Electric Motor
Handbook

Electric Motor Handbook

Edited by
B. J. Chalmers
With specialist contributors

Butterworths
London   Boston   Singapore   Sydney   Toronto   Wellington

First published 1988

© **Butterworth & Co. (Publishers) Ltd., 1988**

British Library Cataloguing in Publication Data

Electric Motor Handbook
  1. Electric Motors
  I. Chalmers, B. J.
  621.46′2   TK2511

  ISBN 0-408-00707-9

Library of Congress Cataloging-in-Publication Data

Electric motor handbook/edited by B. J. Chalmers.
    p. cm.
  Bibliography: p.
  Includes index.
  ISBN 0-408-00707-9
  1. Electric motors–Handbooks, manuals, etc.
  I. Chalmers, B. J. (Brian John)
  TK2511.E423   1988   87-20887
  621.46′2–dc19

Photoset by Butterworths Litho Preparation Department
Printed and bound in Great Britain by Butler and Tanner, Frome, Somerset

Preface

The opportunity to edit a major new handbook, to stand alongside the long established *J & P Transformer Book* and *J & P Switchgear Book,* is at once an honour and a challenge. The J & P Books have been widely used by generations of engineers all over the world and, in setting out to create this *Electric Motor Handbook,* I was conscious that the target was high.

As with J & P books, the aim has been to compile an authoritative handbook which will be of real practical use to engineers working in a wide range of capacities including plant design, equipment specification, commissioning, operation and maintenance. Such a task would have been over-ambitious for most single authors, and certainly for the present editor. I therefore readily acknowledge the expertise contributed by the individual chapter authors, which represents their combined experience accumulated over many decades. The Chapter contents are essentially theirs and I am greatly indebted to them for the manner in which they responded to the tasks I set them. A world-wide rather than parochial view has been sought by including international authors, considerations and illustrations.

While the chapters are, in the main, independent it may be observed that some points are covered in more than one chapter, e.g. under both commissioning and maintenance. I have taken the view that the reader seeking information in a particular context should not be disappointed. In consequence, a little duplication has been accepted.

I must also acknowledge that the original proposal for this form of handbook was made by K. K. Schwarz, Dr M. R. Lloyd and the late J. C. H. Bone, all then with Laurence, Scott and Electromotors Ltd. It was only after force of circumstances prevented them pursuing the project that I became involved. Their initial outline plans gave me a good starting point.

As one who has for about 30 years been mainly concerned with electromagnetic aspects of electrical machines, I confess I have long believed that the majority of their operational problems, and practically all the manifestations of failure or breakdown, are non-electrical, whether they be of the nature of fracture, abrasion, burning or explosion. Accordingly, a significant proportion of this book relates to the causes, occurrence and avoidance of such problems. It is to be hoped that the availability of this Handbook will help many readers avoid such misfortunes.

The vast majority of textbooks on electrical machines concern methods of performance analysis, while a rather smaller number concern the design process. This book is not primarily directed towards either of these activities. Rather it is intended as a user's handbook, written by engineers for engineers, which may provide a useful source of reference.

Although motor size is strictly dependent upon torque rather than power, the scope of the Handbook has been broadly defined as relating to rotating machines of above about 10 kW output (with normal industrial supply frequency). This somewhat arbitrary lower limit was set in order to restrict the coverage to the more common machines and to eliminate the very diverse range of small motors. It was considered that to attempt to include the great variety of motor types of less than 10 kW output within a reasonable limit of total page number would inevitably have made the treatment somewhat formless and superficial. Similarly, linear motors are excluded.

At a time when many senior engineers are retiring and there is serious concern regarding the shortfall in young recruits into electrical power engineering, it is pleasing to be able to participate in activities which may assist in alleviating this deficiency. The Chapter authors are all, in their own spheres, contributing towards promotion of our profession while it is incumbent upon those of us involved in teaching to stimulate the interest of students in what we believe to be a challenging career area. In Manchester, for so long the home of much heavy electrical engineering, we have taken an initiative and created the 'Manchester Machines Research Group', aiming to provide an increased service in research, training and consultancy. I hope the *Electric Motor Handbook* will serve as a useful adjunct to these endeavours, helping to disseminate technical interests and information to a wide readership. If it approaches the popularity of the J & P books it will provide a valuable service to our industry.

B. J. Chalmers
Manchester, 1987

Contents

1 Characteristics

J. E. Brown

Formerly Reader in Electrical Engineering, University of Newcastle upon Tyne

1.1 Introduction

Electric motors convert electrical energy, supplied from an a.c. or d.c. source, to mechanical energy at a rotating shaft. There are several different types of electric motor but here we are concerned with the main ones only, namely: (1) the induction motor and its derivatives which are equipped with a commutator, such as the Schrage motor; (2) the synchronous motor; and (3) the d.c. motor.

All electric motors have certain basic features in common. Each has a stationary member, the stator, and a rotating member, the rotor, separated by an airgap. The stator and rotor each have a magnetic core, usually laminated, although on some high-speed a.c. machines the rotor may be of solid steel. The core carries copper or aluminium windings in slots or on salient poles. The windings are insulated except in the case of cage (squirrel cage) rotors. Details of construction and windings are given in Chapters 5 and 6.

The theory of electric motors is described in many textbooks.[1-8] Only the essentials will be summarized here, for balanced polyphase a.c. machines and d.c. machines in the first instance.

Consider first the principles of operation common to all electric motors. The windings carry currents which may be caused to flow by direct conduction or electromagnetic induction. The currents produce m.m.f. waves of magnitude varying approximately sinusoidally around the airgap, circumferentially (very approximately in the case of the d.c. machine). These space-sinusoidally distributed m.m.f. waves can be represented, as shown in Figure 1.1, by space vectors $\mathbf{F}_s$ and $\mathbf{F}_r$ for stator and rotor, respectively. $\mathbf{F}_s$ and $\mathbf{F}_r$ may be stationary or rotating but, for the production of useful torque, must be stationary relative to each other. They add vectorially in accordance with the equation:

$$\mathbf{F}_m = \mathbf{F}_s + \mathbf{F}_r \tag{1.1}$$

to produce the resultant $\mathbf{F}_m$, called the magnetizing m.m.f. The magnetizing m.m.f. produces the magnetizing flux, which can be represented by the space vector $\mathbf{\Psi}_m$. Flux and current react to produce torque. It can be shown that the total torque developed is given by the expressions:

$$T_T = K \Psi_m F_s \sin\delta_{ms} \tag{1.2}$$

and

$$T_T = K F_r \Psi_m \sin\delta_{rm} \tag{1.3}$$

where K is a constant and $F_s\sin \delta_{ms} = F_r\sin \delta_{rm}$, as can be seen from Figure 1.1.

The relationship between flux and m.m.f. is non-linear because of the well-known B–H properties of the magnetic core. However if, for approximation,

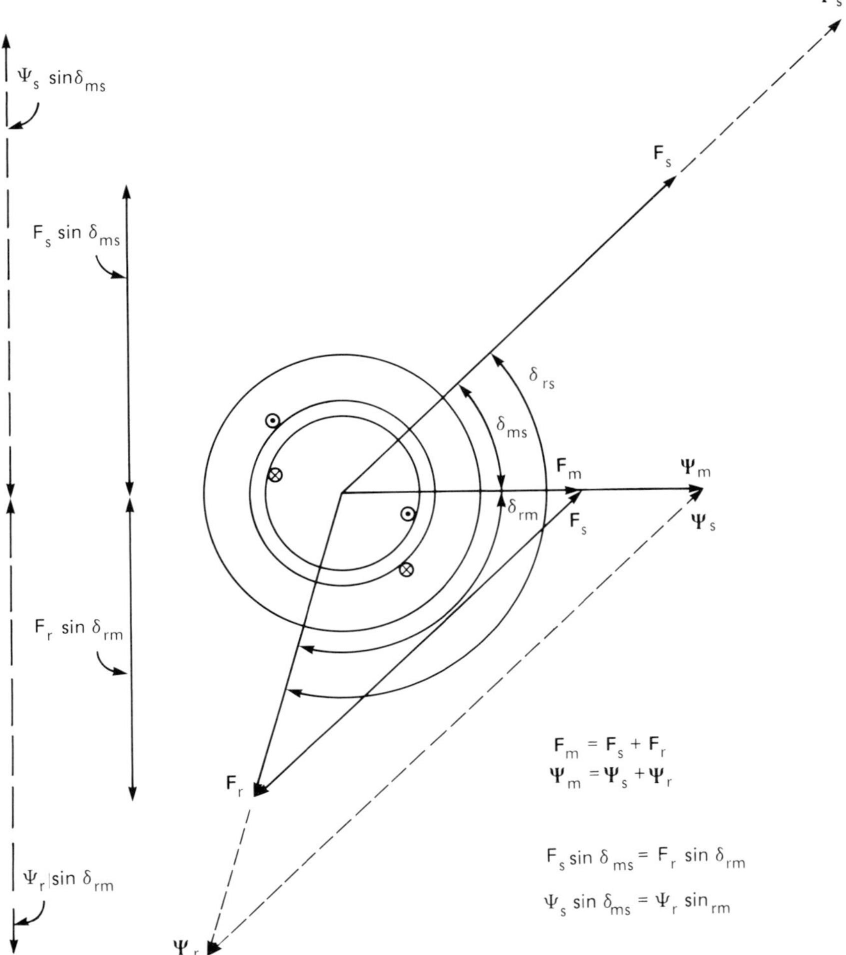

Figure 1.1 The m.m.f. and flux space vectors of an electric motor

magnetic linearity is assumed, i.e. $\Psi_x \alpha F_x$, then the m.m.f. space vectors $\mathbf{F}_s$ and $\mathbf{F}_r$ can be regarded as producing separate space vectors $\mathbf{\Psi}_s$ and $\mathbf{\Psi}_r$, respectively, where:

$$\mathbf{\Psi}_m = \mathbf{\Psi}_s + \mathbf{\Psi}_r \tag{1.4}$$

as shown in Figure 1.1.

It follows that, when magnetic linearity is assumed, other expressions for torque can be derived. These can be summarized in the general equation:

$$T_T = K F_x \Psi_y \sin\delta_{xy} \tag{1.5}$$

where F_x and ψ_y are the magnitudes of m.m.f. and flux space vectors, respectively, and δ_{xy} is the angle between the vectors.

Note that Figure 1.1 includes a representation of a section through the machine in which the directions of current flow in rotor and stator windings are shown. The rotor tends to rotate in the counter-clockwise direction, seeking to maximize the flux linkage by making δ_{rs} equal to zero.

In all polyphase a.c. machines (except the Schrage motor, discussed in Section 1.5.3) the stator m.m.f. wave is developed by alternating currents of supply frequency f_s. It rotates at synchronous speed $N_s = f_s/p$ rev./s, relative to the stator, where p is the number of pole pairs of the winding.

In asynchronous- or induction-type motors, voltages are induced in the rotor windings at frequency $f_r = (N_s-N)p$, where N is the speed of the rotor. These voltages produce currents which develop an m.m.f. wave rotating at speed N_s-N, relative to the rotor, i.e. at speed $(N_s-N) + N = N_s$, relative to the stator. Thus, regardless of the speed of the rotor, the m.m.f. space vectors $\mathbf{F}_s$ and $\mathbf{F}_r$ remain stationary relative to each other.

In a synchronous motor, operating in the steady state, the rotor rotates at exactly synchronous speed and therefore no voltage is induced in the rotor windings. The rotor m.m.f. wave is developed usually by d.c. fed to the rotor windings through sliprings. It is therefore stationary relative to the rotor and $\mathbf{F}_s$ and $\mathbf{F}_r$ are again stationary relative to each other.

In a d.c. motor, the field winding on the stator is supplied with d.c. and therefore the stator m.m.f. vector $\mathbf{F}_s$ is stationary. The armature winding, on the rotor, is supplied with d.c. via brushes bearing on a commutator. The action of the commutator simultaneously converts the external d.c. to a.c. within the armature and maintains the rotor m.m.f. vector $\mathbf{F}_r$ stationary and in quadrature with the $\mathbf{F}_s$ vector.

In addition to their contribution to magnetizing flux, the stator currents produce leakage flux which links with the stator windings only, and the rotor currents produce leakage flux which links with the rotor windings only. The effects of leakage flux can be represented by current flowing through leakage inductances.

Equivalent circuits and phasor and vector diagrams, which take account of all the reactions described above, can be used for modelling the performance of the individual machine types. The general forms of the steady state performance characteristics can be deduced from these models.

In deducing the performance characteristics it will be assumed, unless otherwise stated, that the supply voltage and frequency remain constant, regardless of the power taken by the machine.

The steady state performance characteristics of a motor largely determine its suitability for a particular application. However, they give an incomplete picture of the performance and capability of the motor. The complete picture may require an examination of the transient performance of the motor in the system of which it forms a component. Only brief references to transient performance will be made here.

1.2 Modelling of steady state motor performance

The complex interactions in electrical machines can be modelled by relatively simple equivalent circuits when certain simplifying assumptions are adopted. Thus, in the modelling of polyphase a.c. machines, sinusoidal time variation of applied voltage and sinusoidal space variation of airgap m.m.f. will be assumed. Completely balanced operation will also be assumed, enabling performance to be represented on a per-phase basis.

An equivalent-circuit model for a polyphase a.c. machine, operating in the steady state with voltage applied to both stator and rotor windings, is shown in Figure 1.2. This model can be developed in a way which enables the inherent similarities between different machine types to be appreciated. A proper understanding of the model requires a study of one or more of the standard texts mentioned above, but a fair understanding can be gleaned from careful

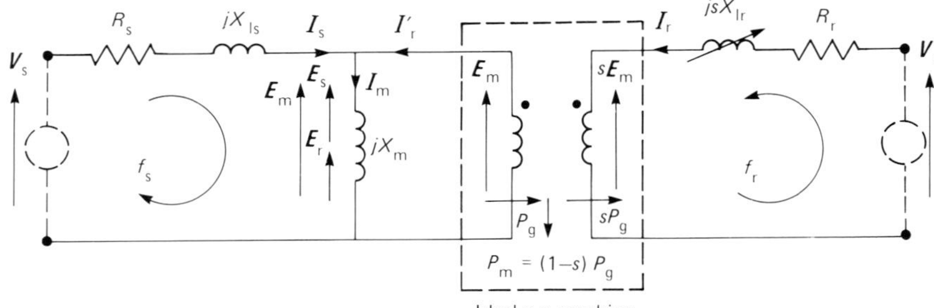

Figure 1.2 Steady state equivalent circuit of a polyphase a.c. machine with voltage applied to both stator and rotor circuits

All stator quantities at stator frequency f_s
All rotor quantities at rotor frequency f_r

V_s voltage applied to a stator phase
V_r voltage applied to a rotor phase

I_r current in a rotor phase
I_r' component of current in a stator phase (at stator frequency) which produces the same magnetizing effect as I_r
I_m $= I_r' + I_s$ magnetizing current (the equivalent current in a stator phase required to establish the magnetizing flux)
I_s total current in a stator phase

jX_m magnetizing reactance of a stator phase

E_m $= jX_m I_m$ magnetizing e.m.f. in a stator phase
E_s $= jX_m I_s$ notional e.m.f. in a stator phase due to stator currents only
E_r $= jX_m I_r'$ notional e.m.f. in a stator phase due to rotor currents only

jX_{ls} leakage reactance of a stator phase
jsX_{lr} leakage reactance of a rotor phase at rotor frequency where jX_{lr} is its value at stator frequency

R_s resistance of a stator phase
R_r resistance of a rotor phase

P_g power transmitted from stator to airgap
sP_g power transmitted from airgap to rotor circuit
P_m $= (1-s)P_g$ mechanical power developed

consideration of the circuit parameters and the phasor and vector diagram shown in Figure 1.3.

The first point to note is that all stator parameters are referred to stator frequency f_s and all rotor parameters to rotor frequency f_r where $f_r = sf_s$ and s is the slip $(N_s-N)/N_s$, as already defined. Furthermore, rotor parameters have been referred to the effective number of turns per phase of the stator winding, e.g.:

rotor winding resistance $R_r = n^2 R_{r\,(\text{actual})}$,

rotor voltage $V_r = n V_{r\,(\text{actual})}$

where n is the ratio of effective turns per phase, stator to rotor.

The equivalent circuit incorporates a representation of an ideal polyphase motor. This conceptual device:

(1) Has no winding resistance.
(2) Has perfect coupling between stator and rotor windings and therefore no leakage flux.
(3) Has infinite permeability in its main magnetic path and therefore requires no magnetizing m.m.f.

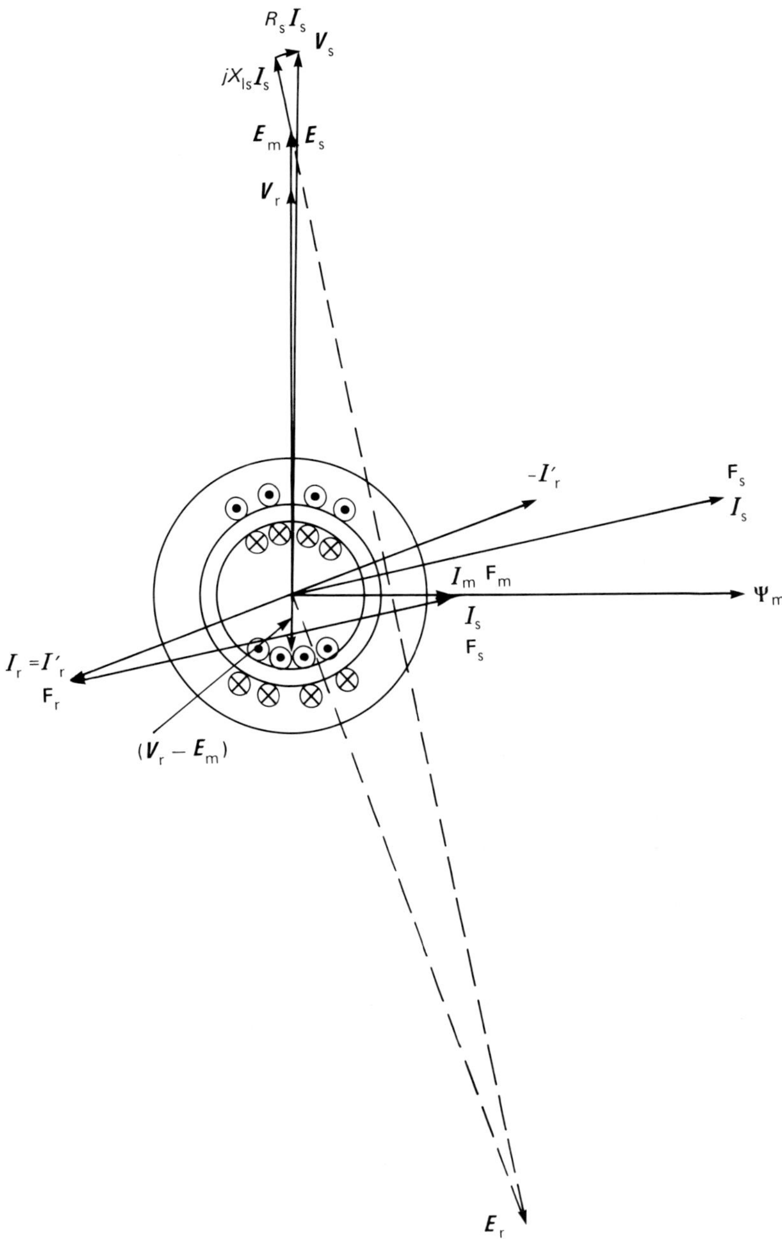

Figure 1.3 Phasor and vector diagram for the equivalent circuit of Figure 1.2 for $s = 1$. (The current direction symbols represent conditions in the reference phases, not resultant current sheets)

(4) Converts voltage, frequency and power from stator to rotor in the ratio $1:s$, but converts current in the ratio $1:1$.

The effects of:

(1) Winding resistances are accounted for in resistors R_s and R_r.
(2) Leakage flux in leakage reactances jX_{ls} and jsX_{lr}.
(3) Finite permeability in magnetizing reactance jX_m carrying magnetizing current I_m.
(*Note:* Losses in the magnetic core are neglected at this stage.)

Figure 1.3 shows a phasor and vector diagram for a special case in which the rotor is locked with the axes of the stator and rotor reference phases aligned, and in which the voltage V_r is in phase with the e.m.f. E_m. To enable 'time' phasors and 'space' vectors to be meaningfully related, the diagram includes a representation of the reference phases in a transverse section through the machine, assuming two-pole construction. For the locked rotor condition, $s = 1$, $f_r = f_s$, and all phasors and vectors rotate in the counterclockwise direction at the same angular velocity $\omega_s = 2\pi f_s$ rad./s. The diagram is drawn for the instant that the magnetizing current $i_m = I_{m(max)} \cos \omega_s t$ has its maximum value, i.e. when $\omega_s t = 0$.

The positive direction of current flow is that shown in the conductors of the stator phase. The so-called 'back e.m.f.' convention is adopted for e.m.f.[9] Back e.m.f. $e = d(Li)/dt$, corresponding to $e = d\Psi/dt$, is an e.m.f. in the direction opposing positive current flow.

The voltage driving rotor current in the positive direction is $(V_r - E_m)$. This voltage produces a current I_r, lagging $(V_r - E_m)$ by the angle arctan X_{lr}/R_r. This current is negative at the reference instant, giving the current directions shown in the rotor conductors. A current I_r' in a stator phase would produce the same magnetizing effect as I_r. The stator current I_s must balance this current and in addition establish the magnetizing current I_m. Thus, $I_s = (I_m - I_r')$ or, otherwise expressed:

$$I_m = I_s + I_r' \tag{1.6}$$

Evidently, the stator current is positive at the reference instant.

The resultant polyphase stator and rotor current 'sheets' which are, of course, displaced from the reference phases, produce the m.m.f. space vectors $\mathbf{F}_s$ and $\mathbf{F}_r$ whose axes are cophasal with the time phasors I_s and I_r', respectively. The m.m.f. space vectors produce the resultant magnetizing m.m.f. space vector $\mathbf{F}_m$ in accordance with Equation (1.1).

Alternatively expressed, the polyphase magnetizing currents produce the magnetizing m.m.f. space vector whose axis is cophasal with the time phasor I_m. Corresponding current phasors and m.m.f. space vectors may be taken as equal in per-unit terms.

The magnetizing m.m.f. produces the magnetizing flux, represented by the space vector $\mathbf{\Psi}_m$. The magnetizing flux induces the magnetizing e.m.f.s E_m in a stator phase and, in general, sE_m in a rotor phase.

E_s and E_r are notional e.m.f.s in a stator phase, associated with notional fluxes $\mathbf{\Psi}_s$ and $\mathbf{\Psi}_r$ produced by the m.m.f.s $\mathbf{F}_s$ and $\mathbf{F}_r$, respectively, where

$$E_m = E_s + E_r \tag{1.7}$$

The space vectors $\mathbf{\Psi}_s$ and $\mathbf{\Psi}_r$ are not shown in Figure 1.3 but can be inferred from the corresponding vectors in Figure 1.1. Corresponding flux space vectors and stator e.m.f. phasors may be taken as equal in magnitude in per-unit terms.

Only the magnetizing e.m.f. can be regarded as having a real existence, in the sense that it could be measured in a shadow winding. E_s and E_r can be regarded as existing, as components of E_m, only under linear magnetic conditions. Such conditions have been assumed in Figure 1.3 and therefore the triangle formed by the m.m.f. space vectors and that formed by the e.m.f. phasors are similar. The latter is displaced from the former by a counterclockwise rotation through 90°. The triangle formed by the flux vectors would be identical to that formed by the e.m.f. phasors, except for the 90° displacement.

In the analysis of induction motors, the separate existence of E_s and E_r is not normally considered. However, in the analysis of synchronous motors it is normal practice to consider the equivalents of E_s and E_r as existing separately, as will be shown later.

Finally, it is worth pointing out that there is more to the equation $f_r = sf_s$ than just a simple numerical relationship. For a given value of f_s there are only two possibilities, either f_r is determined by s or s is determined by f_r. The former leads to asynchronous operation, the latter to synchronous operation.[10]

Asynchronous operation occurs when f_r is determined continuously by the rotor speed, as it is when:

(1) V_r is zero, as in the normal induction motor.
(2) V_r is supplied through sliprings at continuously varying frequency f_r.
(3) V_r is supplied through a frequency changer converting supply frequency to f_r, e.g. a commutator converting from f_s externally to f_r internally.

Synchronous operation occurs when the rotor is fed through sliprings at a frequency which is independent of rotor speed. The rotor must then rotate at a speed such that $s = f_r/f_s$. In practice, if f_s is the a.c. supply frequency the only stable operating condition arises when $f_r = 0$, i.e. when d.c. is supplied to the rotor and it rotates at normal synchronous speed so that $s = 0$. Synchronous operation can also occur with alternative forms of rotor which enable similar magnetic field conditions to be established.

1.3 The polyphase induction motor

1.3.1 Note on construction

The stator of a polyphase induction motor is wound with a polyphase winding, now almost invariably a three-phase winding. The rotor may be wound with a similar winding but of a different number of turns, star- or delta-connected, with the ends of windings brought out to sliprings. Alternatively, the rotor may have a cage winding consisting of bars through the rotor slots, joined together at each end by endrings. An integral cage winding of aluminium can be formed by the die-casting process.

The airgap is regarded as uniform, in contrast to that resulting from the salient pole construction of d.c. machines and some synchronous machines, as discussed later. However, the fact that the windings are carried in slots punched in the laminations of the magnetic core leads to important second-order effects discussed in Section 1.3.6.

1.3.2 Equivalent circuit and phasor and vector diagram

The reactions in a polyphase induction motor, operating under balanced steady state conditions, can be modelled by an equivalent circuit developed from that shown in Figure 1.2. The modelling of a cage rotor by a circuit applicable to a phase-wound rotor is permissible because, in essence, the rotor manifests itself to the stator as a circuit which produces the m.m.f. vector $\mathbf{F}_r$, rotating at synchronous speed. The stator is unaware not only of the number of phases on the rotor but also of its rotation.

The development of the equivalent circuit to its familiar form is illustrated in Figure 1.4. Figure 1.4(a) is similar to Figure 1.2 except that V_r is taken as zero and,

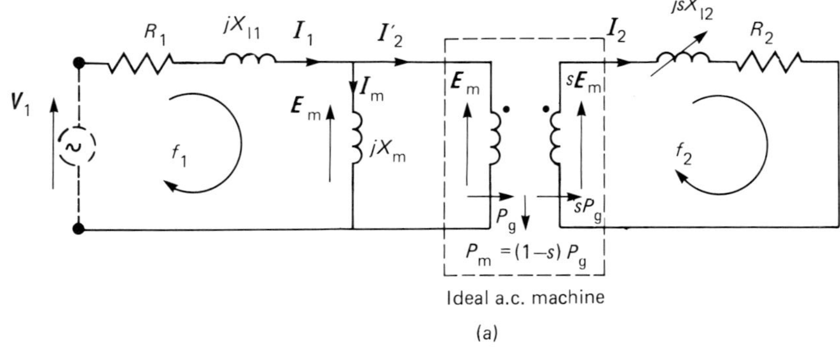

(a)

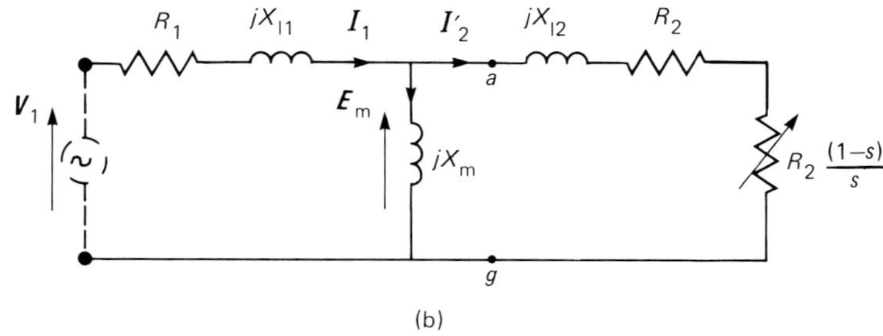

(b)

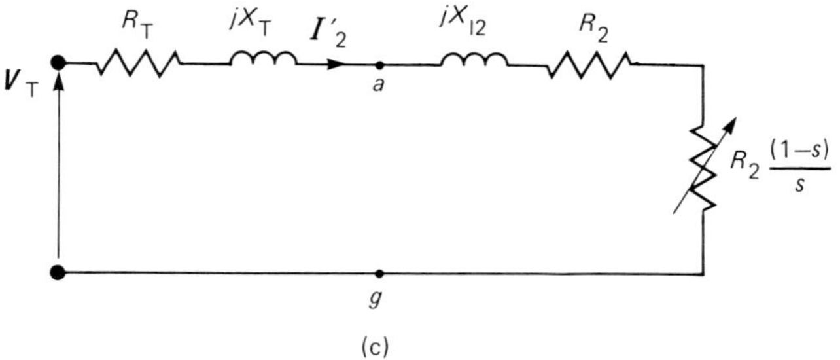

(c)

$$V_T = V_1 \; [jX_m] \, / \, [R_1 + j\,(X_{11} + X_m)]$$

$$R_T + jX_T = [jX_m \,(R_1 + jX_{11})] \, / \, [R_1 + j\,(X_{11} + X_m)]$$

Figure 1.4 Alternative forms of steady state equivalent circuit for a polyphase induction motor. (a) Complete form. (b) Normal form. (c) Thévenin-modified form

in accordance with convention, the reference direction of rotor current is reversed and suffixes s and r are replaced by 1 and 2 respectively. In Figure 1.4(b), the circuit is referred to the stator by displacing the representation of an ideal a.c. machine off the diagram to the right. The resulting resistor R_2/s is replaced by resistors R_2 and $R_2(1-s)/s$, having equivalent total resistance.

Thus, remembering that core losses must be taken into account separately, the electrical losses in the machine are accounted for by the power consumed in the

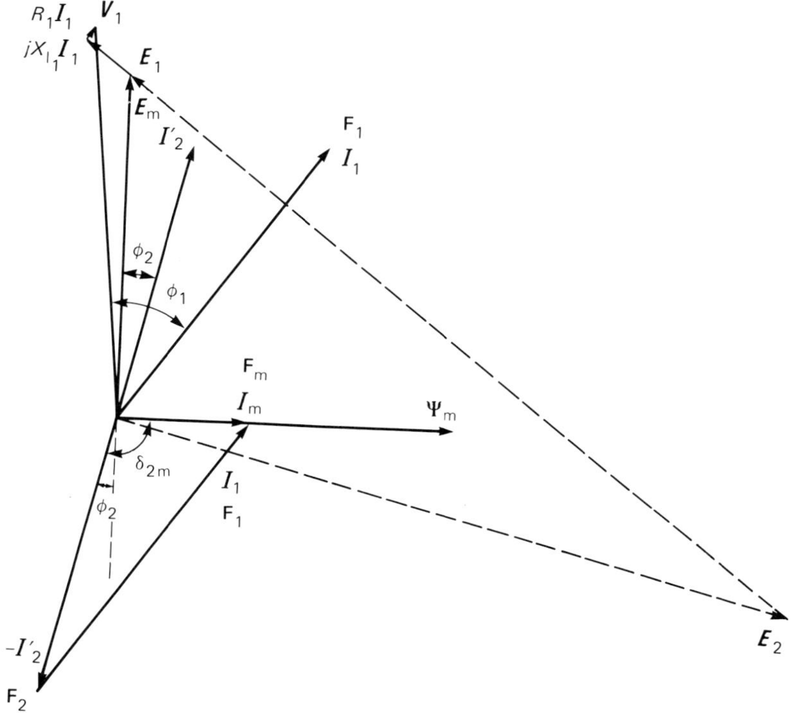

Figure 1.5 Phasor and vector diagram for an induction motor corresponding to the equivalent circuit of Figure 1.4(b)

resistors R_1 and R_2. Therefore, the mechanical power developed is represented by the power consumed in the resistor $R_2(1 - s)/s$.

A typical phasor and vector diagram corresponding to Figure 1.4(b) is shown in Figure 1.5. Note that no rotor–current phasor I_2, equivalent to $-I_r$ of Figure 1.3, appears in the diagram, since this phasor would be required here to represent a current of frequency f_2, where $f_2 \neq f_1$. The phasor $-I_2'$, equivalent to I_r' of Figure 1.3, and the space vector $\mathbf{F_r}$ are correctly shown as cophasal. The phasors $E_1 = jX_m I_1$ and $E_2 = -jX_m I_2'$, notional components of E_m for assumed linear magnetic conditions, are included only for later comparison with the corresponding phasors for a synchronous machine (see Figure 1.13, page 21).

It follows from the equivalent circuit and phasor and vector diagram that:

$$\text{Torque per phase} = R_2 \frac{(1 - s)}{s} I_2'^2 \ \frac{1}{2\pi N}$$

$$= \frac{R_2}{s} I_2'^2 \ \frac{1}{2\pi N_s}$$

$$= \frac{\text{power crossing air gap at a–g}}{2\pi N_s}$$

$$= \frac{E_m I_2' \cos\phi_2}{2\pi N_s} \tag{1.8}$$

where $E_m \ \alpha \ \Psi_m \ \alpha \ F_m$, $F_2 \ \alpha \ I_2$ and $\cos\phi_2 = \sin\delta_{2m}$

Thus, the expression for torque, deduced from essentially power balance considerations, is consistent with the fundamental expression in Equation (1.3).

Given the availability of modern computing aids there can be no justification for introducing the well-known 'approximate' equivalent circuit. However, for our purpose, it is useful to introduce a simplification. By applying Thévenin's theorem to the part of the circuit to the left of a–g in Figure 1.4(b) the simplified circuit of Figure 1.4(c) is obtained. From Figure 1.4(c) the following expressions for the performance characteristics dependent on the current I_2' can readily be deduced:

(1) Stator equivalent of rotor current:

$$I_2' = \frac{V_T}{[(R_T + R_2/s)^2 + (X_T + X_2)^2]^{1/2}} \tag{1.9}$$

(2) Torque developed per phase:

$$T = \frac{R_2}{s} \frac{V_T^2}{(R_T + R_2/s)^2 + (X_T + X_2)^2} \frac{1}{2\pi N_s} \tag{1.10}$$

This has a maximum value at a slip S_{MT} such that:

$$S_{MT} = \frac{R_2}{[R_T^2 + (X_T + X_2)^2]^{1/2}}$$

Evidently, when the other parameters are constant the maximum torque is independent of R_2.

(3) Mechanical power developed per phase:

$$P_m = R_2 \frac{(1-s)}{s} \frac{V_T^2}{(R_T + R_2/s)^2 + (X_T + X_2)^2} \tag{1.11}$$

This has a maximum value at a slip S_{MP} such that:

$$S_{MP} = \frac{R_2}{R_2 + [(R_T + R_2)^2 + (X_T + X_2)^2]^{1/2}}$$

1.3.3 Performance characteristics of wound-rotor motors

In a wound rotor induction motor, the parameters can be regarded as constants if the effects of saturation of the magnetic paths are neglected. Theoretical performance characteristics can then be calculated from Equations (1.9)–(1.11). However, it is convenient to express values in per-unit of the corresponding values at full load, with slip expressed in its normal per-unit of synchronous speed. For this purpose the following equations can be derived from Equations (1.9)–(1.11):

$$\frac{I_2}{I_{2FL}} = \left\{ \frac{\left[1 + \frac{S_{MT}}{S_{FL}}(1 + Q^2)^{1/2}\right]^2 + Q^2}{\left[1 + \frac{S_{MT}}{s}(1 + Q^2)^{1/2}\right]^2 + Q^2} \right\}^{1/2} \tag{1.12}$$

$$\frac{T}{T_{FL}} = \frac{1 + \dfrac{1}{2}(1 + Q^2)^{1/2}\left(\dfrac{S_{FL}}{S_{MT}} + \dfrac{S_{MT}}{S_{FL}}\right)}{1 + \dfrac{1}{2}(1 + Q^2)^{1/2}\left(\dfrac{s}{S_{MT}} + \dfrac{S_{MT}}{s}\right)} \tag{1.13}$$

$$\frac{P}{P_{FL}} = \frac{T}{T_{FL}}\frac{(1 - s)}{(1 - S_{FL})} \tag{1.14}$$

where $Q = (X_T + X_2)/R_T$ and suffix FL denotes full load.

A set of performance characteristics calculated from Equations (1.12)–(1.14) using the values $S_{FL} = 0.0234$, $S_{MT} = 0.15$ and $Q = 7$, which are based on the parameters of a particular 75 kW, 50 Hz, three-phase, six-pole motor, is shown in Figure 1.6.

The ratio of maximum torque to full-load torque differs in machines of different rating and/or pole number. A minimum value of 2 is required by American

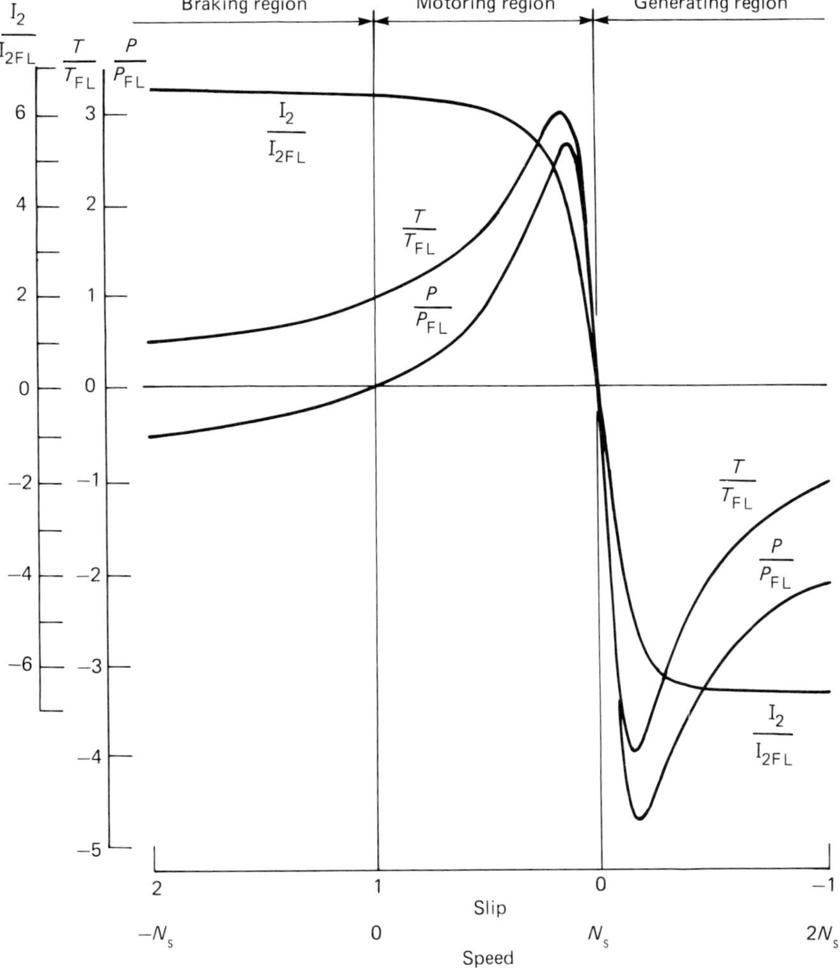

Figure 1.6 Typical performance characteristics of a wound rotor induction motor

Standards, but values of up to 4, for some motors, are claimed in manufacturers' literature. A full-load slip of about 0.025 is typical of a 75 kW motor. A 350 kW machine would have a full-load slip of about 0.01 and a 10 kW machine one of about 0.05. Higher values of full-load slip are associated with higher values of rotor resistance relative to leakage reactance and, consequently, with higher values of slip for maximum torque. However, bearing in mind the variations noted in this paragraph, the characteristics shown in Figure 1.6 can be regarded as typical of those of wound rotor induction motors.

In Figure 1.6 three separate regions can be distinguished, as follows:

(1) The normal motoring region, in which $1 > s > 0$.
(2) A braking region in which $2 > s > 1$. This arises, in so-called 'plugging', when the sequence of the supply voltages to a machine operating at normal speed is deliberately reversed.
(3) A generating region in which s is negative, the machine being driven at super-synchronous speed by external means.

It is evident from an examination of the motoring region that the starting torque is relatively low and the starting current relatively high. However, the characteristics can be modified usefully by the connection of external resistors in series with the rotor circuit, as described in the next section.

1.3.4 Performance characteristics of wound rotor motors with external resistors in rotor circuit

When a balanced polyphase set of resistors is connected in series with the sliprings of a wound rotor induction motor the parameter R_2 in Equations (1.9) and (1.10) is valid for the resistance per-phase of the whole rotor circuit including the external resistance. It is evident from Equation (1.9) that if R_2 and s are changed by the same factor, the current I_2' is unchanged. A similar argument applies to the expression for torque in Equation (1.10). Therefore, the current–slip and torque–slip characteristics for any value of R_2 can easily be deduced from the corresponding characteristics for any other known value of this parameter.

The performance characteristics of the machine of Figure 1.6, for a range of values of $R_2 = R_{2W} + R_{2E}$, where R_{2W} is the resistance of the rotor winding and R_{2E} the resistance added externally, are shown in Figure 1.7. Only the motoring and braking regions need be considered for this aspect of machine performance.

Clearly, the effects of increasing R_2 are to reduce the starting current and increase the starting torque, until a value of R_2 is reached which causes maximum torque to occur at standstill. The use of higher values of resistance then leads to a reduction of starting torque, from the maximum value, and a continued reduction of starting current.

Braking performance over the whole range $2 > s > 1$ can be optimized by suitable choice of R_{2E}. However, in general, good braking performance is associated with relatively high values of torque at standstill and consequently with ideal characteristics for accelerating the machine from standstill in the reverse direction. This is a very unsatisfactory feature of 'plugging', and other methods of braking are generally to be preferred.[11]

1.3.5 Performance characteristics of cage rotor motors

There is no single set of performance characteristics, similar to Figure 1.6, which can be said to be typical of an induction motor with a cage rotor, mainly because a cage rotor does not exhibit constant rotor parameters R_2 and X_2. In virtually all cage motors, to some extent, and with large machines and cages of special design in

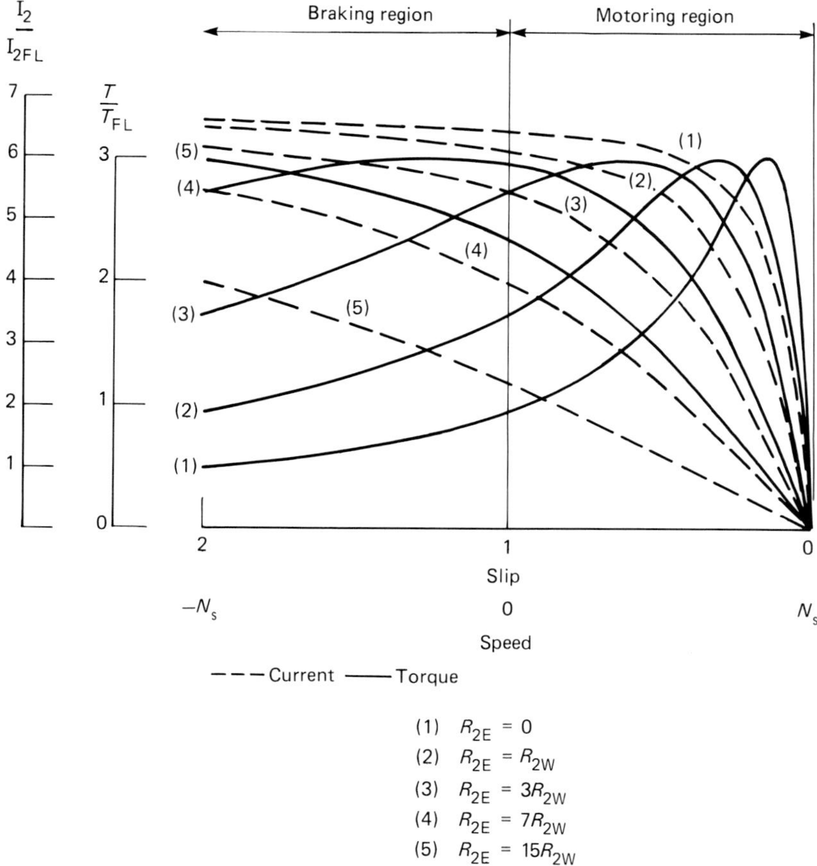

Figure 1.7 Performance characteristics of wound rotor induction motor with external resistors in rotor circuit

particular, these parameters vary with rotor frequency, and therefore speed, because of the so-called 'current displacement' or 'skin' effect. This effect can be explained by a consideration of the leakage flux paths associated with the cages of special design. For example, Figure 1.8(a) shows a cross-section of a 'deep' bar in a rotor slot and the approximate leakage flux paths. The section of conductor at the bottom of the slot is linked by all the leakage flux, that at the top by only a small proportion of this flux. Thus, the leakage inductance of the bottom section is relatively high, that of the top section relatively low. Therefore, when reactance effects are significant, e.g. at standstill where rotor frequency is equal to stator frequency, the impedance of the bottom section of conductor is greater than that of the top section.

Consequently, the current density over the cross-section is non-uniform, current being displaced towards the top. Therefore, the effective resistance of the bar is higher than its resistance to d.c. At rated speed the rotor frequency is very low, the rotor leakage reactance effect is negligible, the current spreads uniformly over the cross-section and the bar exhibits its minimum resistance. This combination of relatively high resistance at standstill and low resistance at rated speed is exactly what is required for good performance, as explained in the discussion related to Figure 1.7 above.

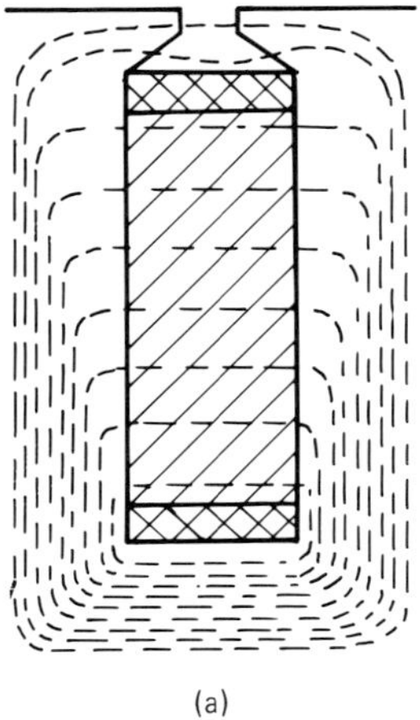

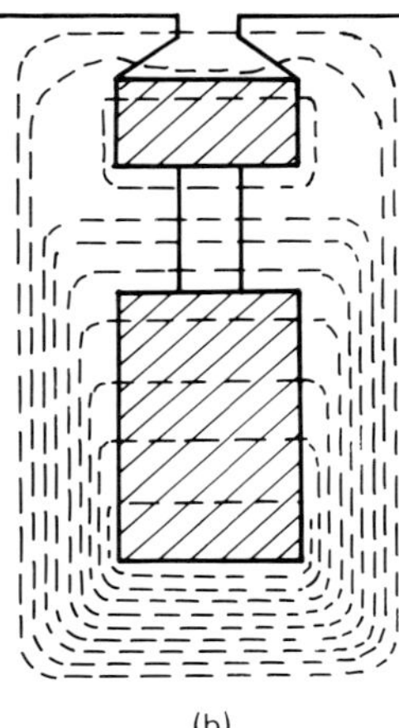

(a)   (b)

Figure 1.8 Diagrammatic illustration of leakage flux paths of induction motor rotors.
(a) Deep-bar. (b) Double cage

The effect can be modified by using bars of various shapes or by using the double-cage construction illustrated in Figure 1.8(b). Evidently, the choice of slot design and bar material, including aluminium in the die-casting process, can lead to a wide range of designs and a wide range of performance characteristics. These are classified in different ways by the various associations responsible for specifying standards.[12,13]

Other factors, affecting the performance characteristics of cage rotor motors more than those of wound rotor motors, should be mentioned. For example, the stator and rotor m.m.f. distributions are not sinusoidal but stepped, because of the concentration of conductors in slots, and the permeance of the gap between stator and rotor is non-uniform because of the slotting and the relative motion of the slotted surfaces. Consequently, the gap flux contains harmonics of the fundamental sinusoid. These harmonics can give rise to both asynchronous and synchronous torques superimposed on that due to the fundamental.[14,15] The design, particularly in the choice of slot combinations, will usually be such as to eliminate the harmonics which lead to synchronous torques, but the asynchronous effects cannot be eliminated completely. The latter can be taken into account in an extended equivalent circuit,[16] but even when this is done there remain other phenomena which can affect performance significantly, as discussed in Section 1.3.6.

Figure 1.9 shows representative characteristics for induction motors with four types of cage rotor. Table 1.1 gives some indication of the range of variation of values at specific points in the characteristics, for a wide range of machines embracing ratings from 1 to 630 kW and pole numbers from two to twelve. The four types correspond to the similarly designated NEMA classifications, but the characteristics, and the information in the table, are based on data from several sources including manufacturers' publications. Reference should be made to the latter for the characteristics of machines of particular type and rating.

1.3.6 Efficiency

The efficiency of an induction motor is best expressed in the form:

Efficiency = 1 − (losses/input)

where the term in brackets can be called the deficiency.

The losses consist of:

(1) Stator and rotor copper losses due to both load and magnetizing currents. These are the only losses taken into account in the simple equivalent circuit of Figure 1.2(a). Strictly, if a.c. values are used for the resistances they include an element for the loss due to pulsation of leakage flux at standstill.

In addition there are:

(2) Stator core losses which are a function of magnetizing flux density and which, at constant supply voltage, are approximately constant.
(3) Rotor core losses which are a function of magnetizing flux density and speed, and which are virtually negligible at rated speed.
(4) Extra stator copper loss associated with the component of stator current required to supply the core losses.
(5) Mechanical losses due to friction and windage.
(6) So-called stray losses, being all the losses additional to the above.

All losses, except stray losses, can be determined with reasonable accuracy from the results of standstill and running-light tests (see Chapter 8).

Table 1.1 Range of variation of particular performance figures for various types of cage induction motor

Designation	(A)	(B)	(C)	(D)
Rotor type	Single cage low-resistance	Deep bar or double cage	Double cage	Single cage high-resistance
Full-load slip	$0.01 > S_{FL} > 0.005$	$0.05 > S_{FL} > 0.005$	$0.05 > S_{FL} > 0.005$	$0.15 > S_{FL} > 0.05$
Breakdown torque	$2.8 > T_{BD} > 2.4$	$2.0 > T_{BD} > 1.9$	$2.0 > T_{BD} > 1.9$	$3.0 > T_{BD} > 2.6$
Pull-up torque	$1.2 > T_{PU} > 0.5$	$1.4 > T_{PU} > 1.0$	$2.0 > T_{PU} > 1.4$	
Standstill torque	$2.0 > T_{SS} > 0.8$	$2.0 > T_{SS} > 1.4$	$2.8 > T_{SS} > 2.0$	$2.7 > T_{SS} > 2.5$
Starting current	$5.0 < I_{SS} < 8.0$	$4.5 < I_{SS} < 6.0$	$4.5 < I_{SS} < 6.0$	$4.0 < I_{SS} < 5.5$

Notes:
(1) All values are in per-unit of corresponding full load value, except slip.
(2) $S_{FL}(A) < S_{FL}(B) < S_{FL}(C) < S_{FL}(D)$ for machines of similar rating.
(3) In all cases, figures on the right apply to machines of higher rating.
(4) Breakdown torque and pull-up torque are designated in Figure 1.9(a).
(5) Breakdown torque may be less than the maximum torque developed by machines of Type C.
(6) Some machines of Type D may not develop a breakdown torque in the region $1 > s > 0$.

Stray losses, as defined in (6) above, include rotor losses associated with the space harmonics of the magnetic field which induce fundamental, i.e. supply–frequency, voltages in the stator windings. To some extent these may be taken into account in an extended equivalent circuit as already mentioned.[17]
There remain:

(1) Losses produced by magnetic fields which link stator and rotor and which induce voltages in the stator at frequencies other than supply frequency; these voltages are short-circuited by the supply network.
(2) Losses due to pulsations of leakage flux additional to those which occur at standstill.

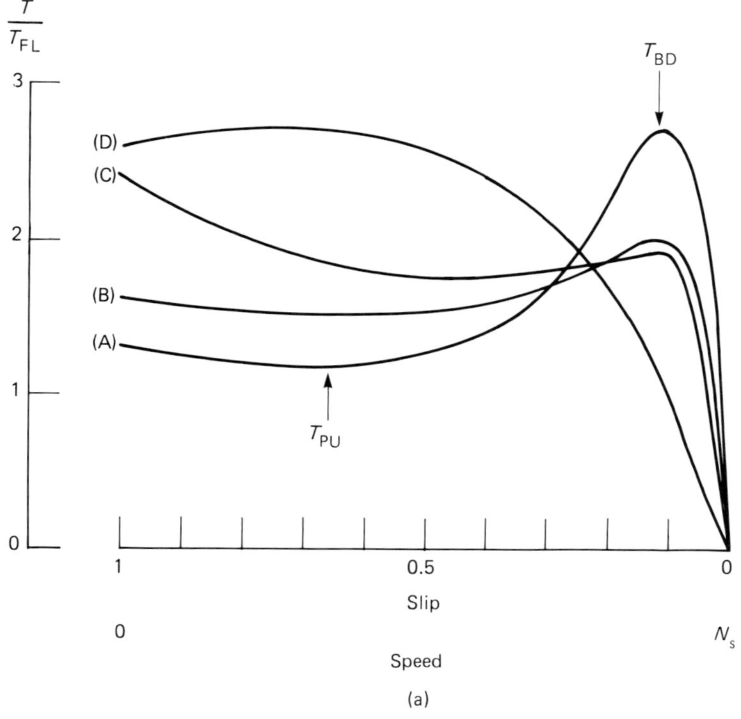

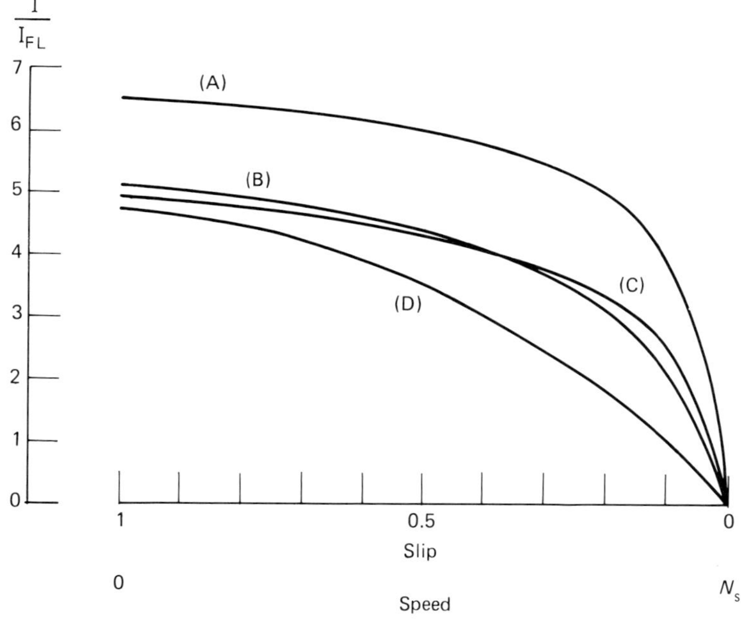

Figure 1.9 Performance characteristics of induction motors with various types of cage rotor. (a) Torque–slip. (b) Current–slip

These arise because of the presence of teeth on stator and rotor, in relative motion, and would exist even with stator windings of notionally perfect sinusoidal distribution.

The interactions causing stray losses are evidently complex.[18,19] These losses can be evaluated at one speed by the reverse-rotation test[20] but this has its limitations.[21] Some designers conclude that, for the purposes of a declared efficiency, it suffices to make an allowance based on experience.[22] However, it must be noted that in the regions $0.9 > s > 0.1$ these losses reduce the motoring torque considerably, and in the region $s > 1.1$ they increase the braking torque considerably.[23]

In an induction motor of full-load efficiency 0.88, the deficiency of 0.12 would be comprised approximately as follows: stator copper loss 0.04, rotor copper loss 0.02, core losses 0.04, mechanical losses 0.01 and stray losses 0.01. The core losses and mechanical losses taken together as rotational losses remain approximately constant; the other losses are load-dependent. Maximum efficiency occurs approximately when these two sets of losses are equal, and usually at a slip less than full-load slip. A typical curve of efficiency against slip is shown in Figure 1.10(b).

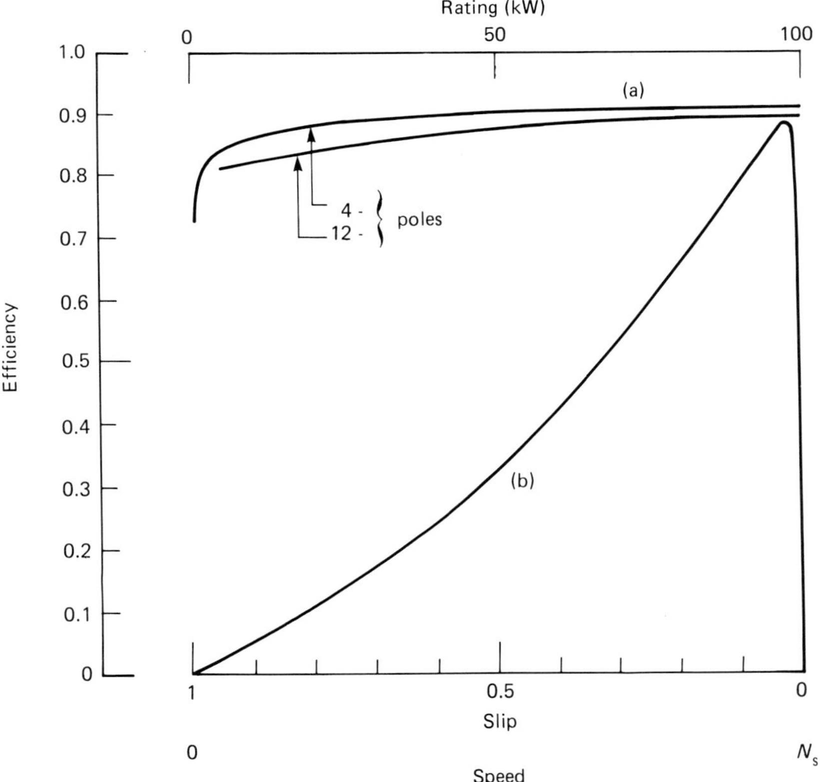

Figure 1.10 Efficiency. (a) Full-load, against rating, for typical cage induction motors. (b) Against slip for a typical motor

Efficiencies of induction motors of a given pole number increase with rating. Those of motors of a given rating decrease with increase in pole number. Curves of full-load efficiency against rating for cage rotor machines of ratings up to 100 kW are included in Figure 1.10. At a rating of 1000 kW, the efficiency would increase by between 0.02 and 0.03.

The efficiencies of wound rotor motors are lower than those of the corresponding cage rotor motors by about 0.02 in the lowest ratings, and by 0.01 or less in the highest ratings.

1.3.7 Power factor

An induction motor operates at a lagging power factor which is a function of slip. Maximum power factor usually occurs at a slip slightly greater than full-load slip. A typical curve of power factor against slip is shown in Figure 1.11(b).

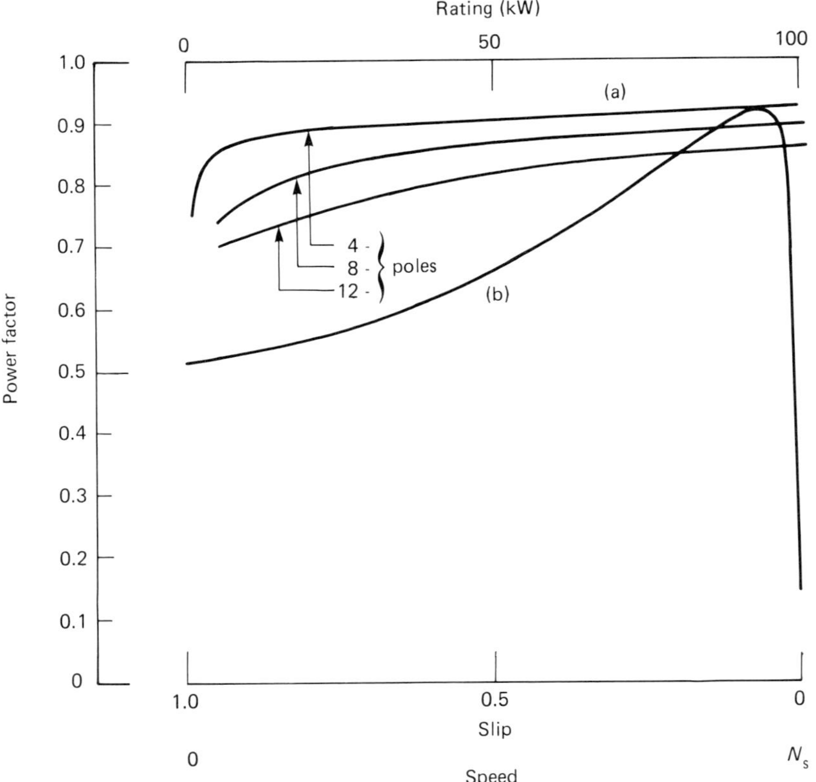

Figure 1.11 Power factor. (a) Full-load, against rating, for typical cage induction motors. (b) Against slip for a typical motor

Power factors of induction motors of a given pole number increase with rating. Those of motors of a given rating decrease with increase in pole number, the decrease being significantly greater than the corresponding decrease in efficiency mentioned in Section 1.3.6. Typical curves of full-load power factor against rating for cage rotor machines of ratings up to 100 kW are included in Figure 1.11(a). At a rating of 1000 kW, the power factor could be expected to rise to 0.91 for a twelve-pole motor and to 0.93 for an eight-pole or four-pole machine.

The power factors of wound rotor motors are lower than those of the corresponding cage rotor motors by about 0.03–0.05 in the lowest ratings, and by about 0.01 or less in the highest ratings.

1.4 The polyphase synchronous motor

1.4.1 Note on construction

The stator of a polyphase synchronous motor is essentially similar to that of a polyphase induction motor. The rotor may be of cylindrical or salient pole form, carrying windings which, when excited with d.c., produce an approximately sinusoidally distributed m.m.f. Alternatively, excitation may be provided by permanent magnets.

A salient pole rotor is capable of synchronous operation without excitation. This mode of operation is utilized in a special type of synchronous motor called the 'reluctance' motor.

Synchronous operating torque is only developed at synchronous speed. The rotor must be accelerated to this speed from standstill by other means. Occasionally, a 'pony' motor coupled to the motor's shaft may be used for this purpose. Normally, the motor is started as an induction motor, utilizing a cage winding embedded in the pole faces. Alternatively, the torque developed by eddy currents in solid pole faces may be used for the starting purpose.

Finally, in the so-called synchronous induction motor, the excitation winding is of the polyphase type, usually three-phase, less often quadrature-phase. External resistors can then be connected to the sliprings to produce good starting characteristics, as described in Section 1.3.4.

1.4.2 Equivalent circuit and phasor and vector diagram of a cylindrical rotor motor

The reactions in a polyphase cylindrical rotor synchronous motor, operating under balanced steady state conditions, can also be modelled by an equivalent circuit developed from that in Figure 1.2. Figure 1.12(a) represents a special case of Figure 1.2 which is suitable for this purpose. Suffixes a for armature and f for field replace s and r respectively. The frequency of the d.c. source, i.e. zero, defines the rotor frequency as zero, and hence the only stable operating speed as the synchronous speed N_s, at which the slip s is zero. I_f' represents an a.c. of supply frequency which, when flowing in the stator, produces the same magnetizing effect as the field current I_f in the field winding.

Strictly, the equivalent circuit is only valid for a cylindrical rotor machine with a polyphase excitation winding. However, all forms of rotor can be assumed to give rise to excitation which could be produced by armature current I_f'. Therefore, the circuit is valid *on the armature side* for all types of cylindrical rotor synchronous motors and, since all of the power which crosses the airgap is converted to mechanical power, it is only the armature side which needs to be considered for steady state performance calculations. Furthermore, as will be shown later, the theory related to the equivalent circuit for the armature can be modified to take the effects of saliency into account.

Significant differences between the performance characteristics of induction motors and synchronous motors arise from the fact that, in the latter, the field (rotor) current I_f, and consequently the notional current I_f', can be controlled independently. For this reason, in modelling the armature circuit, attention is directed to E_f and E_a, the notional components of the magnetizing e.m.f. E_m. It must be emphasized that this separation of E_m into components E_f and E_a, which are directly proportional to I_f' and I_a, respectively, and the implied separation of the true magnetizing flux space vector Ψ_m into notional components Ψ_f and Ψ_a (see Figure 1.13) are valid for linear magnetic conditions only.

Figure 1.12(b) shows a modified equivalent circuit for the armature circuit only. Note that the following voltage equations are valid for the armature circuits of Figure 1.12(a) and (b):

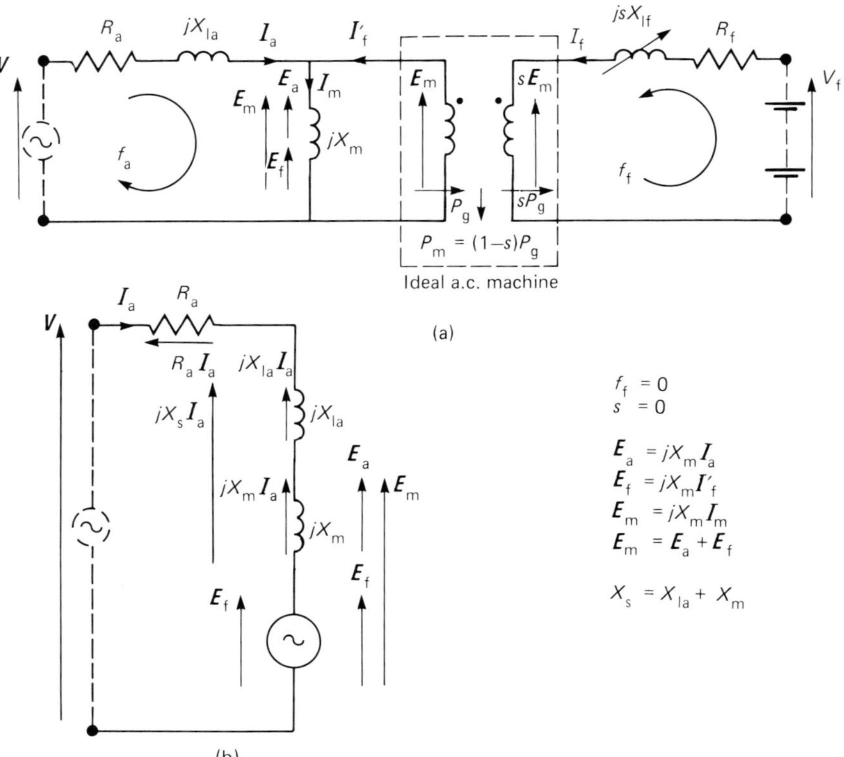

Figure 1.12 Steady state equivalent circuits for a polyphase cylindrical rotor synchronous motor. (a) Complete form. (b) Armature circuit only

$$V = R_a I_a + jX_{la} I_a + E_m$$

$$= R_a I_a + jX_{la} I_a + E_a + E_f$$

$$= R_a I_a + jX_s I_a + E_f$$

where $E_f = jX_m I'_f$ (1.15)

and $jX_s = j(X_{la} + X_m)$ (1.16)

The fictitious, but conceptually convenient, reactance jX_s is called the 'synchronous reactance'.

A phasor and vector diagram for a cylindrical rotor synchronous motor corresponding to Figure 1.12(b) is shown in Figure 1.13. Here, an operating condition similar to that for the induction motor in Figure 1.5 has been chosen deliberately, so that comparisons can be made. A value of magnetizing reactance of 1.25 per unit has been assumed for the synchronous motor in contrast to that of 2.5 per unit assumed for the induction motor of Figure 1.5.

For all except very small machines (or on low-frequency operation), the voltage drop $R_a I_a$ is very small relative to the drop $jX_s I_a$ across the synchronous reactance. Therefore, it is normal practice to ignore the effects of resistance when synchronous motor performance is related to the e.m.f. E_f.

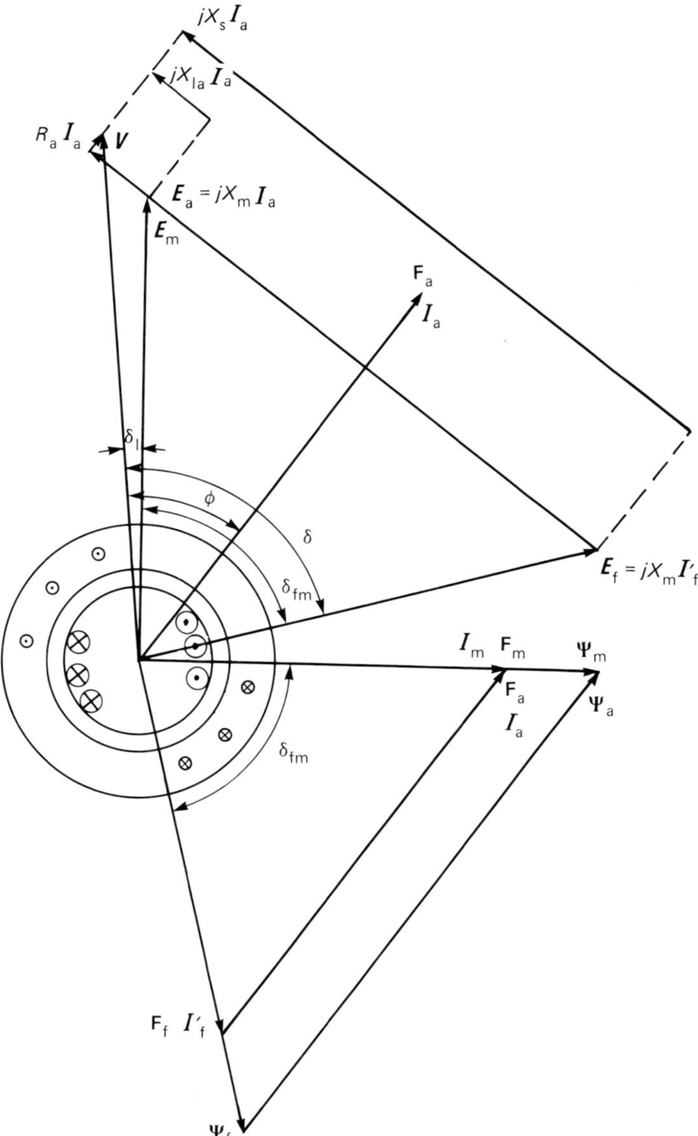

Figure 1.13 Phasor and vector diagram for a cylindrical rotor synchronous motor corresponding to the equivalent circuit of Figure 1.12(b)

Figure 1.14 shows a phasor and vector diagram for a cylindrical rotor synchronous motor in which R_a is assumed to be negligible. This diagram has been drawn for the instant at which the applied voltage is zero, a datum which will be adhered to for subsequent diagrams. A value for synchronous reactance of 1.25 per unit has been adopted, and a relatively high value of excitation has been assumed.

When the losses caused by armature resistance (and all other armature losses) are neglected the electrical input power and mechanical output power are equal. Very simple expressions for power can then be developed. Thus, referring to Figure 1.14:

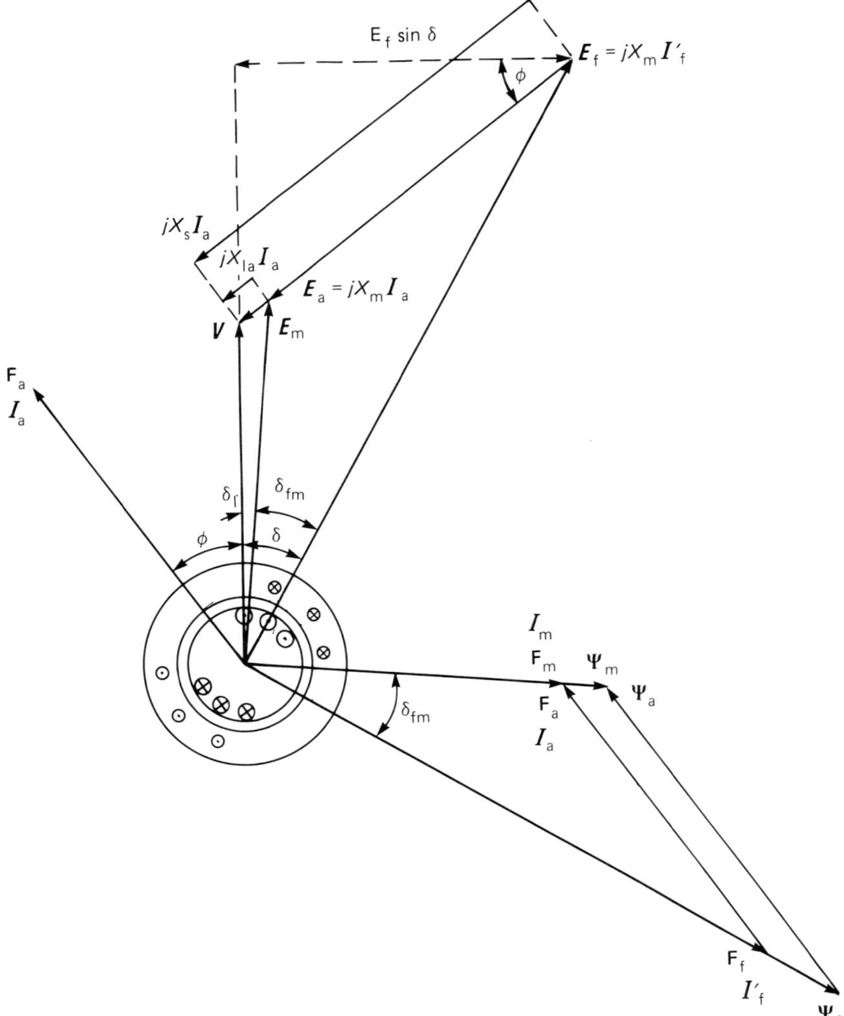

Figure 1.14 Phasor and vector diagram of a cylindrical rotor synchronous motor; machine overexcited and armature resistance assumed to be zero

Input power $= V\,I_a\,\cos\phi$

$$= \frac{V\,E_f\,\sin\delta}{X_s} \tag{1.17}$$

$$= \frac{E_f\,E_m\,\sin\delta_{fm}}{X_m} \tag{1.18}$$

$$= \frac{V\,E_m\,\sin\delta_l}{X_{la}} \tag{1.19}$$

where Equation (1.19) is included only for comment.

The voltage drop $R_a I_a$ is not negligible relative to the drop $jX_{la}I_a$ and Equation (1.19), though correct for zero R_a, is not a useful approximation.

The above expressions are valid for the whole machine when the per-unit notation is adopted and a base value for power equal to the rated VA is utilized. Otherwise, they may be regarded as per-phase expressions in the normal way.

The corresponding expressions for torque are:

$$\text{Torque} = \frac{V\,E_f\,\sin\delta}{2\pi N_s X_s} \tag{1.20}$$

$$= \frac{E_f\,E_m\,\sin\delta_{fm}}{2\pi N_s X_m} \tag{1.21}$$

and it can readily be seen that Equation (1.21) is consistent with Equation (1.3).

The angle δ corresponds to the physical swing of the rotor from its no-load position to its load position. It is called the 'load angle', and in practice can be measured stroboscopically. The angle δ_{fm} is sometimes called the 'torque angle'.

Finally, Figures 1.13 and 1.14 include a representation of a section through the main magnetic circuit, for comparison with that of a salient pole motor to be considered next. Note that in these diagrams the current direction symbols represent conditions in the resultant current sheets.

1.4.3 Phasor and vector diagram for a salient pole motor

Figure 1.15 shows a phasor and vector diagram of a salient pole synchronous motor in which R_a is assumed to be negligible. Arbitrary, but reasonably typical, values have been assumed for all other parameters. Again, the diagram includes a representation of a section through the main magnetic path which should be contrasted with that of the cylindrical rotor machine in the two previous figures.

The field and armature m.m.f.s produce the magnetizing m.m.f. in accordance with Equation (1.1), i.e. $\mathbf{F}_m = \mathbf{F}_f + \mathbf{F}_a$. However, it is evident from Figure 1.15 that, in contrast to the conditions in the cylindrical rotor machine depicted in Figures 1.13 and 1.14, the m.m.f.s $\mathbf{F}_m$ and $\mathbf{F}_a$ are not directed along axes of uniform magnetic reluctance, because the airgap of the salient pole machine is not of uniform length. If saturation is taken into account the conditions are extremely complex. However, if magnetic linearity is assumed, two paths of uniform reluctance can be distinguished. One lies along the axis of the salient poles, called the direct axis. The other lies along an axis at right angles, called the quadrature axis.

The m.m.f.s $\mathbf{F}_m$ and $\mathbf{F}_a$ may be resolved into components along these axes. Thus:

$$\mathbf{F}_a = \mathbf{F}_{ad} + \mathbf{F}_{aq} \tag{1.22a}$$

$$\mathbf{F}_m = (\mathbf{F}_f + \mathbf{F}_{ad}) + \mathbf{F}_{aq} \tag{1.23a}$$

$$= \mathbf{F}_{md} + \mathbf{F}_{mq} \tag{1.24a}$$

Note that the only component of m.m.f. producing excitation in the quadrature axis is the component of armature m.m.f. $\mathbf{F}_{aq} \equiv \mathbf{F}_{mq}$. The component of armature m.m.f. $\mathbf{F}_{ad}$ may be demagnetizing, as in Figure 1.15, or magnetizing, as in cases considered later.

The components of magnetizing m.m.f., $\mathbf{F}_{md}$ and $\mathbf{F}_{mq}$, produce components of flux $\boldsymbol{\Psi}_{md}$ and $\boldsymbol{\Psi}_{mq}$, respectively, where a given value of m.m.f. in the direct axis will produce a bigger value of flux than that produced by an equal value of m.m.f. in the quadrature axis. The resultant flux $\boldsymbol{\Psi}_m$ induces the magnetizing e.m.f. E_m with components E_{md} and E_{mq}.

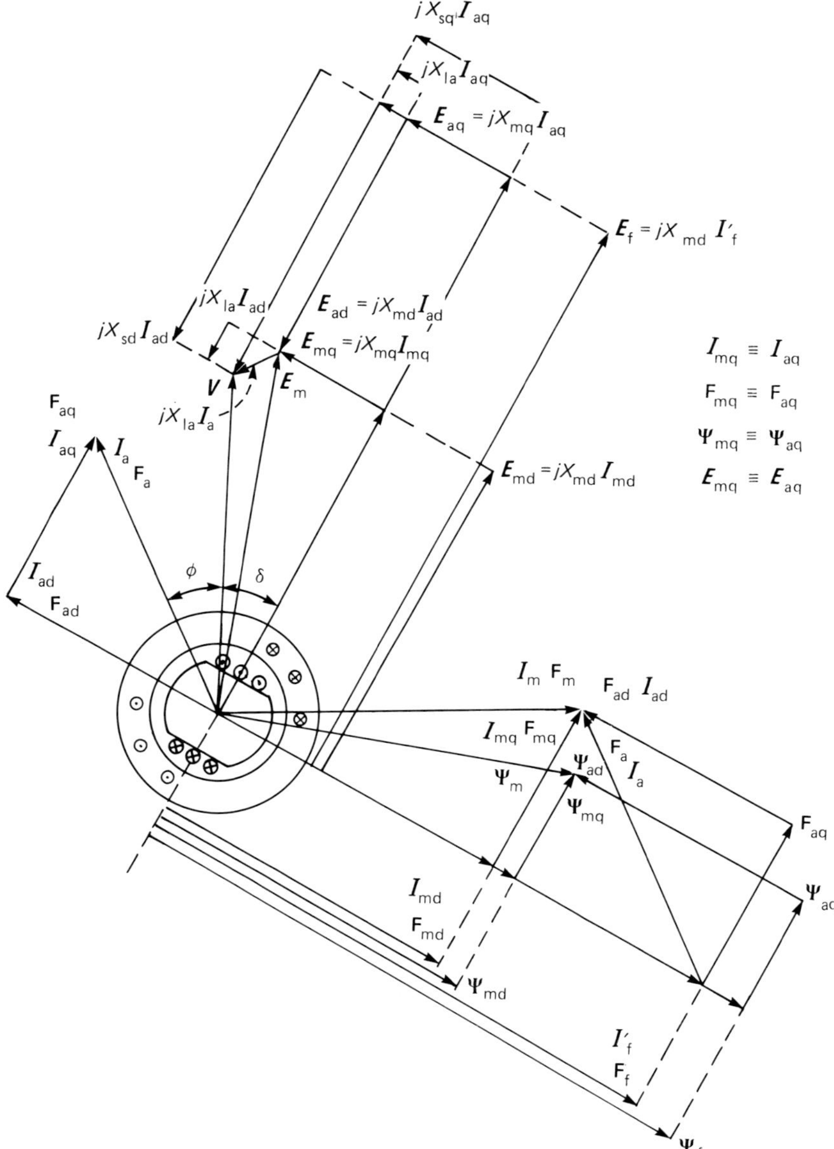

Figure 1.15 Phasor and vector diagram of a salient pole synchronous motor; machine over-excited and armature resistance assumed to be zero

In circuit terms, the current equations corresponding to Equations (1.22a)–(1.24a) are:

$$I_a = I_{ad} + I_{aq} \tag{1.22b}$$

$$I_m = (I'_f + I_{ad}) + I_{aq} \tag{1.23b}$$

$$= I_{md} + I_{mq} \tag{1.24b}$$

and hence the components of the magnetizing e.m.f. are $E_{md} = jX_{md}I_{md}$ and $E_{mq} = jX_{mq}I_{mq}$, where jX_{md} and jX_{mq} are the direct-axis and quadrature-axis magnetizing reactances, respectively.

The above description can be expanded to enable the effects of field current to be considered separately, as for the cylindrical rotor machine. Thus, for the linear magnetic conditions already assumed, the field m.m.f. $\mathbf{F}_f$ produces flux $\mathbf{\Psi}_f$ which induces e.m.f. $\boldsymbol{E}_f = jX_{md}\boldsymbol{I}_f'$. The components of armature m.m.f., $\mathbf{F}_{ad}$ and $\mathbf{F}_{aq}$, produce fluxes $\mathbf{\Psi}_{ad}$ and $\mathbf{\Psi}_{aq}$, which induce e.m.f.s $\boldsymbol{E}_{ad} = jX_{md}\boldsymbol{I}_{ad}$ and $\boldsymbol{E}_{aq} = jX_{mq}\boldsymbol{I}_{aq}$, respectively, where $\mathbf{F}_{aq} \equiv \mathbf{F}_{mq}$, $\mathbf{\Psi}_{aq} \equiv \mathbf{\Psi}_{mq}$ and $\boldsymbol{E}_{aq} \equiv \boldsymbol{E}_{mq}$. Then:

$$\boldsymbol{E}_m = \boldsymbol{E}_f + \boldsymbol{E}_{ad} + \boldsymbol{E}_{aq} \tag{1.23c}$$

Finally, the addition of the voltage drop in the armature leakage reactance gives the terminal voltage. Thus:

$$V = \boldsymbol{E}_m + jX_{la}\boldsymbol{I}_a$$

$$= \boldsymbol{E}_f + \boldsymbol{E}_{ad} + \boldsymbol{E}_{aq} + jX_{la}(\boldsymbol{I}_{ad} + \boldsymbol{I}_{aq})$$

$$= \boldsymbol{E}_f + jX_{md}\boldsymbol{I}_{ad} + jX_{mq}\boldsymbol{I}_{aq} + jX_{la}(\boldsymbol{I}_{ad} + \boldsymbol{I}_{aq})$$

$$= \boldsymbol{E}_f + j(X_{md} + X_{la})\boldsymbol{I}_{ad} + j(X_{mq} + X_{la})\boldsymbol{I}_{aq}$$

$$= \boldsymbol{E}_f + jX_{sd}\boldsymbol{I}_{ad} + jX_{sq}\boldsymbol{I}_{aq} \tag{1.25}$$

where X_{sd} and X_{sq} are called the direct-axis and quadrature-axis synchronous reactances, respectively.*

Figure 1.16 shows a phasor diagram which corresponds to Figure 1.15 and contains all the elements required for calculation purposes. It can be seen from Figure 1.16 that the armature current can be expressed in terms of its 'power' and 'reactive' components as:

$$\boldsymbol{I}_a = \boldsymbol{I}_{ap} + \boldsymbol{I}_{ar} \tag{1.26}$$

and it can be shown that:

$$I_{ap} = \frac{E_f \sin\delta}{X_{sd}} + \frac{V}{2X_{sd}}\left(\frac{X_{sd}}{X_{sq}} - 1\right)\sin2\delta \tag{1.27}$$

$$I_{ar} = \frac{E_f \cos\delta}{X_{sd}} - \frac{V}{X_{sd}} - \frac{V}{2X_{sd}}\left(\frac{X_{sd}}{X_{sq}} - 1\right)(1 - \cos2\delta) \tag{1.28}$$

It is also possible to make a useful distinction (see Sections 1.4.5 and 1.4.7) between the components of armature current attributable to excitation and those attributable to terminal voltage. Thus:

$$\boldsymbol{I}_{aE} = \boldsymbol{I}_{apE} + \boldsymbol{I}_{arE} \tag{1.29}$$

and $$\boldsymbol{I}_{aV} = \boldsymbol{I}_{apV} + \boldsymbol{I}_{arV} \tag{1.30}$$

---

*The equivalent circuit of Figure 1.12(b) is a well-established and useful pictorial aid but it is not a true equivalent circuit because the e.m.f. E_f is not a function of the current I_a only. A circuit representation of Equation (1.25) would contain representations of three voltages in series, each a function of a different current. Such a circuit would not serve a useful purpose.

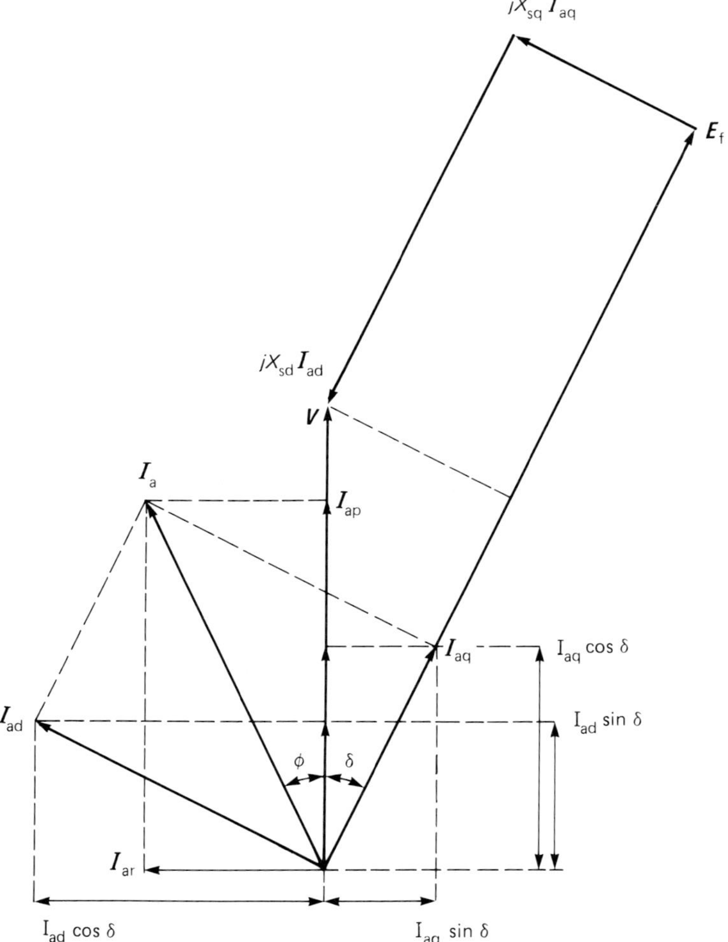

Figure 1.16 Phasor diagram of a salient pole synchronous motor; machine over-excited and armature resistance assumed to be zero

where
$$I_{apE} = \frac{E_f\,\sin\delta}{X_{sd}}$$

$$I_{arE} = \frac{E_f\,\cos\delta}{X_{sd}}$$

$$I_{apV} = \frac{V}{2X_{sd}}\left(\frac{X_{sd}}{X_{sq}}-1\right)\sin2\delta$$

and
$$I_{arV} = -\left[\frac{V}{X_{sd}} + \frac{V}{2X_{sd}}\left(\frac{X_{sd}}{X_{sq}}-1\right)(1-\cos2\delta)\right]$$

It follows from Equation (1.27) that the power and torque developed are given by the expressions:

$$\text{Power} = \frac{V\,E_f\,\sin\delta}{X_{sd}} + \frac{V^2}{2X_{sd}}\left(\frac{X_{sd}}{X_{sq}} - 1\right)\sin2\delta \tag{1.31}$$

$$\text{Torque} = \frac{1}{2\pi N_s}\left[\frac{V\,E_f\,\sin\delta}{X_{sd}} + \frac{V^2}{2X_{sd}}\left(\frac{X_{sd}}{X_{sq}} - 1\right)\sin2\delta\right] \tag{1.32}$$

When $X_{sd} = X_{sq}$, as in the cylindrical rotor machine, Equations (1.31) and (1.32) reduce to the same forms as Equations (1.17) and (1.20) respectively.

The term $(X_{sd}/X_{sq} - 1)$ in the above equations is often expressed in the form $(X_{sd} - X_{sq})/X_{sq}$. The first of these two forms is preferred here because it draws attention directly to the importance of the ratio of X_{sd} to X_{sq}.

1.4.4 Synchronizing torque and limit of steady state stability

When a synchronous machine is operating in the steady state at constant voltage and frequency, two of the parameters in the equations developed above are effectively under the control of the operator. These are the e.m.f. E_f by control of the field current, and the power component of armature current I_{ap} by control of the load applied to the shaft. The effects of varying each of these parameters while the other is held constant will be considered in Sections 1.4.5 and 1.4.6.

When one or other of these parameters is changed there is a momentary change in speed and load angle. If, under the changed conditions, the load is less than the maximum the machine can sustain, torque tending to restore the rotor to synchronism is developed. An expression for the *synchronizing torque* can be obtained by differentiating Equation (1.32) with respect to δ, giving:

$$\text{Synchronizing torque} = \frac{1}{2\pi N_s}\left[\frac{V\,E_f\,\cos\delta}{X_{sd}} + \frac{V^2}{X_{sd}}\left(\frac{X_{sd}}{X_{sq}} - 1\right)\cos2\delta\right] \tag{1.33}$$

per electrical radian of displacement.

The synchronizing torque has a maximum value at zero load, where δ is zero, and a value of zero, representing the limit of steady state stability, when

$$\delta = \delta_L = \text{arc cos}\,\frac{[E_f^2 + 8V^2\,(X_{sd}/X_{sq} - 1)^2]^{1/2} - E_f}{4V\,(X_{sd}/X_{sq} - 1)} \tag{1.34}$$

at which value of δ the power developed is a maximum.

For a cylindrical rotor machine $\delta_L = 90°$, regardless of the value of E_f, and for a salient pole machine (with $X_{sd} > X_{sq}$) $\delta_L < 90°$, each particular value being a function of the corresponding value of E_f.*

The limit of steady state stability described above is an idealized concept for an idealized machine. It assumes the very gradual application of load up to the maximum the machine can sustain. In practice, the machine has a transient response to any of the changes in operating parameters discussed in the next three sections. This response is considered briefly in Section 1.4.8.

---

*For permanent-magnet synchronous motors X_{sd} is often less than X_{sq}. δ_L is then somewhat greater than 90°.

1.4.5 Operation at constant excitation and various values of power

The performance characteristics in this and the next section have been calculated for two specific machines, a cylindrical rotor machine having a synchronous reactance $X_{s(d)} = 1.25$ per unit and a salient pole machine having synchronous reactances $X_{sd} = 1.25$ per unit and $X_{sq} = 0.8$ per unit, giving a ratio $X_{sd}/X_{sq} = 1.56$. The resultant characteristics, expressed in per-unit terms, typify the general form of the characteristics of synchronous motors. The effects of higher values of the ratio X_{sd}/X_{sq} will be considered in Section 1.4.7 on the reluctance motor.

A useful overview of the performance characteristics of synchronous motors operating at constant excitation and various values of power, including those of the reluctance motor, can be obtained from phasor locus diagrams. For this purpose it is advantageous to utilize the expressions for the components of armature current attributable to excitation and terminal voltage, separately, as given by Equations (1.29) and (1.30) respectively. Then, referring to Figure 1.17, the locus of the (tip of) the E_f phasor is a circular arc, between the limits $0 \leqslant \delta \leqslant \delta_L$. The locus of the phasor I_{aE} is a circular arc, of radius E_f/X_{sd} in quadrature with the e.m.f. locus. The locus of the phasor I_{aV} is a circular arc between the limits $0 \leqslant 2\delta \leqslant 2\delta_L$, of radius $(V/2X_{sd})(X_{sd}/X_{sq} - 1)$, having a centre defined by the terms $I_{ap} = 0$ and $I_{ar} = - (V/2X_{sd})(X_{sd}/X_{sq} + 1)$. The locus of the armature current phasor I_a is the locus of the sum of the components. (For clarity, in the general locus diagram of Figure 1.17 only, values of $E_f = 2$ per unit, $X_{sd} = 1.25$ per unit and $X_{sd}/X_{sq} = 2$ have been utilized.)

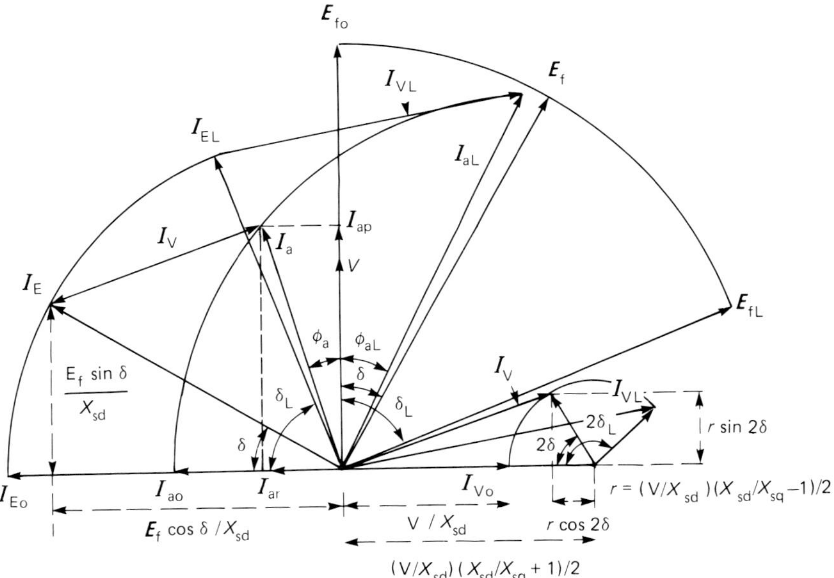

Figure 1.17 General phasor locus diagram for operation of a synchronous motor at constant excitation

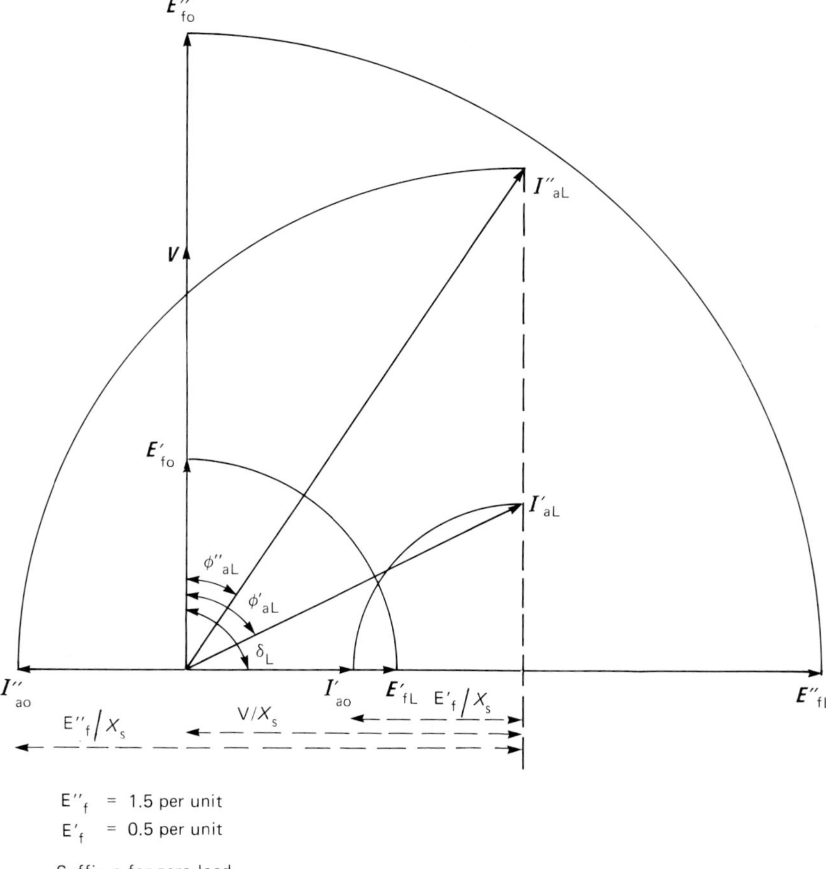

E''_f = 1.5 per unit
E'_f = 0.5 per unit

Suffix o for zero load
Suffix L for limit of steady-state stability

Figure 1.18 Phasor locus diagram for operation of a cylindrical rotor synchronous motor at constant excitation

Figures 1.18 and 1.19 show phasor locus diagrams for the cylindrical rotor and salient pole machines respectively, in each case for values of E_f of 1.5 per unit and 0.5 per unit. The values of δ_L for the salient pole machine, as given by Equation (1.34), are 72.2° and 58.7° respectively, significantly different from the 90° of the cylindrical rotor machine. In the case of the latter only, the locus of the armature current phasor itself, lies on a circular arc. Its radius is E_f/X_s and its centre is defined by $I_{ap} = 0$ and $I_{ar} = -V/X_s$.

Figure 1.20 shows components of power for the two values of excitation, as given by Equation (1.31). The component due to excitation, the first term in the equation, is sinusoidal in δ and is the whole power developed in the case of the cylindrical rotor machine. The component due to saliency, the second term in Equation (1.31), is sinusoidal in 2δ. The power developed by the salient pole machine is the sum of the two components.

Figures 1.21 and 1.22 show complete sets of performance characteristics corresponding to Figures 1.18 and 1.19 respectively. The curves for total power are repeated from Figure 1.20 as curves for I_{ap}.

Further comment on the characteristics is deferred to Section 1.4.9.

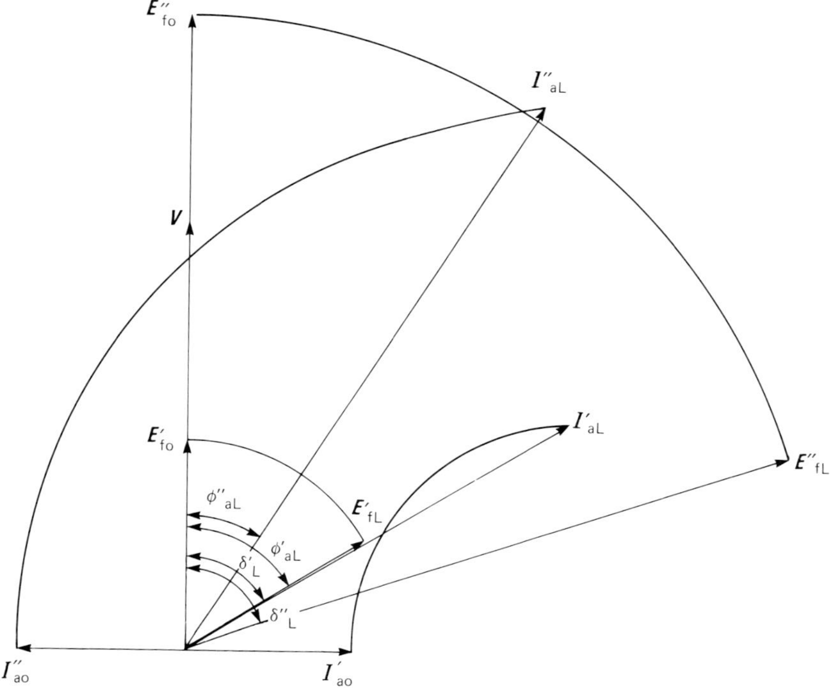

E''_f = 1.5 per unit
E'_f = 0.5 per unit

Suffix o for zero load
Suffix L for limit of steady-state stability

Figure 1.19 Phasor locus diagram for operation of a salient pole synchronous motor at constant excitation

1.4.6 Operation at constant power and various values of excitation

Again, a useful overview of the performance characteristics of synchronous motors operating at constant power and various values of excitation can be obtained from phasor locus diagrams. Figure 1.23 shows such a diagram for the cylindrical rotor and salient pole motors specified previously, when each is operated at powers of 0.8 and 0.225 per unit. The latter value represents the maximum power which can be delivered by the salient pole machine when its excitation is reduced to zero.

The locus of the (tip of) the armature current phasors, for operation at constant power, is a straight line perpendicular to the voltage phasor. The locus of the corresponding e.m.f. phasor, for a cylindrical rotor machine, is a straight line parallel to the voltage phasor and displaced from it by an interval $X_s I_{ap}$ per unit. The loci of the e.m.f. phasors for the salient pole machine, for the two values of power, have been determined by calculation and it is evident that they do not take a simple form.

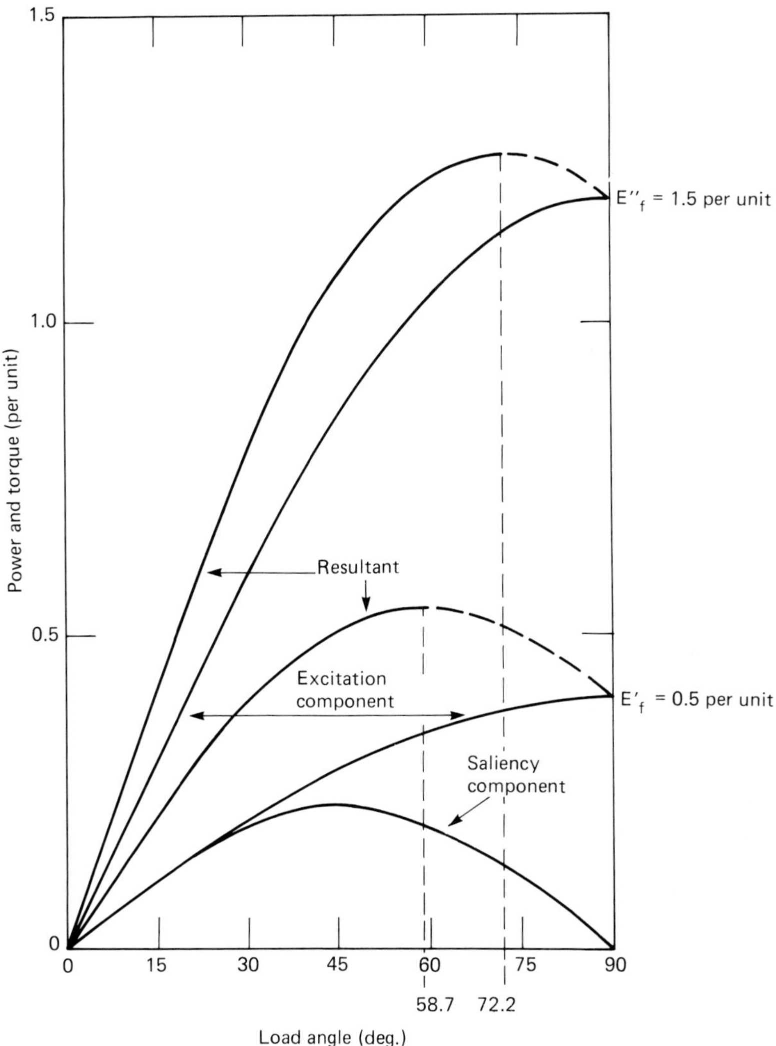

Figure 1.20 Excitation and saliency components of power developed by a synchronous motor at constant excitation

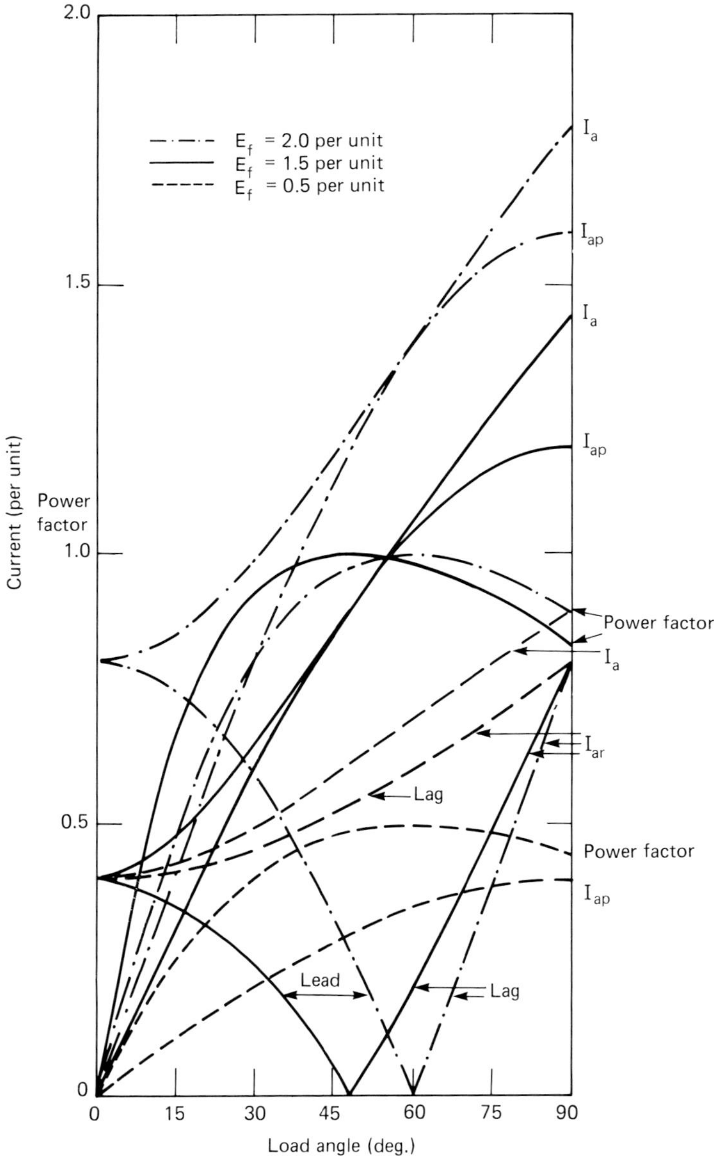

Figure 1.21 Performance characteristics for operation of a cylindrical rotor synchronous motor at constant excitation

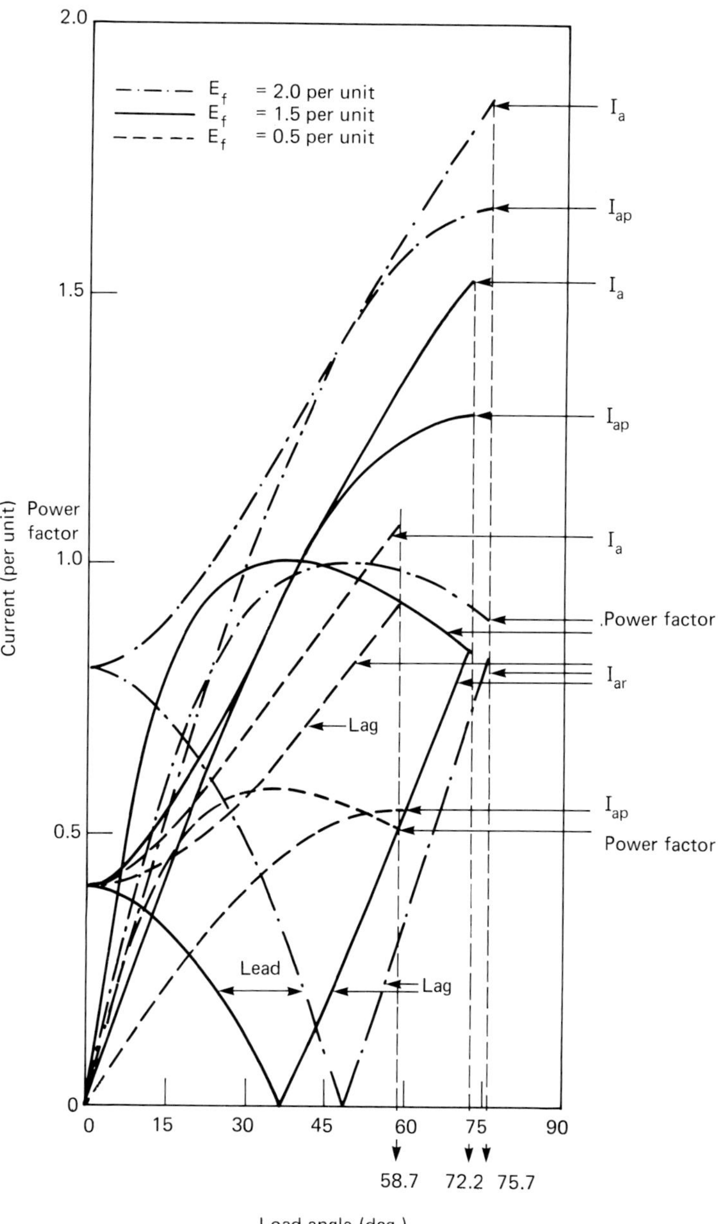

Figure 1.22 Performance characteristics for operation of a salient pole synchronous motor at constant excitation

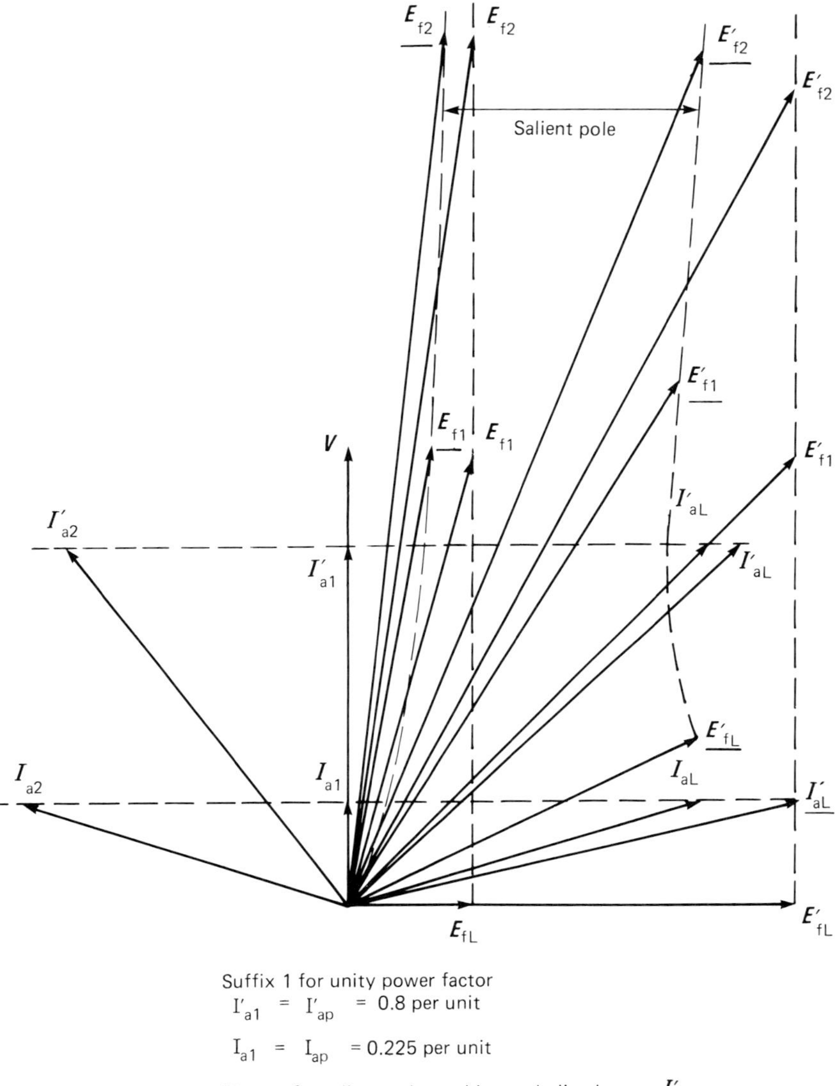

Figure 1.23 Phasor locus diagram for operation of synchronous motors at constant power

In Figure 1.23, current phasors with suffixes 1 (for unity power factor), and 2 are common to the two machines, but those with suffix L (representing the limits of steady state stability) differ for the two machines. Where necessary, phasors for the salient pole machine have been distinguished by underlining.

It can easily be deduced from the phasor locus diagram that the variation of armature current with the magnitude of e.m.f., and therefore with field current, will be of V-shape. The classic V-curves for the two machines, corresponding to the conditions in Figure 1.23, are included in the comprehensive characteristics shown

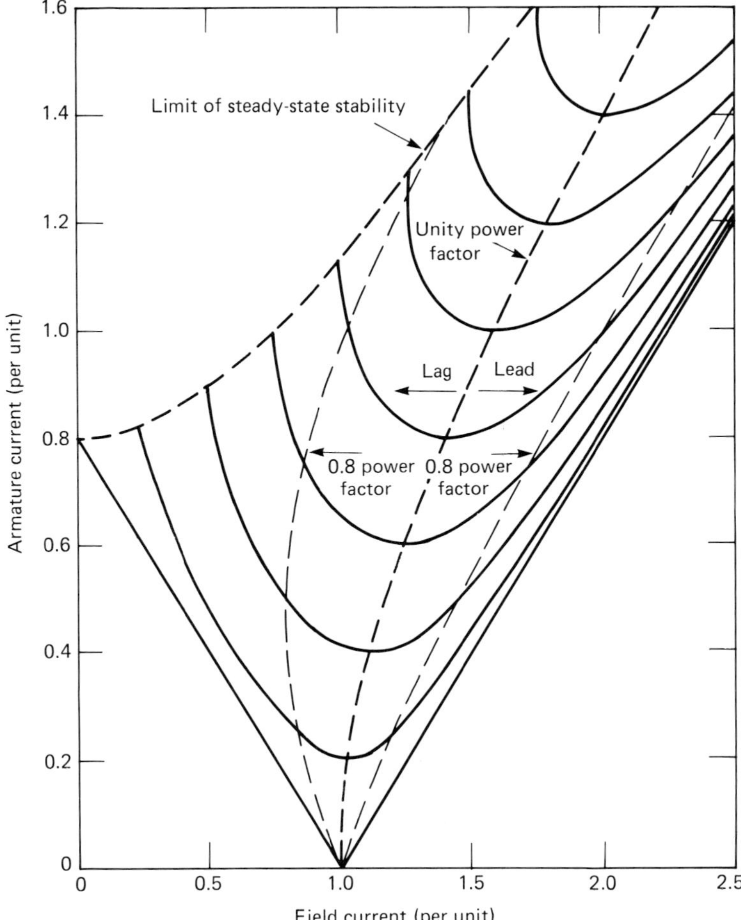

Figure 1.24 V-curves of a cylindrical rotor synchronous motor

in Figures 1.24 and 1.25. Alternatives, showing the reactive components of armature current, are given in Figures 1.26 and 1.27. These are even more distinctly V-shaped, and are complementary to the normal V-curves in giving a clear indication of the reactive capability at a specified load.

1.4.7 The reluctance motor

Figure 1.28 shows a phasor locus diagram for the salient pole synchronous motor of Figure 1.17 ($X_{sd} = 1.25$ per unit, $X_{sd}/X_{sq} = 2$), when it is operated as a reluctance motor, i.e. with $E_f = 0$, at various values of power from zero up to the limit of steady state stability, i.e. when $2\delta = 90°$. The armature current is the current I_{aV} of Equation (1.30). The diagram includes the loci of the phasors of the direct and quadrature axis components of the armature current and of the phasor $jX_{sd}I_{ad}$.

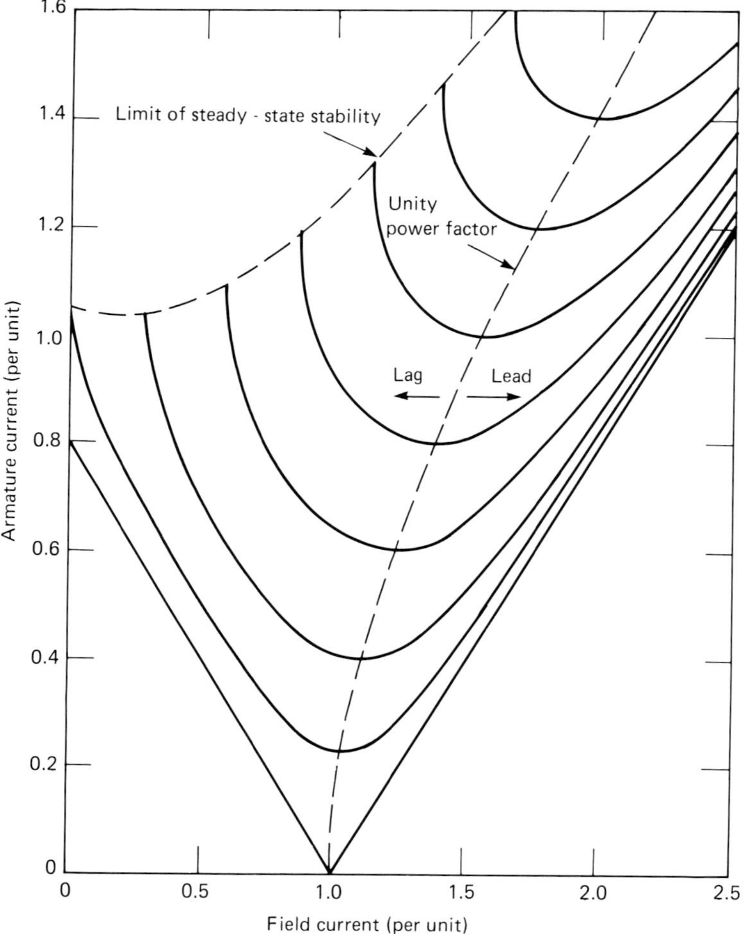

Figure 1.25 V-curves of a salient pole synchronous motor

Figure 1.28 also includes a complete phasor diagram for the maximum power factor condition, at which:

$$\phi_{m.p.f.} = 2\delta_{m.p.f.} = \text{arc}\cos\left(\frac{X_{sd}/X_{sq} - 1}{X_{sd}/X_{sq} + 1}\right) \tag{1.35}$$

In this case the maximum power factor is 1/3.

It can easily be deduced from Figure 1.28 that the machine could be operated at rated current at very nearly its maximum power factor, but at an output of only 0.31 per unit, not a very satisfactory performance. Evidently, for better performance a much higher value of the ratio X_{sd}/X_{sq} is required.

A great deal of attention has been given in recent years to the problem of achieving high values of this ratio in the design of reluctance motors. With conventional salient pole construction the value is unlikely to exceed 4.0, but many commercial machines are still built in this way. For the segmented,[24] flux barrier,[25]

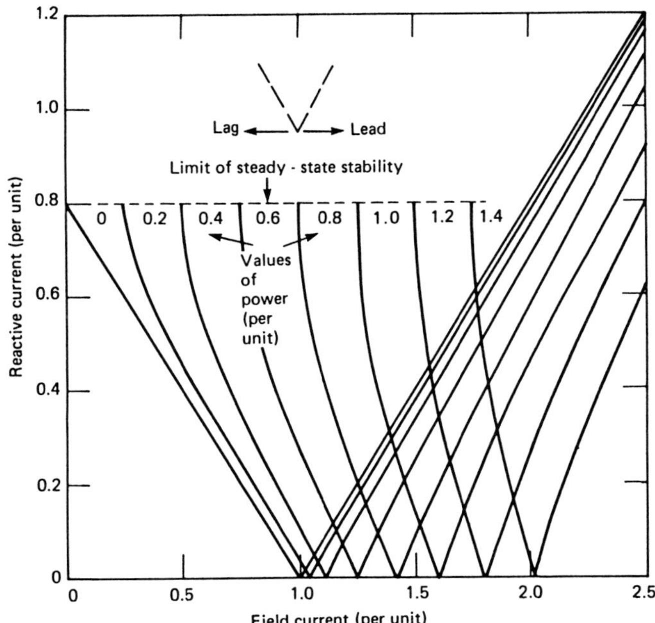

Figure 1.26 Reactive current capability of a cylindrical rotor synchronous motor at constant power

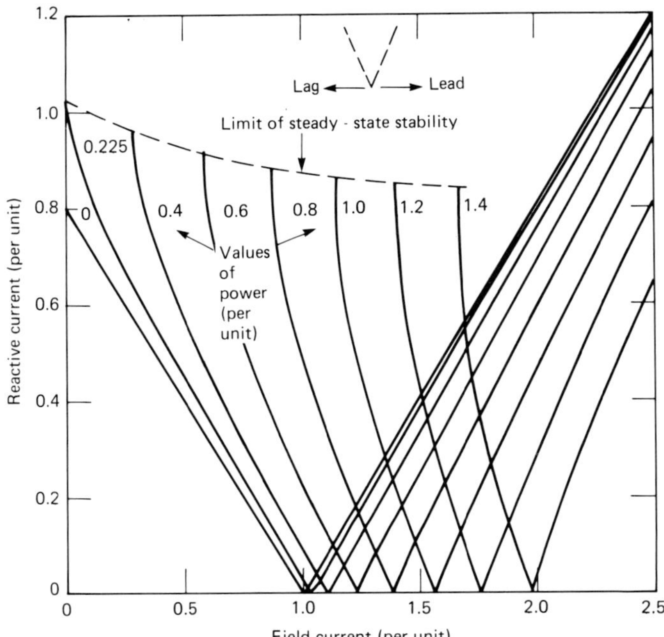

Figure 1.27 Reactive current capability of a salient pole synchronous motor at constant power

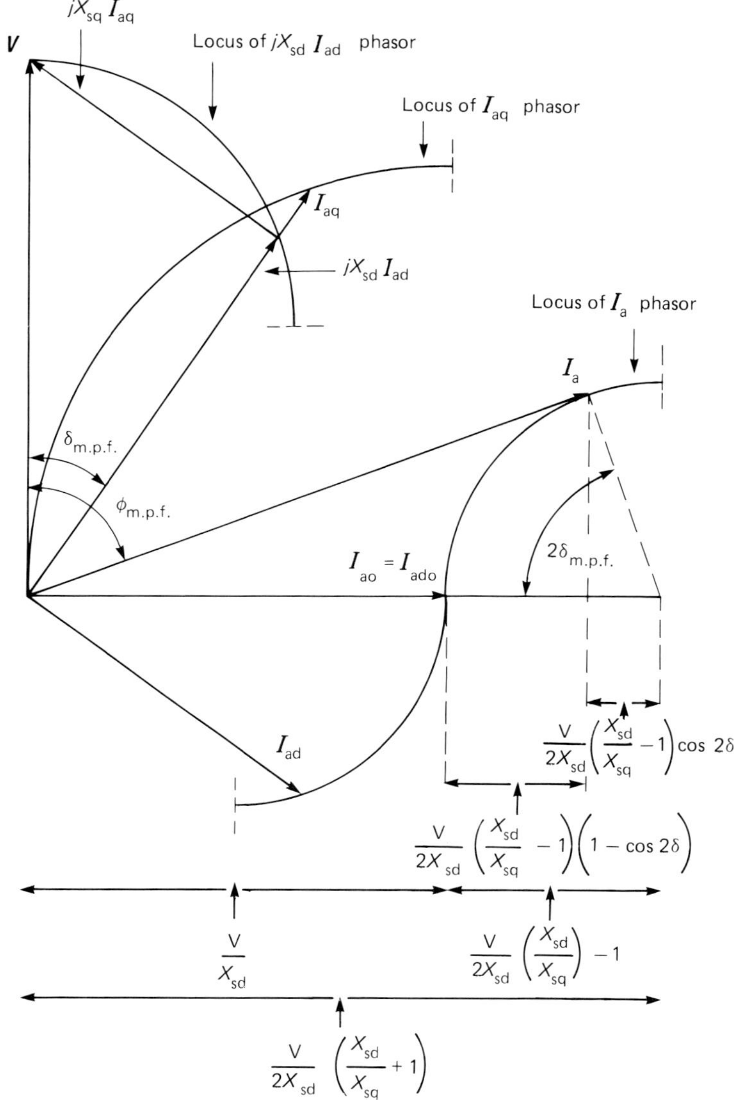

Figure 1.28 Loci of phasors of a reluctance motor between zero load and limit of steady state stability. Complete diagram for condition of maximum power factor

and layer type[26] rotors, values exceeding 7.0 at realistic flux levels are claimed. The axially laminated rotor[27] is probably capable of the highest figure, but it is not easy to construct nor equip with a winding to give good asynchronous performance (see Section 1.4.8).

A value of 9.0 for the ratio X_{sd}/X_{sq} is probably unattainable but would lead to excellent characteristics, as will be shown. A value of 5.0 is certainly attainable, even when saturation is taken into account. These two values will be taken as extremes in what follows.

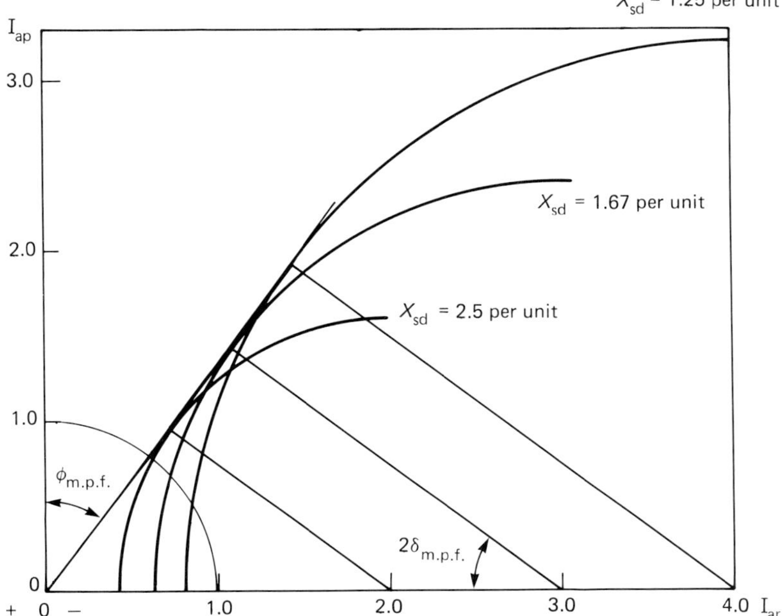

Figure 1.29 Current phasor loci of reluctance motors with $X_{sd}/X_{sq} = 9.0$, for various values of X_{sd}

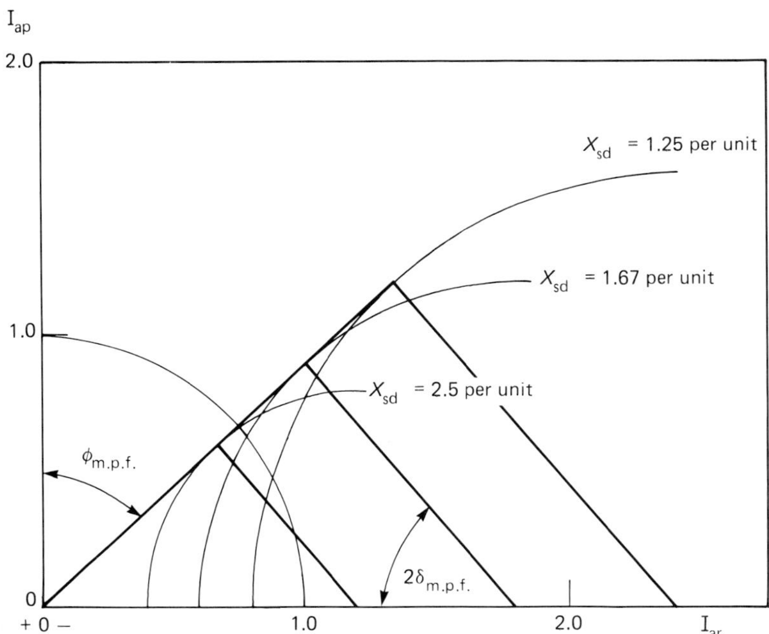

Figure 1.30 Current phasor loci of reluctance motors with $X_{sd}/X_{sq} = 5.0$, for various values of X_{sd}

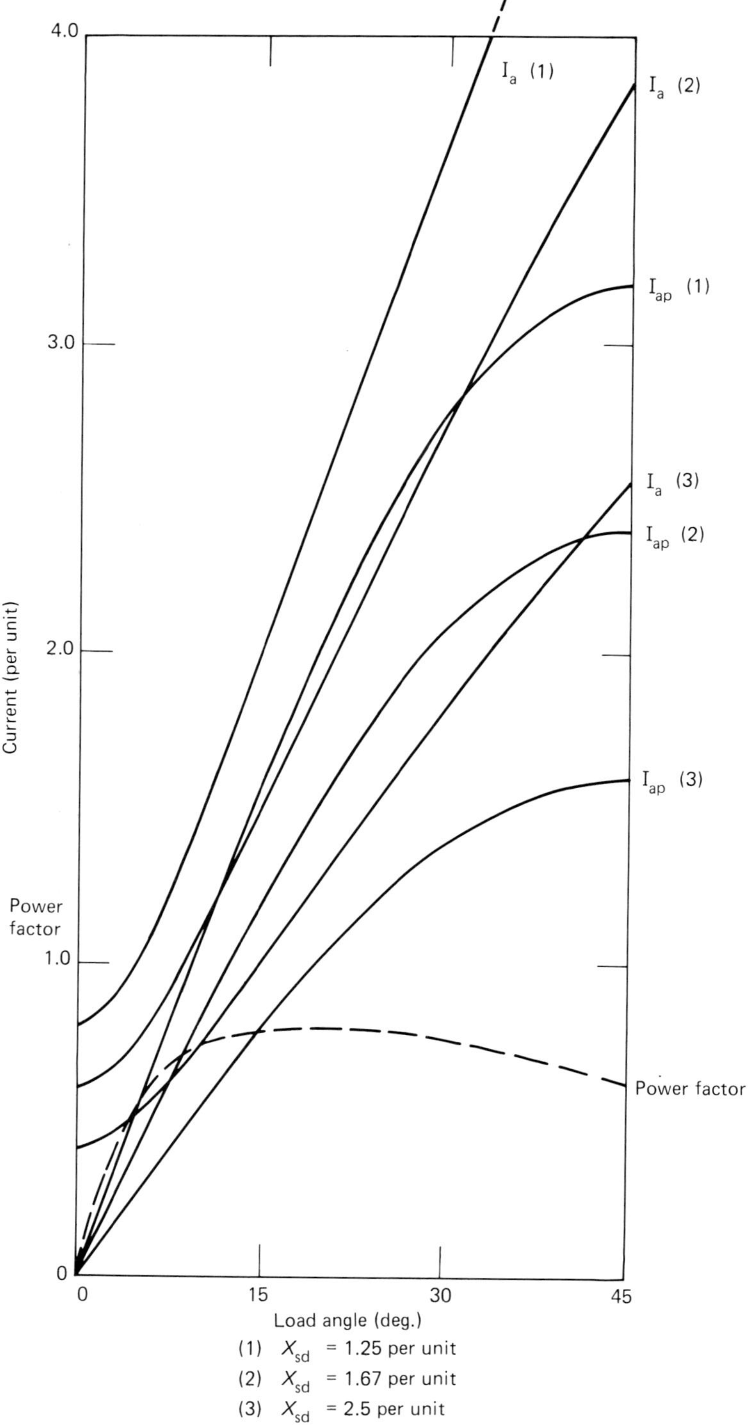

Figure 1.31 Performance characteristics of reluctance motors having a ratio of X_{sd}/X_{sq} of 9.0 and a range of values of X_{sd}

Figure 1.29 shows the loci of the armature current phasors of postulated reluctance motors having values of X_{sd} of 1.25, 1.67 and 2.5 per unit, being reciprocals of 0.8, 0.6 and 0.4 respectively, in all of which the ratio $X_{sd}/X_{sq} = 9.0$. For this ratio the maximum power factor is 0.80.

It is clear from Figure 1.29 that a machine in which $X_{sd} = 1.25$ per unit could not operate in the region of maximum power factor without greatly exceeding rated current. However, a machine in which $X_{sd} = 2.5$ per unit could be operated at rated current at very nearly maximum power factor. In this case the maximum power would be approximately twice the rated power. A machine in which $X_{sd} = 1.67$ could be operated at rated current at a power factor of 0.71 and develop a maximum power of 3.39 times rated power.

Figure 1.30 shows the corresponding loci for machines having the same three values of X_{sd} as previously and a value of the ratio X_{sd}/X_{sq} of 5.0. In this case the maximum power factor is 0.67.

Once again, it is evident that a machine in which $X_{sd} = 1.25$ per unit could not be operated in the maximum power factor region without exceeding rated current. A machine in which $X_{sd} = 2.5$ per unit could be operated at rated current at a power factor of 0.66, almost exactly the maximum, but it would develop a maximum power of only 1.2 times rated power. A machine in which $X_{sd} = 1.67$ could be operated at rated current at a power factor of 0.63 and develop a maximum power of 1.91 times rated power.

Performance characteristics can be calculated using the expressions:

$$I_{ap} = \frac{V}{2X_{sd}} \left(\frac{X_{sd}}{X_{sq}} - 1 \right) \sin 2\delta \tag{1.36}$$

$$\text{and } \phi = \arctan \frac{\left(\dfrac{X_{sd}}{X_{sq}} + 1 \right) - \left(\dfrac{X_{sd}}{X_{sq}} - 1 \right) \cos 2\delta}{\left(\dfrac{X_{sd}}{X_{sq}} - 1 \right) \sin 2\delta} \tag{1.37}$$

Those for the two sets of three machines specified above are shown in Figures 1.31 and 1.32.

It should be noted that, as Equation (1.37) indicates, the power factor for any particular value of δ is independent of X_{sd}, if the ratio of X_{sd}/X_{sq} is fixed.

It can be concluded from the characteristics calculated in this section that the optimum value for X_{sd} is of the order of 1.67 and that the claimed figure of 7.0, for the ratio X_{sd}/X_{sq}, would lead to very good performance characteristics.

1.4.8 Asynchronous capability; the synchronous induction motor

It has already been shown, and it is, in fact, a matter of definition, that the steady state operating torque of a synchronous motor can only be developed at synchronous speed. Therefore, except when the rotor of a synchronous motor is to be brought up to synchronous speed by a separate pony motor, some form of induction motor capability must be provided for starting and run-up purposes. It follows that the accepted, but rather unsatisfactory, term 'synchronous induction motor' could be applied to all synchronous motors except those utilizing a pony motor. However, the term is reserved for a particular type of motor described below.

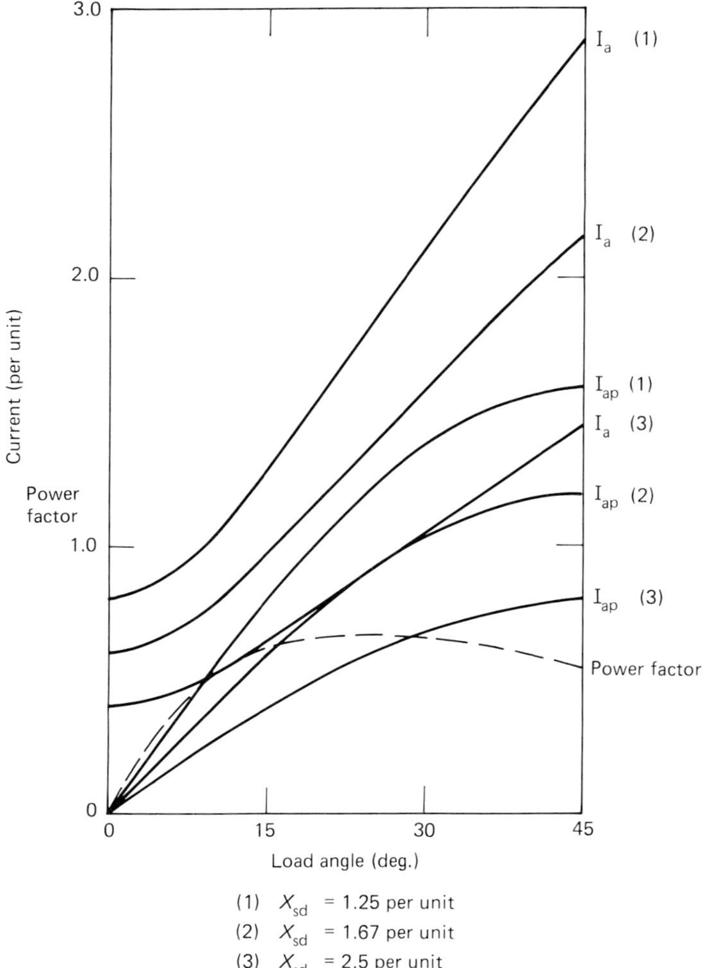

Figure 1.32 Performance characteristics of reluctance motors having a ratio of X_{sd}/X_{sq} of 5.0 and a range of values of X_{sd}

In synchronous motors, other than the synchronous induction motor, the induction motor capability is provided by either cage windings embedded in slots in the pole faces, normally with endrings bridging the interpolar gap, or by the action of eddy currents in the pole faces when these are of solid steel, or both. The torque–speed curve resulting from these reactions would be of the general shape of curves A or B in Figure 1.9(a). In addition, the field winding may be short-circuited, to reduce the danger from the high voltages induced in the winding at low speeds. This winding then develops the typical asynchronous torque of a single-phase rotor winding,[28,29] illustrated qualitatively in Figure 1.33. A similar effect is caused in a salient pole machine by the effect of the saliency itself, but without the torque becoming positive after dipping into the negative region. The resultant torque–speed curve is of the general shape illustrated in Figure 1.33. The reactions are evidently complex[30] and the complexity is increased when there are rectifiers in the excitation circuit.[31]

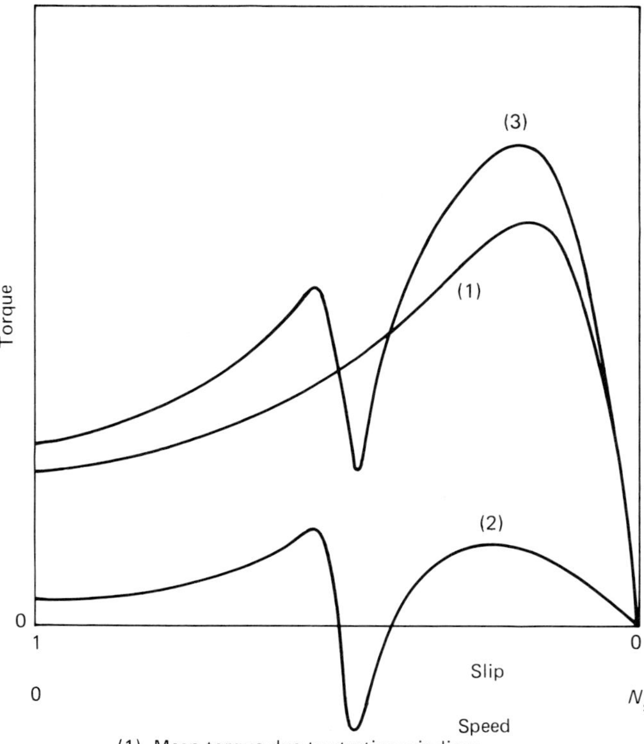

(1) Mean torque due to starting windings
(2) Mean torque due to field winding and saliency
(3) Resultant

Figure 1.33 General form of the asynchronous torque components of a synchronous motor

The synchronous induction motor differs from other synchronous motors in that its rotor is equipped with a polyphase winding which is connected to sliprings. Resistors can be connected to the sliprings to give starting and run-up characteristics similar to those described in Section 1.3.4.

The polyphase winding may be of three-phase or quadrature-phase form. In either case it can be supplied with d.c. in such a way as to simulate the freezing of an instant in an a.c. cycle.[32,33] Its m.m.f. has then the same nearly sinusoidal space distribution as when the winding carries balanced polyphase alternating currents.

The synchronous induction motor seeks to combine the excellent starting properties of a wound rotor induction motor with the desirable operational properties of a synchronous motor, goals which are not totally compatible. Careful examination of the theory and characteristics described in previous sections would indicate that, whereas good induction motor performance demands a relatively high magnetizing reactance, say not less than 2.5 per unit, good synchronous motor performance requires a synchronous reactance of about 1.25 per unit. (In the case of the reluctance motor considered in the previous section, other factors have to be taken into account.)

Its synchronous performance can be improved if a salient pole rotor is used. In this case a conventional excitation winding is wound on the salient poles and the polyphase winding in slots on the pole faces. A compromise on the magnitude of the direct-axis synchronous reactance is still required, but the excitation winding and induction winding can be designed independently for optimum performance.

The construction of the synchronous induction motor is more expensive than that of other comparable synchronous motors but its extra cost is justified whenever its superior starting performance is essential.

The procedure for the synchronizing of a synchronous motor which is run up by means of a pony motor is similar to that for synchronizing a generator to the mains.[34] Self-starting synchronous motors are synchronized by switching-in the d.c. excitation when the rotor approaches synchronous speed. Therefore the asynchronous torque must be capable of accelerating the rotor to a speed at which the maximum excitation current is capable of pulling the rotor into synchronism.

It is convenient to consider here a further function served by the starting windings. When a synchronous motor is subjected to any sudden change such as to cause it to seek to develop a new value of load angle, it is prevented from responding to its natural tendency to oscillate about the new value by the damping action of the asynchronous torque developed by the starting windings. In fact, whereas the polyphase rotor windings of the synchronous induction motor are clearly starting windings which also perform the damping function, the pole-face cage windings of the synchronous motor are regarded as damper windings which also perform the starting function. The damping action also prevents oscillation when the self-starting machine is synchronized by the switching-in of the d.c. excitation.

1.4.9 Some comments on the steady state performance characteristics

When load is applied to the shaft of an induction motor the resulting performance is absolutely fixed by the parameters of the motor, apart from the limited variation available by changing rotor resistance in the case of a motor having a phase-wound rotor. When load is applied to the shaft of a synchronous motor the performance can be modified, usefully, by adjusting the excitation. Therein lies the most important advantage of the synchronous motor.

It can readily be seen from Figures 1.21 and 1.22 that operation at unity power factor at full load, and at near unity power factor over a range of outputs near to full load, can be achieved if required. However, operation at unity power factor may not necessarily be a desired performance condition. A machine may be designed for a specified mechanical output at a leading power factor, thus enabling it to contribute power factor improvement in an otherwise inductive installation. The reactive capability of a synchronous motor is well illustrated by Figures 1.24–1.27. In the limit, the so-called 'synchronous compensator' is a synchronous machine designed for zero mechanical output and near zero power factor operation, leading or lagging.

The neglecting of armature resistance in determining the general form of the performance characteristics, and the complete absence of any consideration of the losses in the excitation circuit, thus far, leaves the efficiency of the synchronous motor open to question. It is not convenient to give efficiency curves of the type shown in Figure 1.10 (page 17) because of the wide range in performance possible for a particular motor. However, it is possible to make some comparisons between the factors which determine the efficiencies of comparable induction and synchronous motors.

The armature windings of comparable machines are similar. Therefore the armature efficiency of the synchronous motor is higher than that of the induction motor when the former is operated in the region of unity power factor. However, as mentioned above, the synchronous motor may be operated deliberately at leading power factor to compensate the lagging power factor of an inductive load, usually an induction motor load. On a hypothetical one-to-one compensation basis the armature efficiencies of the two machines would be equal.

The airgap of the synchronous motor is longer than that of the induction motor.

Therefore a higher magnetizing m.m.f. has to be established, normally from the rotor, i.e. excitation, side, except in the rare case of lagging power factor operation. However, this can be established in a relatively heavy gauge winding, with correspondingly low resistance loss, particularly when the salient pole construction is adopted, allowing more room for the excitation winding. The rotor losses will therefore be lower in the synchronous motor than in the induction motor, but this advantage is partly offset by losses in the excitation circuit external to the winding.

Second order losses in the synchronous motor are lower than those in the comparable induction motor because of the longer airgap of the former.

Overall, the efficiency of the synchronous motor is slightly greater than that of the comparable induction motor, particularly in the case of relatively slow, multi-polar, machines (see Figure 1.10, page 17) of which there are many in practice.

Finally, when the performance of cylindrical rotor and salient pole synchronous motors of equal rating is to be compared it might be expected that the extra term in the output of the latter, as shown in Equation (1.27), gives it a superior performance. However, it can be seen from Figure 1.23 that, for a specified output at unity power factor, or at a particular leading power factor, the two motors considered require approximately the same excitation, the cylindrical rotor machine operating at the greater load angle. Figures 1.21 and 1.22 show that the salient pole motor is a 'stiffer' machine with a greater maximum output for a given excitation. Recalling the fact that the salient pole construction is advantageous for the design of the excitation winding and the consequent efficiency of the machine, it is to be expected that salient pole motors predominate and this is the case. The cylindrical rotor construction would only be preferred for high-speed machines of similar form to the turbo-generator.

1.5 Polyphase a.c. commutator motors

1.5.1 The performance of an induction motor operating with variable voltage applied to its rotor circuit

The induction motor and synchronous motor thus far considered are essentially single-speed machines, when operated from supplies of constant voltage and frequency. The former operates within a narrow range about its nominal rated speed, a little below synchronous speed (see Table 1.1, page 15), the latter at exactly synchronous speed. For many industrial applications variable-speed operation is required and this requirement is considered in detail in Chapter 4. Here, the forms of a.c. motor, specifically designed for variable-speed operation, will be considered. It will be shown that these motors are derivatives of the induction motor, utilizing, in their different ways, the method of varying the speed of this motor by means of a variable voltage applied to its rotor circuit.

The performance of an induction motor operating in this way can be represented by the equivalent circuit of Figure 1.2, reproduced in Figure 1.34(a) with minor changes of notation consistent with that adopted previously for the induction motor. The voltage V_c in the figure is controllable in magnitude and phase and has a frequency, continuously determined by rotor speed, to be exactly equal to the frequency of voltage induced in the rotor from the stator.* It will be assumed

---

*This is a form of non-synchronous operation, to be distinguished from the alternative occurring when the frequency of V_c is determined independently. In general, useful unidirectional torque is developed only when the rotor speed $N = (f_1 \pm f_2)/p$. Therefore, if f_1 and f_2 are independent a particular constant speed is imposed on the machine by the frequencies. This mode of operation can be classified as synchronous. Unfortunately, all such modes are inherently unstable,[35] except that occurring when f_2 is zero, as in the normal synchronous machine.

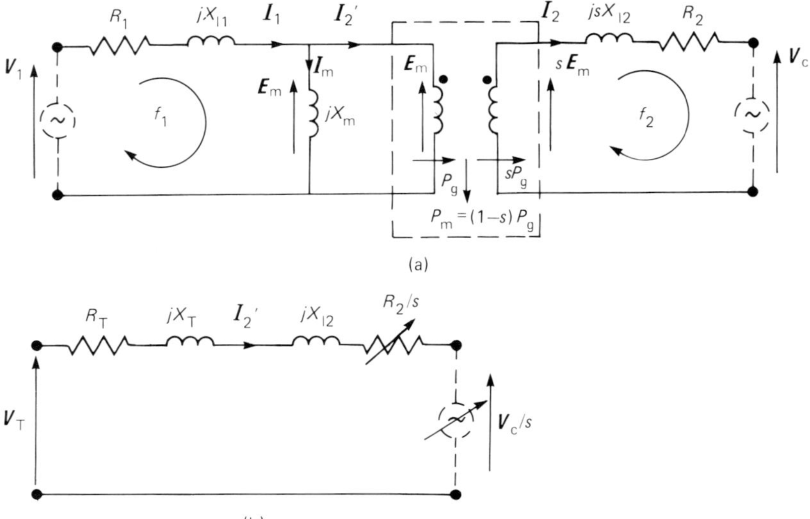

Figure 1.34 Steady state equivalent circuits for a polyphase induction motor with control voltage applied to rotor windings

initially that V_c is obtained from some ideal source having zero internal impedance.

The parameters of the equivalent circuit can be referred to the stator and the circuit simplified by an application of Thévenin's theorem, as was done for the induction motor in Figure 1.4 (page 8). This leads to the equivalent circuit of Figure 1.34(b). It will then be assumed initially that V_c is in phase with V_T, but the effects of varying the phase of V_c will be considered later.

Under ideal no-load conditions I_2' is zero and $V_T = V_c/S_{NL}$, where S_{NL} is the no-load slip. The resulting linear relationship between no-load speed and control voltage is illustrated in Figure 1.35.

For the calculation of performance characteristics on load the further approximation of neglecting the effects of R_1 will be adopted. This is equivalent to assuming Q (see Equation (1.12)) is infinite, and is permissible when, as here, only the general form of the performance characteristics is required. The following equations can then be derived:

$$\frac{I_2'}{I_{2FL}'} = \left[1 - \frac{V_c}{sV_T}\right]\left[\frac{s}{S_{FL}}\right]\left[\frac{S_{MT}^2 + S_{FL}^2}{S_{MT}^2 + s^2}\right]^{1/2} \tag{1.38}$$

$$\frac{T}{T_{FL}} = \left[1 - \frac{V_c}{sV_T}\right]\left[\frac{s}{S_{FL}}\right]\left[\frac{S_{MT}^2 + S_{FL}^2}{S_{MT}^2 + s^2}\right] \tag{1.39}$$

$$\text{Efficiency} = \frac{1 - s}{1 - S_{NL}}$$

where S_{MT} and S_{FL} are the slips for maximum torque and full-load torque, respectively, for normal operation, i.e. for $V_c = 0$. A machine having values of S_{MT} and S_{NL} of 0.25 and 0.046, respectively, giving a maximum torque of 2.81 per unit in normal operation, will be considered for performance calculations.

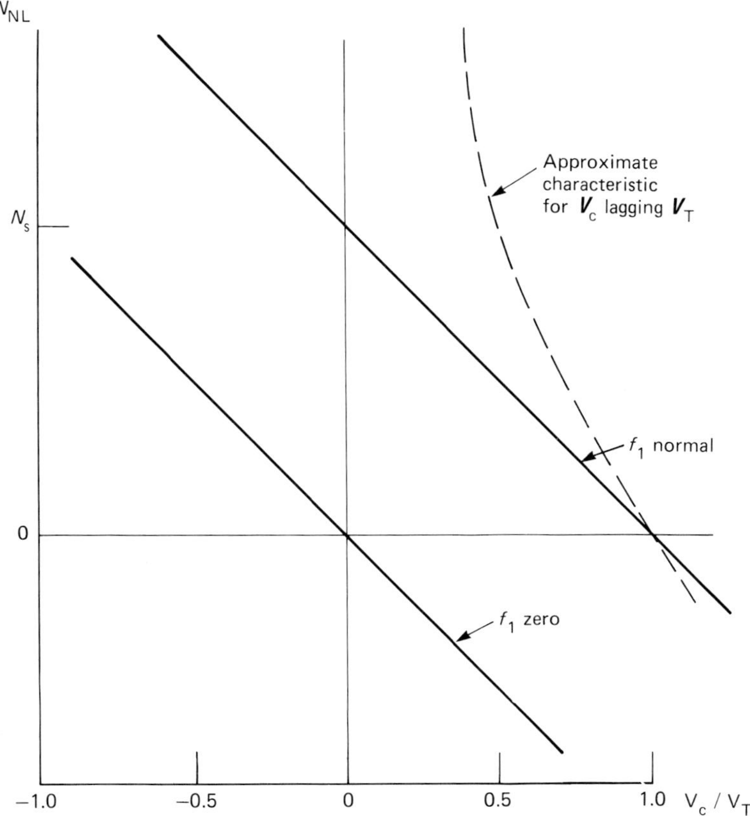

Figure 1.35 Ideal no-load speed characteristic of induction motor with control voltage applied to rotor circuit

Figure 1.36 shows torque–slip and current–slip characteristics for values of the ratio V_c/V_T of 0.5, 0.25, 0, −0.25 and −0.5. The figure also includes the corresponding efficiency curves for the motoring modes only.

It is evident from the approximate characteristics that the speed of an induction motor can be controlled by applying a variable voltage to its rotor circuit, whilst holding constant the voltage applied to the stator windings, and that the performance characteristics on load are potentially useful. However, some caution is required in the interpretation of these characteristics. For example, the efficiencies shown, which of course fall to zero at no-load speed, are idealized values which cannot be attained in practice. Losses in the stator windings have been neglected and it has been assumed that all energy transferred to the source of the control voltage at positive slip, i.e. sub-synchronous speed, is ultimately returned to the constant frequency supply without loss in the control circuit. Similarly, it is assumed that no energy is lost in this circuit when, at negative slip, i.e. super-synchronous speed, energy is supplied to the rotor through this source. Nevertheless, the efficiencies shown are not grossly misleading.

The characteristics appear to be more suitable for loads of the fan type, where torque is proportional to speed squared, than for loads demanding constant torque. If, however, the control voltage has a quadrature component such that V_c *lags* V_T the characteristics are considerably modified. The obvious reason for introducing

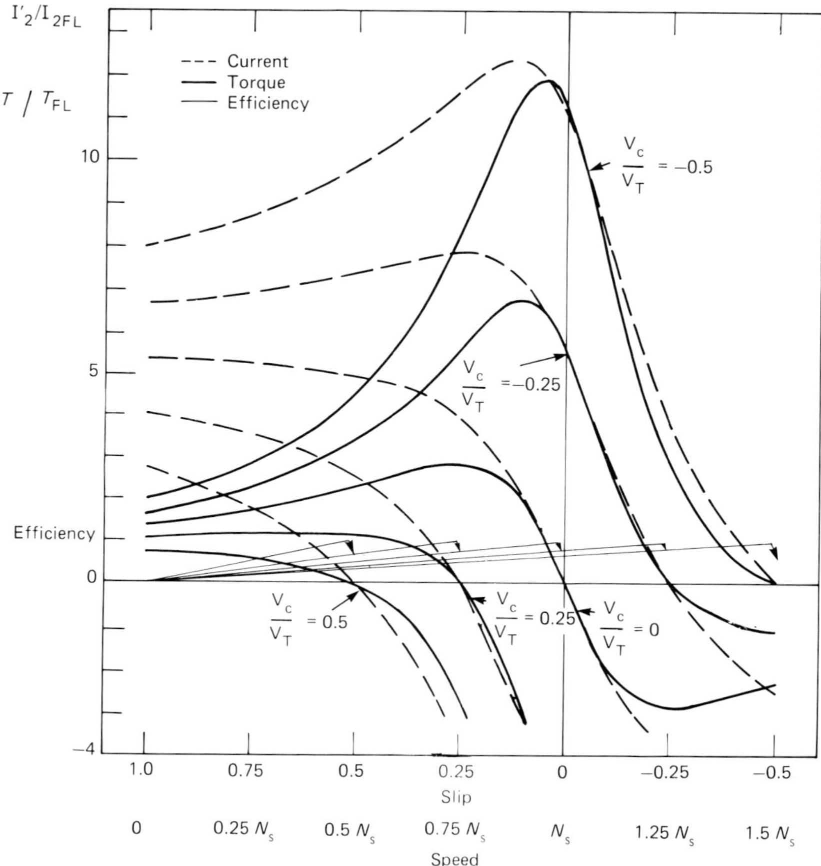

Figure 1.36 Performance characteristics of an induction motor with various values of control voltage applied to the rotor circuit

such a component, in the first place, is to modify the power factor. Power factor does, in fact, need modification.

The phase angle of I_2', as obtained from the equivalent circuit of Figure 1.34(b), gives a very misleading indication of the magnitude of the overall power factor. The magnetizing current of the stator and the effects of the control circuit have to be taken into account. If the impedance of the control source, neglected for the approximate treatment above, were, say, $R_c + jX_c$ then it would appear in the equivalent circuit of Figure 1.34(b) as $(R_c + jX_c)/s$, introducing a large reactance at low values of slip where, normally, the large effective resistance predominates.

When a quadrature component of voltage is introduced there is not only an improvement in power factor, but also a change in the no-load characteristic, indicated qualitatively, in Figure 1.35,[36,37] and an overall improvement in performance characteristics. In particular, the maximum torques for positive values of V_c are increased and the machine becomes capable of developing rated torque over a wide range of speeds.

The two most important types of motor which incorporate the method of speed control discussed above are described in the two sections which follow. For the detailed analysis of the performance of these two machines reference should be

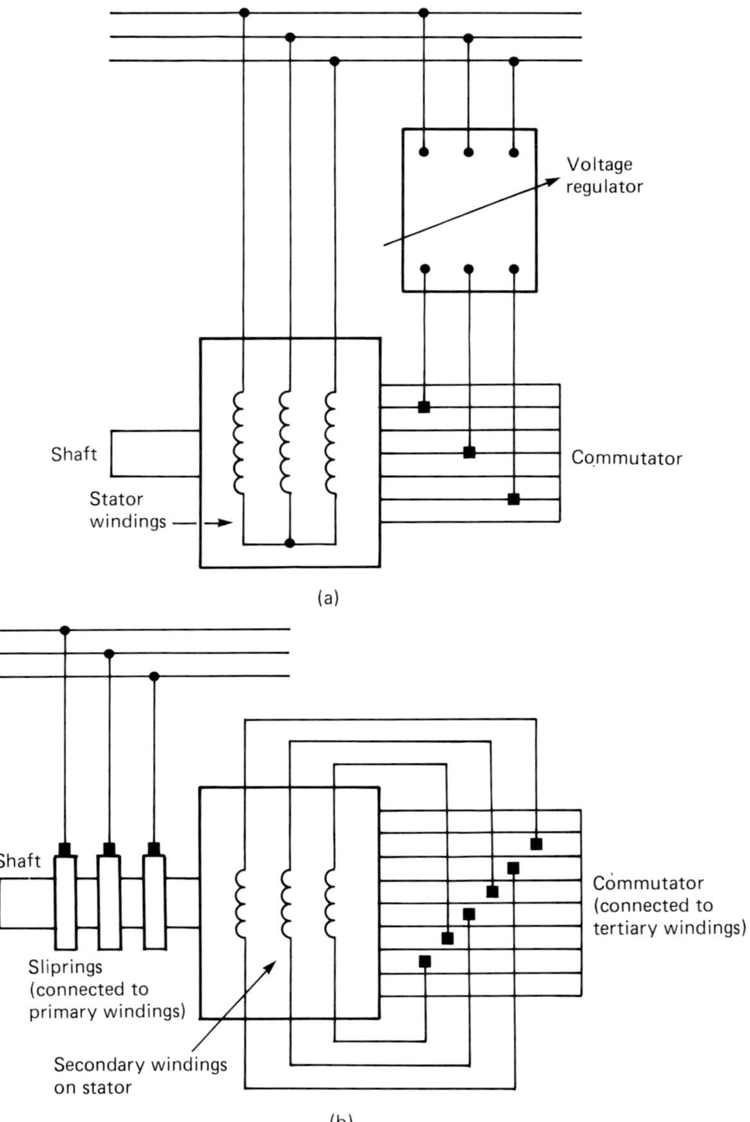

Figure 1.37 Schematic diagrams of connections of two polyphase commutator motors. (a) Polyphase shunt commutator motor. (b) Schrage motor.

made to the specialized texts.[38,39] The representative performance characteristics given in Figure 1.37 are based on those given in these texts.

1.5.2 The polyphase shunt commutator motor

In all practical implementations of the control system described in the previous section the source of energy for the control voltage must be the constant frequency mains, to which the stator windings of the motor are also connected. Some form of frequency changer must therefore be interposed between the mains and the rotor windings.

A convenient form of frequency changer is a commutator connected to the rotor windings. The action of a commutator on a rotating machine changes the internal frequency in the rotor conductors of the machine to an external frequency, appearing at brushes bearing on the commutator, which is exactly equal to the frequency of voltages induced in the stator windings of the machine. The phase of the voltage between a pair of brushes on the commutator, relative to the induced e.m.f. in a rotor phase depends on the angular displacement of the chord between the brushes. Therefore, brush position can be used to ensure that the control voltage is presented to the rotor at the required phase angle. Thus, a motor with three-phase stator windings, having a rotor with a commutator winding, when connected to a voltage regulator in the configuration illustrated in Figure 1.38(a) will, in principle, give a performance similar to that described in the previous

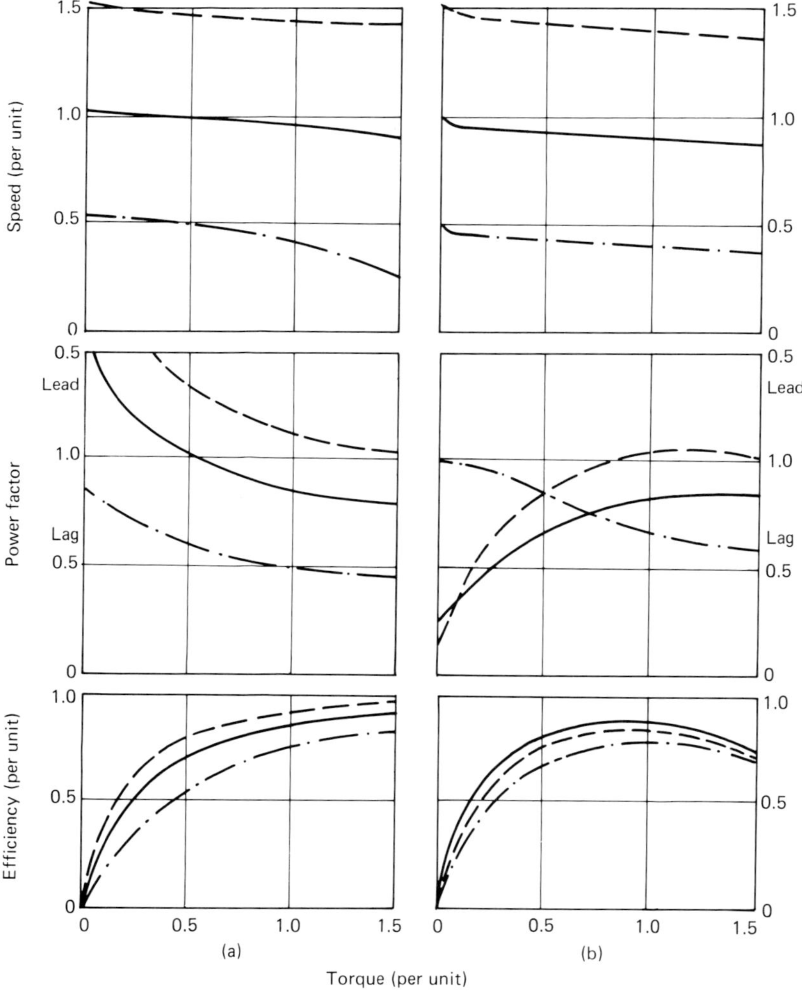

Control voltages having quadrature component and set for approximate no-load speeds of $1.5\,N_s$ --- , N_s —— and $0.5\,N_s$ —·—

Figure 1.38 Representative performance characteristics. (a) Polyphase shunt commutator motor. (b) Schrage motor

section. There are, however, several practical problems. These are discussed in detail elsewhere[40,41] and will only be mentioned briefly here.

Unassisted commutation is best at synchronous speed since the rotational voltage induced in the coils being commutated is then zero. It deteriorates rapidly as slip increases but can be improved by the use of special discharge windings. Even so, speed ranges are best confined to the range $0.5\,N_s < N < 1.5\,N_s$.

It has been stated above that the control voltage should have a quadrature component. This can be achieved by the shift of brush axis on the commutator, but if the shift is correct on one side of synchronous speed it is incorrect on the other. Therefore the phase shift must be achieved in the control circuit. One method is to use a separate auxiliary transformer. A better method is to incorporate the phase shift in the voltage regulator itself. In one type of regulator, the double induction regulator, this can be done in such a way that the phase shift is zero at the extremes of the voltage – and therefore speed – range and a maximum in the middle of the range. This is in keeping with what is required when the range is symmetrical about synchronous speed.

Representative performance characteristics for the shunt commutator motor are given in Figure 1.38(a).

1.5.3 The Schrage motor

The Schrage motor is a self-contained embodiment of the principles described in Section 1.5.1. Referring to Figure 1.37(b) the primary windings, on the rotor, are connected to the mains via sliprings. The secondary windings are on the stator. When the latter are short-circuited the machine is simply an induction motor of inverted construction. The magnetic field rotates at synchronous speed N_s, relative to the rotor, which itself rotates in the opposite direction at 'normal' speed N. Therefore, the magnetic field rotates at speed N_s-N relative to the stator and induces slip frequency voltages in the secondary windings. The speed of the machine can be controlled by voltages of slip frequency applied to these windings.

There is a further winding on the rotor, the tertiary winding, which is connected to a commutator. The frequency of voltages induced in the coils of the tertiary windings is, of course, mains frequency. However, the frequency of voltages appearing between a pair of brushes on the commutator is exactly equal to that induced in the secondary windings on the stator. Thus, the voltage appearing at the commutator satisfies the first essential condition for controlling the speed of the machine.

For the two-pole construction illustrated, each phase of the secondary is connected to a pair of brushes. The brush arms are mounted on two racks, each rack carrying one brush from each phase, the brush arms being displaced from each other by 120°. Operation of the brushgear mechanism causes the brush arms to be displaced by equal angular amounts in opposite directions, and hence the brushes of each pair to be separated by equal amounts. Thus, the magnitude of the voltage appearing between each pair of brushes can be controlled without change of phase, the phase being set by rocking the gear as a whole with the displacement fixed. The basic requirements for speed control have thus been met.

It is desirable that a quadrature component of control voltage be introduced for the improvement of both power factor and the general form of the performance characteristics. On this machine the only method of achieving this is by an overall shift of brush axis, but a constant shift in the phase of the control voltage is unsatisfactory, as already pointed out. However, the brushgear can be so arranged that the displacement of one set of brushes is greater than that of the other. Thus, the control voltage can be varied in both magnitude and phase, the desirable change of phase being relatively small.

For operation at normal values of mains voltage the number of turns on the tertiary winding must be less than that on the primary winding to give satisfactory commutation. However, as on the shunt commutator motor, commutation can be improved by the use of special discharge windings.

Representative performance characteristics for the Schrage motor are given in Figure 1.38(b).

1.6 Direct current motors

1.6.1 Note on construction

The stator of a d.c. motor is usually of salient pole form. The poles, usually solid, carry the field windings which, when excited with d.c. establish the main magnetic field. The rotor is cylindrical, laminated and carries the armature winding which is connected to a commutator, through which it is fed with d.c., externally. In addition, there are usually interpoles. These are poles, of smaller cross-section than the main poles, situated between the latter. They carry windings which are excited by the armature current. Their function is to assist commutation.

In some machines there are compensating windings in the faces of the main poles, which are also excited by the armature current. Their function is to develop an m.m.f. which, ideally, is equal and opposite to that developed by the armature current. In their absence this armature reaction m.m.f. modifies the magnetic field by what is known as 'the demagnetizing effect of cross-magnetizing armature reaction due to saturation'.[42,43]

Finally, it should be noted that in some very small machines the main magnetic field may be established by permanent magnets.

1.6.2 Circuit representation and performance equations

It is possible to regard a simple d.c. motor, with its field and armature windings connected in parallel, as a limiting case of the a.c. shunt commutator motor of Section 1.5.2, when the latter is fed at zero frequency. The no-load speed characteristic for zero frequency operation, included in Figure 1.35 is, in fact, directly applicable to a d.c. machine. It can be shown that the equivalent circuit of Figure 1.2 can be modified to represent a d.c. machine by including a representation of the commutator as a frequency changer and referring all parameters to the rotor side. For our purposes it will suffice to adopt the well-known circuit representation of a d.c. motor, shown in Figure 1.39.

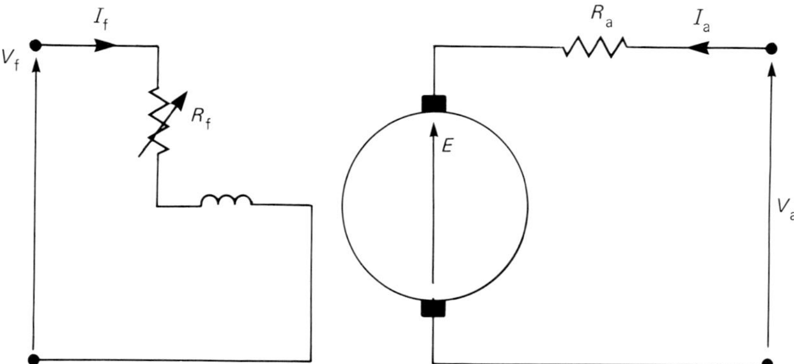

Figure 1.39 Circuit representation of a d.c. motor

For this representation it will be assumed that the magnetic circuit is linear, that the brush axis is exactly in quadrature with the magnetic field and that there is perfect compensation of the armature m.m.f. Thus, the magnitude of the magnetic field can be taken to be linearly related to the field current. The resistance of all windings in the armature circuit and of the brushes (the resistance of the latter actually non-linear) is taken to be represented by the resistor R_a. The following performance equations can be derived:

Voltage equation $\qquad V_a = E + R_a I_a$

where $\qquad\qquad\qquad E = (KI_f)\omega$

whence, power developed $= EI_a$

and torque $\qquad\qquad = \dfrac{EI_a}{\omega} = (KI_f)I_a$

where the speed is expressed in radians per second for the convenience of having the same constant K in the expressions for e.m.f. and torque, and the bracketed term (KI_f) can be treated as a single parameter.

From these equations the following expressions relating the variables can be derived:

Current–speed $\qquad\qquad I_a = \dfrac{V_a - (KI_f)\omega}{R_a}$ $\qquad\qquad$ (1.40)

Torque–speed $\qquad\qquad T = \dfrac{(KI_f)V_a - (KI_f)^2\omega}{R_a}$ $\qquad\qquad$ (1.41)

Power–speed $\qquad\qquad P = \dfrac{(KI_f)V_a\omega - (KI_f)^2\omega^2}{R_a}$ $\qquad\qquad$ (1.42)

The parameters in these equations can be treated as per-unit values if they are expressed in terms of base values derived from rated values of voltage, current and the parameter (KI_f). The normal maximum value of (KI_f) is taken as its rated value. It follows that the base value of power is an input power and the base value of speed is the ideal no-load value corresponding to zero armature current and rated voltage.

1.6.3 Separate and shunt excitation

In shunt excitation, from a source of zero internal resistance, and separate excitation the field current is independent of the armature current.

Figure 1.40 shows a comprehensive set of steady state theoretical performance characteristics for a machine having an armature resistance of 0.05 per unit, a typical mid-range value. Characteristics are depicted for the following three types of operation:

(1) 'Normal', $V_a = 1$ per unit, $(KI_f) = 1$ per unit.
(2) With field weakening, $V_a = 1$ per unit, $(KI_f) = 0.5$ per unit.
(3) At reduced armature voltage $V_a = 0.5$ per unit, $(KI_f) = 1$ per unit.

Figure 1.40(a) shows the characteristics for (1) and (2) over the complete theoretical range. Figure 1.40(b) shows characteristics for (1), (2) and (3) in the region of full load.

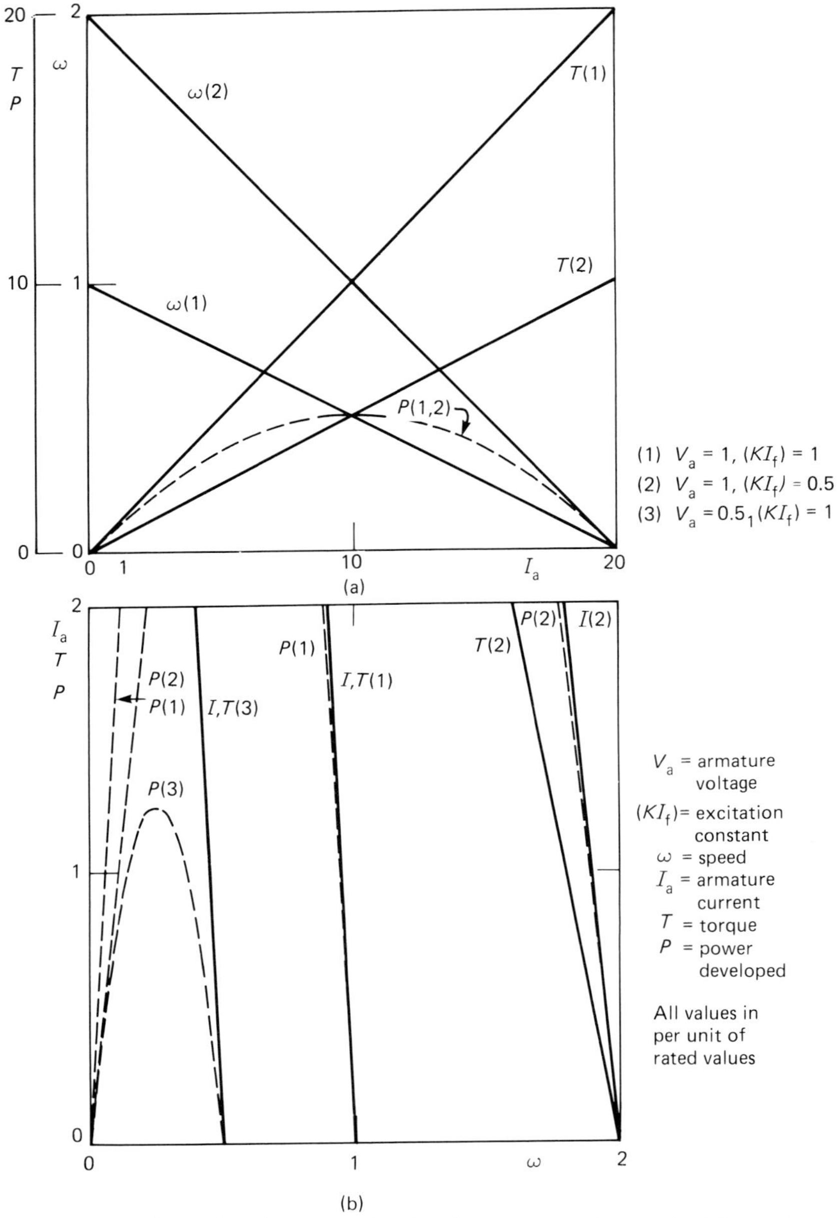

Figure 1.40 Performance characteristics of a shunt or separately excited d.c. motor.
(a) Complete range. (b) Full-load region

It is evident that starting current would be excessively high unless steps are taken to reduce it. The normal method is to increase the resistance in series with the armature by means of a starter incorporating a series of reducing steps in resistance. This is unnecessary when wide-range armature voltage control is adopted.

In the region of full load the motor can be regarded as a nearly constant speed machine, even when the field current is reduced to half its normal value.

It can be seen that the curve of output power is symmetrical about a maximum at the mid-point of the theoretical current range in Figure 1.40(a) or at the mid-point of the actual speed range in Figure 1.40(b). It can easily be shown that this maximum value is equal to $V_a^2/4R_a$, giving values of power of 5 per unit at a current of 10 per unit in cases (1) and (2), and 1.25 per unit at a current of 2.5 per unit in case (3). Thus the maximum power is not relevant to the practical limits in performance of the machine under consideration. (See, however, the characteristics in Figure 1.42 for the motor considered in the next section.)

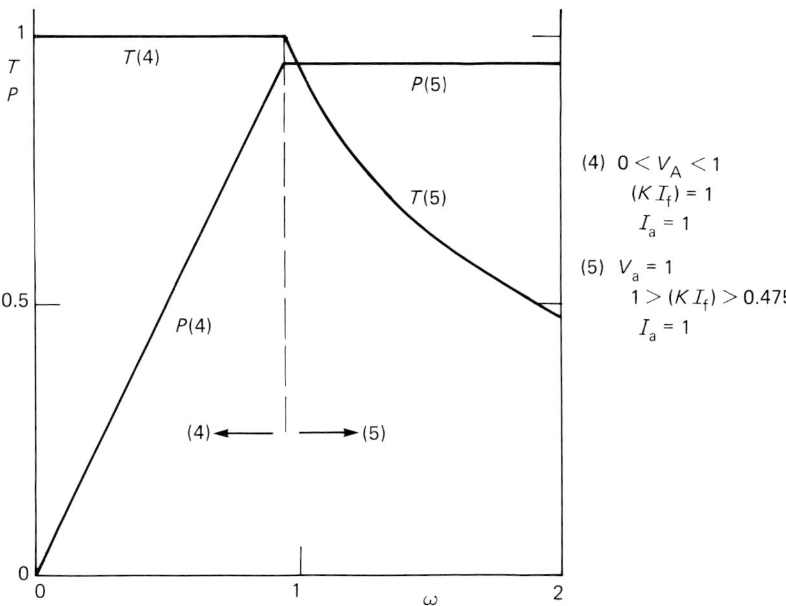

Figure 1.41 Torque and power characteristics of a separately excited d.c. motor at rated current. (Notation as for Figure 1.40)

Figure 1.41 shows the curves of torque and power for the above motor when:

(4) With the field current at its normal value the armature voltage is varied from zero to the rated value.
(5) With the armature voltage at the rated value, the field current is varied between the normal value and about half of this value.

1.6.4 Permanent-magnet excitation

Permanent-magnet machines are similar to separately excited machines in operating with the equivalent of a constant value of (KI_f), but only one. They are manufactured in smaller ratings down to less than 5 W. Consequently, they have relatively high per-unit values of R_a.

Figure 1.42 shows a set of performance characteristics for a motor for which $R_a = 0.2$, when it is operated at rated voltage. It can easily be seen that, with an appropriate change of scale, these characteristics can be adapted to apply to a motor having any value of R_a. Compare, for example, Figure 1.40(a).

A machine of 5 W rating might have an armature resistance of 0.3 per unit. Such machines can be operated over the full range shown in Figure 1.42.

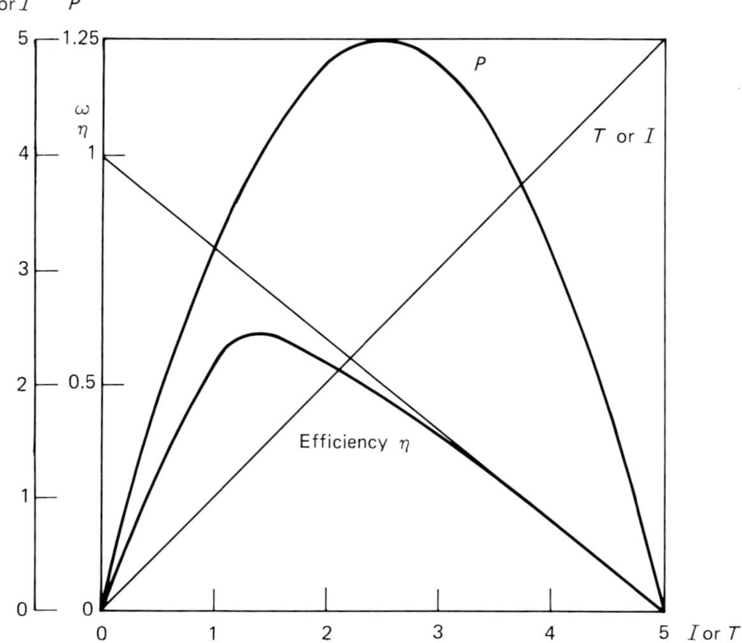

Figure 1.42 Performance characteristics of a permanent magnet d.c. motor. (Notation as for Figure 1.40 ($V_a = 1$))

1.6.5 Series excitation

In a series-excited motor the field and armature are connected in series across the supply. Consequently, the performance equations given above require modification.

The supply voltage and current will be denoted V and I respectively. The total resistance of the armature and field circuits in series, will be denoted R. The performance equations can then be expressed in the following form:

$$V = E + RI \tag{1.43}$$

$$E = KI\omega \tag{1.44}$$

$$T = KI^2 \tag{1.45}$$

$$\text{whence} \quad I = \frac{V}{K + K\omega} \tag{1.46}$$

$$T = K\left[\frac{V}{R + K\omega}\right]^2 \tag{1.47}$$

$$P = \omega K\left[\frac{V}{R + K\omega}\right]^2 \tag{1.48}$$

Again, the parameters in these equations will be expressed as per-unit values based on three base values: the rated voltage and current, as before, and the base value of

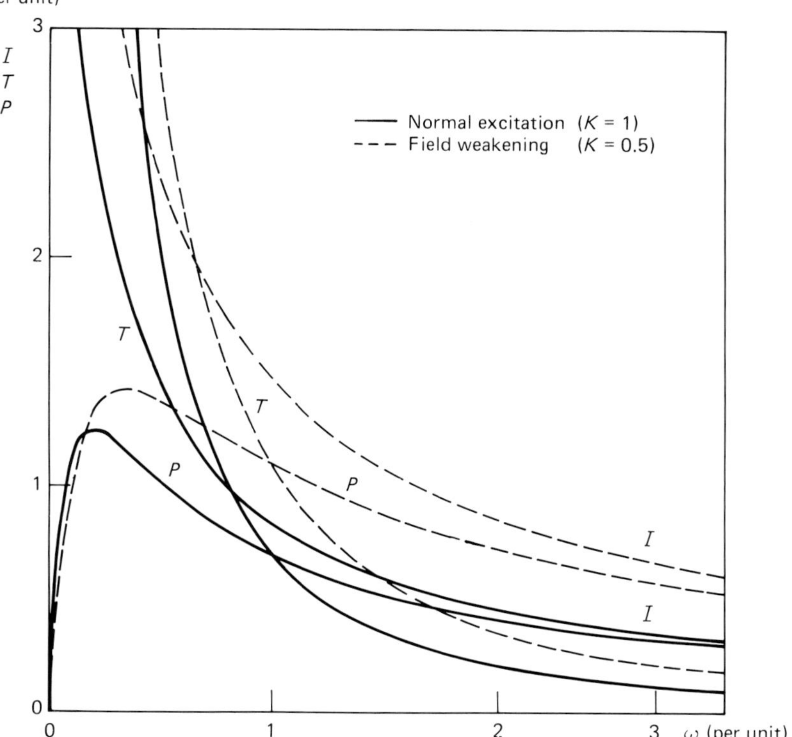

Figure 1.43 Performance characteristics of a series-excited d.c. motor

the e.m.f. and torque coefficient (KI), taken as the value corresponding to rated current.

As a consequence of the series excitation, theoretically, the speed approaches infinity as the current approaches zero. Thus, the ideal no-load speed is not 1 per unit. Some caution is therefore required when interpreting the per-unit values of speed in the calculated performance characteristics which follow. However, it is a simple matter to relate any per-unit speed to the per-unit speed when, for example, rated torque is developed.

Figure 1.43 shows two sets of performance characteristics for a series motor. The first set, for normal excitation, corresponds to a motor having a resistance R of 0.2 per unit and normal excitation, effectively $K = 1$. For the second set, a value for K of 0.5 has been taken and a reduced value for R of 0.175 has been used.

Field weakening can be achieved in practice by, either, a diverting resistor connected across the field winding, or, tapping connections to the winding itself. In each case there is a reduction of the resistance of the field circuit.

The outstanding feature of the characteristics is the very large starting torque which falls off rapidly as speed increases. This is the feature which makes the motor suitable for traction and for hoists. The outstanding problem, or cause for caution, is the very high no-load speed. Where there is a danger of losing the load, a relatively weak shunt field may be incorporated to limit the no-load speed.

The practical methods for controlling the excitation are much less convenient than those for the shunt- or separately excited machine, but the principle is clear. There is, however, the further alternative of voltage control. The performance equations are presented in a way which enables the effect of a change in voltage at a given speed to be estimated readily.

1.6.6 Efficiency

An estimate of efficiency based only on the armature I^2R loss in the simple model would be misleading, particularly at low values of current.

For a shunt- or separately excited machine operating over a small range of speeds, there are three main components to the losses, as follows:[44,45]

(1) Brush contact losses which are proportional to armature current.
(2) Losses in the armature circuit, including stray-load losses, which are proportional to the square of armature current.
(3) 'Constant' losses due to friction and windage, no-load iron losses in the armature iron, and the power consumed in the excitation circuit.

These may be summarized as:

$$P_{\text{loss}} = K_1'I + K_2'I^2 + K_3'$$

Whence the efficiency is:

$$\eta = 1 - \frac{P_{\text{loss}}}{P_{\text{in}}}$$

$$= 1 - \left(\frac{K_1'I + K_2'I^2 + K_3'}{VI}\right)$$

$$= 1 - (K_1 + K_2I + K_3/I)$$

where the term V is absorbed in the modified constants.

The three components of the *deficiency* are recorded in Figure 1.44, leading to the graph of efficiency shown. The general order of the various terms is reasonably typical but higher efficiencies are claimed for some modern machines. The efficiency curve of a relatively inefficient machine is included in Figure 1.44.

For the series motor the estimation of efficiency is more approximate because both speed and current vary significantly with load. Friction and windage may be taken as inversely proportional to the square of current, and the deficiency takes the form $K_1 + K_2I + K_3/I + K_4/I^3$.[46,47] The extra term is included in Figure 1.44 to yield the approximate form of the efficiency curve of a series-excited motor.

1.7 The single-phase series commutator motor

Although, strictly, it does not fall within the scope of this handbook it is convenient at this point to consider briefly the performance of the single-phase series commutator motor. This is, in effect, the motor of Section 1.4.6 operated on a.c. However, it should be noted that, for a.c. operation, both stator and rotor should be of laminated construction, in the case of the stator to reduce the effects of eddy currents induced by the a.c. excitation.

For satisfactory power factor the armature reaction m.m.f. must be fully compensated. This is best achieved by transformer action from the armature to the

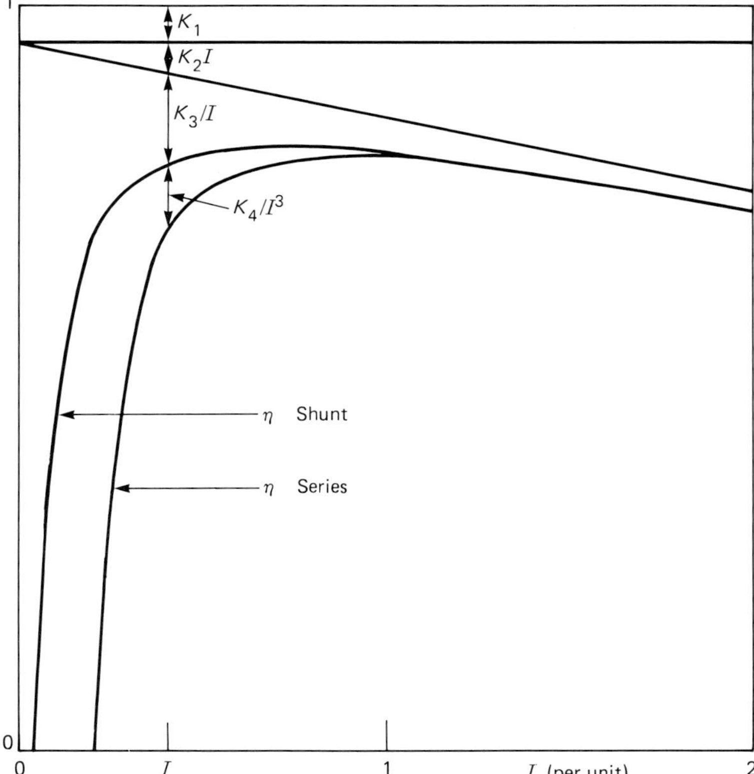

Figure 1.44 Efficiency and components of deficiency of d.c. motors

pole face compensating winding, rather than by series connection of the armature and compensating windings, as in the d.c. case.[48,49] Simple, uncompensated, series motors are, however, used in large numbers in handtools and in other applications of motors of low rating.

The performance characteristics of fully compensated a.c. series motors are generally similar to those of the d.c. motor shown in Figures 1.43 and 1.44. In addition, the power factor may exceed 0.8 over most of the speed range. As already noted these characteristics are ideally suited to the traction application.

When compared with its d.c. counterpart the a.c. motor has a special commutation problem, arising from the transformer e.m.f. induced from the excitation winding into the coils undergoing commutation. This e.m.f. is frequency-dependent and cannot be compensated, at all speeds, by the normal use of interpoles. It can be reduced by reducing the operating frequency, and there are large-scale traction systems in Europe and the US utilizing frequencies of 16.66 Hz and 25 Hz, respectively.* It should, however, be pointed out that successful traction motors, for operation at 50 Hz, have been built in France.

---

*It would seem to be logical to reduce the operating frequency of the motor to zero, as is done in the modern UK mainline traction system. However, full-wave rectification of a single-phase, 50 Hz supply introduces a 100 Hz component which contributes to the commutation problem.

Detailed analysis of the single-phase series commutator motor can be found in the specialized texts.[50,51]

1.8 Transient operation

It has been emphasized that the characteristics presented in this chapter are the steady state characteristics of the individual machines, derived from circuit models which cannot be utilized for the calculation of transient performance. For this purpose the differential voltage equations of the machine and the equation of motion of the machine and its coupled mechanical load are required. Furthermore, if the motor forms part of a complete control system the equations of other elements of the system have to be taken into account. In some cases it is necessary to take into account non-linear effects such as that of saturation.

The required theory is beyond the scope of this chapter. It is covered, in part, in some of the references already cited[52–54] and in others.[55,56] These texts themselves contain many references to the literature.

Nomenclature

$\mathbf{F}, \mathbf{\Psi}$	space vectors
F, ψ	magnitudes of space vectors
V, E, I	phasors
V, E, I	magnitudes of phasors
V, E, I	d.c. quantities

References

1. SAY, M. G. *Alternating current machines* (5th edn). Pitman (1983)
2. SAY, M. G. and TAYLOR, E. O. *Direct current machines.* Pitman (1980)
3. CLAYTON, A. E. and HANCOCK, N. N. *The performance and design of direct current machines.* Pitman (1959)
4. TAYLOR, E. O. *The performance and design of a.c. commutator motors.* Pitman (1958)
5. HINDMARSH, J. *Electrical machines and their applications* (3rd edn). Pergamon (1977)
6. JONES, C. V. *The unified theory of electrical machines.* Butterworth (1967)
7. FITZGERALD, A. E., KINGSLEY, C. and KUSKO, A. *Electric machinery* (3rd edn). McGraw-Hill (1971)
8. SLEMON, G. R. *Magnetoelectric devices.* Wiley (1966)
9. SAY (1983) op. cit.
10. JONES, B. L. and BROWN, J. E. 'Electrical variable-speed drives', *Proc. Inst. Elec. Engrs*, A **131**, 7, 516 (1984)
11. BROWN, J. E. and VAS, P. 'Review of methods of braking three-phase induction motors', *Proceedings, Drives, Motors and Controls*, **85**, p. 169 (1985)
12. NATIONAL ELECTRICAL MANUFACTURERS' ASSOCIATION. *Motors and generators.* NEMA Publication No. MG 1. (1978)
13. BRITISH STANDARDS INSTITUTION. *Specification for general requirements for rotating electrical machines.* BS 4999:Part 42 'General characteristics' (superseded by Part 112). BSI (1976)
14. ALGER, P. L. *The nature of induction machines.* Gordon and Breach (1965)
15. CHRISTOFIDES, N. 'Origins of load losses in induction motors with cast aluminium rotors', *Proc. Inst. Elec. Engrs*, **112**, 2317 (1965)
16. ALGER (1965) op. cit.
17. ALGER (1965) op. cit.
18. SAY (1983) op. cit.
19. CHRISTOFIDES (1965) op. cit.

20. MORGAN, T., BROWN, W. E. and SCHUMER, A. J. 'Reverse-rotation test for the determination of stray-load loss in induction motors', *Trans. Am. Inst. Elec. Engrs,* **58,** 319 (1939)
21. CHALMERS, B. J. and WILLIAMSON, A. C. 'Stray losses in cage induction motors', *Proc. Inst. Elec. Engrs,* **110,** 1773 (1963)
22. SCHWARZ, K. K. 'Survey of basic stray losses in cage induction motors'. *Proc. Inst. Elec. Engrs,* **111** (Part IX), 1565–1574 (1965)
23. BARTON, T. H. and AHMAD, V. *The measurement of induction motor stray loss and its effect upon performance.* Institution of Electrical Engineers Monograph 255 (1958)
24. LAWRENSON, P. J. and GUPTA, S. K. 'Developments in the performance and theory of segmental-rotor reluctance motors'. *Proc. Inst. Elec. Engrs,* **114,** 5, 645 (1967)
25. FONG, W. and HTSUI, J. S. C. 'New type of reluctance motor'. *Proc. Inst. Elec. Engrs,* **117,** 3, 545 (1970)
26. CHALMERS, B. J. and MULHI, A. S. 'Design and performance of reluctance motors with unlaminated rotors'. *Inst. Elec. Electronic Engrs. Trans,* **PAS–91,** 4, 1562 (1972)
27. CRUIKSHANK, A. J. O., MENZIES, R. W. and ANDERSON, A. F. I. 'Axially laminated anisotropic rotors for reluctance motors'. *Proc. Inst. Elec. Engrs,* **113,** 12, 2058 (1966)
28. JONES (1967) op. cit.
29. BARTON, T. H. and DOXEY, B. C. 'Operation of three-phase induction motors with unsymmetrical impedance in the secondary circuit'. *Proc. Inst. Elec. Engrs,* **102,** 71 (1955)
30. WIDGER, G. F. T. and ADKINS, B. 'Starting performance of synchronous motors with solid salient poles'. *Proc. Inst. Elec. Engrs,* **115,** 1471 (1968)
31. CHALMERS, B. J. and RICHARDSON, J. 'Steady state asynchronous characteristics of salient pole motors with rectifiers in the field circuit'. *Proc. Inst. Elec. Engrs,* **115,** 987 (1968)
32. SAY (1983) op. cit.
33. RAWCLIFFE, G. H. 'The secondary circuits of synchronous induction motors'. *Inst. Elec. Engrs J.,* **87,** 282 (1940)
34. SAY (1983) op. cit.
35. JONES and BROWN (1984) op. cit.
36. TAYLOR (1958) op. cit.
37. JONES and BROWN (1984) op. cit.
38. TAYLOR (1958) op. cit.
39. JONES (1967) op. cit.
40. TAYLOR (1958) op. cit.
41. SCHWARZ, B. 'The stator-fed a.c. commutator machine with induction regulator control'. *Proc. Inst. Elec. Engrs,* **96** (Part II), 755 (1949)
42. SAY (1980) op. cit.
43. CLAYTON (1959) op. cit.
44. SAY (1980) op. cit.
45. CLAYTON (1959) op. cit.
46. SAY (1980) op. cit.
47. CLAYTON (1959) op. cit.
48. TAYLOR (1958) op. cit.
49. JONES (1967) op. cit.
50. TAYLOR (1958) op. cit.
51. JONES (1967) op. cit.
52. SAY (1983) op. cit.
53. SAY (1980) op. cit.
54. JONES (1967) op. cit.
55. ADKINS, B. and HARLEY, R. G. *The general theory of alternating current machines.* Chapman and Hall (1975)
56. THALER, G. J. and WILCOX, M. L. *Electric machines.* Wiley (1966)

2 Environment

M. G. Simms*, P. N. Williams† and B. Pomfret‡

* Engineering Director, Brush Electrical Machines Ltd.

† Engineering Manager AC Machines, Brush Electrical Machines Ltd.

‡Deputy Chief Designer, GEC Turbine Generators Ltd. Formerly Engineering Manager, DC Mining and Turbogenerators, Brush Electrical Machines Ltd.

2.1 Ambient conditions

2.1.1 Introduction

There are two main categories of ambient condition: (1) those due to geographic conditions; and (2) those which are man made. The former embraces temperature, altitude and the effects of weather, e.g. lightning strikes. The latter are those produced by the supply system, which embraces system faults and voltage surges produced by switching.

In designing an electric motor the designer must cater for the effect of both categories. In the following section, each condition will be taken in turn and the procedures or precautions taken by the designer to cater for these conditions will be shown. For nearly every type of ambient condition, there is an international, a national or a large user or industry specification or standard to provide guidelines or even the complete procedure.

There are two main bases for national standards: (1) countries using 60 Hz supply systems tend to follow the ANSI or NEMA standards; whilst (2) the remaining countries, mainly those using 50 Hz systems, tend to follow the recommendations of the IEC. In the following text the main differences between NEMA and the IEC are noted.

2.1.2 Temperature

One of the factors which determines the rating of an electric motor is the temperature rise at full load of the active materials, i.e. windings and iron cores. The permissible temperature rises for various classes of insulation are specified in the standards. Both NEMA and the IEC use 40°C as the basic ambient temperature. If a motor is to be installed in an ambient temperature in excess of 40°C, the permissible temperature rise will then be reduced by the amount by which the site ambient temperature exceeds 40°C. By agreement with the user, a correction may also be made up to maximum of 10°C for machines operating at temperatures below 40°C on a degree-for-degree basis down to 30°C. Below 30°C, the IEC-based standards do not permit any further correction.

If the cooling medium employed is a liquid then the temperature rise permitted is increased by 10°C and is measured relative to the cooling liquid, providing the liquid temperature is less than 30°C. For higher coolant temperatures, 1° is deducted from the permissible temperature rise for every degree in excess of 30°C. This basis is used where the liquid is used to cool the air circulating through the machine windings and core.

Where motors employ naturally cooled oil-lubricated sleeve bearings, the designer must also ensure that bearings will not overheat when operating in elevated ambient temperatures. The common solution for this problem is to provide additional cooling for the oil, which often necessitates providing an oil-circulating system with pumps, filters and cooler which may use either liquid or air as the cooling medium. If the temperature of the oil is only marginally above the ambient temperature, a water-cooled finned tube may be fitted in the oil sump to maintain the bearing at an acceptable temperature.

High ambient temperatures affect the rating of motors, but at very low temperatures it is necessary to consider the effect on mechanical components. Where motors are fitted with oil-lubricated sleeve bearings, the effect of low temperature on lubricants should be considered. An important consideration is the selection of a lubricant which can operate at reduced temperatures or, alternatively, it may be necessary to fit thermostatically controlled heaters in the oil. Where circulated oil is used, it is common practice to circulate the heated oil for a period before putting the motor into operation.

Where motors are operated in temperatures of the order of $-20°\,C$ or lower, it is necessary to ensure that any steel components which are operated under stress are checked to ensure that the steel used is an alloy capable of maintaining its mechanical properties at reduced temperature. Many steels become brittle at low temperatures and an alloy must be chosen which remains ductile at the minimum operating temperature.

2.1.3 Altitude

Owing to the fact that air density reduces with increase in altitude, it is necessary to allow for this for any motors operating at altitudes in excess of 1000 m. When a motor which is to be operated at an altitude over 1000 m is tested at a location at an altitude of less than 1000 m, the allowable temperature rise at the test location must be reduced. The IEC reduces the permissible temperature rise by 1% per 100 m above 1000 m. The adjustments for the various insulation systems is shown in Table 2.1, reproduced from BS 4999:Part 32. The corrections specified in NEMA MG1 are the same as those listed in Table 2.1.

Table 2.1

Item	Class of insulation				
	A	*E*	*B*	*F*	*H*
Temperature rise in degrees centigrade of stator winding	60	70	80	100	125
1% of temperature rise per 100m	0.6	0.7	0.8	1.0	1.25
Altitude increment in metres per degree centigrade	167	143	125	100	80

2.1.4 Power supply system

The supply system factors affecting the design of motors are both man made and caused by natural phenomena. The obvious factor is supply voltage and windings must be designed to cater for this, but there are also a number of constraints imposed by the supply conditions. Where the motor starter does not have fuse protection, it is necessary to select a terminal box which is compatible with the fault capacity of the supply system.

The constraint frequently imposed by the supply is the maximum current or kilovolt-amperes which may be drawn during starting, a condition which may be met either by a lower starting current design of motor or by use of some form of soft-start device to reduce the current drawn during starting. When considering the supply constraints at start, the source impedance must also be considered to ensure that there is sufficient voltage at the machine terminals at start to overcome the load torque, leaving sufficient torque in hand to accelerate the motor against load.

Voltage surges are caused mainly by switching and lightning. Where motors are connected to a supply system fed by overhead lines, electric storms can cause very high surges to be seen by the motor windings; this voltage occurs between line and earth and therefore is seen by the main insulation. The intensity of voltage surges caused by switching varies with the type of switchgear used and the system impedance.

A number of user and industry standards quantify the surge withstand levels required by windings, the best known of these in the UK being the ESI standards issued by the Electricity Supply Authority (ESI 44–5) and the standard issued by the oil and chemical industries (OCMA Elec–1).

In OCMA Elec–1, two voltage impulses are considered. The first with a rise time of 1.2 μs is a form generally experienced in lightning strikes which would normally appear across the main wall insulation. The other with rise time of 0.2 μs is at the other extreme and is the type normally generated by switching surges; this type would normally show between turns.

The formulae for calculating the withstand levels for different authorities are summarized in Table 2.2.

Table 2.2

	To earth (1.2–50 μs)	*Between turns* (0.2 μs front)
OCMA normal level	$5 \dfrac{\sqrt{2}}{\sqrt{3}} U_n$	$3 \dfrac{\sqrt{2}}{\sqrt{3}} U_n$
OCMA special level	$5 \dfrac{\sqrt{2}}{\sqrt{3}} U_n$	$5 \dfrac{\sqrt{2}}{\sqrt{3}} U_n$
IEC (draft)	$4 U_n + 5$	—
ISI 44–5	—	$3 \dfrac{\sqrt{2}}{\sqrt{3}} U_n$

Note: U_n = rated line voltage r.m.s.

2.1.5 Effects of supply waveform harmonics

Static frequency converters are now quite commonly used to vary the speed of both induction and synchronous motors. The two most common types of converter give quasi-square and pulse width modulated output waveforms. A well-designed pulse width modulated converter with a varying pulse width produces a waveform close to a sine wave but, if a constant pulse width is employed, the harmonic content can be greater than that of a quasi-square output. Harmonic distortion causes extra heating in the motor and may necessitate derating to avoid overheating.

A six-pulse, three-phase thyristor system gives rise to harmonics of orders $m = -5, +7, -11, +13, -17, +19, -23, +25$, etc. where the sign indicates direction of harmonic field rotation. Even and triplen harmonics do not occur. Typical harmonic voltage magnitudes, expressed as a fraction of the fundamental voltage, are equal to $1/m$, i.e. 0.20, 0.14, 0.09, 0.076, 0.058, etc. for the above series of harmonics.

The harmonic currents produced in the stator winding cause additional eddy current losses in the stator conductors. These are not significant with wire windings but with bar windings in motors above, say, 300 kW rating at four-poles, it may be necessary to subdivide the conductors into thinner laminations and to transpose the laminations.

A six-pulse, three-phase converter produces additional rotating fields in the airgap. These rotate in space at m times the synchronous speed, the negative harmonic fields, e.g. the -5 and -11 rotating backwards at 5 and 11 times synchronous speed in space but backwards at 6 and 12 times synchronous speed relative to the rotor. The positive harmonics rotate forwards in space, e.g. harmonic fields 7 and 13 rotating forwards at 6 and 12 times synchronous speed relative to the rotor. The rotor will therefore have currents at 6, 12, 18, 24, etc. times stator frequency induced into cage windings, pole faces and field windings, and these produce extra rotor losses, with associated deep-bar skin effect.

On larger thyristor-supplied motors, it is now common to employ twelve-pulse systems feeding motors having two three-phase windings displaced from each other by 30 electrical degrees. The order of the airgap harmonics is then reduced to -23, $+25$, -35, $+37$, etc. resulting in significant reductions in extra losses.

Harmonic fields which are seen to be rotating relative to each other, produce torque pulsations. Quite significant pulsating torques can be produced by interaction between a harmonic field and the fundamental flux. The resulting torque pulsations, at multiples of supply frequency, may need to be taken into account in designing the torsional characteristics of the complete shaft system.

2.1.6 System faults

The most common system fault is short circuit which can be either a balanced short circuit that is across all three phases or a single-phase short circuit either between two phases or from one phase to earth. In the majority of cases any short circuit will fail also to earth.

At the instant of short-circuit the motor, whether it is an induction or a synchronous machine, will operate as a fully excited generator and consequently will feed current into the fault. In addition, there will be a torque reaction transmitted as an impulse load to the foundations; this impulse load will also appear at the motor coupling but the level here will be reduced depending on the inertias of motor and load. The values of these torques and currents are dependent on the machine reactances, typical values being of the order of 6 times full load torque. At one side of the motor, this torque reaction will add to the static load and on the other side will subtract from it, thus putting the foundation in tension if the foundation bolts are too short.

Other faults which can occur are sudden dips in voltage to a level which causes the motor to pull out of step, followed by a reinstatement of the supply. If this reinstatement occurs fairly rapidly before the motor voltage has collapsed and if the motor voltage and supply voltage are then in excess of 120° out of phase, this is equivalent to starting the motor on a voltage approaching twice the rated value.

To cater for such conditions, it is necessary to reinforce the overhang bracing and the winding to limit movement of the winding, since any excessive overhang displacement can cause insulation to fracture leading to a premature breakdown.

Voltage unbalance can produce serious overheating in motors due to the high negative sequence current which flows with a relatively small out-of-balance component. It is therefore important to ensure that protection will operate to trip the motor if the sustained voltage unbalance exceeds approximately 5%. Modern motor protection relays include a negative sequence trip function which can be adjusted to a level which will trip the motor when the negative sequence current exceeds a critical value.

A derating curve given by NEMA should be applied to motors operated on an unbalanced supply. This standard also recommends that motors should not be operated on supplies with a voltage unbalance in excess of 5%. The NEMA derating curve is reproduced in Figure 2.1.

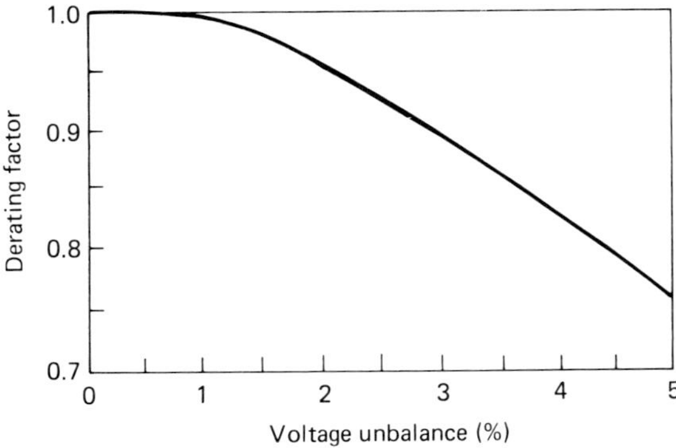

Figure 2.1 Derating curve applied to motors operated on unbalanced supply

2.1.7 Transients and stability

Disturbances in the power supply system of an industrial plant, even though momentary and transient in nature, may lead to a shutdown in plant, resulting in unacceptable outage times. To minimize the risk of such outages, it is possible to carry out stability studies of the complete system including the generators, drive motors, and associated driven equipment.

The major factors affecting transient stability may be listed as follows:

(1) The motor $GD^2 \times$ (rev./min)^2. The greater this quantity, the lower the acceleration factor.
(2) The system impedance, which must include the transient reactances of all machine units. This affects phase angles and the flow of synchronizing power.
(3) The duration of the fault chosen as the criterion for stability. The duration will be dependent upon the circuit-breaker speeds and the relay schemes used.
(4) The machine loadings prior to any fault will determine the internal voltages and the change in output.
(5) The system loading will determine the phase angles of the internal voltages of the various machines contributing to the overall system.

Transient stability, as treated in system design, involves mainly an investigation of the foregoing items and the final result is a balancing of the effects obtainable from the various items to obtain the minimum overall cost.

Item (1) can be considered only in new plants or extensions to existing units.

Item (2) is very important and is best studied by means of a system impedance diagram in which all quantities are expressed on a common voltage or kilovolt-ampere basis. The proportions of transient reactances of machines, as parts of the total impedances to the flow of synchronizing power, can be seen and the effects of possible changes evaluated.

Item (3). Circuit breakers and relays are the principle factors in bringing system stability to the present high levels. Fast clearing of faults is considered to be one of the most important factors in maintaining transient stability.

Item (4). The synchronous machine loadings, and hence the system characteristics, are generally fixed by the generation requirements. The most advantageous loadings from a stability standpoint cannot therefore be specified by the design engineer.

Item (5). The system loads, together with the machine loadings, fix the phase angles in the system and determine the angle margins available for swings.

Study of the steady state and transient stability of rotating machines connected to a system forms an integral part of power system planning and, of the two, the greater importance is usually placed on the transient performance. It is possible to make assumptions on the system side as to the nature and duration of disturbances and, with the assumed conditions, the performance of the connected motors may be analysed for particular load conditions.

The response of a.c. machines following a system disturbance is of great importance. Computer programs, to which the system and machine parameters are input, can be used to simulate various fault conditions and, hence. determine the stability of both the machines and the system. The information required for such a study can vary, but may include the transient, subtransient, saliency, and saturation parameters, and AVR and speed governor control characteristics.

Transient stability studies are essentially a study of the momentary speed changes of rotating equipment, which are functions of their inertia. The inertia characteristics of rotating machines may be expressed by their 'inertia constant' H. The inertia constant of a machine is defined as the stored energy in kilowatt-seconds at rated speed divided by the rated kilovolt-amperes of the machine. It should be noted that the value of H applies to the rated speed of the machine. This value is sometimes used in calculations for machine acceleration, and if these calculations are carried into appreciable speed changes, H should be changed in proportion to the square of the speed.

The transient stability of the synchronous motor can be determined by a step-by-step procedure based on the equal area criterion. This approach superimposes the transient torque–rotor load angle characteristic, for a particular voltage condition, on to the load characteristic. From this the net accelerating and decelerating torque is determined, and the change in rotor load angle determined for increments of time. Knowing the inertia of the system, it is possible to calculate the critical rotor angle from which recovery is not possible. It is usual to make a few simple assumptions when this type of approach is used for determining machine stability. Firstly, it is usually assumed that any voltage loss or recovery is instantaneous about the declared time interval. Secondly, any damping effect produced by load swings is ignored and any changes in excitation are considered on a steady state step-by-step basis.

Load-induced instability may arise through cyclic variation of the load which will cause sympathetic changes in load angle. A reciprocating compressor is a typical example of this type of load, which may have cyclic changes at multiples of the shaft frequency as well as at rotational speed. This subject is well documented in various papers and in a concise form in BEAMA Publication No. 214, which also covers the same problem with induction motor drives.

Non-cyclic load variations arise with loads such as those associated with rubber mills, Banbury mixers and woodchippers where wide fluctuations in load occur due to the nature of the load and method of operation. It is then necessary to take account at the design stage of the maximum load likely to be encountered, to ensure that the high torques will not cause the motor to pull out of synchronism.

The maximum power developed by the synchronous machine increases with increase of load angle until the machine loses synchronism at approximately 90° load angle. Under transient conditions the machine capability is higher.

Control of the excitation system on synchronous machines provides a means for improving stability limits for transient conditions and also for steady state conditions. Excitation systems that are effective from a standpoint of stability are commonly termed quick-response excitation systems. Exciter response is the rate of change of the main exciter voltage when resistance is suddenly removed or inserted in a main exciter field circuit by the action of a voltage regulator. The characteristics which an excitation system must have in order to obtain benefits of quick response include a high ceiling voltage as well as a high rate of build-up. The ceiling voltage of an exciter varies through quite a wide range, depending upon the particular design.

The exciters for quick-response excitation systems are designed so that the field circuits have low time constants, usually obtained by sub-dividing the fields and using external resistances in series with them together with other changes in design proportions. With quick-response excitation systems, it is important that the systems have a reliable source of power for the exciter, sufficiently ample to supply the heavy demands that arise when a fault occurs.

Quick-response excitation systems tend to improve stability limits of power systems in three ways:

(1) Maintaining or increasing machine flux against demagnetizing action of fault currents.
(2) Offsetting deficiency in system excitation due to loss of other sources of excitation.
(3) Increasing steady state stability limits by facilitating action in the region of dynamic stability.

Another feature of quick-response excitation systems is the ability to increase the excitation to meet the requirements of a system arising from the loss of other sources of excitation owing to disconnection of other generators or synchronous compensators.

The importance of quick-response systems has to be balanced against the effectiveness of high-speed circuit-breakers and relays, which limit the duration of fault currents and their demagnetizing effects.

During a system disturbance two factors are of prime importance: (1) the transient infeed to a fault; and (2) the speed reduction of motors connected to the system. The fault infeed is determined by the subtransient reactances and time constants of the machines connected to the system. The correct current–time characteristic is obtained by arithmetic summation of the currents from the separate machines.

Methods are available for simulating multiple machines, using a single equivalent machine/load unit to determine their behaviour in respect of fault infeed and speed reduction, in order to produce a generalized solution. This is of particular value for use at an early stage of a project when precise details of motors and a system are not available.

The process of stability recovery of an induction motor is quite different from that of the transient stability of synchronous machines and, because of the longer times usually involved in the former problem, it has been customary to use the steady state characteristics of induction motors for such investigations.

The stored magnetic energy of the induction motor rotating field causes the motor to contribute to any short-circuit in the system. The system fault level which determines the switchgear rating is increased materially if allowance is made for the fault contribution of the induction motor load.

When disconnecting an induction motor from the supply, its voltage decays according to its open-circuit time constant. Reconnection after a brief supply interruption can lead to dangerous transient torques and severe end-winding forces if the residual voltage is substantially out of phase with the incoming supply.

The induction motor torque varies as the square of the supply voltage. Care must therefore be taken to match the motor characteristic to the expected transient voltage dips of the supply system. This is particularly important if the motor is expected to reaccelerate its load after severe voltage dips or a complete supply interruption.

2.2 Enclosure, cooling and loss dissipation

2.2.1 Enclosures

Electric machine enclosure is commonly specified by an 'IP' number in accordance with BS 4999:Part 20 (IEC 34–5), where the first digit describes the degree of protection against the ingress of solids and the second digit defines protection against harmful ingress of water. The suffix 'S' or 'M' indicates whether the degree of protection applies with the machine stationary or running, whilst a prefix 'W' shows that the machine is weather-protected and that the ingress of rain, snow and airborne particles is reduced to a safe level. The standards also lay down test criteria to establish compliance and give tables of the most frequently used degrees of protection (Tables 20.3, 20.4, and 20.5 of the standard).

Related to the subject of enclosure, but not synonymous with it, is the method of cooling. Broadly speaking, it is possible to group the cooling methods into those where the ambient air is in direct contact with the machine core and windings and those where the heat transfer is indirect, either by heat conduction through the totally enclosed frame or by the provision of closed internal cooling circuits operating in conjunction with heat exchangers. It is the purpose of the cooling process to stabilize the operating temperature of the machine winding within the limits set by the thermal classification of the machine insulation under the specified ambient conditions or within more stringent limits set by the motor user.

The enclosure is usually determined by the motor environment while the cooling arrangements are frequently chosen to minimize machine costs for the particular application. Other factors, such as the availability of cooling water, the need to minimize airborne noise, or space limitations, may influence the eventual choice. Different solutions are appropriate to certain ranges of motor size, but there is a considerable overlap, so that a variety of alternative cooling arrangements are available for a given drive. The following brief notes outline the various options.

2.2.2 Cooling

Direct contact of the cooling air with the motor active parts and the use of shaft-driven fans will generally result in the simplest and most economical arrangement. The degree of protection can be varied from open machines (IP 11S) via screen-protected (IP 21S), drip-proof (IP 22S) and spray-proof machines (IP 24S), to weather protection with filters (IP 44) where the machine is suitable for outdoor installation with the airpath designed to minimize the ingress of harmful substances. Another approach is the use of pipe ventilation. Here the cooling air is taken via ducts from a clean source and is possibly also discharged from the motor in a controlled manner. When the ducting is long and where there are many bends, it may be necessary to incorporate motorized blowers for the cooling air. Forced ventilation may also be needed for variable-speed drives in order to ensure adequate ventilation over the whole speed range.

In spite of their higher cost, totally enclosed machines are being used increasingly for industrial drives. Totally enclosed air-cooled machines are substantially larger than equivalent open-ventilated machines. In order to achieve effective material utilization, it has been found necessary to improve cooling efficiency as the motor size increases by the adoption of more complex cooling circuits. Fractional-power motors often rely on convection cooling from ribbed or smooth frames. Integral-power motors up to 400 kW are usually built in ribbed frames with shaft-driven fans blowing external air over axial cooling ribs on the motor frame. Internal air circuits are incorporated by some manufacturers and one recent development uses a frame made from folded sheet metal which forms hollow ribs, to provide a wrap-round heat exchanger with internal air circulating inside the hollow ribs while external air passes over the outside of the ribs.

On larger machines, closed air circuits with separate tube-type heat exchangers are adopted. The external air is forced by a shaft-driven fan through cooler tubes, which are assembled in a top-mounted air-to-air heat exchanger. The internal air is circulated round the outside of the cooler tubes and returns to the machine after having given off its heat to the external air via the heat exchanger.

If cooling water is available, some economies in motor size can be obtained by the use of water-cooling. On small machines, the cooling is achieved by fitting a water jacket to the frame of totally enclosed motors and, for difficult applications, cooling water is also applied to the rotor via rotating seals. On larger machines, air–water heat exchangers are provided. The cooling air gives off its heat to the cooling water and returns to the motor at a low temperature to resume its cooling function. As the water temperature is usually less than $25°C$, the recooled air temperature can be less than $40°C$ and the motor size need not be larger than for an open-ventilated machine. Water cooling may raise a number of special problems which must not be overlooked. Thus, corrosion risks must be avoided by the use of suitable cooler materials and there is the danger of frost damage in cold climates, which may necessitate the use of anti-freeze.

There are many variations and combinations of cooling arrangements. British Standard 4999:Part 21 (IEC 34–6) uses a code of 'IC' numbers to provide a shorthand definition.

The heat which has to be removed by the cooling process is produced by the mechanical and electrical losses. These are often grouped into fixed losses which occur at no-load and variable losses which vary approximately as the square of the motor load. The fixed losses include friction, windage and iron losses as well as a small stator copper loss associated with the no-load current. The variable losses consist of the stator and rotor copper losses and the stray load loss. As with all design decisions, the balance of losses is the result of a compromise.

2.2.3 Losses

Friction losses occur in the machine bearings. The losses in anti-friction bearings are generally very small so that, apart from adequate lubrication, there is little need to provide special means of loss dissipation although ribbed end-covers are sometimes employed. As the motor speed and size becomes larger, sleeve bearings are employed. For moderate speeds and sizes, self-cooling is possible with ribs on the oil reservoir helping to dissipate the bearing losses. At high speeds and large powers, oil is circulated via an oil-cooler and removes the frictional heat, thus ensuring stable bearing temperature.

Windage losses are produced by the machine fans and also by the movement of air resulting from the rotation of the rotor. Although the cooling air is heated by this process, windage losses rarely add a significant amount to the motor heating and an adequate quantity of cooling air is, of course, essential to effective cooling. Windage losses can be minimized by careful fan design, unidirectional fans being generally more efficient than radial fans, and the use of volutes and careful baffling can ensure an adequate cooling airflow with reduced windage loss.

Iron losses occur primarily in the stator cores. For a given iron grade, they depend on the flux density and supply frequency and comprise hysteresis and eddy–current components. The choice of lamination material and thickness is largely made on economic considerations. To some extent the iron losses can be affected by the manufacturing process and the quality of the core-plate insulation.

There is an element of no-load loss which appears as an iron loss, although it is caused by high-frequency permeance variations which produce tooth pulsation iron losses in wound rotors and high-frequency copper losses in cage rotors.

These losses can be controlled by close attention to the choice of slot combination, airgap and slot openings.

Stator copper losses are proportional to the product of stator current density and ampere wires per unit length of stator bore periphery. The space available for the stator copper, when taking account of the rival requirements of winding insulation and magnetic flux, has a critical influence on the stator copper loss, as has the correct balance between electric and magnetic loading. Loss dissipation depends greatly on the thermal resistance of the winding insulation. For this reason, the use of few large slots, leading to lower current densities due to better space factors, is not always the route to lower winding temperatures.

Rotor copper losses in cage motors with uninsulated windings are not directly related to the temperature rise of the insulated stator winding, but excessive rotor copper losses can have significant adverse effects. In order to achieve the desired speed–torque characteristics, cage rotors are designed with various degrees of current displacement, i.e. skin effect or deep-bar effect, so that the cage resistance during starting is adequate while the running resistance is kept low. Rotors with insulated windings are limited by the thermal classification of the insulation materials and their design is governed by similar considerations as those for the stator windings.

Stray load losses are by definition difficult to define and to measure. It is well known from calorimetric tests that additional losses are present in most machines, which are not covered by the above-mentioned categories and which are approximately proportional to the square of the motor load. Various test methods have been proposed for their isolation, but there is no universal agreement about the validity of these tests. It is generally agreed, however, that the stray load losses can be divided into fundamental-frequency and high-frequency losses. The fundamental-frequency losses can be accounted for by the difference between a.c. and d.c. resistance of the primary winding and by induced currents in the core end-regions due to end-winding leakage fluxes. The high-frequency losses are caused by space harmonics which can produce significant additional rotor losses. Although skewing should theoretically minimize the effect of the slot harmonics, it can actually cause an increase in stray load loss unless the rotor cage is effectively insulated. For medium and large machines, the present trend is towards unskewed, uninsulated cages.

2.2.4 Loss dissipation

Loss dissipation depends critically on effective ventilation, taking into account the various factors which affect heatflow. Thermal equivalent circuits are being used to model the cooling process. With such a complex phenomenon, it is not possible to obtain exact answers. Small changes in air baffling or thermal conduction can have significant effects on nominally duplicate machines. However, certain general guidelines emerge.

(1) It is advantageous to minimize the resistance to heatflow. This is demonstrated by the benefit of an interference fit between stator cores and case on fan-cooled blow-over machines, minimizing the thermal gradient at this interface. The lower temperature rise of vacuum pressure impregnated windings can be attributed to the better heatflow, particularly in the end-winding region, as a result of the greater homogeneity of the insulation. Advances in motor rating are due partly to the thinner insulation thicknesses of modern, synthetic resin systems.

(2) If hotspots are to be avoided, it is necessary to provide sufficient surface pick-up area near the source of the heat and to ensure that the cooling air has access to the cooling surface. If these points are kept in mind, it is possible to manage with lower air quantities.

(3) In the past, 2800 l/min of primary air per kilowatt loss was considered normal but half that quantity is now quite common and, although the air temperature rise has doubled to 40° C, adequate cooling of Class F insulation is still assured.

In general, it is more difficult to dissipate copper losses than iron losses. For this reason there has been a trend to work with higher flux densities in the airgap and to limit the electric loading.

The cooling circuit is of great importance for the effective utilization of the active material of the machine. On small- and medium-size machines, axial airflow over cores without radial ducts is almost universal. On large machines, radial ducts are almost always used and this feature is being extended downwards so that most ventilated motors down to size 315 now use radial ventilation or one of its variants. The series of sketches in Figures 2.2–2.4 illustrate the general principles.

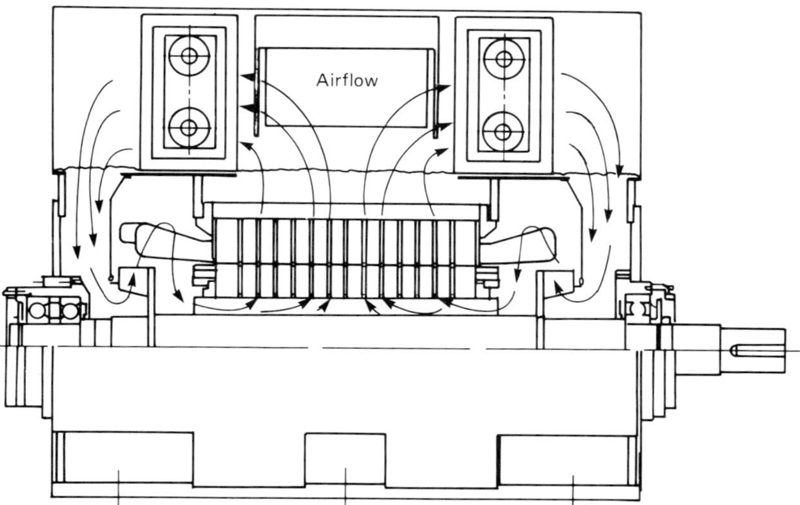

Figure 2.2 Enclosure protection IP44, cooling method ICW37A81, mounting arrangement B3

These notes may give the impression that heating and loss dissipation are of primary importance in motor design. They are important factors, but other parameters such as operating efficiency, speed–torque and speed–current characteristics may set the limits. It is, nevertheless, a fact that the advent of improved insulation materials, greater awareness of the causes of stray load losses, and greater efforts to improve ventilation and to understand heat flow problems, have resulted in significant frame size reductions.

2.3 Potentially hazardous atmospheres

2.3.1 Introduction

Manufacturing processes in many sectors of industry may be described as hazardous by the nature of the operating environment, where the potential

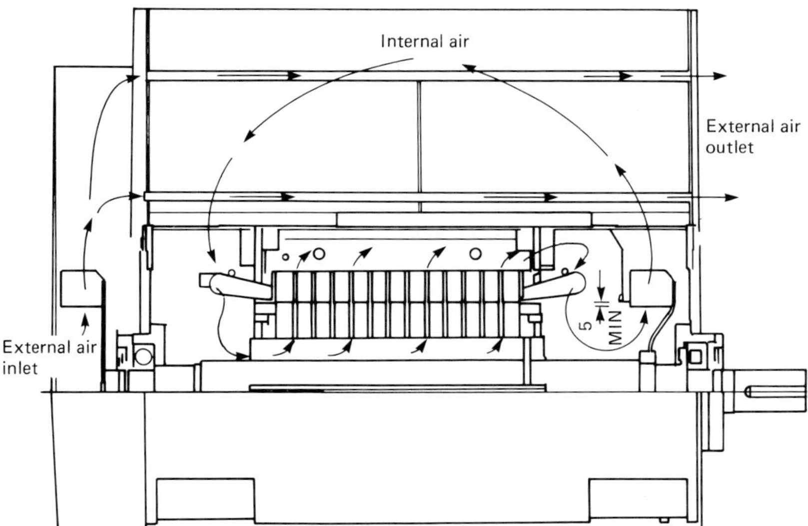

Figure 2.3 Enclosure protection IP44, cooling method IC0161, mounting arrangement B3

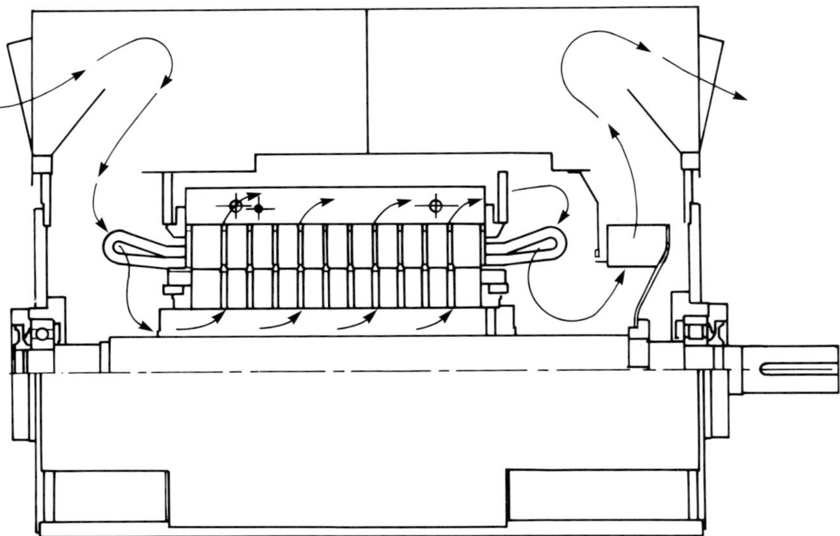

Figure 2.4 Enclosure protection IP22, cooling method IC01, mounting arrangement B3.

presence of an explosive gas–air mixture makes it imperative that formal steps are taken to protect personnel and equipment from the dangers of explosion or fire.

From the very beginning it was realized that the degree of hazard could be variable and this ultimately led to the concept of area classification and the development of design techniques to ensure that electrical equipment would operate safely in specified hazardous area zones. The earliest form of protection for electrical apparatus was the flameproof enclosure first developed in the UK for use in coalmines and later extended to include chemical and petrochemical industries.

Against this background the need for formal specifications to define properly the required standards of safety became essential, together with the establishment of national authorities to provide a test and accreditation service. The earliest specification for a flameproof enclosure was BS 229:1926 and the first explosion tests to determine the maximum experimental safe gap were conducted in Sheffield, although testing was later transferred to the well-known test and certification facility at Buxton. Due in the main to the way in which electrical safety standards have developed, compatibility between the specifications of industrialized nations became extremely difficult and formed a barrier to trade. With this in mind, the first (CENELEC) harmonized standards were introduced in 1977 covering electrical equipment for potentially explosive atmospheres, which greatly assisted European understanding of the subject and eased the technical problems of exporting across the borders of member states. In North America, emphasis is also placed on test and accreditation, but the nomenclature employed to determine area classification, ignition temperature, gas grouping and enclosure types are different to European practice.

The seven commonly recognized methods of explosion protection, as published by CENELEC, have become national standards in force throughout all member countries. Table 2.3 identifies the national versions of these harmonized standards which have been published without change by CENELEC member countries. Dates of issue are included for German and Italian standards to avoid confusion with standards having the same designation but of earlier issue.

2.3.2 Area classification zones

In all locations where flammable liquids or gases are processed, handled or stored, all sources of ignition, including arcs, sparks and hot surfaces associated with electrical equipment should be excluded unless they are contained by use of explosion-protected electrical apparatus. The degree of protection required is dependent upon the presence of ignitible concentrations of flammable gas or vapour in relation to the length of time the explosive atmosphere may exist, and this is defined by a system of area classification (Table 2.4). Thus, the user may select appropriate apparatus according to the needs of a particular operating environment, although it must be stressed that only persons with sufficient knowledge and experience should be involved in the selection and installation of explosion-protected electrical equipment. Those without this technical expertise should seek advice (see BS 5345:Part 2:1983). Whilst comprehensive guidance on the selection of explosion-protected electrical apparatus is available in the BS 5345 series, the information to be considered falls into four categories:

(1) The type of protection of the apparatus in relation to the zonal classification of the hazardous area.
(2) The temperature classification of the apparatus in relation to the ignition temperature of the gases and vapours involved.
(3) The apparatus sub-group (if applicable) in relation to the relevant properties of the gases and vapours involved.
(4) The suitability of the apparatus for the proposed environment.

Explosion-protected apparatus appropriate to a particular zone is readily identified by reference to Table 2.7.

2.3.3 Temperature classification

Electrical apparatus must be selected properly to ensure that the maximum surface temperature is below the ignition temperature of the specified gas. This is determined easily by reference to the temperature classification (Table 2.5)

Table 2.3 European standards

CENELEC standard	EN 50 014 General	EN 50 015 (Ex)o	EN 50 016 (Ex)p	EN 50 017 (Ex)q	EN 50 018 (Ex)d	EN 50 019 (Ex)e	EN 50 020 (Ex)i
UK	BS 5501: Part 1	BS 5501: Part 2	BS 5501: Part 3	BS 5501: Part 4	BS 5501: Part 5	BS 5501: Part 6	BS 5501: Part 7
Belgium	NBN C23–001	NBN C23–104	NBN C23–105	NBN C23–106	NBN C23–103	NBN C23–102	NBN C23–101
Denmark	AFSNIT 50–0	AFSNIT 50–1	AFSNIT 50–2	AFSNIT 50–3	AFSNIT 50–4	AFSNIT 50–5	AFSNIT 50–6
Finland	SFS 4094	SFS 4095	SFS 4096	SFS 4097	SFS 4098	SFS 4099	SFS 4100
West Germany	VDE 0170/0171: Part 1/5.78	VDE 0170/0171: Part 2/5.78	VDE 0170/0171: Part 3/5.78	VDE 0170/0171: Part 4/5.78	VDE 0170/0171: Part 5/5.78	VDE 0170/0171: Part 6/5.78	VDE 0170/0171: Part 7/5.78
Italy	OEI 31–8:1978	OEI 31–5:1978	OEI 31–2:1978	OEI 31–6:1978	OEI 31–1:1978	OEI 31–7:1978	OEI 31–9:1978
Netherlands	NEN–EN 50 014	NEN–EN 50 015	NEN–EN 50 016	NEN–EN 50 017	NEN–EN 50 018	NEN–EN 50 019	NEN–EN 50 020
Norway	NEN–110	NEN–111	NEN–112	NEN–113	NEN–114	NEN–115	NEN–116
Spain	UNE 21814	UNE 21815	UNE 21816	UNE 21817	UNE 21818	UNE 21819	UNE 21820
Switzerland	SEV1068	SEV1069	SEV1070	SEV1071	SEV1072	SEV1073	SEV1074

Table 2.4 Area classification: zones

Zone 0 A zone in which an explosive gas–air mixture is continuously present, or present for long periods

Zone 1 A zone in which an explosive gas–air mixture is likely to occur in normal operation

Zone 2 A zone in which an explosive gas–air mixture is not likely to occur in normal operation, and if it occurs, it will exist only for short time

Table 2.5 European and American temperature classifications for electrical apparatus other than mining

EN 50–014:1977	NEC NFPA 70:1984	Maximum surface temperature (°C)
T1	T1	450
T2	T2	300
	T2A	280
	T2B	260
	T2C	230
	T2D	215
T3	T3	200
	T3A	180
	T3B	165
	T3C	160
T4	T4	135
	T4A	120
T5	T5	100
T6	T6	85

Table 2.6 Relationship between 'T' classification and maximum surface temperature with a selection of representative gases

Example of compound	Ignition temperature (°C)	Suitable equipment regarding temperature classification						APP. group and sub-group
Acetone	535	T1	T2	T3	T4	T5	T6	11A
Butane	365		T2	T3	T4	T5	T6	11A
Hydrogen sulphide	270			T3	T4	T5	T6	11B
Diethyl ether	170				T4	T5	T6	11B
Carbon disulphide	100					T5	T6	*

Note: *Not yet allocated a sub-group

Table 2.7 Selection of apparatus according to zone or risk

Zone	Type of protection
0	(Ex)ia; (Ex)s (specifically certified for use in Zone 0)
1	Any of the above and (Ex)d; (Ex)i(b); (Ex)p; (Ex)e*; (Ex)s
2	Any of the above and (Ex)N or (Ex)n; (Ex)o; (Ex)q

Note: For qualifications regarding the permissible degrees of protection of enclosure for (Ex)e apparatus in Zones 1 and 2 see BS 5345:Part 6.

together with the examples (Table 2.6) which also shows how apparatus certified for one temperature classification can be used in areas with a higher ignition temperature. A more complete list of the properties of some flammable gases, vapours and liquids in relation to T-classification and apparatus grouping is contained in BS 5345:Part 1:1976. Further reference to Table 2.5 shows the American NEC to be different to the CENELEC standards. Although the basic temperature classification is retained, interpolation has occurred between some T classifications.

2.3.4 Apparatus grouping

Explosion-protected electrical apparatus is divided into two main groups: (1) Group I apparatus is only concerned with mining; whilst (2) Group II relates to all non-mining applications.

For some types of protection, notably flameproof enclosures, it is necessary to sub-divide Group II according to the properties of the gases, vapours or liquids, since apparatus certified and tested for, say, a pentane–air mixture will not be safe in a more easily ignitible hydrogen–air mixture. This has led to apparatus sub-grouping, i.e. Groups IIA, IIB and IIC, although care must be exercised in this regard with older standards.

Explosion-protected apparatus should only be used in the environment for which it is certified, although apparatus bearing a higher grouping may be used in

Table 2.8 European and North American testing authorities

Name	Country	Marks used
(1). European Economic Community approved testing authorities		
INIEX paturages	Belgium	**Ex**
DEMKO Herlev	Denmark	**Ex**
Cerchar Verneuil	France	
LCIE Paris	France	**MS AE**
BVS Dortmund-Derne	West Germany	(Sch)
PTB Braunschweig	West Germany	(Ex)
CESI Milan	Italy	**AD-PE**
BASEEFA Buxton	UK	(Ex) (FLP)
All	—	(Ex)
(2). Main North American testing authorities		
Factory Mutual Norwood	US	(FM) / ◀F M▶
Underwriters Laboratory Northbrook	US	(UL)
CSA Toronto	Canada	(CSA)

applications having a lower classification, provided the ignition temperature of the gas–air mixture is correct for the T-classification assigned to the certified apparatus. In practical terms, temperature classification and apparatus grouping must always be considered separately, owing to the wide range of ignition temperatures in gas Groups IIA and IIB.

In the US the system of area classification and gas grouping is again different to European practice. Here, the hazardous area is divided into flammable gases or vapours and combustible dusts, with a further division according to the degree of risk. Hence, Class 1, Division 1 would be approximately equivalent to Zone 0/1 in Europe and Class 1, Division 2 approximating to Zone 2. Similarly, ignitible dust is assigned Class 2, Division 1 and Class 2, Division 2. It is also recognized that the explosion characteristics of gas–air mixtures, vapours or ignitible dusts vary with the specific material involved and an upper-case letter is employed to define this requirement. Hence, Class 1, Division 1, Group D would be the approximate equivalent of the European Group IIA with temperature classification to suit the specific explosion hazard.

2.3.5 Recognized types of protection for electrical apparatus in potentially explosive atmospheres

The recognized types of protection are as follows:

(1) Oil-immersed apparatus   o
(2) Pressurized apparatus   p
(3) Powder-filled   q
(4) Explosion-proof apparatus   d
(5) Increased safety   e
(6) Intrinsically safe apparatus   i(a) or i(b)
(7) Special protection   s
(8) Type 'N' apparatus   N

2.3.5.1 Oil-immersed apparatus (Ex)o

In this type of protection, all or part of the electrical apparatus is immersed in oil to prevent the ignition of an explosive atmosphere which may be present above the oil or outside the enclosure.

2.3.5.2 Pressurized apparatus (Ex)p

Since it is not practical to manufacture explosion-proof motors in very large sizes, it is common practice to employ pressurized motors for Zone 1 applications.

The motors must be totally enclosed and may be cooled by either air-to-air or air-to-water heat exchangers.

The motors are designed to normal industrial standards except that special attention is paid to the sealing of all removable covers and to shaft seals, to keep air leakage to a minimum. Before a motor is energized, it must be purged with at least 5 times its own volume of clean air to remove any flammable gases which may be present. The purging is monitored by proving the airflow rate and maintaining it without break for a predetermined time. On completion of purging, the purge exhaust valve is closed and the machine enclosure maintained at a preset pressure using a supply of clean air to compensate for leakages. The monitoring control unit also provides interlocking with the starter so that the motor cannot be energized until the purge is complete. A failure to maintain pressure will operate the interlock circuit and cause the motor to be de-energized.

Air for purging and pressurizing may be obtained from either high-pressure sources (instrument air) through a pressure regulator or from a fan at the machine pressure, with air drawn from a clean source.

Terminal arrangements may be of standard industrial type if included in the pressurized enclosure or, if outside, the pressurized circuit may be any type certified for use in a Zone 1 environment.

All auxiliary devices which can be energized when the motor is not pressurized must be certified for the zone in which the motor is installed.

2.3.5.3 Powder-filled (Ex)q

In this protection concept the enclosure of the electrical apparatus is filled with powder, usually sand, to prevent the transmission of arcs or sparks which may occur in the enclosure from igniting the surrounding atmosphere. This form of protection is normally confined to small enclosures and is rarely used in the UK.

2.3.5.4 Explosion-proof apparatus (Ex)d

The basic concept of the explosion-proof motor is to ensure that an explosive gas–air mixture ignited within the motor will not be transmitted to a potentially explosive surrounding atmosphere. It is recognized that unless additional steps are taken to seal the joints of an explosion-proof motor it will breathe through the flamepaths, during the normal process of heat cycling, and if the surrounding atmosphere is potentially explosive, the interior of the motor will be contaminated. Thus, a spark or hot surface developed within the motor may initiate an internal explosion and, in these circumstances, flame or hot gases escaping through the flamepaths will be cooled sufficiently to prevent ignition of the surrounding atmosphere without permanently distorting the motor's enclosure. This is achieved by ensuring that the flameproof enclosure is extremely robust and by controlling the length of path and closeness of all joint surfaces. The same rules also apply to the terminal box if the enclosure is designed to be explosion-proof, although it must be stressed that some (Ex)d specifications allow the use of an (Ex)e terminal box.

2.3.5.5 Increased safety (Ex)e

This type of motor is permissible in Zone 1, with some qualifications, and in Zone 2 areas. It is required that all surface temperatures are kept to within the ignition temperature of the specified gas under all conditions of operation and that by design/construction techniques are safeguarded against the production or arcs or sparks.

All internal and external surface temperatures of (Ex)e motors must be kept below the ignition temperature of the specified gas under both normal and abnormal operating conditions. The majority of gases come within the T3 class (200°C) and it is common practice to adopt this classification for standard ranges. The T3 classification also covers the T2 (300°C) and T1 (450°C) requirements; therefore, if equipment is designed for safe installation in a potentially hazardous atmosphere having an ignition temperature of 200°C, it would also be acceptable for use with gases with a higher ignition temperature.

The worst abnormal condition which can occur without damaging a motor permanently (provided it is suitably protected against overload) is a stalled condition which, is of course, the most severe overload condition. In keeping with this specification, therefore, the associated overload device must, in addition to its normal function, operate and trip the motor supply before any surface temperature exceeds the limit of the T3 classification (200°C).

Most motor designs are rotor-critical, the rotor temperature increasing more rapidly than the stator temperature under stalled conditions, and accordingly it is the surface temperature of the rotor conductors that are the critical and limiting factors of this feature.

The t_E characteristic is an additional characteristic peculiar to the increased safety (Ex)e design which must be known, to enable adoption of the requisite overload device that will trip before any dangerous surface temperature is reached.

The t_E time is defined as the time taken for an a.c. winding (rotor) when carrying the starting (locked rotor) current to be heated up from the temperature reached in rated service (under full load conditions) to the limiting temperature.

Every increased-safety motor must bear the t_E characteristic on the nameplate. The motor starter must incorporate an overload device which will trip the motor under stalled conditions within the t_E time, which must never be less than 5 s.

Terminations are required to meet the clearance and creepage distances appropriate to the working voltage. The terminals must be dimensioned generously and of a design whereby they can be fixed in their mountings without the possibility of self-loosening and must not have sharp edges which could damage conductors.

2.3.5.6 Intrinsically safe apparatus (Ex)i

The concept of intrinsic safety is based upon restricting electrical energy within the apparatus and its associated wiring to prevent the occurrence in normal service of incendive arcs, sparks or hot surfaces. In this regard, it is extremely important to ensure that voltages at a higher potential cannot be induced into the intrinsically safe circuit.

Shunt diodes usually are employed as barriers between the safe area and the hazardous area to ensure that the energy within the wiring is brought to a safe level before entering the hazardous location.

Intrinsically safe apparatus is divided into two categories: i(a) and i(b). Equipment manufactured to the i(a) standard is less likely to ignite an explosive mixture under fault conditions than i(b) and can be used in Zone 0 hazardous areas.

2.3.5.7 Special protection (Ex)s

This protection concept provides for the certification of electrical apparatus which does not comply with the specific requirements of more established forms of protection.

Apparatus in this category may be certified for use in designated hazardous areas if suitability can be demonstrated, where necessary by test, for operation in the prescribed hazardous areas zones.

2.3.5.8 Type-N apparatus

Non-sparking motors which are suitable for use in Zone 2 hazardous areas are supplied in the UK to BS 5000:Part 16. These motors are designated type N. The standard requires that the machines must not contain any sparking components and specified minimum clearance must be maintained between stationary and rotating parts. No part of the motor must exceed a temperature of 200°C during normal operation; it is, however, permissible for this temperature to be exceeded during starting. This temperature limitation is also applied to any accessories fitted to the machine, such as anti-condensation heaters.

Minimum creepage and clearance distances are laid down for terminations, and terminal boxes may be phase-insulated, phase-segregated, phase-separated, or air-insulated. Motors for operation in Zone 2 hazardous areas may be either totally enclosed or ventilated to suit customers' requirements. It is common practice, particularly in the US, for NEMA II weatherproof ventilated machines to be operated in petrochemical plants.

In the US, the national electrical codes require that motors operating in Class 1, Division 2 areas shall be non-sparking, which enables the standard cage motor to be used in this environment without any special considerations regarding clearances or creepage distances between live parts. When exposed to flammable atmospheres, however, motors for this duty are subject to temperature limitations on internal and external surfaces.

In the designation of type-N apparatus, the upper-case N is used in the UK, but the lower-case n has been proposed as the symbol for a European standard having a similar concept.

2.3.6 Selection, installation and maintenance of electrical apparatus for use in potentially explosive atmospheres (other than mining applications or explosives processing and manufacture)

Correct selection of electrical apparatus for hazardous areas together with installation and maintenance is an extremely complex subject, requiring expert guidance. In this regard, the codes of practice published by the British Standards Institution are invaluable. Published in twelve parts under BS 5345, the codes embrace all the principle forms of protection in a precise and authoritative manner. A complete list is given below.

BS 5345:Part 1:1976 'Basic requirements for all parts of the code'.
BS 5345:Part 2:1983 'Classification of hazardous areas'.
BS 5345:Part 3:1979 'Type of protection d: explosion-proof enclosures'.
BS 5345:Part 4:1977 'Type of protection i: intrinsically safe electrical apparatus and systems'.
BS 5345:Part 5:1983 'Type of protection p: pressurization and continuous dilution'.
BS 5345:Part 6:1978 'Type of protection e: increased safety'.
BS 5345:Part 7:1979 'Electrical apparatus with type of protection N'.
BS 5345:Part 8:1980 'Type of protection s: special protection'.
BS 5345:Part 9:* 'Type of protection o: oil-immersed apparatus and with type of protection q: sand-filled apparatus'.
BS 5345:Part 10:* 'Electrical apparatus for use with combustible dusts'.
BS 5345:Part 11:* 'Specific industry applications'.
BS 5345:Part 12:* 'The use of gas detectors'.

A further subject for concern is the control of repairs when certified electrical apparatus fails in service. While it is expected that the repairer will be qualified to carry out the work, the consistency of the repair procedure employed would be largely unknown. The problem has now been resolved with the publication in 1985 of a joint BEAMA/AEMT code of practice for the *Repair and overhaul of electrical apparatus for use in potentially explosive atmospheres* (other than mining applications or explosives processing and manufacture).

The code covers all the principle types of protection and is again an invaluable guide to all who are authorized to carry out repairs to this type of equipment.

2.4 Submerged motors

2.4.1 General

There are a number of advantages of submerged drives, which range from the elimination of rotating seals for nuclear power station gas circulators and other high-pressure equipment, to the avoidance of long line shafts in deep borehole

*Not yet issued.

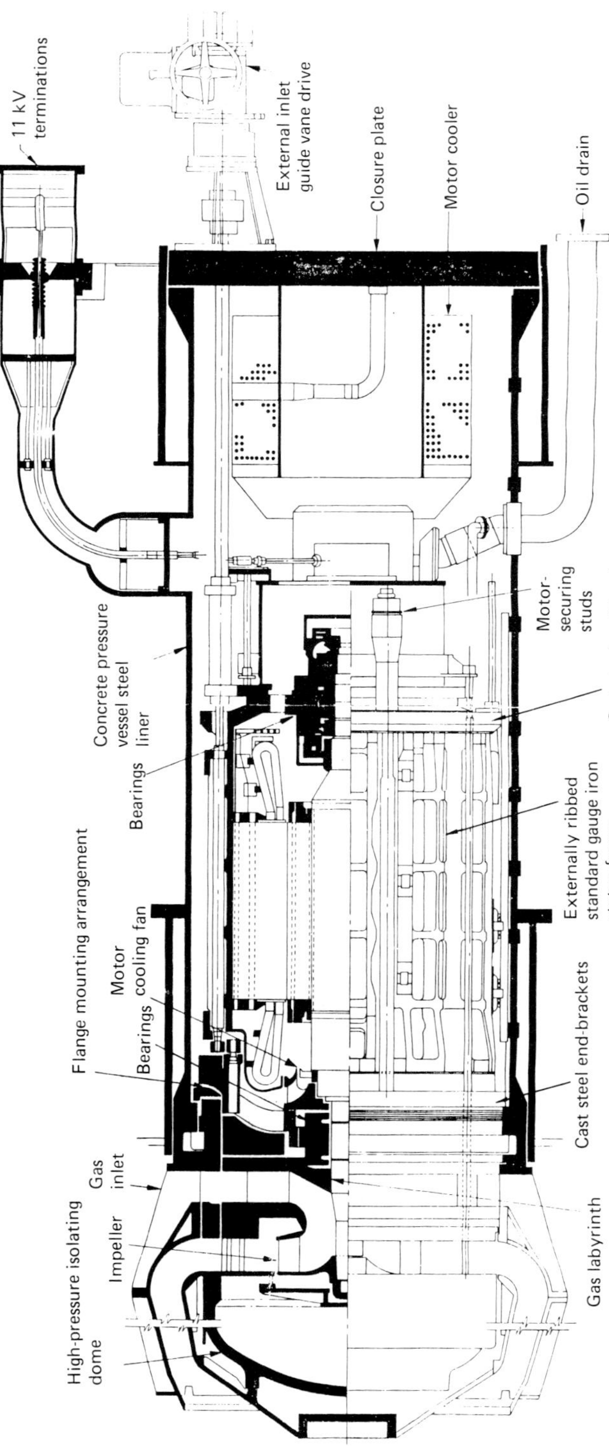

Figure 2.5 Sectional arrangement of boiler circulating unit. (*Courtesy:* Sulzer Bros (UK) Ltd and Laurence, Scott and Electromotors Ltd, Norwich.) *Copyright:* Institution of Mechanical Engineers

pumps. Squirrel cage motors can be designed for immersion in gases and liquids, subject to the level of toxicity of the fluid actually in contact with both stator and rotor windings. For special cases such as circulating pumps in nuclear water reactors, a 'canned' motor is used, where a non-magnetic, generally steel, skin is attached to both stator and rotor to isolate the windings and cores from the cooling fluid.

2.4.2 Water-submersible pump motors

One particular solution[1] used in high- and low-pressure water circuits of power stations, e.g. booster and heater extraction pumps, is based on wet windings (Figure 2.5). The stator winding consists of a continuous length of extra-flexible cable insulated with cross-linked polyethylene insulation, each phase being terminated by two ceramic terminals. This jointless winding uses a quasi-normal winding configuration with a wedge on the top of the slot, but some distance away from the bore to allow cooling fluid to pass through the top of the slot. The bracing of the overhang is designed to sustain the substantial forces created by the fluid movement. The rotor utilizes a bridged slot which, together with the capping of the cage overhang, provides a smooth surface to minimize friction losses. Cooling is provided by a separate water-to-water heat exchanger, a shaft-mounted impeller circulating the water through the enlarged airgap and stator holes. The overhung impeller is separated from the motor compartment by a labyrinth, to minimize the fluid exchange from the main water circuit which tends to be contaminated. This provides a thermal barrier as well as improving the reliability of the water-lubricated bearings.

2.4.3 Submerged gas circulator

An integral part of the original advanced gas-cooled reactor concept was the submerged gas circulator[2] which, by eliminating all rotating seals, gave a major improvement in the simplicity, reliability, and safety of the overall pressure circuit (Figure 2.6). A current design of motor is rated at 5.2 MW, 3000 rev./min, 11 kV, nose-suspended in a horizontal penetration of the concrete reactor wall and operates between the normal cooling level of 40 atm of CO_2 down to 1 atm of air.

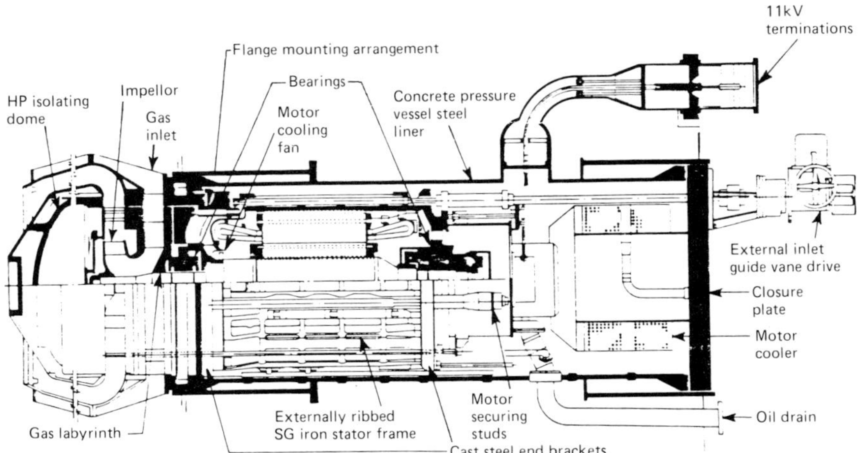

Figure 2.6 Basic section of circulator motor for Heysham II/Torness AGR power stations. (*Courtesy:* James Howden Ltd and Laurence, Scott and Electromotors Ltd, Norwich.) *Copyright:* Institution of Electrical Engineers

Flow control is achieved by inlet guide vanes, and there is an auxiliary variable frequency supply for limited auxiliary and emergency duty, thus avoiding the necessity for a pony motor system.

The stator winding is based on a resin-rich mica paper design. Special motorette testing techniques are used to prove reliability under the particular conditions of pressure cycling and water contamination in case of boiler tube leaks. The rotor is completely smooth, with bridged slots and a quasi-sealed overhang structure achieved by means of a combination of a shrinkring and end-plate of non-magnetic (turbo-end-bell) steel. The high-integrity cage-winding, manufactured from copper–chromium to achieve long fatigue life, uses a 'keyhole'-shaped bar, turned down to a round section to enter the drilled endring. The performance is computed by finite difference methods.

Both for thermodynamic and mechanical reasons, the stator frame is an externally ribbed standard gauge iron casing. This, together with cast steel end-brackets, and standard gauge iron bearing housings, acts as a 'secondary containment' system in case of a depressurization accident. This structure has been analysed fully by finite element methods. The main impeller is overhung on the shaft; a labyrinth separates the motor compartment, with its own separate water-to-gas heat exchanger, from the main reactor gas circuit to provide the necessary thermal and oil-ingress barrier. The oil-lubricated journal and double-thrust bearings are provided with appropriate seals and pressure-control arrangements to minimize the egress of oil into the motor compartment, the duplicated oil supply being derived from a separate submerged pump–motor system.

Full power testing is carried out on each complete circulator to ensure that the quality requirements have been achieved. Strict quality control is exercised throughout to maintain the integrity of the structure, whilst the whole design is based on quality assurance (see Chapter 3, Section 3.3.4) level 1.

2.5 Noise

2.5.1 Introduction

Noise and vibration are both unwanted cyclic oscillations and vibration can be considered in this context as structure-borne noise, as opposed to airborne noise.

There is an increasing awareness of the effect of noise on the well-being of individuals[3] and of the effect of low-frequency vibration on the life of the motor and the driven machine.

The transmission of high-frequency vibration can be a serious problem in special locations such as on board ship. Keeping such vibration down to a minimum level can have an important effect on the security of an active ship in time of war, as these vibrations are readily traced by enemy shipping.

This type of vibration is unlikely to have any effect on the operation or life of the motor and therefore would be largely ignored for normal industrial drives.

The threshold between airborne sound and objectionable noise is subjective and depends on the age, sensitivity and interests of the individual. What is certain is that exposure to high levels of sound is damaging to the ear and is cumulative. Each exposure to such sound causes irreversible impairment of the individual's hearing.

A sound pressure level of 90 dBA is generally considered to be the highest level that an individual should be exposed to over an 8-h working day.[4] In practice, many people would find this level objectionable and tiring and often lower levels are specified for those motors operating in areas requiring frequent attendance by personnel.

Achieving specified noise levels can involve significant expenditure, and the required noise level to be specified needs careful assessment.

2.5.2 Definitions

2.5.2.1 Decibel (dB) is 10 times the logarithm of the ratio of two powers W_1 and W_2:

$$dB = 10\log_{10}\left(\frac{W_2}{W_1}\right) \tag{2.1}$$

To have any meaning, therefore, a decibel scale must be referred to an agreed power level W_1.

For the sound power scale, this reference is taken as $10^{-12}\,\text{W}$.

2.5.2.2 Sound power level:

$$\text{Sound power level} = 10\log_{10}\left(\frac{W_2}{10^{-12}}\right)dB \tag{2.2}$$

where W_2 is the radiated sound power in watts.

2.5.2.3 Sound pressure level:

$$\text{Sound pressure level} = 10\log_{10}\left(\frac{P_2}{P_1}\right)^2 \tag{2.3}$$

In this case the reference pressure P_1 is taken as $2 \times 10^{-5}\text{N/m}^2$.

$$\text{Sound pressure level} = 20\log_{10}\left(\frac{P_2}{2 \times 10^{-5}}\right)dB \tag{2.4}$$

2.5.2.4 Sound measurement

It is normal to measure sound pressure using a precision noise meter. The noise meter is fitted with filters which accept noise within set frequencies. In all cases the r.m.s. of the sound pressures in the filter pass band are indicated.

The noise meter or analyser can be one of several differing types:

(1) Wide band analyser: a single bandwidth, normally covering 20 Hz–20 kHz.
(2) Octave band analyser: an analyser with several filters centred on a frequency (X) with a pass band of $1\sqrt{2}\ (X)$ to $\sqrt{2}\ (X)$. The internationally accepted values of (X) are: 31.5, 63, 125, 250, 500, 1000, 2000, 4000 and 8000 Hz.

A precision instrument incorporating wide band, octave bands and A, B and C weighting networks is usually used for noise measurements on machines (see BS 4197).
(3) For more specialized analytical work, half, third or tenth octave analysers can be useful.

2.5.2.5 Weighting networks

The ear does not respond equally to sound pressures of different frequency. To give an approximate loudness level, the sound pressures recorded are adjusted in each frequency band by differing amounts. Three weighting networks, A, B and C are normally included in a noise analyser but only the A-weighted network is in general use and the adjusted pressure reading is referred to as dBA (see Section 2.5.2.6 following).

2.5.2.6 Calculation of dBA from octave band sound pressure levels

This is carried out by summating the antilogs of the pressure levels after they have been corrected by application of the A weighting characteristic for the octave concerned (Figure 2.7 and Table 2.9).

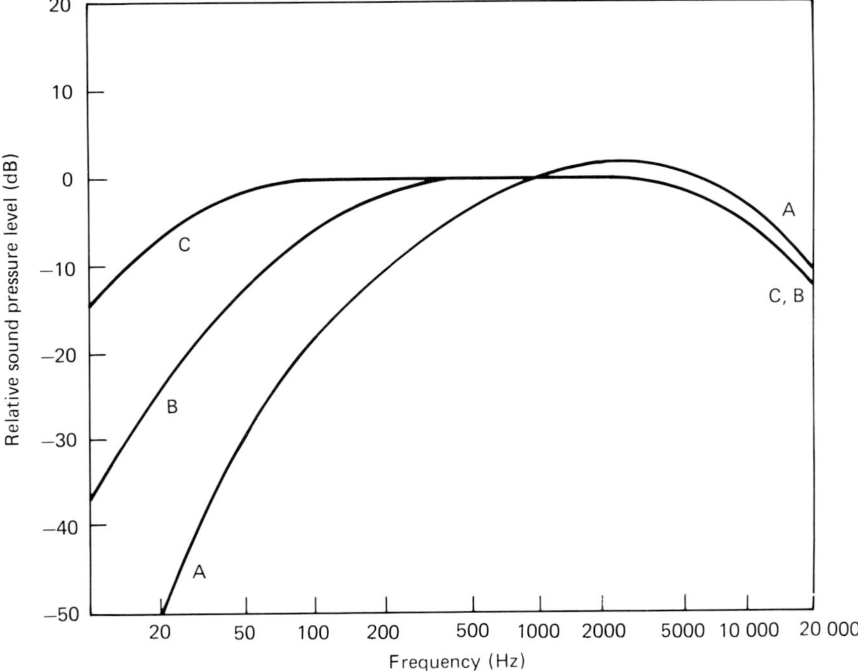

Figure 2.7 International standard A, B and C weighting curves for sound level meters

Table 2.9 Example of calculation of human ear decibels from octave band sound pressure levels

Octave band	Measured sound pressure level (dB)	'A' band correction (dB)	Revised sound pressure level	Antilog dB/10
63	88	−26	62	0.02×10^8
125	88	−16	72	0.16×10^8
250	94	−9	85	3.16×10^8
500	96	−3	93	19.95×10^8
1000	96	0	96	39.81×10^8
2000	92	1	93	19.95×10^8
4000	89	1	90	$10.0 \ \times 10^8$
8000	76	−1	75	0.32×10^8
Total				93.37×10^8

Mean noise level = $10 \log_{10}$; $93.37 \times 10^8 = 99.7$ dBA.

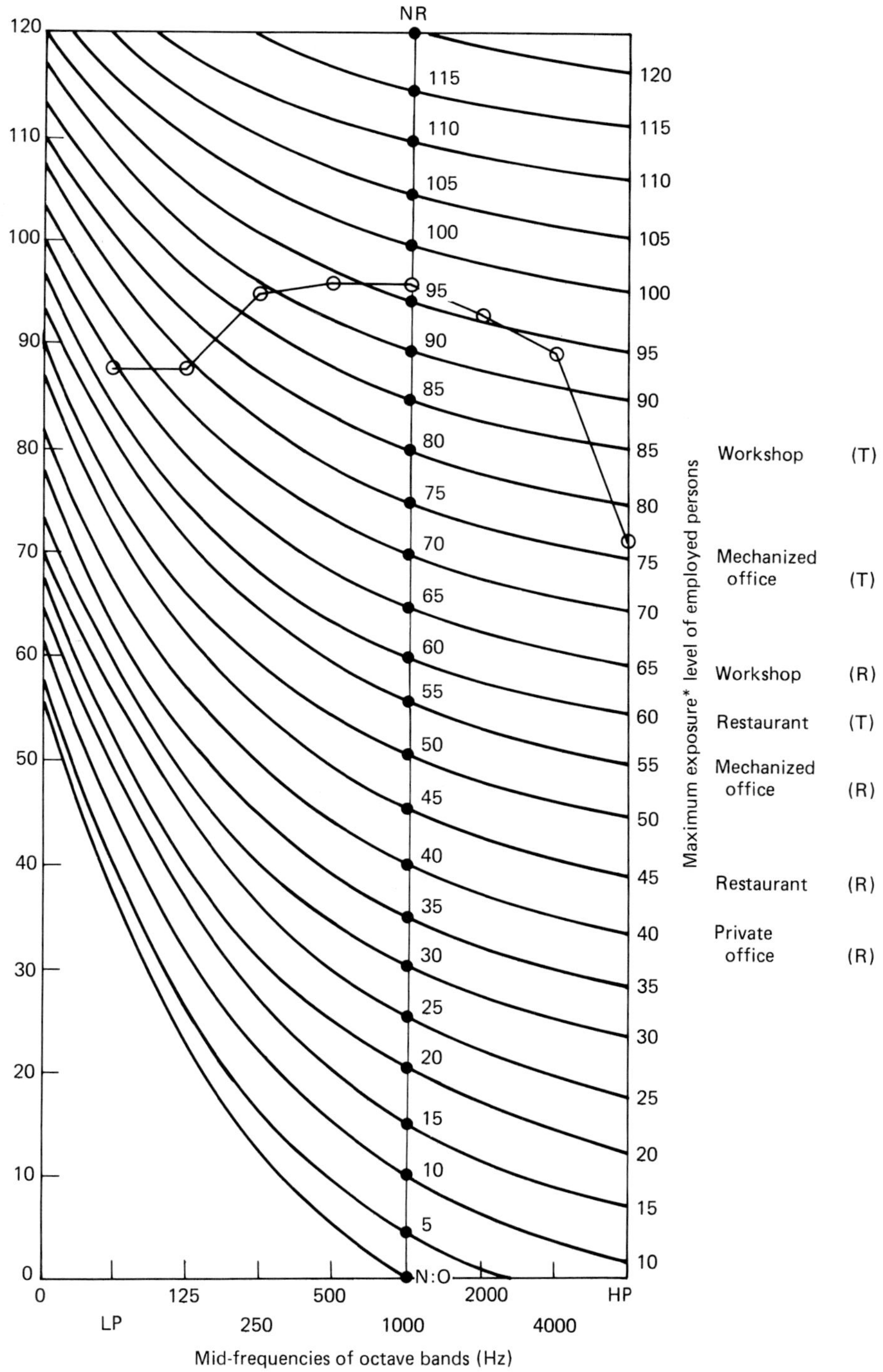

Figure 2.8 Noise rating curves with recommended *R* and typical *T* noise ratings for various common environments

2.5.2.7 Noise rating curves

The ear does not respond equally to sound pressures of different frequency but, equally important, the difference the ear can discern changes with the intensity of the sound.

Various curves of equal loudness have been proposed but the one of interest to engineers is the ISO noise rating curves shown in Figure 2.8. To use the curves, the measured octave band sound pressure levels are plotted on the curve and the point which touches the highest curve number gives the noise rating. The example plotted in Figure 2.8 is that used in Section 2.5.2.6 and it will be seen that, for this set of measured readings, the noise rating is 96 and the noise level is 99.7 dBA.

2.5.2.8 Addition of sound pressures

The use of a logarithmic scale for sound measurement means that the summation of two sound pressures requires the summation of the antilog of the values, as shown in Section 2.5.2.6. In fact, it can be shown that the answer is equal to the highest individual reading plus a value dependent upon the difference between the two readings.

Table 2.10 Addition of sound pressures (in decibels)

Difference	0	1	2	3	4	5	6	7	8	9	10+
Add to higher level	3.0	2.5	2.0	2.0	1.5	1.0	1.0	1.0	0.5	0.5	0

Example 2.1

Two sound sources give pressure levels of 70 and 74 dB. The difference in readings is 4 dB which requires from Table 2.10 the addition of 1.5 dB to the highest individual reading. The combined pressure therefore is 75.5 dB.

2.5.2.9 Variation in sound level with distance

It can be shown that, over short distances, sound pressure varies inversely with the square of the distance from the source.

As one moves further away from the sound source, this equation shows that there is a decay of 6 dB per doubling of the distance.

2.5.2.10 Radius of equivalent hemisphere

Referring to Figure 2.9, for the purposes of calculation it is assumed that the noise readings taken have been made over a hemisphere of radius r_s:

$$\text{where } r_s = \tfrac{1}{2}\sqrt{(ac = bc)} \tag{2.5}$$

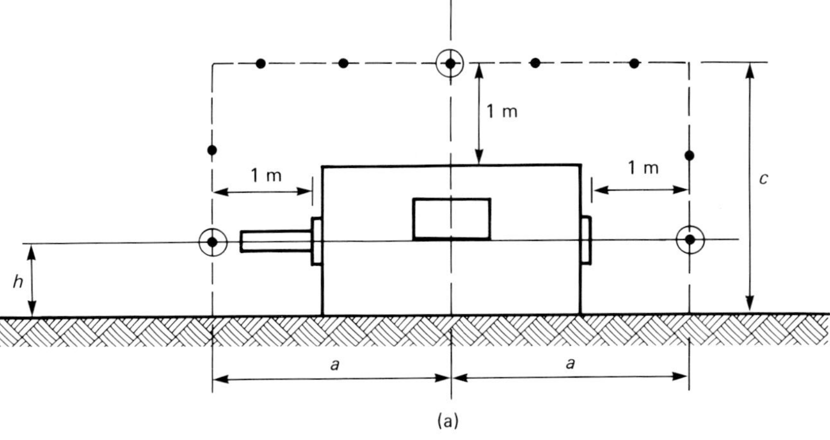

(a)

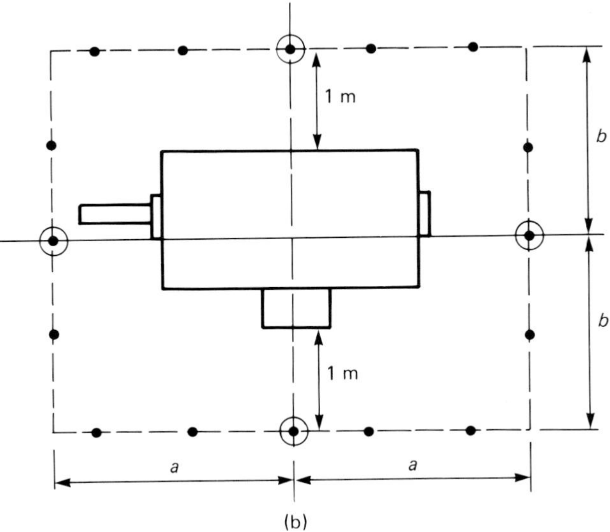

(b)

Figure 2.9 Location of measuring points and prescribed paths for horizontal machines with axial length greater than or equal to the width. (a) Vertical plane. (b) Horizontal plane

2.5.3 Specifying noise requirements

The project engineer responsible for the design of a new installation must take into account the effect of that installation on the environment. He will be particularly concerned with the noise level emanating from the plant, especially if it is located close to a residential area or if personnel need to work adjacent to it.

He will no doubt be aiming to meet a specified or recommended noise level in a given place of work or residential area, taking into account the increase in noise in that area likely to result from the new installation.

Table 2.11 indicates the recommended noise rating for various environments. The requirements for residential areas is more complicated and reference should be made to prescribed standards such as BS 4142:1967 *Method of rating industrial noise affecting mixed residential and industrial areas.*

Table 2.11 Recommended noise ratings for various environments

Location	Noise rating (dB)
Workshops	60–70
Mechanized offices	50–55
Gymnasia, sports hall, swimming baths	40–50
Restaurants, bars, cafeterias	35–45
Private offices, libraries, courtrooms	30–40
Cinemas, hospitals, churches, small conference rooms	25–35
Class rooms, television studios, large conference rooms	20–30
Concert halls, theatres	20–25
Diagnostic clinics, audiometric rooms	10–20

Table 2.12 Upper limit of normal noise rating on no-load

Rating (kW or kVA)		Noise rating numbers		
		Rated speed (rev./min)		
Above	Up to	3000–1501	1500–1001	1000 and below
0.0	2.5	60	60	55
2.6	6.3	70	65	60
6.4	16.0	75	70	65
16.1	40.0	80	75	70
40.1	100.0	85	80	75
100.1	250.0	85	85	80
250.1	630.0	90	90	85
630.0	1100.0	90	90	85

Note: An upward tolerance of 3 in the noise rating number is allowed.

Source: BEAMA Publication No. 225:1969, Appendix C.

Working back from the desired noise level and taking into account the distance of the plant from the area in question, a maximum noise level for the motor may be determined. Whilst a margin will no doubt be applied, it should be borne in mind that specifying low noise levels for the motor may have an adverse effect on cost, particularly for motors of high output.

Appendix C of the BEAMA Publication No. 225:1969 *Measurement and classification of acoustic noise from rotating electrical machines* is reproduced as Table 2.12. This gives the upper noise rating that can be expected from a standard motor. Noise ratings specified substantially below these figures may involve the manufacturer in taking special steps such as fitting air silencers or, in exceptional cases, acoustic enclosures which will increase the cost.

The noise ratings quoted in Table 2.12 refer to a reference radius of 3 m from the sound source and reference should be made to BEAMA Publication No. 225:1969 for details of how the test measurements are to be taken and how the noise rating should be calculated. A general procedure is outlined in Section 2.5.4.

Mean (A) weighted sound level and noise rating are still the most frequently used means of specifying the standard required but suffer from the requirement to specify the equivalent distance from the noise source to which they refer. Although in practice this causes no real problem, more recent specifications such as IEC 34.9 and BS 4999:Part 51 favour the specification of human ear decibel power levels, which is independent of distance.

Appendix 51D of BS 4999:Part 51:1978 gives the limiting mean sound power level in human ear decibel.

Although the room conditions have an important effect on the measured sound pressure levels and, hence, the calculated power level of the machine, it can be taken as a guide that for a source above a horizontal reflecting surface the sound power level is numerically 17.5 dB greater than the sound pressure level referred to 3 m.

Inspection of Table 2.12 and Appendix 51D of BS 4999:Part 51:1973 will indicate a general lowering of noise levels in the later specification, reflecting the improvements manufacturers have been able to make to combat noise at source.

2.5.4 Noise testing

Noise testing follows the procedure laid down in the relevant specification. There are detailed differences between them which makes it necessary to refer to the specifications themselves. In principle, however, referring to Figure 2.9, a series of background sound pressure readings are taken at the location points specified. The readings taken will be either single reading giving the sound level in human ear decibels or octave band sound pressure readings, or both. The motor will then be run on no load and at full speed. Alternating current machines will be supplied at rated voltage and frequency. Synchronous machines will be run at unity power factor.

Table 2.13 Correction factors

Increase in level produced by the machine (dB)	*Subraction from the measured value* (dB)
3	3
4–5	2
6–9	1
10+	0

A new set of noise readings will be taken with the machine running at the test conditions and if any reading is less than 10 dB higher than the background reading a correction is applied as given in Table 2.13. The corrected readings can now be used to compute the mean sound level in human ear decibels and the mean octave sound level in human ear decibels which, when plotted on a noise rating curve, will give the noise rating number after correcting the readings to 3 m reference radius. Procedures outlined in Sections 2.5.2.6 and 2.5.2.9 can be used for these calculations.

If the mean sound power level in human ear decibels is required, this can be computed from the equation:

$$L_{WA} = L_{AM} + 10\log_{10}\left(\frac{2\pi r_s^2}{S_0}\right) \tag{2.6}$$

where L_{WA} is the A-weighted sound power level; L_{AM} is the mean sound level in human ear decibels; r_s is the radius of the equivalent hemisphere; S_o is the reference area of $1\,m^2$.

It may be questioned why noise tests are specified as being carried out at no load. The basic reason is the difficulty of separating out the noise generated by the motor under test from that of the driven load machine. Only the smaller industrial machines can be tested economically inside an anechoic chamber.

This is accepted in practice as the increase in noise level due to load is often less than $3\,dB$, particularly when ventilation noise predominates.[5]

If it is necessary to obtain the increase in noise level due to load, this can be estimated from near field measurements as specified in Appendix 51B of BS 4999:Part 51:1973.

2.5.5 Noise reduction

The three main sources of noise emanating from electric motors are: (1) magnetic; (2) aerodynamic; and (3) bearing noise. The extents to which these three contribute varies with the speed and output and also with the type of enclosure.

Totally enclosed water-cooled machines are by their nature quieter than open-ventilated machines of the same speed and output. Ventilation noise probably predominates in two-pole machines, magnetic noise in machines of higher polarity, and when both of these are sufficiently low the rolling element bearing noise becomes noticeable.

2.5.5.1 Bearings

Oil-lubricated sleeve bearings are by comparison very quiet and, when fitted, eliminate bearings from the noise spectrum.

Rolling element bearings are used whenever possible because of their reduced cost and various types of assembly are used dependent upon the application.

Ball and roller bearings used in electric motors require radial clearances between moving parts and, if these clearances are excessive, increased noise will result. Inaccuracies in the track and ball geometry also lead to increased noise; Figure 2.10 illustrates an increase of $6\,dB$ in the $8\,kHz$ band due to a drive-end bearing. An increase in structure-borne noise caused by wear can be used to predict the remaining life of a bearing using special monitoring equipment. Cage clearance and construction can have a noticeable effect and a particular problem occurs with roller bearings used at the drive end of vertical motors.

Expansion of the shaft on a vertical machine is generally downwards with the thrust bearing located at the upper non-drive-end location. If a ball-bearing is used at the drive end, then it must be axially free to slide in its housing. On occasion, even though this axial movement is planned for, the bearing can cross-lock in the housing and shaft expansion is upwards, unloading the upper bearing. This can have very serious effects on vibration and noise and can lead to rapid bearing failure.

A roller bearing is often chosen to permit thermal expansion of the shaft without the risk of unloading the upper non-drive end-bearing. The shaft diameter required to transmit the motor fault torques, however, requires the choice of a roller bearing which is underloaded compared with the bearing maker's design value. This leads to skidding of the rollers in the bearing and excessive cage noise.

Using a light series bearing with minimum acceptable internal clearances and changing to a bearing with a different cage construction can minimize this problem.

Some authors claim that the choice of lubricant can affect the noise level emanating from bearings. It is known that only in exceptional cases has lubrication been a positive factor in the noise level of industrial machines. In one case a bearing manufacturer had changed the grade of material used on a particular roller-bearing cage. This resulted in the cage being excited at its natural frequency due to the changed boundary lubrication effect. The frequency was a pure tone within the audible range and was unacceptable. Changing the grease to a lighter grade effected a cure but, as this grease was not acceptable for the application, the bearing was changed to another manufacturer's design with no further problem.

Faulty lubrication causing premature bearing failure is, of course, a separate but important subject.

All major bearing manufacturers offer products of high quality manufactured to internationally accepted standards. In practice, the user has little choice but to accept these bearings and reduces noise by using minimum internal clearances and rigid housings, with controlled tolerances on the diameter and truth of the bearing seatings. Axial preloading of ball-bearings, using some design of springing, can be beneficial in effectively eliminating internal clearance.

2.5.2.2 Ventilation

Ventilation noise can be a major problem and the increased specific output achieved by manufacturers in recent years has largely been achieved by increased ventilation of machines. Ventilation noise can be minimized by careful design of the machine air circuits. Significant improvements have been made in the detail design of air circuits and the use of computers has enabled much more accurate modelling of the heat transfer system.

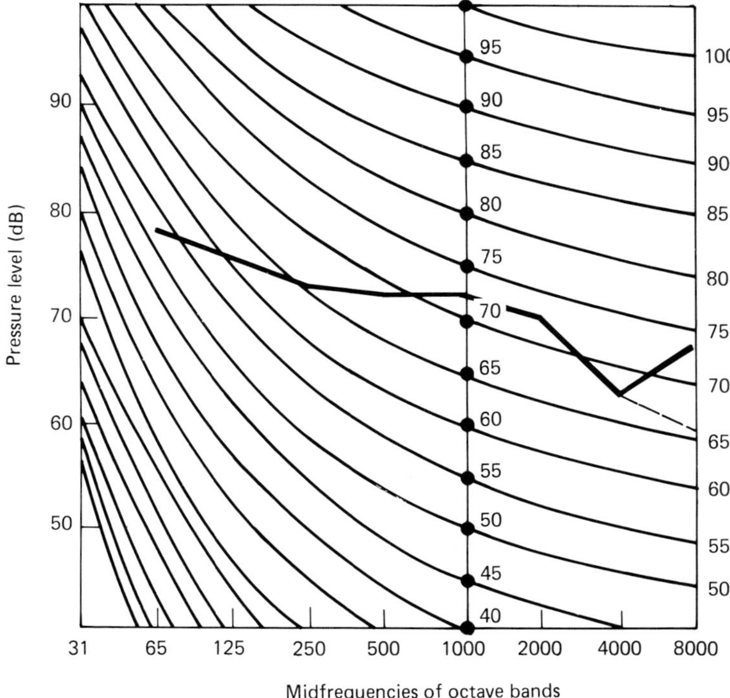

Figure 2.10 Noise spectrum of 3000 rev./min motor illustrating high-frequency bearing noise

Ensuring that cooling air is directed to those parts of the machine where it will be used most usefully, and in the correct proportion to the total volume of air, achieves the best balance of stator and rotor winding temperatures with minimum airflow. Reducing the volume of air has an important effect on both machine efficiency and noise.

The type of cooling circuit used and its overall resistance to airflow is important and explains why manufacturers use different air circuit designs for different applications. Figure 2.11 illustrates the characteristics of a typical motor resistance

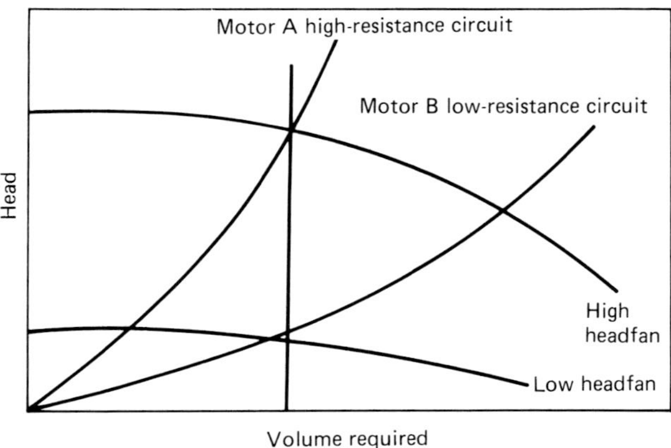

Figure 2.11 Head–volume characteristics for machine resistance and fan performance

curve and the associated fan curve. Assuming the two motors A and B require the same volume of air to dissipate losses, then the duty point (intersection of the fan curve and machine resistance curve) demands a fan for motor A which can achieve a much higher head of pressure than that required for motor B. This will, for a similar design of fan, result in a higher sound level.

The overall fan noise is dependent upon such parameters as blade angle, tip speed, blade dimensions and detail of the fan enclosure. Empirical relationships between generated noise and size, static pressure and speed for a family of fans can be derived. The difference in mean sound power level between fans 1 and 2 is:

$$70\log_{10}\left(\frac{\text{tip diameter 2}}{\text{tip diameter 1}}\right) + 50\log_{10}\left(\frac{\text{tip speed 2}}{\text{tip speed 1}}\right) \tag{2.7}$$

$$\text{or } 20\log_{10}\left(\frac{\text{tip diameter 2}}{\text{tip diameter 1}}\right) + 25\log_{10}\left(\frac{\text{pressure 2}}{\text{pressure 1}}\right)$$

$$\text{or } 10\log_{10}\left(\frac{\text{volume 2}}{\text{volume 1}}\right) + 10\log_{10}\left(\frac{\text{pressure 2}}{\text{pressure 1}}\right)$$

If the fan speed was increased by a factor of 2, the fan laws would predict a doubling of the volume of air and 4 times the generated load. Substituting these factors into the third Equation (2.7) mean sound power level given above will predict an increase in sound power of 15 dB.

If the sound power of fan 1 is not available, an approximate value can be predicted using Equations (2.8):

$$SWL = 77 + 10\log_{10} kW + 10\log_{10} P \, dB$$

$$SWL = 25 + 10\log_{10} Q + 20\log_{10} P \, dB \tag{2.8}$$

$$SWL = 130 + 20\log kW - 10\log Q \, dB$$

where SWL = mean sound power level;
    kW = motor power;
    P = static head in millimetres water gauge; and
    Q = volume flow in cubic metres per hour.

If it is necessary to predict the mean sound power level into octave band sound power levels then the mean sound power level calculated from one of the Equations (2.8) can be corrected using the factors in Table 2.14.

Table 2.14 Corrections to obtain octave band spectra for various fan types

Octave band centre frequency (Hz)	*Add to overall sound power level* (dB)							
	63	*125*	*250*	*500*	*1000*	*2000*	*4000*	*8000*
Centrifugal:								
backward curved blades	−4	−6	−9	−11	−13	−16	−19	−22
forward curved blades	−2	−6	−13	−18	−19	−22	−25	−30
radial blades	−3	−5	−11	−12	−15	−20	−23	−26
Axial	−7	−9	−7	−7	−8	−11	−16	−18
Mixed flow	0	−3	−6	−6	−10	−15	−21	−27

2.5.5.3 Electromagnetic noise

The driving torque of an electric motor is produced by interaction between the main flux wave and the m.m.f. wave produced by the ampere–conductor distribution. These forces and their harmonics can excite the stator teeth but in general do not result in unacceptable noise being generated. This is because, to be significant, the amplitude of vibration of the teeth needs to be magnified by a tooth natural frequency.

The natural frequency of teeth usually is too high to be of importance and, perhaps more significantly, the inertia of the main winding acts as an effective damper limiting vibration amplitude.

Induction motors.   Radial forces distort the stator frame and, depending on the proximity of a core natural frequency, the radial vibration of the core can set up objectionable pure-tone noise.[6] The airgap flux density is represented by:

$$B_g = \Sigma_n B_n \cos(n\theta\omega_n t - \phi_n) \tag{2.9}$$

where n is the order of harmonics;
    B_n is the maximum amplitude of the nth wave;
    ω_n is the harmonic angular frequency;
    ϕ_n is the harmonic phase angle; and
    θ is the angle from a reference point.

The radial magnetic force F_g per unit area at the airgap is

$$F_g = B_g^{2}/2\mu_0 \tag{2.10}$$

B_g may be substituted from Equation (2.9) and the squared terms of the harmonics often ignored, either because their frequencies are low or the harmonic numbers are high and therefore contribute little to audible noise.

The various magnetic force waves distort the stator producing discrete tones in the low- and high-frequency ranges. Low-frequency noise is mainly a problem in two-pole induction motors and is produced by the main rotating flux wave. For this condition:

$$B_g = B \sin\omega t$$

$$F_g = (B^2/2\mu_0) \sin^2 \omega t$$

$$= (B^2/4\mu_0)(1 - \cos 2\omega t) \tag{2.11}$$

The second term of Equation (2.11) is a force that varies at twice line frequency and will produce an audible hum. The autensity of this hum depends upon the amplitude of the core deflection which is, without resonance, proportional to $1/m^4$.

For a two-pole motor the force wave mode number $m = 4$ and for a four-pole motor $m = 8$; the noise-exciting force for a four-pole machine is 1/16 that of a two-pole machine. Furthermore, at twice line frequency it falls into a band to which the ear is not particularly sensitive (see Figure 2.7, page 86).

High-frequency noise in the much more sensitive 1–4 kHz range is produced by the interaction of the stator and rotor harmonics. The lower modes are more objectionable than the higher ones and are caused by harmonics in the winding m.m.f. wave and the airgap permeance, the former being caused by the winding distribution and the latter by slot openings.

To avoid these problems, Jordan[7] recommended the following rules for stator and rotor slot numbers S_1 and S_2:

$S_1 - S_2$ must not equal $\pm 2 \times$ pole pairs   (0 mode)
$S_1 - S_2$ must not equal $\pm 2 \times$ pole pairs ± 1 (mode 1)
$S_1 - S_2$ must not equal $\pm 2 \times$ pole pairs ± 2 (mode 2)
$S_1 - S_2$ must not equal $\pm 2 \times$ pole pairs ± 3 (mode 3)

In addition, odd rotor slots should not be used and S_2/S_1 should fall outside the range 0.85–1.15. (Note that this conflicts with a desirable condition for low harmonic losses.)

The prediction of the actual sound intensity is extremely difficult since it depends not only upon the forces but also on the stator stiffness, natural frequency and damping. Liwschitz-Garik and Whipple[8] give a general method which, if not precise, does give by comparison with successful designs a good guide to whether or not a proposed design is likely to be equally successful.

The airgap permeance harmonics can be reduced by bridging the slots, using magnetic closing wedges, or increasing the airgap length. Skewing the rotor slots can be beneficial in eliminating the lowest frequency slot harmonics but can, on large machines, result in torsional oscillations with associated noise.

Chording the winding to reduce the troublesome fifth and seventh harmonics is desirable but sometimes difficult on two-pole motors due to the large coil span.

Several of these factors which lower noise, such as reducing the flux densities, have an adverse effect on the machine performance and cost. Increasing the airgap may adversely affect performance and, whilst other factors are equal, machines with more rotor, than stator, slots tend to be quieter than the reverse, it is at the expense of higher stray load losses and parasitic torques. As always, a compromise has to be made.

Variable-speed a.c. commutator motors have smaller airgaps but, in these machines, an auxiliary commutating winding in parallel with the main rotor winding acts as an efficient damper winding which provides a low impedance path for harmonic currents. This reduces harmonic noise significantly.

Direct current machines. Although, as previously stated, the d.c. motor is generally less troublesome than an induction motor there are applications such as submarine drives that require special attention. The armature slotting interacts with the main poles, commutating poles and slotting in the pole shoe if the machine has a compensating winding.

Current ripple produced by thyristor supplies results in a ripple in the magnetic forces in the machine. In extreme cases this has caused serious noise problems.

With the generally low noise level of a d.c. machine, the noise generated by brush friction on the commutator can be noticeable. Tangential harmonic forces on the poles can cause frame vibration and it is unfortunate that steps to reduce the magnitude of the tangential forces increase the radial forces.

The following measures can be made to reduce noise:

(1) Use a reduced magnetic loading.
(2) Increase the number of armature slots.
(3) Skew the armature slots (or, less commonly, the pole shoes).
(4) Use continuously graded main pole gaps or flare the gaps at the edges of the main pole.
(5) Increase the airgap.
(6) Brace the commutating poles against the main poles.
(7) Use semi-closed or closed slots for the compensating winding.
(8) Select the pitch of the compensating winding slots to give minimum variation in airgap permeance.
(9) Use a twelve-pulse rather than a six-pulse thyristor supply or fit a choke in series with the machine to reduce the current ripple.

Salient pole machines. The ability to use fractional slot windings in multipole synchronous machines introduces the possibility of local noise due to the presence of long pole pitch harmonics in the force wave produced by the stator m.m.f. The resultant hum which occurs at twice supply frequency is dependent on the load current. By applying suitable design techniques, these harmonics may be reduced to an acceptable level.

The most difficult noise source to allow for at the design stage is the magnetic noise that can be generated by flux pulsations in the airgap region. There are many factors that influence the magnitude of these pulsations. For instance, the radial airgap length, the stator slot opening/slot pitch, the airgap shaping and the pole arc:pole pitch ratio.[9–11]

On multipole laminated-pole machines, where the rotor slot opening is significant compared with the airgap length, the rotor slots can introduce appreciable harmonics into the flux wave. These may appear in the voltage waveform and may give rise to magnetically induced noise in the stator core. Consideration must then be given to offsetting the rotor slots on alternate poles.

The dynamic response of the stator core pack and the way in which it is supported in the stator frame influences the noise level generated, as there is little damping in the system.

2.5.6 Noise attenuation

After the designer has taken whatever steps were open to him to minimize noise generation at source, it may still be higher than specified. To achieve the specified value it will now be necessary to apply external silencing. This may be in the form of inlet or outlet air duct silencers or even the fitting of a complete enclosure. If this situation is reached then it is essential to know at least the octave band sound pressure levels in order that effective silencing can be achieved.

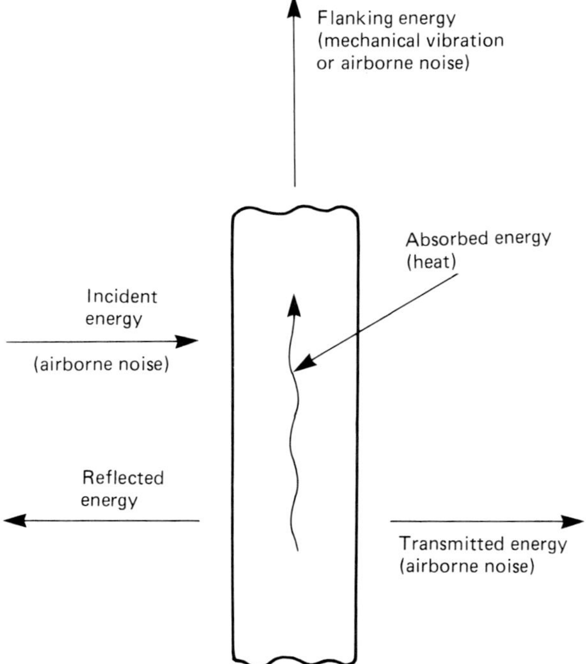

Figure 2.12 Energy flow in an acoustically excited partition

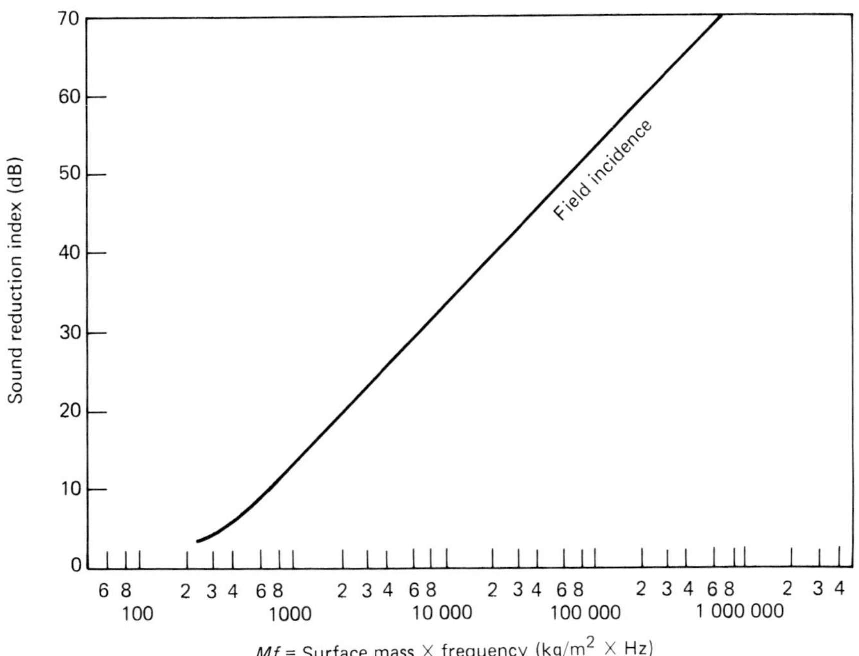

Figure 2.13 The mass law for partitions in a semi-reverberant room

Airborne noise striking a surface will, like other forms of energy, be dissipated in various ways. Some will be reflected, some will be converted into heat and some will be transmitted through the surface to emerge as airborne noise (Figure 2.12).

The ratio of the sound energy passing through a panel to that approaching it is called the transmission coefficient τ and the sound reduction index R is a measure of the panel's ability to prevent acoustic energy passing through it:

$$R = 10 \log_{10} (1/\tau) \text{ dB} \tag{2.12}$$

R is dependent upon both the mass of the panel and the frequency of the applied force from the sound waves. The resistance to panel movement, or its inertia, increases both with mass and frequency (Figure 2.13). The ratio of the sound energy retained or absorbed by the panel to the energy approaching the panel is called the absorption coefficient. An important objective in controlling noise is to increase the absorption coefficient of surrounding panels, thus converting more sound energy into heat. The inside surface of an acoustic panel is often lined with a porous blanket. This allows the incident sound energy to enter the blanket, cutting down the reflected energy, and the internal damping of the material absorbs this energy converting it to heat, ideally without increasing the transmitted energy.

Low frequencies are difficult to treat without increasing the mass of the panel but some gain can be made by using the principle of a multiple resonant absorber.

If air flows in a duct it has a natural inclination to resonate and the greater the amplitude of oscillation the more energy is absorbed. By the suitable combination of a perforated cover plate (multiple resonant absorber) and absorbent blanket in the air space between the coverplate and panel, a higher absorption at lower frequencies can be obtained than using an absorbent blanket on its own.

Figure 2.14 gives the typical performance of an acoustic panel both with and without a perforated cover plate.

Inlet and outlet silencers consist of ducts with splitters, with the walls and the splitters lined with acoustic panels consisting of a combination of perforated sheet and absorbent blanket. The attenuation obtained by these silencers obviously

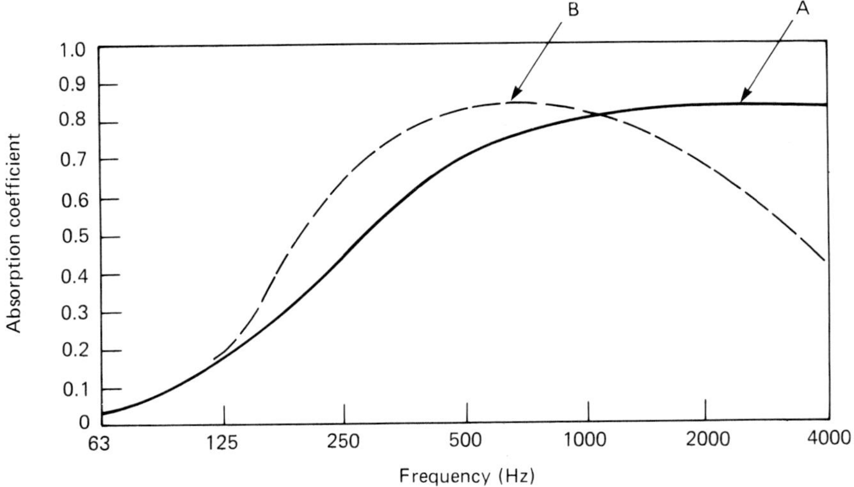

A   Typical absorption of 50-mm thick blanket.
B   As A but with the addition of a perforated cover plate 3 mm thick with 10% open area created by 5-mm diameter holes.

Figure 2.14 Absorption coefficient of a typical multiple resonant absorber

depends upon the dimensions of the ducts and the make-up of the perforated sheet and absorbent lining. A typical curve for such an assembly is shown in Figure 2.15.

A photograph of an induction motor fitted with both inlet and outlet silencers is shown in Figure 2.16. This machine achieved the noise spectrum quoted in Table 2.15 from which the noise rating in Figure 2.17 was obtained.

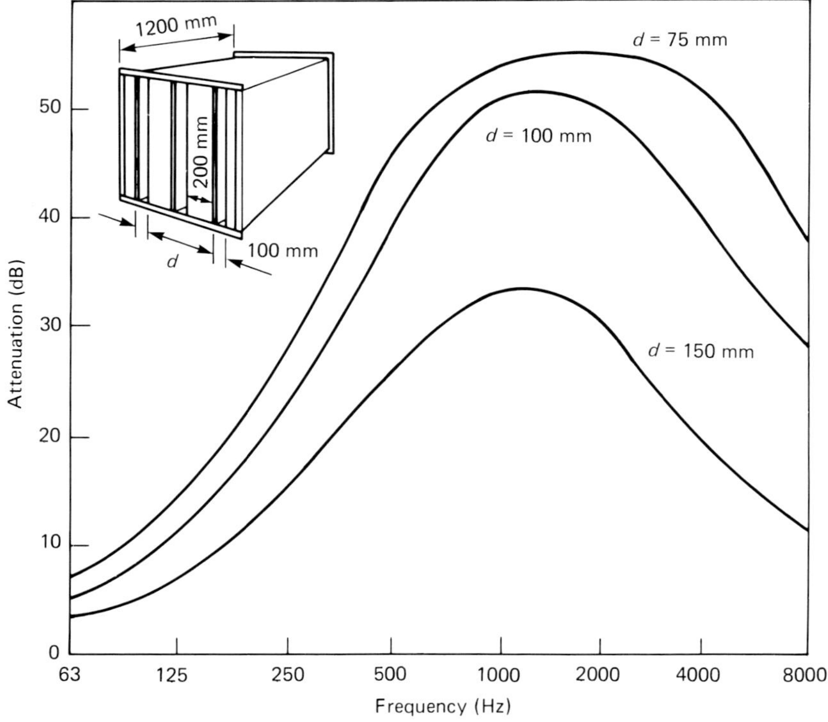

Figure 2.15 Typical attenuation from a 1200 mm splitter silencer

Table 2.15 Effect of inlet and outlet silencers on a CACA four-pole induction motor, 7460 kW, 60 Hz, three-phase and 6.6kV

		Frequency							
		63	*125*	*250*	*500*	*1000*	*2000*	*4000*	*8000*
Sound pressure (dBA)	A	91.8	67.3	79.5	74.9	87.6	85.8	85.9	83.9	77.7
Reference 3 m radius	B	84.7	68.2	75.9	74.7	78.7	76.6	77.4	74.5	68.0

Noise rating
  A = without silencers NR 89
  B = with silencers   NR 80

Sound power level
  A = without silencers = 109.3
  B = with silencers   = 102.2
  Limit BS 4999:1973   = 116.0

Figure 2.16 7460 kW, NR80 CACA induction motor with inlet and outlet silencers

2.6 Vibration

2.6.1 Introduction

There is every reason to suppose that vibration has had more attention than many other subjects over the years. Numerous papers are read at conferences each year and yet the average practising engineer has little practical understanding of the subject. A manufacturer may be permitted a wry smile as he runs a four-pole motor on his test bed, expecting to meet a specified vibration level of 0.041 mm peak-to-peak (BS 4999:Part 50:1978), whilst on the next stand a generator of basically similar construction is also being tested. In service, the generator will be driven by a diesel engine and, hence, is expected to withstand continuously imposed vibration at a level of up to 9 mm/s (BS 5000:Part 3:1980). Assuming a sinusoidal wave at a frequency of 25 Hz, this equates to an amplitude of 0.16 mm peak-to-peak, 4 times that of the motor.

As with noise, achieving low vibration levels does have a cost penalty, and below a certain level it is doubtful if anything is to be gained with regard to improved service life of the motor. Specifying vibration levels below those agreed in national or international standards would be difficult to justify in terms of improved field service.[12,13] There are some exceptions where structure-borne noise (vibration) is considered of prime importance and must be kept to a minimum regardless of cost, and normal standards do not apply. Motors destined for use on submarines are a

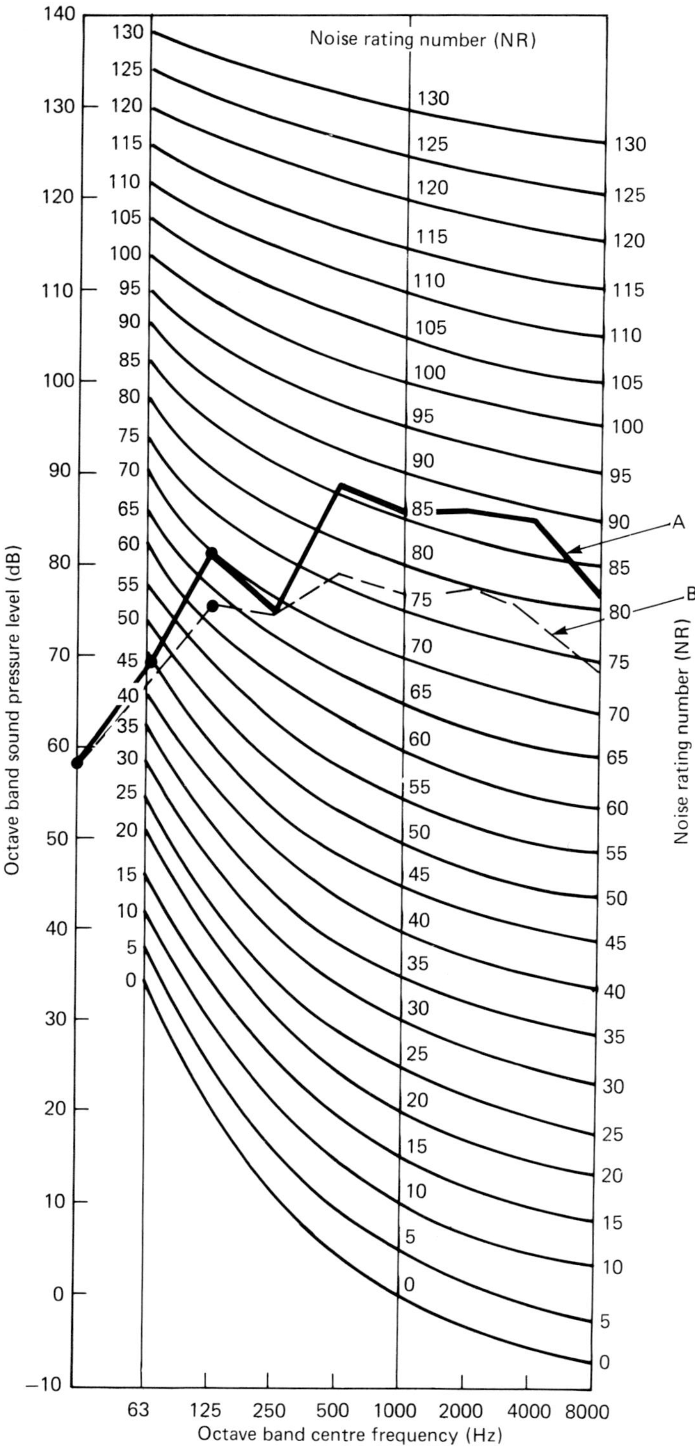

Figure 2.17 Noise rating for 7460 kW, 60 Hz CACA induction motor. (B) Without silencers. B – With inlet and outlet silencers

case in point, where great efforts are made to minimize vibration over a wide frequency spectrum. Isolation techniques are then applied to reduce to even lower levels the residual vibration transmitted to the hull.

Vibration control may be exercised for a number of reasons. Reducing the forces induced into the motor itself, or into its supporting structure or driven machinery, may be the prime motive in order to increase the service life or availability. This certainly is a main purpose of specifications for industrial motors.

Other considerations, such as using vibration as a measurable quantity for quality control purposes or to minimize the forces imposed on human beings[14] who may be in close proximity to the machinery, may equally be valid.

The project engineer purchasing a motor from one supplier, a pump from a second, and then mounting them on a flexible steel foundation designed and supplied by a third party can be faced with a major problem if the combination at site results in unacceptable vibration levels. Is the motor or the pump adequately balanced, or is the foundation too responsive to the imposed forces? Without attempting to supply an answer at this point, it is perhaps not surprising that the project engineer may be tempted to specify for future projects lower vibration levels for works acceptance tests on the motor.

2.6.2 Vibration measurement

Vibration is the cyclic or random oscillation of a component or assembly, and just as noise can be analysed into its harmonics and measured by using a broad or narrow band analyser so, indeed, can vibration. In fact, a noise analyser is a convenient laboratory tool for measuring vibration using specially calibrated transducers instead of a microphone.

Fortunately, in many applications, the predominant frequencies are multiples of the shaft rotational speed and it is convenient to have calibrated instruments designed specifically for analysing vibration. For reasons that will become apparent, the more sophisticated instruments have the facility for recording the phase difference between the peak of the vibration wave and some reference plane on the shaft.

Consider the simple case of a pendulum swinging through an amplitude $2x$ as in Figure 2.18. If C is taken as the reference point, the amplitude varies between plus and minus x at A and B respectively.

A marker on the pendulum in contact with a moving strip of paper would trace out a sine wave, as shown in Figure 2.19. The amplitude x is referred to as the zero-to-peak amplitude (or single amplitude or half amplitude) whilst the amplitude $2x$ is referred to as peak-to-peak amplitude or double amplitude. Great care must be taken when making reference to vibration specifications that one correctly defines what is meant by the amplitude, as there is often no consistency of approach.

If one measured the velocity of the pendulum, the wave would be displaced in time when compared to the amplitude wave, with the peak velocity occurring when the amplitude was zero.

As the velocity is changing, clearly the pendulum is accelerating and decelerating. The velocity and acceleration waves are also shown in Figure 2.19.

The instantaneous displacement wave can be written as:

$$x = X \sin \omega t \tag{2.13}$$

where ω = angular velocity, in radians per second.

Velocity is:

$$v = dx/dt = \omega X \cos \omega t \tag{2.14}$$

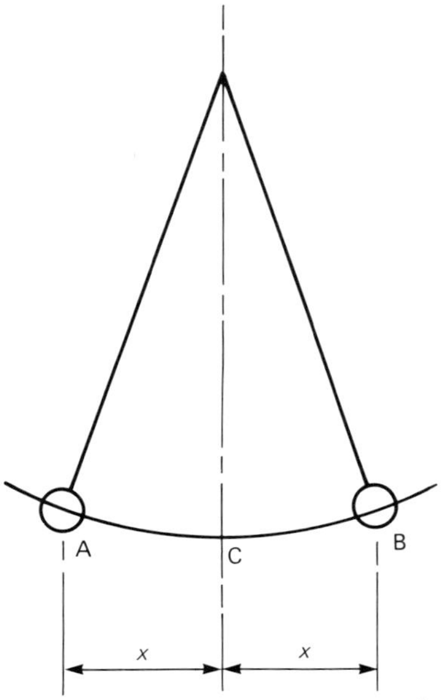

Figure 2.18 Pendulum motion

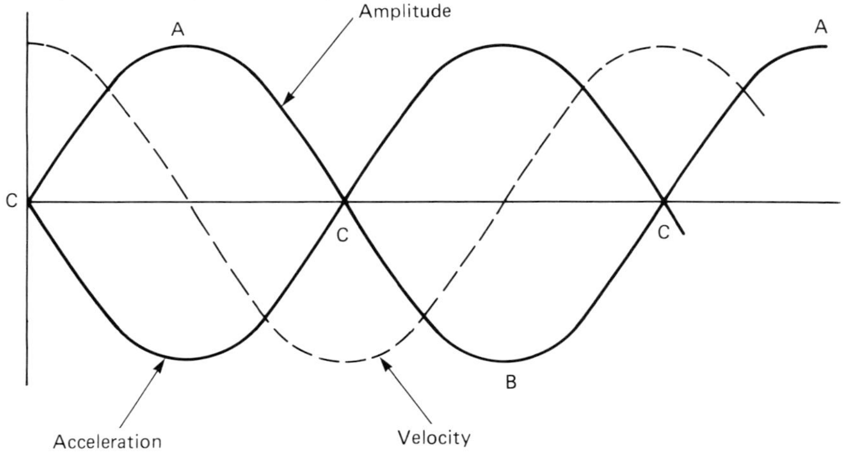

Figure 2.19 Variables of pendulum motion

(1) Case
(2) Springs
(3) Damper
(4) Magnet
(5) Wound coil
(6) Output leads

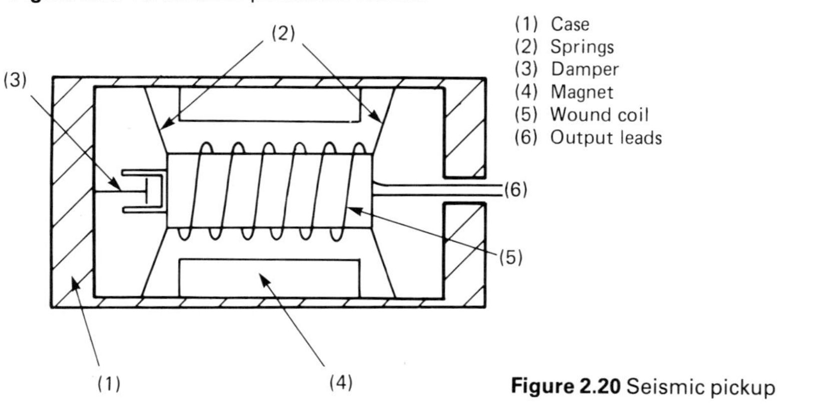

Figure 2.20 Seismic pickup

acceleration is:

$$a = dv/dt = -\omega^2 X \sin \omega t \qquad (2.15)$$

Conversion factors for the relationship between velocity and amplitude for sinusoidal vibration are as follows.

$$A = \text{amplitude (peak-to-peak)} = V_p/\pi f$$

$$= \frac{\sqrt{2}V_{r.m.s.}}{\pi f} = 0.45\ V_{r.m.s.}/f$$

where V_p = peak velocity $= \pi f A;$

$V_{r.m.s.}$ = r.m.s. velocity $= \dfrac{V_p}{\sqrt{2}}$ $= 2.22 f A;$ and

f = frequency

Consistent units must be used.

In practice, the direct measurement of amplitude is restricted to those cases where the amplitude is large and the frequency is low. Vibration monitoring of industrial machines invariably involves measuring an electrical signal which is then processed electronically in order to derive the equivalent amplitude of, say, a bearing housing.

One such method would be to use a seismic pickup, as illustrated in Figure 2.20. This is a moving-coil transducer consisting of a casing to which a magnet is rigidly connected and which vibrates with the bearing housing. A wound coil supported on diaphragm springs tends to remain stationary, with the magnets vibrating round it. This movement induces a voltage in the coil, proportional to the rate at which the flux surrounding the core is cut by the coil. The output voltage of the transducer is therefore proportional to the velocity of the vibrating housing. The amplitude may be required in order to compare, say, with an appropriate standard such as BS 4999:Part 50:1978, Table 50.9.1. This is achieved by passing the signal through an integrating circuit in a suitable vibration measuring instrument and the reading on a calibrated meter will then give the amplitude. Several instruments are available which will give both velocity and amplitude and some include a differentiating circuit which will process the transducer signal to give acceleration.

The advantage of a seismic transducer is that it gives a high signal:noise ratio. This has proved useful in field use where, if the balance is good, it can be difficult to differentiate machine-induced vibration from industrial structure-borne noise. The pickups are, however, bulky and on small industrial machines cannot be bolted directly to the machine without some doubt arising as to whether or not the mass is large enough to affect the behaviour of the component it is mounted on. Below about 500 rev./min the output signal falls and requires correction.

The use of an accelerometer as shown in Figure 2.21 can overcome these problems, taking advantage of its lower mass and physical size. The signal generated is less than that of a velocity transducer and requires greater amplification before it can be displayed on a meter. The signal:noise ratio can be a problem but for most applications the use of this type of measuring equipment is quite acceptable. This instrument will integrate the signal to give velocity and again to give amplitude. It is important to check that the stray fields from the motor do

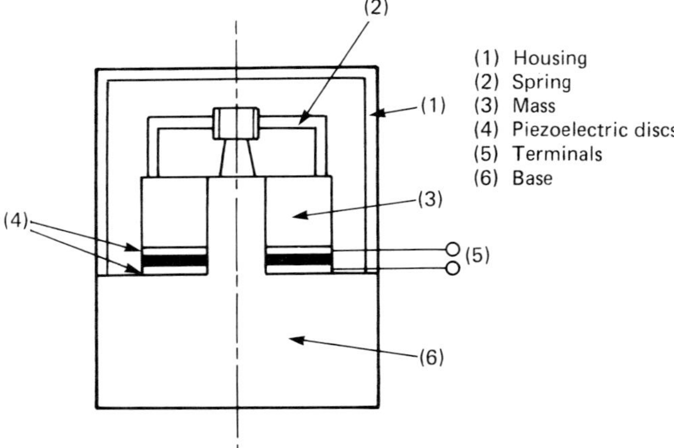

Figure 2.21 Piezoelectric accelerometer

not generate a spurious signal in the leads. It may be necessary to use screened leads and the transducer itself may require to be electrically insulated from the motor housing.

If the vibration signal is to be analysed over a wide frequency spectrum, the characteristics of the transducers over the frequency band must be available for the type of mounting to be used. Figure 2.22 shows typical response curves for an accelerometer subjected to constant acceleration at various frequencies but mounted in alternative ways.

If only the lower harmonics such as half, full and twice operating speed are of interest then the type of mounting is unimportant provided the signal is filtered or, if the instrument is a wide-band instrument, there are no significant higher harmonics.

If octave-band analysis is carried out, as would be the case on machines destined for naval vessels, then the type of mounting is very important.

Laboratory investigation of hydrodynamic sleeve bearing characteristics requires measurement of the shaft location and dynamic response to imbalance. This has led

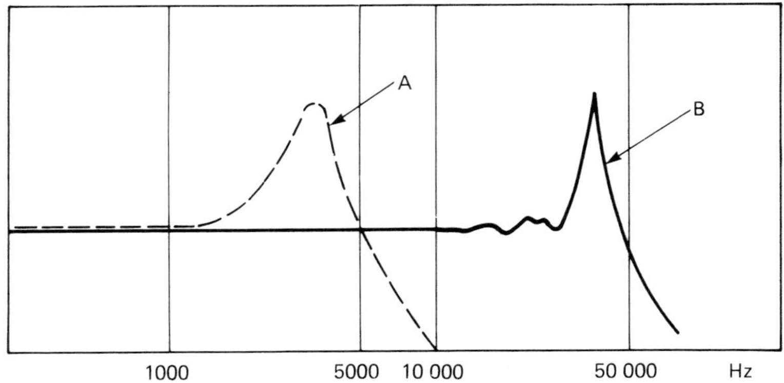

A – Handheld
B – Stud-mounted

Figure 2.22 Accelerometer output: constant acceleration, varying frequency

to the development of non-contacting proximity probes for use with such bearings. These instruments use the eddy–current principle to measure the distance from a coil mounted on the tip of the probes to the surface of a shaft. A radio frequency wave is generated by the probe and is transmitted into the air space between the tip and shaft. As a conductive surface (the shaft) approaches the tip, the radio frequency wave sets up eddy currents in the surface which produce a measurable power loss in the circuit. This power loss can be analysed to indicate both the average displacement of the shaft from the probe and the amplitude of shaft vibration about this displacement.

Some authorities claim that a direct reading of increased shaft vibration will lead to identification of incipient failure earlier than would be possible with indirect readings of the bearing housing vibration. There is an equally strong opinion that this is not so and that points to some serious disadvantages in using this type of probe in the industrial field. The main disadvantage is that the signal is not only dependent upon the airgap but upon the surface condition of the shaft. The track at which the tip is pointing must be smooth and circular, because out-of-roundness or other geometrical discontinuities will generate a signal which can be incorrectly diagnosed as vibration. A scratch or slight corrosion on the surface will give an increased signal and it is true that some apparent vibration problems in the field have eventually been diagnosed as a deterioration in the surface of the track and not a deterioration in the balance of the rotor.

The manufacturer can have difficulty achieving a suitable track, not so much in controlling the geometric tolerance specified but because the physical characteristics of the shaft surface can vary, giving rise to a signal even with a perfectly round shaft running with zero vibration. This is the so-called electrical runout.

Earlier disadvantages of non-linearity and temperature sensitivity have been largely overcome and proximity probes are being used increasingly for health monitoring of important drives.

It is important to appreciate that, whilst the velocity transducer or accelerometer measures the absolute vibration of the housing, the proximity probe measures the movement of the shaft relative to the housing. Other things being equal, a light shaft running in heavy rigid bearing supports could begin to vibrate more heavily with little or no change in housing vibration level. The use of proximity probes would certainly be of benefit in this case.

A comparatively heavy shaft, however, supported on light flexible pedestals could also see a rise in vibration. In this case the bearing housing will also suffer increased vibration and the relative shaft-to-housing movement could change only marginally. In this case the absolute measurement of housing vibration is the better safeguard. Most of the disadvantages can be overcome by fitting both absolute and relative vibration detectors, and the problem of electrical and mechanical runout can also be overcome by additional electronic circuitry which cancels these signals. The increased cost, however, could only be justified on the most important drives or in support of a research or development programme.

As relative movement in a rolling element bearing is restricted to the internal clearance of the bearing, a proximity probe is not a suitable basis for measuring vibration of electric drives with this type of bearing. Instrumentation either permanently or periodically attached is used to measure bearing housing vibration. This instrumentation is of two types: (1) the first as described above using accelerometers or moving coil sensors concentrates on the lower harmonics; and (2) that which essentially looks at the shock pulses generated in the rolling element bearing due to imperfections on the tracks and/or rolling element surfaces. As the damage in the latter type extends, due to fatigue, the signal increases and by monitoring the rate of deterioration a decision can be taken as to when the bearing should be replaced. This can be an important factor in the planned maintenance of important drives.

2.6.3 Rotor dynamics

The vibration level observed in operation is the result of imposed cyclic forces either from residual imbalance of the rotor or from some other cyclic force, such as from a diesel drive, and the response of the machine or component to these forces.

The designer needs to determine the source and level expected of the forces and should design each component of the motor to absorb these forces without impairing performance and without reaching unacceptable levels of vibration. Some of the more important factors to be taken into account are described below.

2.6.3.1 Sub-synchronous vibration

This problem, often referred to as oil whirl, can occur on rotating machines fitted with sleeve bearings when an adverse combination of bearing load, rotational speed and shaft design is met. The problem is complex and is usually associated with the use of flexible shafts, i.e. a shaft whose critical speed is lower than operating speed. Oil whirl can occur at shaft operating frequency but more frequently at a lower frequency near to half operating speed or the shaft critical speed. The problem can be eliminated by careful choice of bearing loading, lubrication and profile. Vertical motors which require the use of sleeve bearings pose a particular problem, as the load on these bearings is low and largely unpredictable. The use of tilting-pad bearings in such cases is normal, taking advantage of the inherent stability of this design. With horizontal motors, the use of tilting-pad bearings is not normally required and simpler fixed-geometry bearings such as lemon bore, multiland or offset halves will be chosen if cylindrical bearings cannot be designed to suit. This problem is unlikely to develop during service as the adverse combination of factors, should they exist, will occur during initial testing and commissioning and will require immediate correction. A cure is likely to involve a change in bearing or shaft geometry and will be permanent.

On induction motors, electromagnetic forces occur at frequencies equal to, or a multiple of, the supply frequency and any imbalance will result in a cyclic force on the rotor. The rotational speed, however, is slightly less than the supply frequency and mechanical imbalance forces will be cyclic at this reduced frequency. This is the classic case of two combined cyclic forces at frequencies relatively close together and results in a noticeable beat at low frequency. A vibration meter indicating the vibration level of, say, an induction motor pedestal will show gradual increases and decreases in vibration at the slip frequency of the motor. In the extreme case of a rotor with broken rotor bars, the variation in amplitude at slip frequency may be excessive.

2.6.3.2 Synchronous vibration

The most likely sources of vibration at shaft rotational frequency are mechanical imbalance or shaft misalignment. Imbalance may be residual mechanical imbalance or shaft bending owing to unequal cooling or heating of the rotor, due to some rotor winding or ventilation problem.

Thermal problems are usually time-dependent and a symptom is the gradual change of vibration with time and load and, where the load can be controlled, the vibration change reverses with load reduction (see Section 2.6.5).

An elliptical bore or a rotor grossly offset with respect to the axis of the stator will cause cyclic magnetic forces at shaft rotational speed. This usually can be identified by a sudden fall in vibration level when the motor supply is disconnected.

2.6.3.3 Super-synchronous vibration

Out-of-roundness of journals can be a source of vibration, with oval or three-lobed journals giving rise to vibration at 2 and 3 times rotational speed respectively. If the shaft stiffness is asymmetric along two major axes due to, say, a single large keyway along the length of the shaft, this can give rise to vibration at twice rotational frequency. It is only in exceptional cases that either of these sources gives rise to unacceptable vibration levels.

Owing to wear or damage to either track or rolling element, rolling-element bearings can give rise to high-frequency vibration, the level of which increases as damage becomes more extensive. The energy levels are initially small and would be undetected by a wide-band vibration meter. Special shock pulsemeters are used, specially designed for monitoring these higher frequencies.

The initial increase in bearing vibration is low but the rate of wear gradually increases and, if undetected, will lead to a sudden catastrophic failure of the bearing, probably resulting in major shaft damage and even stator core damage.

2.6.3.4 Vertical motor bearing problems

In vertical motors, the choice of upper thrust bearing depends upon the maximum speed and total thrust loads that have to be accommodated. The simplest arrangement is a deep-groove ball-bearing which will take thrust both in its upward or downward directions.

Angular contact bearings and spherical roller bearings are used where larger thrust-carrying capacity is required but, in isolation, will not permit upward thrust loads in excess of the weight of the shaft, as the rolling elements lift clear of the track and radial location of the shaft no longer exists. This causes shaft whirl and rubbing between shaft and housing. This excessive upward thrust may arise from a pump upthrust higher than predicted but can also occur with thermal expansion of the motor shaft.

Shaft expansion is accommodated by allowing the shaft to expand towards the driven machine. If, for reasons of high rotational speed or quietness, a ball-bearing has been chosen for the lower guide bearing, then it must be allowed to move axially in its housing. In certain instances cross-locking of the bearing occurs in the housing and the thermal expansion of the shaft is then upwards, lifting the rotating elements of the upper bearing off their seating, causing vibration and possible damage as a result. This tends to be less of a problem with a roller-guide bearing but other problems can result from using a roller-bearing. In this application the relatively unloaded bearing, which requires increased internal clearances compared with the equivalent ball-bearing, can result in noise and skidding of the rollers, which may lead to premature failure.

A vertical flange-mounted motor must inevitably result in the upper bearing being less rigidly supported radially than the lower bearing mounted close to the foundation. As this upper support stiffness is lower than that which would be applicable to the bearing support on horizontal machines, it is not surprising that the vibration level measured on the upper bearing will in general be higher than that expected on a horizontal motor manufactured and balanced to the same standards.

It is common practice to permit the level of vibration measured at the upper bearing to reach 1.5 times that specified in standards referring to horizontal motors.

2.6.3.5 Critical speeds

As the physical size of motors increases and their use with inverters to provide variable-speed drives at speeds as high as 8000 rev./min becomes more common, motors are being manufactured with a first critical speed below the maximum operating speed. This so-called 'flexible shaft' arrangement is inevitable but needs careful assessment if smooth operation is to be achieved.

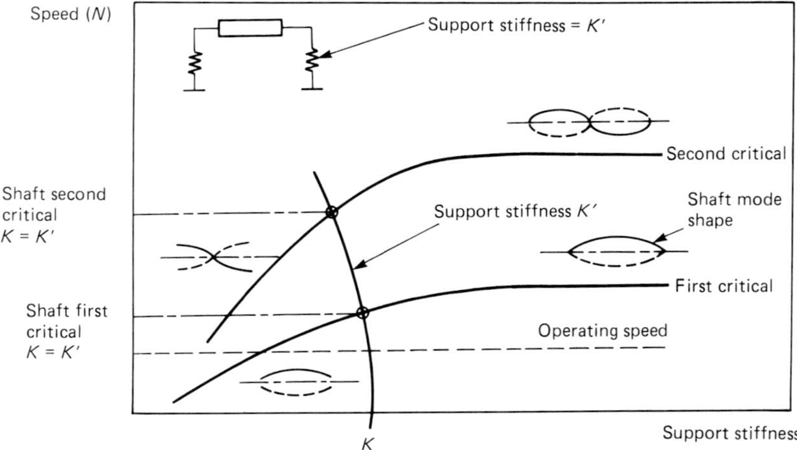

Figure 2.23 Shaft critical speed as a function of support stiffness K. (K' is the actual bearing stiffness curve, including oil film stiffness

The critical speed of a shaft is not dependent solely on its geometry and mass but is also very much affected by the stiffness of its bearing supports (Figure 2.23). The modal shape of the shaft at any speed is also dependent upon the relative stiffness of shaft and supports. If one considers the first critical speed of a shaft on rigid bearings, shaft deflection at the bearings is low and the mode shape will be similar to that shown to the right in Figure 2.23. If the journal support is very flexible then the mode shape will have the form shown to the left in Figure 2.23.

The actual amplitude of shaft and bearing housing vibrations for a given imbalance, without damping, is dependent upon the resonance curve for the shaft and the closer the shaft is running to the critical speed, the higher the vibration level (Figure 2.24).

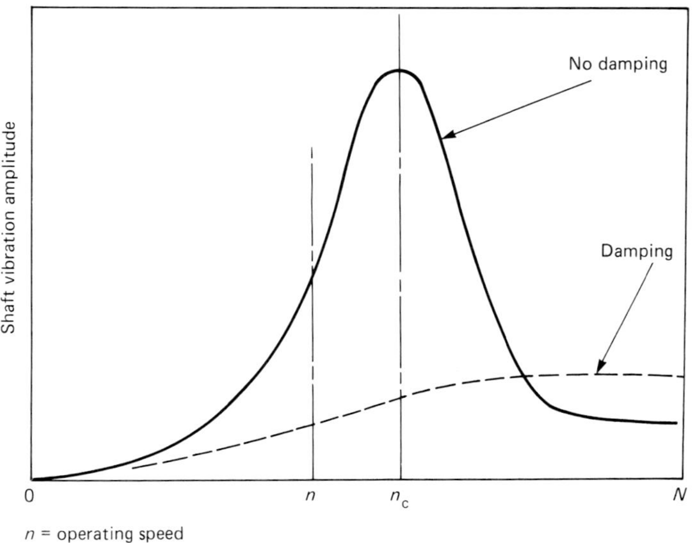

n = operating speed
n_c = undamped critical speed

Figure 2.24 Shaft resonance curve

It is probably impractical to run a shaft supported by rolling element bearings near to its critical speed without introducing some means of providing damping to reduce effectively the dynamic magnification factor. Fortunately, sleeve bearings provide a means not only of supporting the shaft but also of introducing significant damping via the oil film. This makes it entirely feasible to run a shaft close to, or even actually at, its undamped critical speed whilst still maintaining acceptable vibration amplitudes.

Inverter-driven motors of large output running at speeds significantly above normal synchronous speed may often have one or more critical speeds within the operating speed range. Choosing suitable bearing designs with the optimum stiffness and damping coefficients enables safe operation to be maintained at acceptable vibration levels throughout the speed range. Although at the time of writing the number of such high-speed motors is small, the problem has existed on two-pole turbogenerators for many years and design information and analytical procedures are well established.

It has been the practice to specify that the operating speed should be at least 20% away from the undamped critical speed. This is, in fact, good practice for an undamped system such as a shaft supported by rolling elements. It can, however, lead to an uneconomic design if rigidly applied to shafts supported on oil-lubricated sleeve bearings. Present practice is for the designer to calculate the response of the shaft system to imbalance and use this as the criterion for acceptability. Vibration levels specified in international standards must, and can be, met.

2.6.4 Balancing

Balancing, whether carried out by the manufacturer at his works or in the field, requires a means of measuring some factor proportional to imbalance and the angular displacement of that imbalance factor with respect to some fixed point on the shaft. Several methods exist for obtaining a reference signal from a shaft, such as a photocell viewing a painted mark or a magnetic pickup and a bolthead. These signals can also be processed to give a speed signal; this is a useful feature, as maintaining a constant speed is important whilst taking readings. An older method, used less frequently now but useful to illustrate the point, is to couple a sine wave generator temporarily to the shaft. The output wave from the transducer measuring force, amplitude or whatever other factor is being taken as a measure of imbalance is compared with the output from the generator (Figure 2.25).

The balancing instrument will read out the peak amplitude of the transducer signal, say pedestal displacement, and the phase angle shift between the two signals. These are plotted by the operator on a polar chart. If a modern balancing machine is being used, the signals will be processed electronically to advise the operator what size of balance weight is required in a particular balancing plane and the angular position that is required in that plane.

To balance a shaft demonstrating a simple static imbalance at points A and A$'$ (Figure 2.26) after a trial run, a trial weight W_1 is added to the rotor. The points B and B$'$ indicate a possible result with vectors **AB** and **A$'$B$'$** being the effect of W_1.

If W_1 is removed and is replaced by a second weight W_2 where

$$W_2 = W_1 \times \frac{\textbf{OA}}{\textbf{OB}} \tag{2.16}$$

and is located at angle α from the position of W_1, then the rotor should be in balance. If not, the process is repeated.

In practice, only the simplest of balance problems can be solved by single-plane balancing and multi-plane balancing is frequently necessary.[15,16]

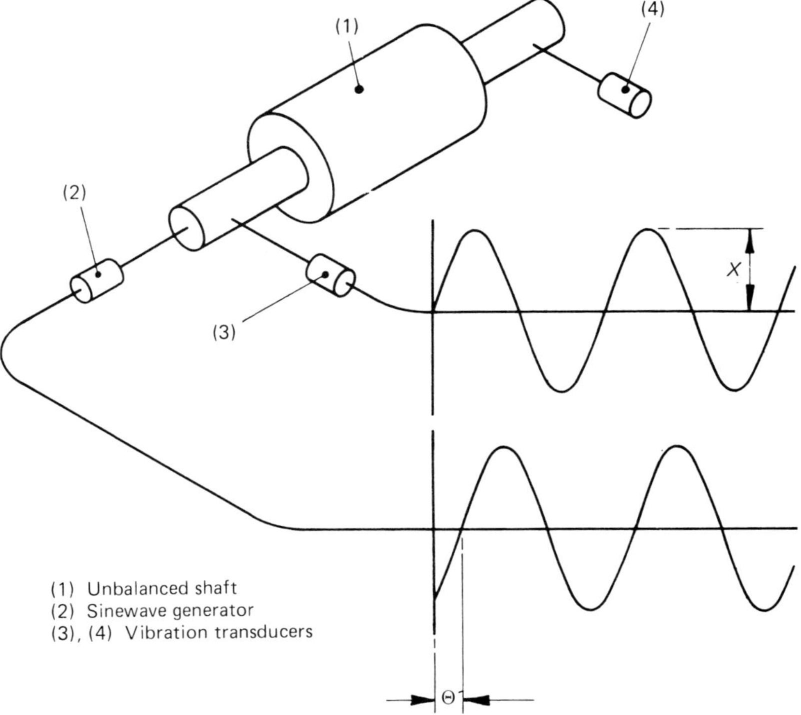

Figure 2.25 Measurement using a sinewave generator

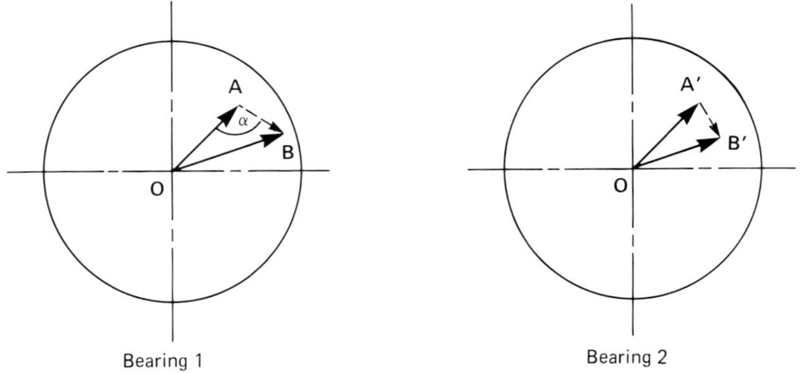

Figure 2.26 Polar diagram showing effect of balance weight W_1

2.6.5 Thermal imbalance

Portable balancing instrumentation as described above is most useful for analysing site vibration problems, particularly time-dependent symptoms. As an example, consider the case of an open-ventilated synchronous motor which, after some years of service, indicates a vibration level increasing with load to an unacceptable value. The symptoms are shown in Figure 2.27. On reaching operating speed, the vibration level is well within the accepted value. The machine is allowed to run at a

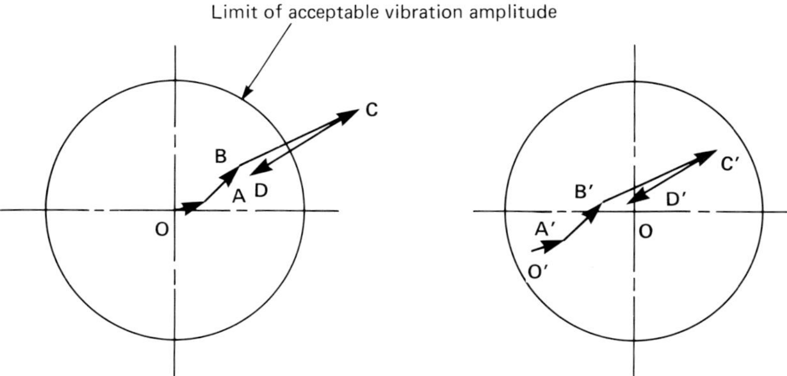

OA vector change due to residual imbalance
AB vector change due to windage heating
BC vector change due to increased load
CD vector change due to reducing load

Figure 2.27 Polar diagram showing balance change during run-up OA and at operating speed due to motor heating

low load held constant and the vibration vector changes to point B, where it then stabilizes, the change **AB** probably being the effect of the rotor heating-up due to the electrical losses and windage. The load is then gradually increased and the vibration vector changes to point C, outside acceptable limits. Removing the load causes a vector change to point D, sensibly the same as point B. This phenomenon would indicate unequal heating or cooling of the rotor, causing it to bend and indicate increasing static imbalance. Assuming electrical tests do not indicate a winding fault, then the cause is probably partially blocked ventilation ducts causing

Table 2.16 Vibration severity ranges and examples of their application to small machines (Class 1), medium-size machines (Class 2), large machines (Class 3) and turbo machines (Class 4)

Ranges of vibration severity		*Examples of quality judgement for separate classes of machines*			
Range	*r.m.s.–velocity v at the range limits (mm/s)*	*Class 1*	*Class 2*	*Class 3*	*Class 4*
0.28					
0.45	0.28				
0.71	0.45	A			
1.12	0.71		A		
1.8	1.12	B		A	
2.8	1.8		B		A
4.5	2.8	C		B	
7.1	4.5		C		B
11.2	7.1	D		C	
18	11.2		D		C
28	18			D	
45	28				D
71	45				

Source: BS 4675:Part 1:1976, *Mechanical vibrations in rotating and reciprocating machinery.*

uneven cooling. If the rotor cannot be removed for cleaning for operational reasons, one alternative procedure would be to add balance weights to move the initial start point O to O'. This will cause the vibration amplitude level to remain within the accepted value and will permit continued operation until such time that the rotor can be released. A chart commonly referred to is illustrated in Figure 2.28. This philosophy is built into several vibration specifications and reference to Table 2.16 shows an extract from BS 4675:Part 1:1976 *Mechanical vibrations in rotating and reciprocating machinery.* Class 2 includes electric motors up to 75 kW, Class 3 large prime movers on heavy foundations, and Class 4 large prime movers on low-tuned foundations.

In practice, it does not matter whether amplitude or velocity is used as a criterion provided the level observed is known to be normal for the type of particular equipment and usage. One motor mounted on rigid foundations driving a

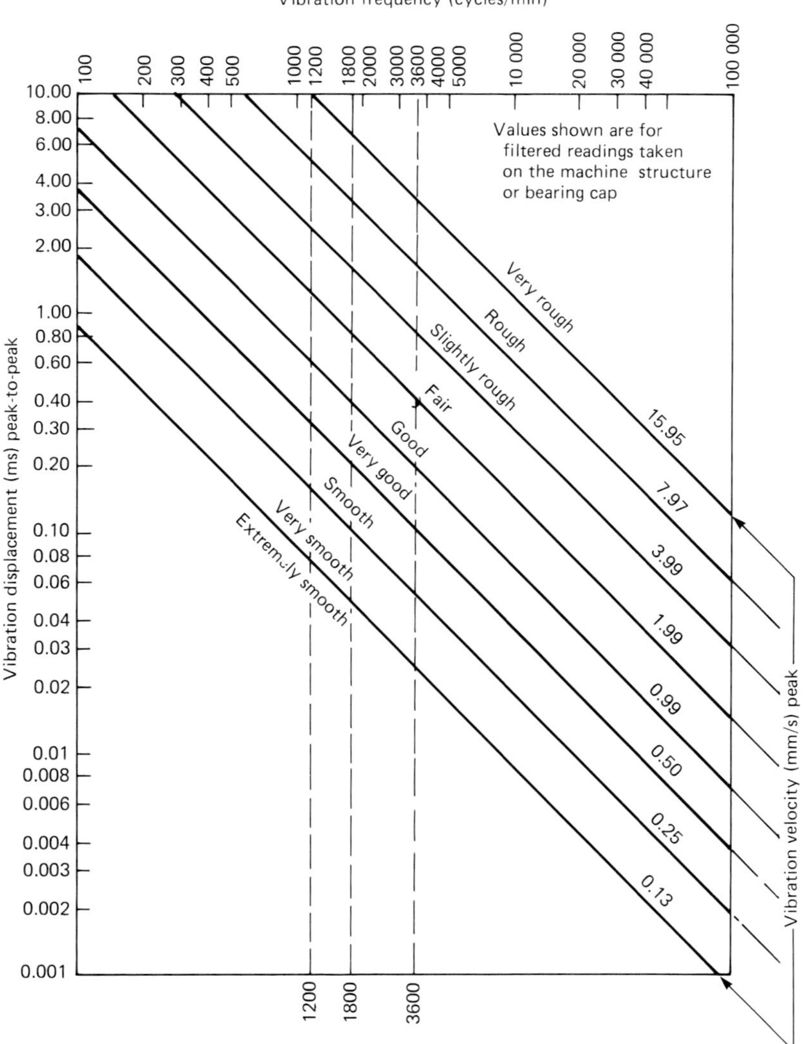

Figure 2.28 Vibration severity chart

Table 2.17 Comparison of vibration specifications

| Speed | | BS 4999:Part 50:1978[a] | | | | BS 4675b | | IEC 2373:1974 Normal grade | | | | | | NEMA[c] | |
| | | A | | B | | Class 3A | | 80≤H≤132 | | 132≤H≤225 | | 225≤H≤400 | | MG1 20.52:1980 | |
(Hz)	(rev./min)	Amp	Vel.	Amp	Vel.	Amp	Vel.	Amp	Vel.	Amp	Vel.	Amp	Vel.	Amp	Vel.
50.00	**3000**			**0.024**	2.66	0.016		0.016		0.025		0.040		**0.025**	2.8
25.00	**1500**	As IEC 2373		**0.041**	2.28	0.032	1.8	0.032	1.8	0.050	2.8	0.080	4.5	**0.050**	2.8
6.67	**1000**			**0.055**	2.04	0.049		0.049		0.076		0.123		**0.064**	2.4
12.50	**750**			**0.065**	1.80	0.065		0.065		0.101		0.163		**0.076**	2.1

Notes: Values in bold type refer to the specific value.

Values in light type are values equivalent to the specified value.

(a) Column A agrees with IEC 2373:1974. The value H is the height in millimetres from the foot to shaft centre height.

(b) Class 3 includes large prime movers on heavy foundations and Grade A would be considered good state of balance. Grade B would be considered satisfactory balance and the velocity level would then be 4.5 mm/s r.m.s.; the equivalent amplitude would then be as given in IEC 2373:1974 225≤H400.

(c) Amplitude converted from equivalent inch units specified.

Amplitude unit is millimetres peak-to-peak.
Velocity unit is millimetres per second r.m.s.

high-frequency generator would be expected to record significantly lower vibration levels than a second identical motor driving pulverizing plant mounted on a low-tuned foundation.

In service, it is changes in vibration level which are important and form the basis of health monitoring.

National standards are used primarily to measure the overall quality of the machine being offered and a comparison of some commonly applied standards is given in Table 2.17. Inspection reveals some interesting variations.

2.6.6 Vibration standards

Section 2.6 was introduced by referring to two similar machines, a four-pole motor and a four-pole generator, being works-tested and probably both being required to meet the vibration level of 0.041 mm peak-to-peak as in the *Specification for general requirements for rotating electrical machines,* BS 4999:Part 50:1978 (superseded by Part 143:1987). In service, this same specification recognizes that a higher vibration level is likely but should not exceed 0.052 mm peak-to-peak, which equates to an r.m.s. velocity of 2.88 mm/s, assuming sinusoidal vibration at the operating speed.

The generator may well be subjected in service to a vibration of 9 mm/s, over twice that for the motor, and still meet the appropriate standard BS 5000:Part 3:1980. More importantly, a long trouble-free service life will also be expected. This raises the question of what the maximum vibration level for rotating equipment should be.

Fortunately, we are not concerned with determining the level at which failure would occur or service life would be seriously affected, as this would be extremely difficult. We are more concerned with what experience dictates is normal for equipment and hence, if this level is exceeded, when it becomes necessary for an investigation to identify and correct the problem before failure takes place.

Rathbone's[17] early work in vibration resulted in assessment criteria being established based on subjective experience which suggested that vibration velocity could be used to judge whether a machine was running smoothly or roughly.

Appendix 2.1

Influence of user requirements and power supply on motor specification

	Aspect	*Relevant British Standard*	*Comments*
Electrical supply	Supply voltage	BS 4999: Part 30	Choice of supply voltage is related to motor output power. Above 1100V, the motor size and cost increases with voltage
	Voltage and frequency fluctuations	BS 4999: Part 31	Motors can generally cope continuously with voltage fluctuations of $\pm 6\%$. Other variations by special agreement
	Voltage transients		Voltage transients due to switching or lightning may impose additional stress on main and interturn insulation
	Neutral earthing	BS 4999: Part 31	The period when operating with one line at earth potential should be limited to prevent the danger of corona attack
	System fault level		It is good practice to supply motor terminal boxes which can deal with the system fault level
	Waveform	BS 4999: Part 31	The instantaneous values of phase voltage should not differ by more than 5% from the fundamental sine wave
	Phase balance	BS 4999: Part 31	The negative-sequence and zero-sequence components should not exceed 2% of the positive sequence component of voltage
User requirements	Noise level	BS 4999: Part 51	Describes test methods and gives tabulation of sound power level limits
	Vibration	BS 4999: Part 50	Gives tables of vibration velocity limits for shaft heights up to 400 mm and vibration amplitude limits for larger motors
	Dimensions and mounting	BS 4999: Part 10 BS 4999: Part 22	Leading dimensions are given for foot-mounted and flange-mounted machines. Standard codings for mounting arrangements are presented
	Cooling methods	BS 4999: Part 21	A classification of cooling methods is presented, using coded nomenclature
	Duty type	BS 4999: Part 30	Different duty types are defined to represent actual operating cycles including stopping, starting and speed changing
	Torque and current limits	BS 4999: Part 41	Design letters define combinations of torque and current at starting. Low starting current and high starting torque need larger motors
	Load inertia		Drives with large load inertia, such as fans, require special design consideration in order to incorporate sufficient thermal capacity
	Starting frequency	BS 4999: Part 41	Cage motor drives involving frequent starting require special design consideration. Normal drives can withstand two consecutive starts
	Overloads	BS 4999: Part 41	Momentary overloads for all types of motor are tabulated. Unless otherwise specified, motors have no sustained overload capacity
	Overspeed	BS 4999: Part 41	Motors are designed to withstand 1.2 times the maximum rated speed

Appendix 2.2

Effect of environmental factors on motor specification and performance

	Aspect	*Relevant British Standard*	*Comments*
Climatic	Ambient temperature	BS 4999: Part 31 BS 4999: Part 32	Up to 40°C without temperature rise adjustment. Precautions needed concerning lubrication and material at very low temperatures
	Cooling medium temperature	BS 4999: Part 31 BS4999: Part 32	Up to 30°C without temperature rise adjustment. Cooling water assumed not to exceed 25°C, unless higher temperatures are specified
	Altitude	BS 4999: Part 31 BS 4999: Part 32	Up to 1000 m without temperature rise correction. Special arrangements for altitudes above 4000 m
	Relative humidity		For high relative humidity and wide temperature fluctuations, consider the use of anti-condensation heaters
Enclosure	Protection against solids	BS 4999: Part 20	Standard specification indicates degree of protection and describes tests to establish effectiveness
	Protection against water	BS 4999: Part 20	Standard specification indicates degree of protection and describes tests to establish effectiveness
Coolants	Cooling air contamination		Dusty location may need the use of filters. Extra impregnation and special painting for chemical contamination and abrasion
	Properties of cooling water		Cooler design is affected by cooling water analysis. Anti-freeze may be needed in cold climates
Hazardous locations	Hazardous Areas zone 2 Ex(n)	BS 5000: Part 16	Specification lists design features for this type of motor
	Hazardous Areas Zone 1 Ex(e)	BS 5000: Part 15	Specification lists other relevant specifications
	Hazardous Areas Zone 1 Ex(d)	BS 5000: Part 17	Specification lists other relevant specifications
	Hazardous Areas Zone 1 Ex(p)	BS 5001: Part 3	Specification lists design features for pressurizing systems and lists other relevant specifications
Mechanical forces	Axial forces from drive		Incorporate thrust bearing, taking note of magnitude and direction of thrust. This is specially important on vertical motors
	Radial forces from drive		High bearing loads due to large overhung weight, belt pull, etc. may need larger shafts and bearings
	Vibration from outside		Imposed vibration, especially if motor is stationary, can result in bearing damage due to brinelling
	Shock loads from outside		For marine duty, excavator drives, etc. or where shock loads are expected, it may be necessary to reinforce the design

References

1. WELDON, R. and KELLET, R. 'Glandless boiler circulating pumps', *Proc. Inst. Mech. Engrs,* **184/3K,** 52–63 (1969)
2. SCHWARZ, K. K. 'Three generations of submerged circulator motors for AGR power stations', *Institute of Electrical Engineers Colloquium Digest.* pp. 5.1–5.4 (1985)
3. DEPARTMENT OF EMPLOYMENT, *Code of practice for reducing the exposure of employed persons to noise.* HMSO (1972)
4. DEPARTMENT OF EMPLOYMENT (1972) op. cit.
5. BROZEN, R. J. 'No-load to full-load airborne noise level change on high-speed polyphase induction motors, *Inst. Elec. & Electronic Engrs Trans.,* **IA.9,** 2 (1973)
6. TSIVITSE, P. J. and WEISHMANN, P. R. 'Polyphase induction motor noise', *Inst. Elec. & Electronic Engrs Trans.,* **IGA-7,** 3 (1971)
7. JORDAN, H. and ROTHERT, H. 'Rules of slot numbers and their relationship with the magnetic noise of induction machines', *Elektrotech. Z(ETZ) A,* **74,** 637–642 (1953)
8. LIWSCHITZ-GARIK, M. and WHIPPLE, C. C. *Alternating current machines.* Van Nostrand Reinhold Co.
9. WALKER, J. H. and KERRUISH, N. 'Open-circuit noise in synchronous machines', *Proc. Inst. Elec. Engrs,* Part A, **107,** 36 (1960)
10. LARTER, F. W. 'Magnetic noise in dynamo-electric machines', *Engrg* (1972)
11. WALKER, J. H. and KERRUISH, N. 'Noise in salient-pole machines', *AEI Engrg,* **1,** 12 (1961)
12. BRITISH STANDARDS INSTITUTION BS 4999:Part 5D: *Mechanical performance – vibration.* BSI, Milton Keynes (1978)
13. INTERNATIONAL ORGANIZATION FOR STANDARDIZATION. *Mechanical vibration of certain rotating electrical machines with shaft heights between 80 and 400 mm. Measurement and evaluation of vibration severity.* ISO R2373 (1974)
14. VEREIN DEUTSCHE INGENIEURE. *Assessing the effects of vibration on human beings.* VDI Publication No. 2057 (1963)
15. POMFRET, B. 'The dynamic analysis of rotors for high-speed rotating machines'. Indian Electrical Manufacturers' Association ELROMA 83 Conference, Session III, Paper 2 (1983)
16. BISHOP, R. E. D. and PARKINSON, A. G. 'Vibration and balancing of flexible shafts'. *Applied Mech. Revs,* **21,** 5 (1968)
17. RATHBONE, T. C. *Power Plant Engrg,* **43,** 721 (1939)

Bibliography

BRITISH ELECTRICAL APPARATUS AND MANUFACTURERS' ASSOCIATION. *Measurement and classification of acoustic noise from rotating electrical machines.* BEAMA Publication No. 255 (1969)

BRITISH STANDARDS INSTITUTION. *Specification for general requirements for rotating electrical machines.* BS 4999:Part 51, 'Noise levels'. BSI (1973)

ELECTRICAL RESEARCH ASSOCIATION. *Heat transfer in electrical machines, a critical review.* ERA Report No. 71–76 (1971)

GORDON, C. G. 'Fan noise and its prediction', Institution Mechanical Engineers conference on fan technology (1972)

INTERNATIONAL ELECTROTECHNICAL COMMITTEE 34–9. *Rotating electrical machines,* IEC 34–9: Part 9, 'Noise limits'. IEC (1972)

JACKSONE CLARK. 'An iterative technique for computing temperature distributions in rotor and stator cores'. Institution Electrical Engineers Power Division Colloquium, 22 January (1976)

KING, A. J. *The measurement and suppression of noise.* Chapman and Hall (1965)

POMFRET, B. 'Noise of electric motors', LSE Bulletin, **12,** 3 (1973)

RENTZCH, H. 'Systems for cooling electrical machines from the point of view of noise generation', *Elektrotech. E. (ETZ)A,* **87,** 450–453 (Electrical Research Association translation IB 2412) (1966)

SHARLAND, I. *Woods' practical guide to noise control.* Woods of Colchester Limited (1973)

3 Selection

M. R. Lloyd* and K. K. Schwarz†

* Engineering Director, GEC Large Machines Ltd.

† Consultant. Formerly Technical Director, Laurence Scott and Electromotors Ltd.

3.1 Electrical supply systems

3.1.1 General

3.1.1.1 Introduction

There are two principal differences when designing (rather than choosing) supply systems for motor drives. For normal land-based installations, there is a choice of 50 or 60 Hz, with a wide-ranging highly non-standardized choice of voltage levels in conjunction with grid supplies and apparatus designs. However, for self-contained systems, such as ships, offshore structures or installations not connected to a grid, there is a choice of frequency as well as voltage. There have been a number of attempts at optimizing the frequency choice, by adopting higher frequencies, such as 400 Hz, whilst traditionally railway networks in Europe use 16.66 Hz on a large scale, originating from the limitations of single-phase commutator motors. In practice, the choice of frequency is no longer an active subject, though in principle a choice in the region of 100–150 Hz might well be optimal for a combination of reasons based on a total system evaluation.

There are still active moves to standardize voltage levels, accepting the two standard frequencies, and this is of considerable economic impact with respect to motor installations. Table 3.1 summarizes the current international situation.

Table 3.1 Standard voltage (up to 15 kV nominal) in accordance with IEC 38 (1983)

	Highest voltage (kV)	Nominal voltage (kV)	Alternative nominal voltage (kV)	
High voltage	3.6	3.3	3	
Series I	7.2	6.6	6	
(50 and 60 Hz)	12.0	11.0	10	Table III
	17.5	—	15	
High voltage	4.40	4.16	—	
Series II	13.20	12.47	—	
(60 Hz)	13.97	13.20	—	Table III
	14.52	13.80	—	
		(V)		
Low voltage	—	380	—	
	—	400	—	
	—	415	—	
	—	480	—	
	—	660	—	Table I
	—	1000	—	

Whilst it is true that voltage levels can always be changed by transformers, this is costly in an area where standardization of apparatus should bring economic benefits. Moreover, it is convenient to consider low voltage ($< 1000\,\text{V}$) and high voltage ($1000-17000\,\text{V}$) separately, these voltage ranges being derived from further international standardization on the subject of insulation.

As electricity is 'unsafe' above about $50\,\text{V}$ r.m.s. a.c. to ground, the choice of voltage level is dictated by the system short-circuit capacity, which has to be dealt with in a manner 'as safe as is reasonably practicable' at all times. It is advisable, therefore, to limit this to some $40-50\,\text{kA}$ as high fault currents, or more precisely high fault energy, presents potentially 'dangerous' situations, as was demonstrated so clearly by the investigations into high-voltage terminal boxes[1] (see Section 3.3.3.5). In this connection both magnitude and time are of importance.

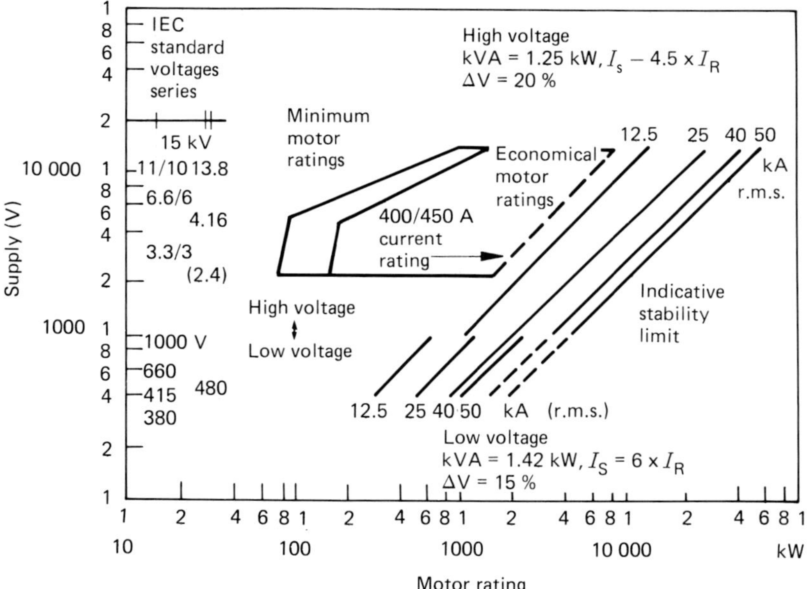

Figure 3.1 Economic motor ratings and systems stability limits for system voltages 380–15 000 V and system short-circuit capacities of 12.5–50 kA (r.m.s.)

The actual detailed design of the supply system has to deal with the transient voltage regulation due to normal overcurrent, e.g. starting currents of motors, and fault conditions, e.g. short circuits. The situation is summarized graphically in Figure 3.1 illustrating the various parameters involved.

3.1.1.2 Low-voltage practice in Europe

The apparently minor difference of British practice using 415/440 V, against the continental standard of 380 V, 50 Hz prohibits the interchangeability of motors. A proposal to compromise on 400 V, with a voltage variation tolerance increased from the current norm of ±5% to ±10% as an interim measure, is being pursued. Moreover, during the last decade, a new standard of 660 V has been promulgated, which has found wide application, particularly in Eastern Europe, but has not been found acceptable in the UK. A number of studies have been carried out to prove the economical advantages of this voltage level, which makes little impact on the cost of motors, contactor control gear or cables, whilst widening the range of motor ratings, individually or in total, by 50% for a given short-circuit current level.

The reasons advanced for not changing are often connected with spares holdings, which is, of course, relevant, but it is important to look ahead and take advantage of the potential of this system. This is one of the areas where British practice has pushed up the short-circuit current to 80 kA in special situations, e.g. marine applications.

3.1.1.3 High voltage practice in Europe

Most European countries have standardized on 6 kV against the British practice of 3.3 kV as the first level above low voltage. The second level is 10 kV and 11 kV respectively, with some 6.6 kV British installations, whilst a higher level of 15 kV with insulation level 17 kV is the proposed extension for larger installations. This has to be compared with US practice of 2.4 kV (obsolescent) and 4.16 kV followed by 13.2/13.8 kV, whilst some offshore operators have chosen 6.6 kV at 60 Hz. For the future, a two-tier system of 660 V and 10 kV is generally economically attractive or 660 V/6 kV/15 kV for the largest installations (Figure 3.2).

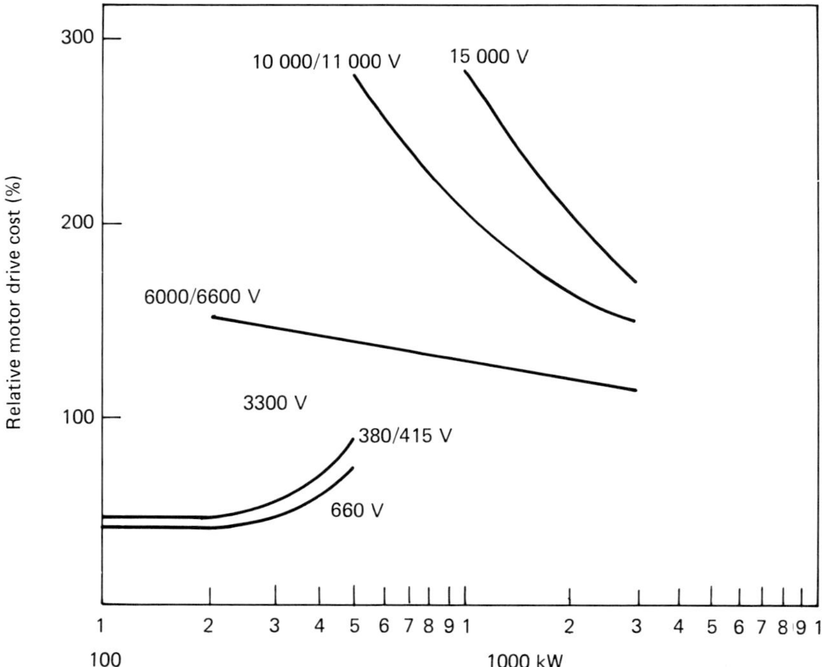

Figure 3.2 Motor drive cost comparison: four-pole, 50-Hz, IP44 squirrel cage motors and starters

Permissible starting current is found by the maximum voltage reduction which can be sustained, normally 15–20%, including cable voltage drops, and whether the complete installation has to recover from a short-circuit fault without a complete restart. This is an area where the system requirements have to be specified clearly and the overall performance computed. Table 3.2 indicates these stability limits against short-circuit faults. Other parameters, e.g. cable costs, are indicated in Figure 3.3 related to a notional 2.5% voltage drop. To match this, acceptable motor ratings are shown in Table 3.3.

Table 3.2 Typical parameters

System voltage	Fault level for 40/50 kA fault current	Induction motor load-stability approximate limit
(kV r.m.s.)	(MVA)	(a) (MW)
15	1040/1300	46/58
13.8	955/1200	42/53
11	760/950	34/42
6.6	455/570	20/25
3.3	230/285	10/13
(V r.m.s.)	(MVA)	(b) (kW)
660	46/57	1350/1650
440	30/38	850/1100
380	26/33	750/950

Notes: (a) Assumes power factor × efficiency = 0.8 per unit and starting current = 4.5 × full load current.

(b) Assumes power factor × efficiency = 0.7 per unit and starting current = 6.0 × full load current.

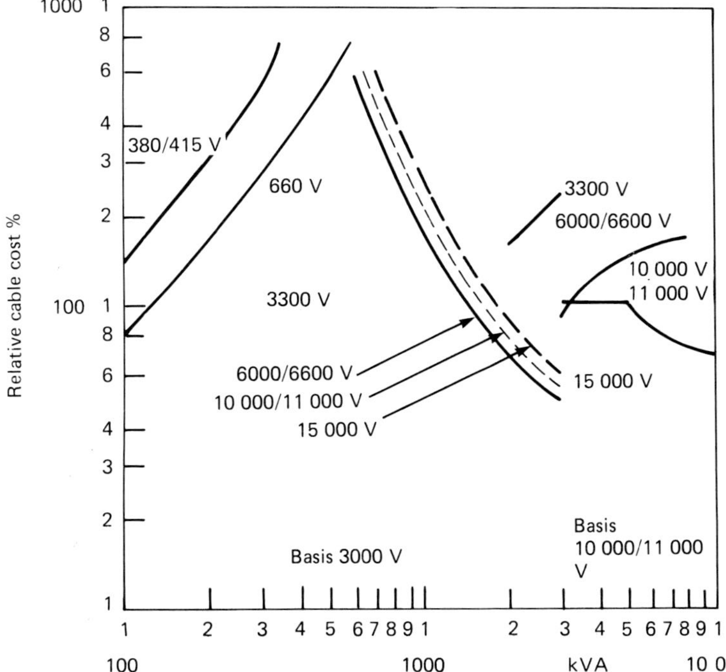

Figure 3.3 Indicative relative cable costs (380–15 000 V). (Stranded copper PVC-insulated three-core 3300 V and below; stranded copper XLPE-insulated core 6000 V and above (also above 550 A))

Table 3.3 Indicative economic ratings for induction motors

Voltage	Minimum (kW)	Normal (a) (kW)	Large (b) (kW)
380/660 V	—	0.1 – 500	1000
2–4.2 kV	75	150 – 2000	3000
5–7.2 kV	150	300 – 4000	6000
10–11 kV	500	750 – 7000	10 000
13.8 kV	750	1000 – 9000	14 000
15 kV	1000	1500 –10 000	15 000

Notes: (a) Maximum current 400–450 A.
   (b) Maximum current 600–700 A.

3.1.1.4 Self-contained systems

The performance parameters discussed so far are all based on a supply which is connected to a stable grid and will not, therefore, fluctuate in frequency, except in extreme fault conditions. However, there are many applications where this is not applicable such as self-contained systems, ships and offshore applications where emergency supplies may have to be installed for safety reasons, resulting in more onerous design requirements. In addition to the voltage regulation, it is necessary to introduce frequency regulation, which may manifest itself in a collapse of the prime mover's capability to produce the necessary power.

Whilst it is well understood how an a.c. generator performs with respect to wattless (kilovar) output, being dependent on its own internal impedance and the reaction of the automatic voltage regulator, power can only be sustained by the prime mover, except for the rotating inertia of the generating system. As a motor has a non-linear characteristic and inherently demands somewhere between 150 and 200% input power as it passes over its maximum torque point, a critical situation can arise, unless a proper design study has been carried out. The

Table 3.4 Typical a.c. generator transient performance parameters utilizing semiconductor automatic voltage regulator

Type of generator	Low power factor load switching on for 15% V dip		High power factor load switching off for 15% V rise		Sub-transient short-circuit current
	(% FLC)	(s)	(% FLC)	(s)	(% FLC)
(1) Standard normal performance	50	1.5	35	2	1200
(2) Standard high performance including brushless	120	0.2–0.5	100	0.3–0.7	1000
	100	0.2–0.5	100	0.3–0.7	800
	80	0.2–0.5	80	0.3–0.7	650
	60	0.2–0.5	60	0.3–0.7	500
(3) Compounded	125	0.2–0.5	125	0.3–0.7	950
	100	0.2–0.5	100	0.3–0.7	750
	75	0.2–0.5	80	0.3–0.7	600

Notes: (a) Recovery times given for recovery to ±2% accuracy for typical four- and six-pole machines with steady state accuracy of ±1%.
   (b) Types 2 and 3 have separately excited exciters, but Type 1 may be self-excited.
   (c) Range of performance given in Types 2 and 3 for standard frames. Performance outside this range can be obtained by adjustment of frame size.

generator performance, moreover, has to be obtained at the expense of the sub-transient reactance, using either a compounded design and/or high-speed voltage regulator, as shown in Table 3.4. The better the performance of the a.c. generator, the more onerous the duty of the prime mover, and an overall performance computation becomes essential.

3.1.2 International overview

The variety of supply systems worldwide is detailed in Table 3.5. To this should be added that the majority of ships and offshore installations operate at 60 Hz with the mixture of voltages already mentioned. Standards are available (see Section 3.3), to help this confused position, but it is still a long way from a satisfactory situation. However, most principal subjects are covered, as illustrated by the selected references to British and international standards in Table 3.6.

Table 3.5

Country	National installed capacity total (MW)	Frequency (Hz)	Industrial three-phase supply voltages in common use below 15 kV
Afghanistan	394	50	380/220 V
Algeria	2235	50	15 kV, 10 kV, 6.6 kV, 380/220 V
Angola	600	50	380/220 V
Antigua and Barbuda	29	60	400/230 V
Argentina	13 480	50	13.2 kV, 6.88 kV, 380/220 V
Australia	27 543	50	11 kV, 6.6 kV, 3.3 kV, 415/240 V
Austria	14 243	50	10 kV, 5 kV, 380/220 V
Bahamas	350	60	415/240 V
Bahrain	655	50, 60	11 kV, 400/230 V
Bangladesh	1025	50	11 kV, 400/230 V
Barbados	94	50	11 kV, 3.3 kV, 380/220 V
Belgium	12 046	50	15 kV, 6 kV, 380/220 V
Belize	21	60	6.6 kV, 380/220 V
Benin	91	50	15 kV, 380/220 V
Bermuda	118	60	4.16 kV, 2.4 kV, 208/120 V
Bolivia	508	50	400/230 V
Botswana	89	50	11 kV, 6.4 kV, 380/220 V
Brazil	38 904	60	13.8 kV, 11.2 kV, 380/220 V
Brunei	240	50	415/240 V
Bulgaria	9220	50	15 kV, 380/220 V
Burkina Faso	40	50	15 kV, 5.5 kV, 380/220 V

Table 3.5 (Continued)

Country	National installed capacity total (MW)	Frequency (Hz)	Industrial three-phase supply voltages in common use below 15 kV
Burma	636	50	11 kV, 6.6 kV, 3.3 kV, 400/230 V
Burundi	3.7	50	6.6 kV, 380/220 V
Cameroon	531	50	15 kV, 10 kV, 5.5 kV, 380/220 V
Canada	83 000	60	12.5 kV, 7.2 kV, 208/120 V
Canary Islands	4	50	380/220 V
Cape Verde Islands	7	50	380/220 V
Cayman Islands	6	60	208/120 V
Central African Republic	30	50	380/220 V
Chad	38	50	15 kV, 380/220 V
Chile	3358	50	13.8 kV, 6.6 kV, 380/220 V
China	76 000	50	12 kV, 11 kV, 5.6 kV, 380/220 V
Colombia	5820	60	13.2 kV, 11.4 kV, 208/120 V,
Comoros	5	50	380/220 V
Congo	149	50	380/220 V
Costa Rica	657	60	13.2 kV, 4.2 kV, 415/240 V
Cuba	2704	60	380/220 V
Cyprus	333	50	11 kV, 415/240 V
Czechoslovakia	18 109	50	15 kV, 6 kV, 3 kV, 380/220 V
Denmark	5768	50	15 kV, 10 kV, 6 kV, 380/220 V
Djibouti	38	50	380/220 V
Dominica	7	50	400/230 V
Dominican Republic	960	60	12.5 kV, 4.16 kV, 2.5 kV, 380/220 V
Ecuador	1315	60	13.8 kV, 6.6 kV, 4.4 kV, 2.2 kV, 220/127 V
Egypt	5145	50	11.0 kV, 6.6 kV, 3 kV, 380/220 V
Eire	3932	50	10 kV, 380/220 V
El Salvador	500	60	14.4 kV, 2.4 kV, 208/120 V
Ethiopia	319	50	15 kV, 380/220 V
Faroe Islands	7	50	415/240 V
Falkland Islands	20	50	415/240 V
Fiji	113	50	11 kV, 415/240 V

Table 3.5 (Continued)

Country	National installed capacity total (MW)	Frequency (Hz)	Industrial three-phase supply voltages in common use below 15 kV
Finland	11 135	50	10 kV, 660/380 V
France	73 984	50	15 kV, 10 kV, 5 kV, 380/220 V
French Guiana	31	50	380/220 V
French Polynesia	69	60	220/127 V
Gabon	175	50	5.5 kV, 380/220 V
Gambia	12	50	11 kV, 400/230 V
Germany, East	21 183	50	10 kV, 6 kV, 660/380 V
Germany, West	85 769	50	10 kV, 6 kV, 380/220 V
Ghana	1060	50	11 kV, 400/230 V
Gibraltar	30	50	415/240 V
Greece	5979	50	15 kV, 6.6 kV, 380/220 V
Greenland	80	50	380/220 V
Grenada	9	50	400/230 V
Guadeloupe	103	50, 60	380/220 V
Guam	302	60	13.8 kV, 4 kV, 208/120 V
Guatemala	473	60	13.8 kV, 220/127 V
Guinea-Bissau	8	50	10 kV, 6 kV, 380/220 V
Guyana	165	50, 60	13.8 kV, 11 kV, 380/220 V
Haiti	126	60	12.5 kV, 7.2 kV, 4.2 kV, 2.4 kV, 380/220 V
Honduras	240	60	13.8 kV, 4.2 kV, 2.4 kV, 380/220 V
Hong Kong	3475	50	11 kV, 380/220 V
Hungary	4865	50	10 kV, 6 kV, 3 kV, 380/220 V
Iceland	795	50	380/220 V
India	38 808	50	11 kV, 6 kV, 400/230 V
Indonesia	5016	50	6 kV, 380/220 V
Iran	5300	50	11 kV, 380/220 V
Iraq	1200	50	11 kV, 380/220 V
Ireland, Northern	4500	50	11 kV, 6.6 kV, 3.3 kV, 415/240 V
Israel	3477	50	12.6 kV, 6.3 kV, 400/230 V
Italy	50 023	50	15 kV, 10 kV, 380/220 V,
Ivory Coast	1163	50	15 kV, 380/220 V

Table 3.5 (Continued)

Country	National installed capacity total (MW)	Frequency (Hz)	Industrial three-phase supply voltages in common use below 15 kV
Jamaica	740	50	15 kV, 4 kV, 2.3 kV, 380/220 V
Japan	154 811	50, 60	11 kV, 6.6 kV, 3 kV, 380/220 V
Jordan	540	50	11 kV, 380/220 V
Kampuchea	40	50	380/220 V
Kenya	556	50	415/240 V
Korea, North	5500	60	380/220 V
Korea, South	11 597	60	6.6 kV, 5.7 kV, 380/220 V
Kuwait	3446	50	415/240 V
Laos	250	50	380/220 V
Lebanon	668	50	380/220 V
Lesotho	1	50	380/220 V
Liberia	306	60	12.5 kV, 7.2 kV, 415/240 V
Libya	1180	50	400/230 V
Luxembourg	1305	50	15 kV, 5 kV, 380/220 V
Macau	104	50	380/220 V
Madagascar	100	50	5 kV, 380/220 V
Malawi	111	50	3.3 kV, 400/230 V
Malaysia	2550	50	11 kV, 415/240 V
Mali	42	50	15 kV, 380/220 V
Malta	152	50	11 kV, 6.6 kV, 3.3 kV, 415/240 V
Martinique	65	50	380/220 V
Mauritania	55	50	380/220 V
Mauritius	244	50	6.6 kV, 400/230 V
Mexico	21 574	60	13.8 kV, 13.2 kV, 220/127 V
Monaco	2	50	380/220 V
Mongolia	450	50	400/230 V
Montserrat	2	60	400/230 V
Morocco	1593	50	8 kV, 380/220 V
Mozambique	1800	50	380/220 V
Namibia	406	50	380/220 V
Nauru	1	50	415/240 V

Table 3.5 (Continued)

Country	National installed capacity total (MW)	Frequency (Hz)	Industrial three-phase supply voltages in common use below 15 kV
Nepal	139	50	11 kV, 400/230 V
Netherlands	18 139	50	10 kV, 6 kV, 5 kV, 3 kV, 380/220 V
Netherlands Antilles	340	50, 60	380/220 V
New Caledonia	381	50	380/220 V
New Guinea	715	50	15 kV, 11 kV, 6.3 kV, 5.5 kV, 415/240 V
New Zealand	6842	50	11 kV, 415/240 V
Nicaragua	400	60	13.2 kV, 7.6 kV, 415/240 V
Niger	50	50	15 kV, 380/220 V
Nigeria	2770	50	15 kV, 11 kV, 400/230 V
Norway	22 119	50	10 kV, 5 kV, 380/220 V
Oman	419	50	415/240 V
Pakistan	4239	50	11 kV, 400/230 V
Panama	744	60	12 kV, 208/120 V
Papua New Guinea	322	50	11 kV, 415/240 V
Paraguay	370	50	380/220 V
Peru	3237	60	10 kV, 6 kV, 380/220 V
Philippines	5003	60	13.8 kV, 4.16 kV, 2.4 kV, 415/240 V
Poland	25 988	50	15 kV, 10 kV, 6 kV, 380/220 V
Portugal	5073	50	15 kV, 10 kV, 5 kV, 380/220 V
Puerto Rico	4100	60	8.32 kV, 4.16 kV, 415/240 V
Qatar	695	50	415/240 V
Reunion	159	50	415/240 V
Romania	17 232	50	10 kV, 6 kV, 5 kV, 380/220 V
Rwanda	39	50	15 kV, 6.6 kV, 380/220 V
Sabah and Sarawak	2	50	415/240 V
St Helena	4	50	415/240 V
St Kitts, Nevis	13.5	60	400/230 V
St Lucia	16	50	11 kV, 415/240 V
St Thomas	3	60	208/120 V
St Vincent and the Grenadines	8.5	50	11 kV, 6 kV, 3.3 kV, 400/230 V

Table 3.5 (Continued)

Country	National installed capacity total (MW)	Frequency (Hz)	Industrial three-phase supply voltages in common use below 15 kV
Samoa	14.5	50	6.6 kV, 415/240 V
São Tomé and Principe Is.	5	50	30 kV, 6 kV, 380/220 V
Saudi Arabia	9145	50, 60	13.8 kV, 11 kV, 6.6 kV, 380/220 V
Senegal	165	50	5.5 kV, 220/127 V
Seychelles	16	50	415/240 V
Sierra Leone	95	50	11 kV, 400/230 V
Singapore	2106	50	11 kV, 6.6 kV, 400/230 V
Solomon Islands	8	50	11 kV, 415/240 V
Somalia	30	50	380/220 V
South Africa	23 074	50	11 kV, 6.6 kV, 3.3 kV, 380/220 V
South Yemen	108	50	15 kV, 11 kV, 10 kV, 400/230 V
Soviet Union	285 492	50	15 kV, 10 kV, 6.0 kV, 3.0 kV, 380/220 V
Spain	29 900	50	15 kV, 11 kV, 380/220 V
Sri Lanka	562	50	11 kV, 400/230 V
Sudan	313	50	11 kV, 415/240 V
Surinam	400	50, 60	220/127 V
Swaziland	31	50	11 kV, 6 kV, 400/230 V
Sweden	29 684	50	10 kV, 6 kV, 380/220 V
Switzerland and Liechtenstein	14 100	50	11 kV, 6 kV, 380/220 V
Syria	1104	50	380/220 V
Taiwan	11 869	60	11.4 kV, 380/220 V
Tanzania	258	50	11 kV, 6.6 kV, 3 kV, 400/230 V
Thailand	5057	50	11 kV, 3.3 kV, 380/220 V
Togo	35	50	15 kV, 5.5 kV, 380/220 V
Tonga	9	50	11 kV, 6.6 kV, 415/240 V
Trinidad and Tobago	760	60	12 kV, 400/230 V
Tunisia	949	50	15 kV, 10 kV, 380/220 V
Turkey	6638	50	15 kV, 6.3 kV, 380/220 V
Uganda	163	50	11 kV, 415/240 V
United Arab Emirates	1510	50	11 kV, 6.6 kV, 415/240 V
United Kingdom	69 191	50	11 kV, 6.6 kV, 3.3 kV, 415/240 V

Table 3.5 (Continued)

Country	National installed capacity total (MW)	Frequency (Hz)	Industrial three-phase supply voltages in common use below 15 kV
United States	666 405	60	12 kV, 4.16 kV, 2.4 kV, 208/120 V, etc.
Uruguay	1364	50	15 kV, 6 kV, 380/220 V
Venezeula	9312	60	12.47 kV, 4.16 kV, 2.4 kV, 208/120 V
Vietnam	900	50	15 kV, 380/220 V
Virgin Is (UK)	12	60	208/120 V
Yemen	104	50	15 kV, 11 kV, 10 kV, 400/230 V
Yugoslavia	14 800	50	10 kV, 6.6 kV, 3 kV, 380/220 V
Zaire	1716	50	6.6 kV, 380/220 V
Zambia	1728	50	11 kV, 400/230 V
Zimbabwe	1192	50	11 kV, 415/240 V

Table 3.6 Standards (IEC and BS) relating to apparatus and systems

IEC38:1983	Standard voltages	BS77:1962
IEC439:1985	LV switch and controlgear	BS5486:1986
IEC298:1981	HV switch and controlgear	BS5227:1984
IEC56:1971–1981	HV circuit-breakers	BS5311:1976–1980
IEC157–1:1983	LV circuit breakers	BS4752:1977
IEC183:1984	HV cables	
IEC502:1984	HV cables	
IEC227:1979–1985	LV cables	BS6004:1984
IEC158–1:1983	LV contactors	BS775:1984
IEC292:1970–1983	LV control gear	BS5486:1977–1986
		BS4941:1977–1979
IEC470:1975	HV contactors	BS775:1984
IEC282–1:1985	HV fuses	BS2692:1976–1986
IEC269:1978–1986	LV fuses	BS88:1982
IEC 34:1968–1985	Rotating electrical machines (see Table 3.8)	BS4999:1972–1987
		BS5000:1972–1985
IEC71:1976–1982	Insulation co-ordination	BS5622:1979
IEC79:1967–1984	Hazardous areas (see Table 3.17)	BS5501:1980–1986
IEC76:1967–1981	Power transformers	BS171:1970–1981
CISPR:1967–1986	Radio interference	BS1597:1985
IEC363:1972	Short-circuit calculations	

3.1.3 Controlgear interface

Control, in its widest sense, encompasses three principal aspects:

(1) Power control.
(2) Monitoring and protection.
(3) Closed-loop (automatic) control where applicable to the process.

The design of controlgear has a major impact on safety in connection with fault energy, resulting in the generally accepted use of fused contactor equipment to limit short-circuit current effects. Provided the motor, system voltage and short-circuit ratings are within the capabilities of contactors, the former being a thermal rating, whilst the latter is a function of the fuse, this form of switching is preferred, as it matches the switching capability of motors. Circuit-breakers, being designed for interrupting short-circuit, i.e. fault currents, have a much lower switching capability in terms of number of operations. Air-break interruptors for low voltage and vacuum or SF6 interruptors for high voltage are eminently suitable for contactors.[2] The advantage of vacuum and SF6 lies in the 'non-sparking' mode of switching, leading to greater reliability and a widening scope of application, where contaminated or even potentially hazardous environments are present. As detailed later (see Section 3.3.3.9 and Chapter 7) close attention has to be paid to protection co-ordination, which is the link between the electrical supply system and the motor, as well as the motor and its load. Equally important is the recognition of the demands of the load machine in the context of these overall system requirements.

On the electrical side, standards stipulate the testing of the combined equipment in terms of basic current co-ordination between the contactors, fuses and overcurrent relays to ensure proper circuit interruption at all times. The same applies to the special case of (Ex)e 'increased safety' motors (see Section 3.3.3.4). An aspect which has become increasingly recognized is the subject of controlgear safety under short-circuit fault conditions, internal as well as external. Two approaches are recognized, one of which is to ensure that short-circuit current clearance is initiated in a sufficiently short time to avoid a dangerous occurrence, using fuses or high-speed circuit-breakers operated by a signal from the arcing fault, usually light. The other approach is to provide pressure relief for the products of the fault, the hot gases being exhausted in a direction away from personnel, allowing normal circuit-breaker clearance to proceed. The co-ordination of the safety arrangement with plant requirements is paramount.

3.1.4 Closed loop/process control

Single-, two- or variable-speed motors may be incorporated in a closed-loop system, controlled by some physical parameter of the process. Effective transducers coupled with modern microprocessors or computers permit such controls to be carried out on a large scale with relatively little expenditure in terms of hardware. However, this may necessitate sequential start-up, repeated starts for so-called 'on–off' controls or selective tripping on reduction or loss of supply or switching over from one supply to another. These control actions all have an effect on the motor design in terms of performance, reliability and life and clearcut numerate evaluation is essential.

In many cases, the attempt to use fixed-speed motors in these situations leads to reduced life. For instance, with a nominal switching life of a medium-sized squirrel cage motor of 20 000 operations, based on 1000 switching cycles per year, even three starts per hour (if permissible) would use up the notional life in less than 1 year.

3.1.5 External system performance

Designs for stability and recovery after faults need studying in terms of the electromechanical parameters of drive and supply system on the one hand, and the effect of transients on the other.

Switching of motors always produces substantial torque transients, which are transmitted mechanically to the driven machine and reflected electrically in the supply system. Furthermore, these transient currents impose major forces on the motor windings and the foundations, to mention but two of the most important points. The dielectric design of windings must take full account of switching surges.[3-4] In addition to following the rules based on experience, effective evaluation is both practical and necessary for important applications. The additional cost of appropriate system-orientated design should be cost effective in the overall scheme of things and is vital in connection with reliability[5] and safety,[6] and has to be an inter-disciplinary design effort to ensure 'fitness for the purpose'.

3.2 Motor drives

3.2.1 Types of motors

3.2.1.1 General

As all types of rotating electrical machines operate in accordance with the same electromagnetic laws, they all behave in a generally similar manner. Contrary to popular belief, there is no basic difference between 'a.c.' and 'd.c.' machines; the so-called 'd.c.' machines appeared first because the only source of electrical supply in 1831 was the primary cell. It follows, therefore, that it is possible to categorize all types of machines on a simple basis, realizing that d.c. is the special case of 'zero-frequency a.c.' This makes all machines a.c. machines, leading to two groupings: (1) stationary field machines; and (2) rotating field machines, as seen from the stationary part. This then determines where collectors (sliprings or commutators) are required and, ignoring single-phase a.c. machines, fits every type of motor into convenient categories as illustrated by Table 3.7. This avoids confusing terminology, like 'brushless d.c.'. Moreover, all types of variable-speed motors with 'static' or 'rotating' frequency changers fit into this pattern. The purpose of any sliprings is to make external connections to the rotor winding whilst the purpose of any commutator is to convert the frequency in the rotor winding to the frequency of the stator winding, be this d.c. or a.c.

Direct current and a.c. collector machines are speed-regulated by voltage injection into the rotor, typically S1, S3, R2 and R4 (Table 3.7) which is effected by low loss, static or rotating equipment or by energy loss, i.e. resistance.

3.2.1.2 Squirrel cage motors

The squirrel cage motor is, without doubt, the basic workhorse on which our electrical civilization is based. After a century of development, it now dominates all motor drives, large and small. The advantages are clear. If properly designed and applied it is compact, robust, adaptable and cost effective. It is not, however, indestructible nor unsophisticated or 'low technology' if cost effectiveness, properly defined, is considered. Other types of drive will be compared with it in appropriate categories of electrical and mechanical execution and performance characteristics.

The basic aspects of designing and manufacturing reliable stator windings covers a wide range of disciplines, as illustrated in Chapter 6. Starting from the steady state and transient system parameters outlined in Section 3.1.5, dielectric and mechanical design, performance and test parameters are derived, long-term to ensure reliability, short-term to ensure adequate test demonstration.

Table 3.7 Principal features of basic motor configurations

Reference nomenclature		Stator	Rotor	Sliprings	Commutator
S	**Stationary field**				
S1	d.c.	Salient pole winding d.c. excited	Distributed winding d.c. excited	—	a.c./d.c.
S2	Synchronous	Salient pole winding d.c. excited	Distributed winding a.c. excited	a.c./a.c.	—
S3	Permanent magnet d.c.	Permanent magnet	Distributed winding d.c. excited	—	a.c./d.c.
R	**Rotating field**				
R1	Squirrel cage	Distributed winding a.c. excited	Distributed winding internally short-circuited	—	—
R2	Slipring	Distributed winding a.c. excited	Distributed winding	a.c./a.c.	—
R3	Synchronous	Distributed winding a.c. excited	Salient pole, or distributed winding, d.c. excited, additional damper winding	d.c./d.c.	—
R4	Stator-fed commutator*	Distributed winding a.c. excited	Distributed winding a.c. excited	—	a.c./a.c.
R5	Rotor-fed commutator*	Distributed winding excited from rotor	Distributed winding a.c. excited Distributed winding to excite stator	a.c./a.c.	a.c./a.c.
R6	Permanent magnet synchronous	Distributed winding a.c. excited	Permanent magnet	—	—
R7	Reluctance	Distributed winding a.c. excited	Salient pole unexcited	—	—
R8	Switched reluctance*	Salient pole sequentially excited	Salient pole unexcited	—	—
R9	Solid rotor	Distributed winding a.c. excited	'Eddy current' rotor	—	—

Operational characteristics

	'Lossfree' speed variation	'Lossful' speed variation	Variants
S1	Variable d.c. rotor voltage series connection Field Control (Limited)	Resistance control	Multifield, including compensated
S2	Variable frequency	NA	—
S3	Variable d.c. rotor voltage		Resistance control
R1	Variable frequency	Variable stator voltage	2 speed etc. stator winding
R2	Variable d.c. rotor voltage	Resistance control	—
R3	Variable frequency	NA	'Brushless' design with rotating semiconductor/exciter system
R4	Induction regulator variable mains frequency rotor voltage	NA	—
R5	Brush shifting	NA	—
R6	Variable frequency	NA	Hybrid PM/reluctance rotors
R7	Variable frequency	NA	Hybrid PM/reluctance rotors
R8	Designed for sequential switching	NA	—
R9	NA	NA	—

Notes: (a) 'Distributed winding a.c. excited' refers to 3 or multiphase configurations
(b) NA: not applicable
*(c) Specifically variable speed motors

The same considerations apply to squirrel cage rotor windings[7] though with somewhat different degrees of emphasis. The versatility of the design lies in the choice of rotor bar shape which is in principle unlimited but in practice requires careful design decisions on standardization, for cost reasons. It is clearly outside the scope of this chapter to do more than summarize what is involved. There are two principal variants: (1) the 'cast' rotor, usually an aluminium alloy; and (2) the 'wound' rotor, usually using copper and/or alloy bars and short-circuiting rings. Figure 3.4 shows some of the variety of shapes used, generally from the point of view of torque–speed characteristic. This is greatly influenced by the number of

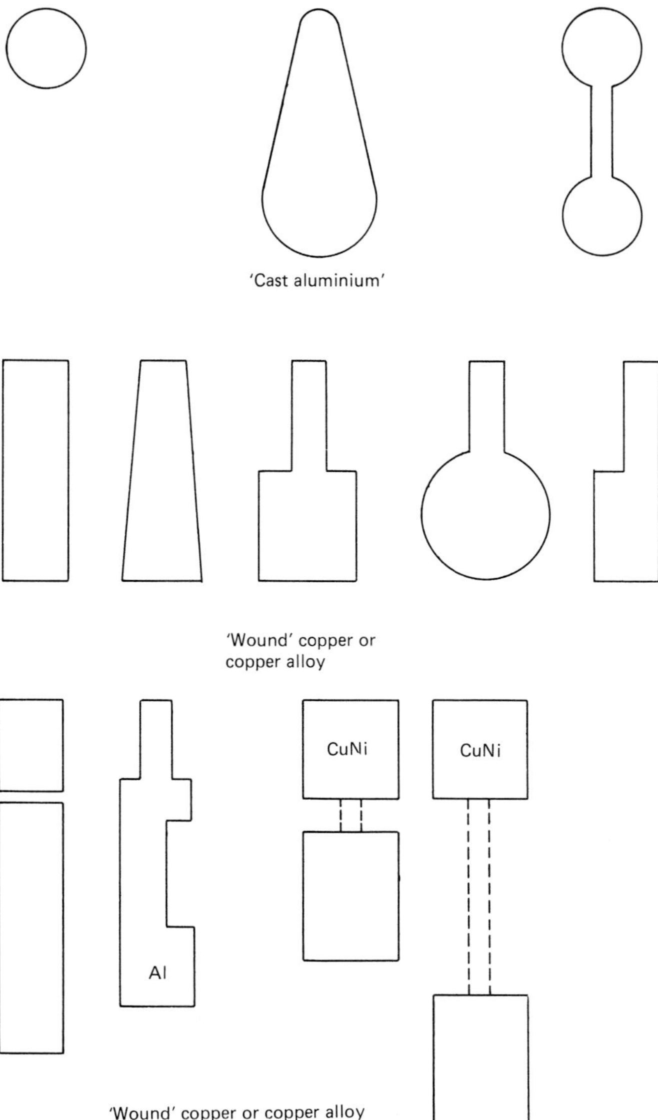

Figure 3.4 Typical squirrel cage bar shapes

slots owing to the effect on reactance which, in conjunction with the number of stator slots, determine noise, vibration and stray losses. The last phenomenon is also influenced in skewed constructions by the insulation level between core and bars. However great the importance of these subjects is judged to be, the most important criterion is the reliability of the squirrel cage.[8] Contrary to popular belief, only the smallest motors, below, say, 100 kW, are thermally limited by the stator winding and termed 'stator-critical'. However, it is more than just temperature which dictates the reliability of the rotor; it is the actual stress history which reflects the steady state and transient performance, structurally related to the specific materials and jointing technique used. Whilst torch brazing is usual, more highly stressed joints are preferably induction or 'puddle'-brazed.

Except for relatively low ratings, below, say, 200 kW, drives requiring life expectancies of 20 000 starts with 1000 starts per year (including a limited number, say, 10%, of two hot starts in quick succession) require special rotor designs. As an example, Figure 3.5 shows how a special shape of bar (keyhole) adapted for a joint configuration of a round bar end, designed for appropriate flexibility as an encastré beam, is brazed into the round hole in the short-circuit ring. The objective of this design is to relieve the braze from the normal first order centrifugal and thermal radial stresses, only having to deal with the much lower axial stresses. Moreover, as copper is not a suitable structural material, high-conductivity (80% IACS) copper–chromium or copper–zirconium is incorporated in both bars and short-circuit rings. Additionally, shrinkrings of turbo-endbell steel (typically

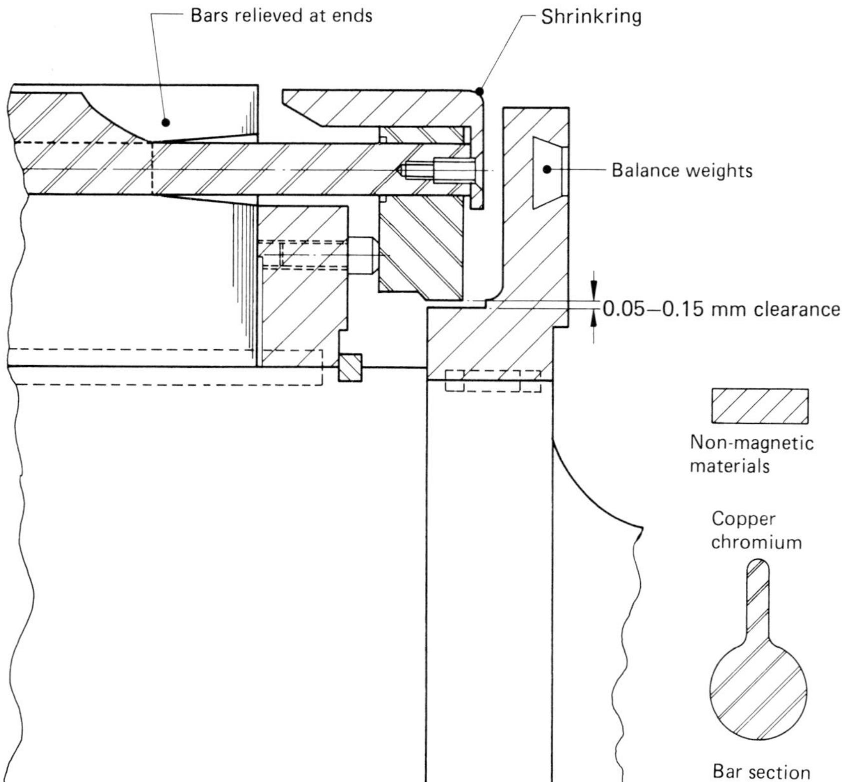

Figure 3.5 Typical high-integrity, high peripheral speed rotor cage overhang support system

non-magnetic stainless steel 18:4 magnesium:chromium) are applied to the complex structure, which is fully designed by appropriate computer programs covering all aspects of mechanical and thermal parameters.

Whilst there are ultimate limits to what can be achieved by proper design, the combination of shapes, multicage arrangements and materials of the squirrel cage, together with special stator winding arrangements and cooling system configurations, makes for great flexibility which is achievable at a finite predictable cost.

The two-speed option is widely used, either with separate or, preferably, where appropriate, tapped reconnectable windings. In view of the impact of stator:rotor slot ratios on most design parameters, it becomes more difficult to finesse various requirements. Consequently, more than two-speed windings are rarely useful. The tapped winding technique is greatly and effectively extended by 'pole amplitude modulation'.[9] Not only does this allow a reconnection using only six terminals, but it also introduces a design methodology, which makes the whole subject most attractive. The reconnection is conveniently carried out by contactors, air for low voltage and vacuum for high voltage.

With all these possibilities, it is only rarely that the torque–speed–system limitation requirements cannot be fulfilled. However, if restrictions are too severe, the solution of the closed-circuit autotransformer starter (see Chapter 7) is to be preferred to reverting to a slipring motor which, by definition requires the complication of collectors and brushes with their inherent maintenance requirements.

The design of squirrel cage motors for use with static frequency converters[10] requires special attention. If losses are to be reduced, and if reliability is to be maintained, designs 'fit for the purpose' must be used. The rotor cage design needs to be based on different requirements, primarily that of reducing harmonic losses and torques to a low level, when supplied by a non-sinusoidal waveform. This means generally rotor bar configurations with the lowest possible eddy current effects, e.g. 'square' bars. There is, of course, normally no direct on line (DOL) starting requirement, except for group drives or special designs, so that the starting current is no longer a lead parameter. The effect of different types of frequency converters varies widely and needs to be included in the efficiency declaration (see Section 3.3.3.10). This is additional to the effect of the extra losses, which usually require a reduction in rating, which also applies to the effect of harmonic torques. Overload capacities have to be matched to the converter design. In practice, the motor performance at low frequencies (below, say, 10 Hz) needs a voltage boost to overcome the internal voltage drop in the stator resistance. At high frequencies (above, say, 100 Hz), eddy currents in iron and windings need particular attention.

As all types of static frequency converter waveforms include high-frequency harmonics, the winding insulation, particularly interturn, has to be designed to cope with these additional dielectric stresses to ensure the appropriate life expectancy.

3.2.1.3 Slipring motors

These are only needed for special purposes. For some drives a sloping torque–speed characteristic is utilized to control torque and to obtain the maximum torque for a given supply current, typically in manoeuvring drives.

'Electrical shafts', i.e. two motors operating in synchronism, have peculiar stability and torque transient problems and use a common rotor resistance. The practice of specifying slipring motors for start-up purposes, and also requiring brush-lifting and short-circuiting gear, is expensive in first cost, not conducive to reliability, and rarely achieves a better overall performance than a properly designed squirrel cage motor drive with autotransformer start, especially for falling torque, i.e. pump drives. One potentially useful feature is short-time, first-order loss, small-speed regulation by resistance control, which can be economical compared with a cascade system especially for a pump drive.

3.2.1.4 Commutator motors

Whilst variable speed is described in Chapter 4, motor design is included here as a subgroup of the total range of motor types. Contrary to many suggestions in the past, commutator motors, particularly d.c., are unlikely to be displaced completely in the foreseeable future, notwithstanding the progress in variable-frequency drives.

Direct current motors are generally associated with variable-speed drives, as there are few, if any, advantages, in a d.c. distribution system, even a local one, although there is still the special case of traction (see Section 3.5.15). Whilst limited loss-free speed control is possible with d.c. motors, any major speed variation requires the supply of a variable voltage (see Table 3.7), which is now usually provided by a semiconductor converter. The d.c. motor design has to cope with two important problems: (1) current collection; and (2) commutation. The former is largely connected with the manufacture and performance of commutators, carbon brushes and brushgear, whilst the latter requires specific electromagnetic features,

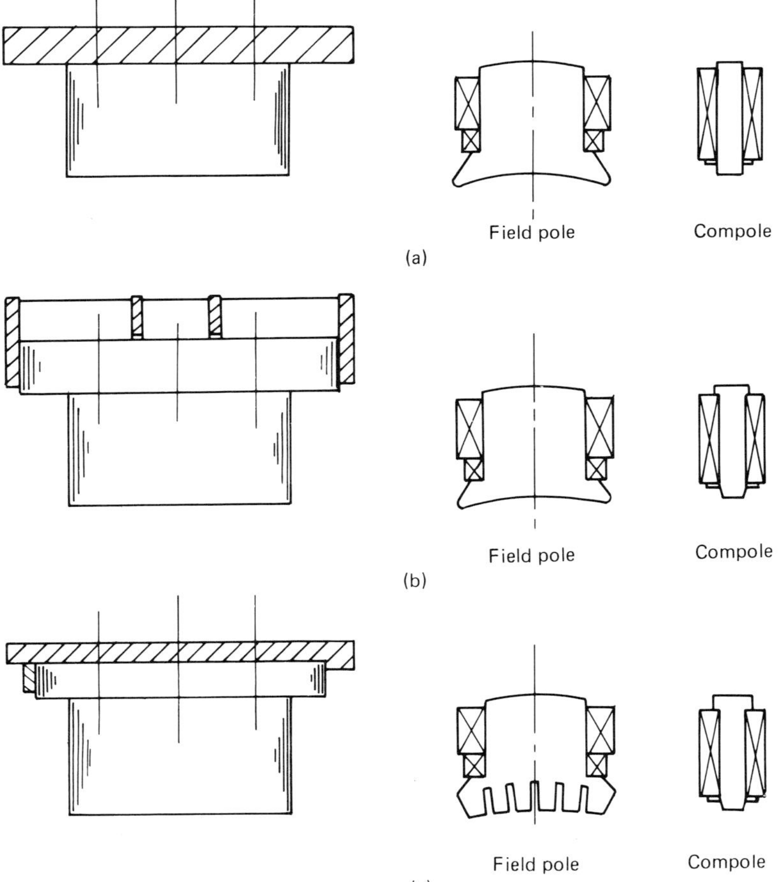

Figure 3.6 Typical d.c. motor pole configurations. (a) Unlaminated yoke. (b) Partly laminated yoke. (c) Fully laminated yoke, compensated field pole

e.g. interpoles. To reduce the induced voltages at the commutator, multiplex windings are occasionally used, whilst for high ratings it is necessary to provide compensating windings in the main pole face to counteract the effects of armature reaction. Typical pole configurations are illustrated in Figure 3.6 from the simplest basic design with unlaminated yokes and interpoles to the high-performance, fully laminated, compensated machine.

The equivalent a.c. motor is the stator-fed version with induction regulator control,[11] the induction regulator being a substantially loss-free, variable-voltage device based on a stationary induction machine with flexibles for the rotor, i.e. a brushless device. As interpoles are not practicable for rotating fields, and as there is an additional problem due to direct transformer action between stator and rotor windings, the motor armature carries a secondary commutating winding. This is separated from the main winding by iron strips and connects each turn of the armature winding by transformer action to other non-commutating turns of the armature winding. The electromagnetic effect is to suppress the voltages arising from the current reversal (commutation) by distributing an amount of circulating current throughout the armature. This mechanism also deals with the transformer-induced voltage. In order to limit the voltage per bar at the commutator, multiplex armature windings, up to quintuplet, are used. This type of motor, known as the NS motor, has been widely applied, particularly to fans and pumps up to 3 MW, but is now generally superseded by variable-frequency drives.

Another type of a.c. commutator motor, the rotor-fed design, known as the Schrage motor, is still manufactured in Europe, providing a simple drive up to some 200 kW. The primary winding is carried by the rotor and connected to the supply by sliprings. A secondary rotor winding with a commutator feeds a tertiary stator winding via movable brushgear to achieve the variable secondary injection. As usual, the commutator acts as a frequency changer, in this case operating at slip frequency.

3.2.1.5 Synchronous motors

European, unlike North American practice, does not favour the synchronous motor. However, this type of motor fulfils several important functions, all connected with its inherent characteristics. It is clearly more complex than a squirrel cage motor, requiring a fully wound, insulated rotor as well as a stator winding, together with some form of double-wound exciter connected to the rotor (excitation) winding by a starter (semiconductor) switching arrangement (see Chapter 7), if brushless operation is to be achieved. It inherently provides the function of a synchronous condenser, i.e. it can generate leading kilovars into the system. This property is used in various ways: (1) as power factor correction for other machines; (2) as intrinsic power factor compensation for high magnetizing current, i.e. high poleage motors; and (3) to provide reduced magnetizing current requirements and/or commutation leading kilovars for static frequency converters. The choice of configuration ranges from permanent-magnet motors up to, say, 100 kW to turbo-type machines, as found in the highest ratings. In the medium power range, the most usual configuration is basically R3 of Table 3.7 with a salient-pole, partially laminated rotor construction with interconnected pole-face squirrel cage windings. Large, high-speed motors are commonly made with a solid-pole construction and there is every variant of laminated–solid pole motors with bolt-on pole tips. A small practical selection of configurations is given in Figure 3.7.

Lastly, the requirements for 'synchronism' as such are rarely used. Moreover, this only refers to steady state operation, as the load angle exhibits itself as a transient speed variation when the torque varies. Where synchronism in space is really needed, e.g. plastic film or fibre manufacture, this phenomenon has to be part of the design calculation.

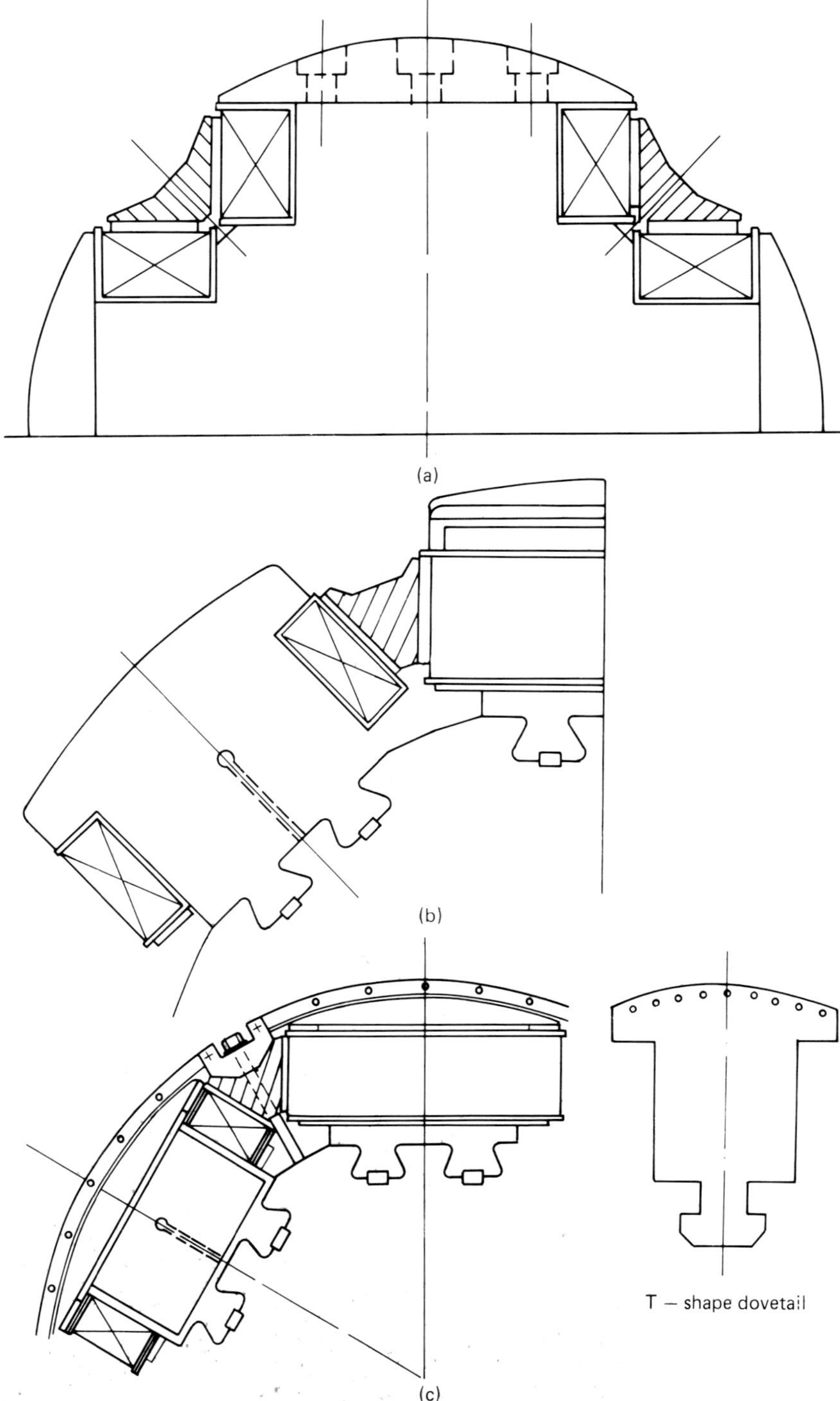

Figure 3.7 Typical synchronous motor pole configurations. (a) Solid pole/shaft with bolted-on pole caps. (b) Solid pole with dovetails. (c) Laminated pole with dovetails and interconnecting starter winding

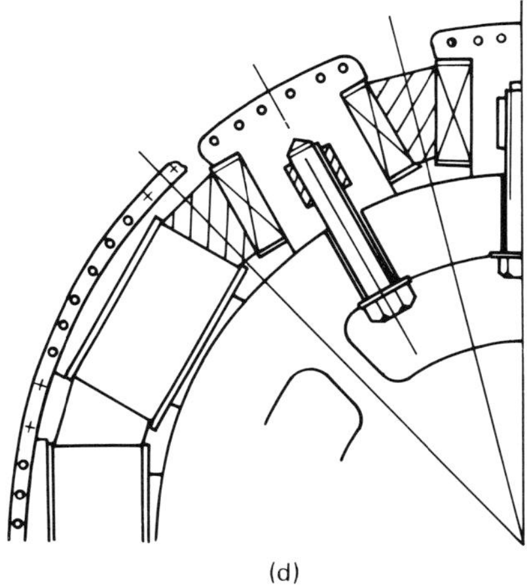

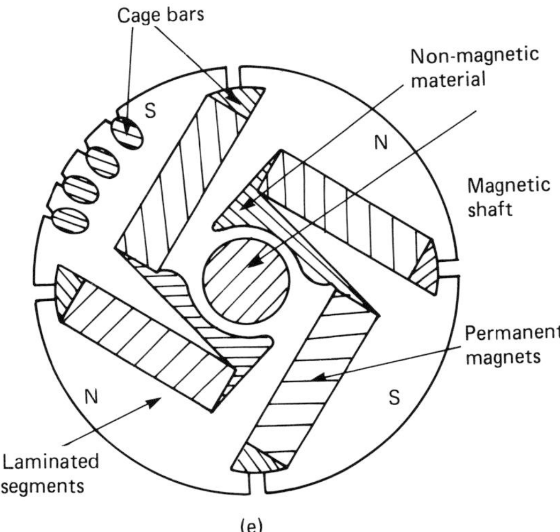

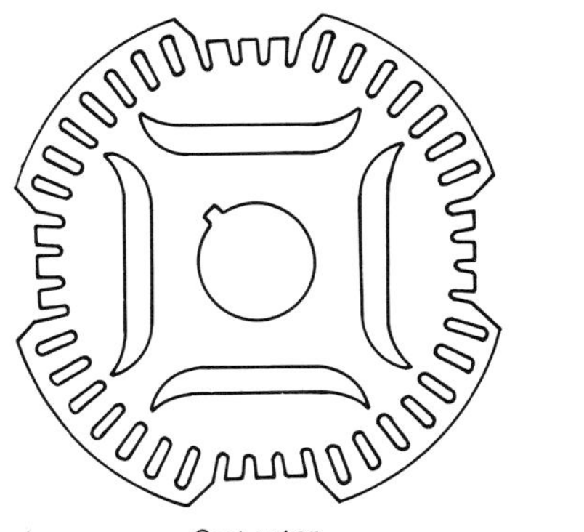

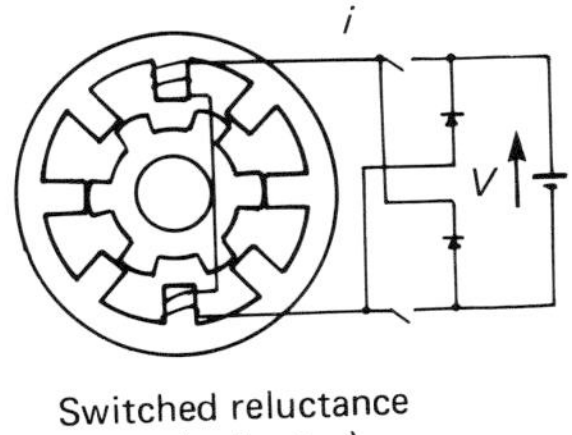

Figure 3.7 (continued) (d) Laminated pole bolted on rotor hub. (e) Permanent magnet. (f) Reluctance machine

In order to take advantage of a synchronous motor's kilovar generation capability, an excitation regulator is essential, which should preferably control to constant kilovar (see Chapter 7). A constant kilovar performance gives control of losses and ensures overload capability at all times, which is not possible with a power factor regulator.

Except for small ratings, the starting performance in terms of torque-amperes is relatively worse than the equivalent squirrel cage motor. A convenient starting method is either an autotransformer, or alternatively a series reactor starter (see Chapter 7).

In order to limit the voltage in the field windings at start, this circuit must always be closed via a correctly designed resistance. A semiconductor switch in conjunction with a resistance and appropriate automatic control system is therefore required as part of the rotating assembly. If collectors and brushes are permitted, the excitation and/or starting system can be fed via sliprings, but this is rarely accepted in modern practice. In fact, a slipring winding with a distributed or salient-pole design has been used, but normally would not be considered, unless there were some exceptional requirements.

The use of permanent magnets to replace separately excited fields is gaining ground both in terms of wider applications for small- and medium-rated motors, with many different configurations to deal with stability for variable-frequency applications and, additionally, with starting performance for fixed-frequency supplies. A variant in this group is the reluctance motor, using no excitation, and also many hybrid designs exist. The so-called switched reluctance motor is another variant designed specifically as part of a variable-speed, variable-frequency package. The interest in permanent-magnet synchronous motors is increasing with the introduction of the improved permanent magnet materials, which are becoming available. Current technology[12] indicates that special designs for variable-frequency drives of 1 MW at 230 rev./min are viable.

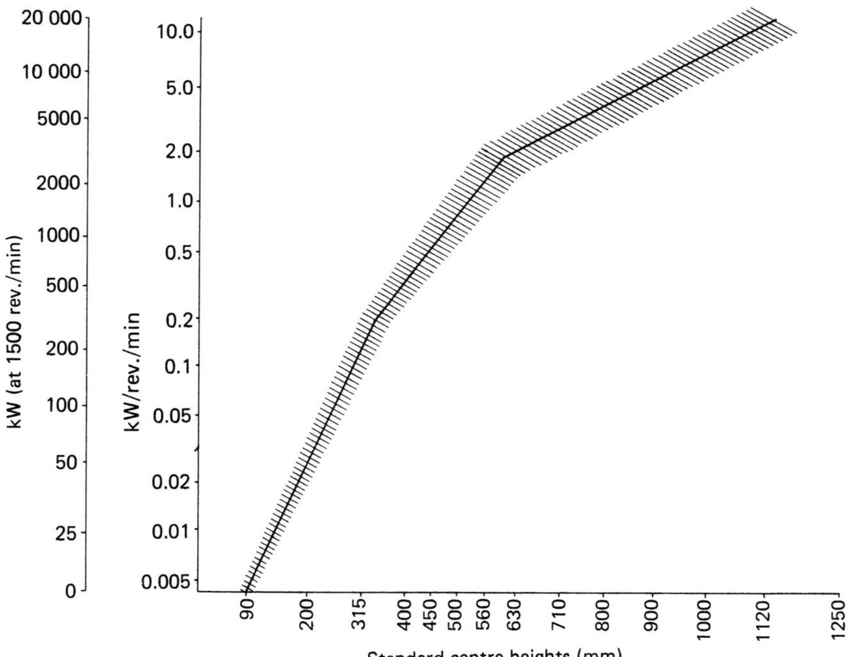

Figure 3.8 Indicative power indices for motors with centre height 90–1250 mm

3.2.1.6 Performance characteristics

Whilst any motor characteristic, such as efficiency or power factor, is, in practice, to be derived from the requirements of the driven machine and environmental and supply system requirements, it is nevertheless useful to indicate ranges of performance in broad terms. Figures 3.8–3.10 illustrate induction motor ratings in

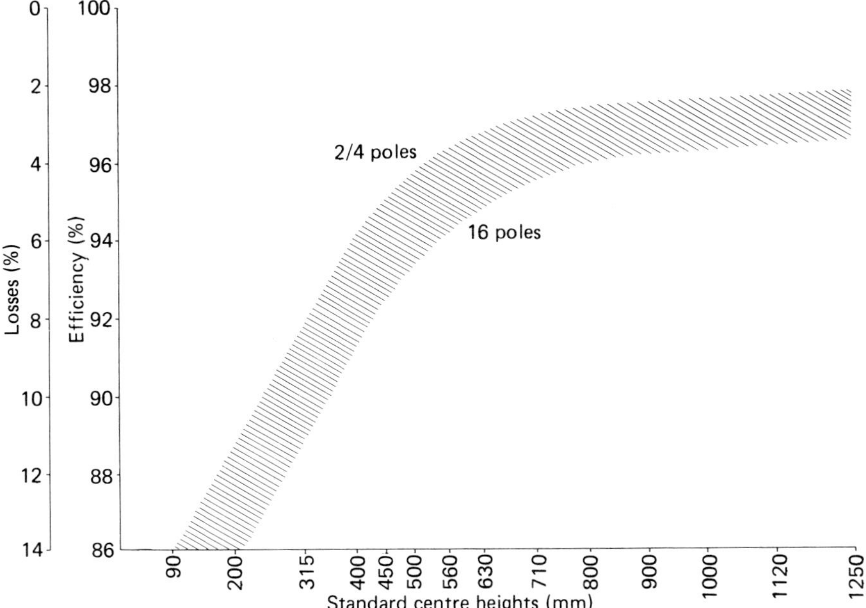

Figure 3.9 Indicative efficiency levels for motors with centre height 90–1250 mm

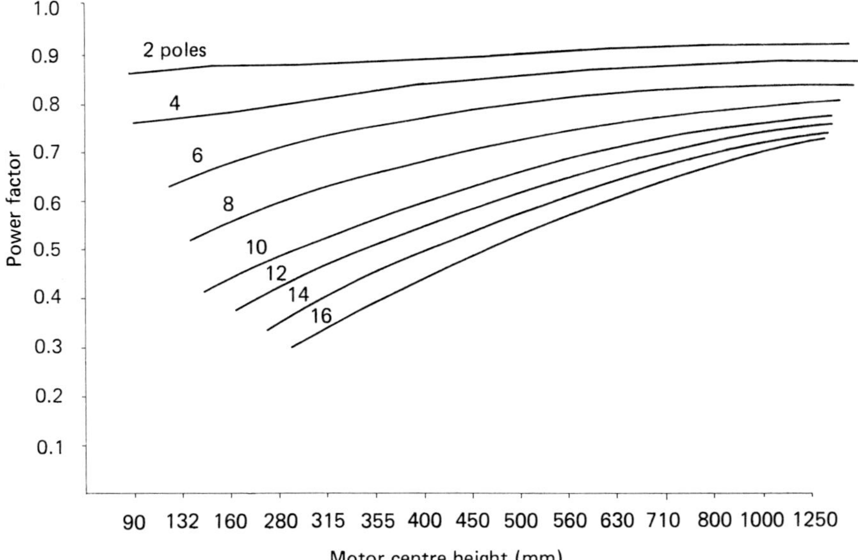

Figure 3.10 Indicative full-load power factors for motors with centre heights 90–1250 mm

conjunction with standard dimensions over a range of poleages; Figure 3.8, output per rev./min; Figure 3.9, efficiency, or more usefully losses; Figure 3.10, power factor. Torque–speed curves have been normalized to some extent by international, British and other national standards; however, this only suffices for simple drive problems. For complex, important, high-integrity applications, the actual starting performance, including permissible starting frequency, must be calculated, based on the actual load torque and inertia, and heating and cooling performance must be considered.

Furthermore, such computations may have to cover not just bulk effects, but also the temperature distribution throughout the motor. An example of such an analysis is presented in Figure 3.11 for a 2 MW motor. These performance parameters are closely connected with the cooling system and enclosure (see Section 3.3.3.3).

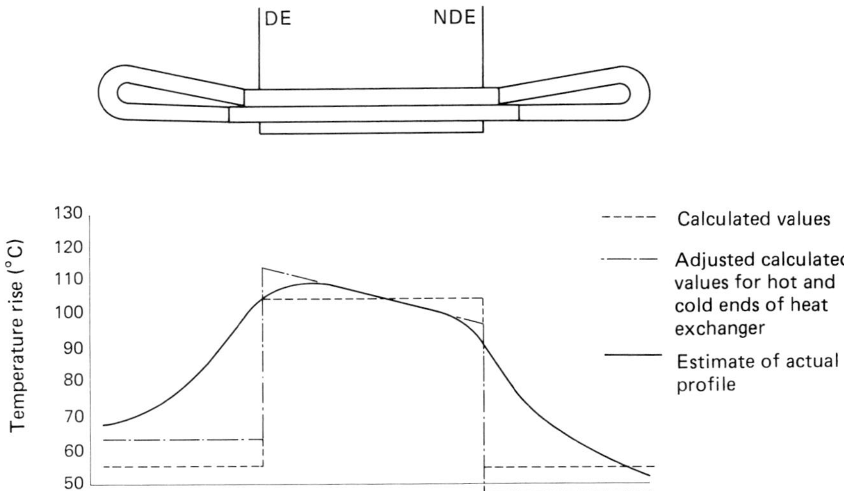

Figure 3.11 Typical temperature distribution in a 2-MW, four-pole, 11-kV CACA motor

The environment should be specified as accurately as possible. There is no point in building in temperature or other margins at random. Class B temperature rise with Class F insulation (see Chapter 6) is a useful guide, but is not the same as knowing what the ambient temperatures are or what torque excursion may be required short term. All information about requirements is useful to the motor drive designer, especially cost information, such as capitalized value of losses, space restrictions, mechanical behaviour of the driven machine and the proposed monitoring system.

3.2.2 Drive requirements

The foregoing paragraphs are intended to indicate that, for economical solutions, the maximum amount of information about the drive must be given, except for routine, relatively small, drive applications. The extra time and, therefore, cost spent on design at the planning stage invariably makes a substantial contribution to overall cost reduction.

Looked at in some detail, Table 3.8, based on the British Standards, can be used as an appropriate checklist. Specification of the need to the constructor, together with any local preferences, should be followed by a proposal from the constructor,

Table 3.8 Selected British and international standards dealing with specific aspects of motors

Reference	Subject	BS 4999	BS 5000	Other BS	European	International
3.2.2	General	0 (1987)				
3.3.3.1	Dimensions	103 (1987)				IEC 72 (1981)
		141 (1987)				IEC 72A (1970)
3.3.3.2	Ratings	101 (1987)	10 (1984)		CENELEC HD 231	IEC 34-1 (1983)
3.1.1	Voltage	101 (1987)				IEC 34-1 (1983)
						IEC 38 (1983)
3.3.3.3	Enclosure	105 (1988)			CENELEC HD 53.5	IEC 34-5 (1981)
3.3.3.3	Cooling	106 (1987)				IEC 34-6 (1969)
3.3.3.3	Mounting	107 (1987)				IEC 34-7 (1972)
3.3.3.5	Terminations	145 (1987)				
	Markings	108 (1987)			CENELEC HD 53.8.S3	IEC 34-8 (1972)
3.3.3.4	Hazardous areas		15 (1985)	5501 series (1980–1986)	EN 50.015-020	IEC 79 series (1967–1984)
			16 (1985)			
			17 (1981)			
3.2.1.6	Torque/speed	112 (1987)	40 (1984)			IEC 34-12 (1983)
	(Steady state and transient)					IEC 892 (1987)
3.2.1.6	Efficiency	102 (1987)				IEC 34-2 (1972–74)
3.3.3.6	Vibration	142 (1987)			CENELEC HD 347	IEC 34-14 (1982)
3.3.3.6	Balancing			5265-1 (1986)		ISO R5406 (1980)
3.3.3.7	Noise	109 (1987)			CENELEC HD 539	IEC 34-9 (1972)
3.3.3.10	Testing (general)	143 (1987)				IEC 34-1 (1983)
3.3.3.10	Testing (specific)	101 (1987)				IEC 34-1 (1983)
		102 (1987)				IEC 34-2 (1972–1974)
3.3.3.10	Testing tolerances	101 (1987)				IEC 34-1 (1983)
3.3.4	Quality assurance			5750 (1979–1981)		ISO 9000 series (1985)
				5882 (1983)		
3.3.3.9	Protection	111 (1987)				IEC 34-11 (1978–1984)
3.3.3.8	Electromagnetic			1597 (1985)		IEC 533 (1977)
	Compatibility					
3.5.7	Marine			2949 (1977)		IEC 92 (1965–1985)
3.2.1.5	Turbine-type machines		2 (1973)			
3.5.8	D.c. mill auxillary motors		25 (1983)			IEC 34-13 (1980)
3.5.11	Motors for driving power station		40 (1984)			
	auxiliaries					
3.5.15	Traction			173 (1984)		IEC 349 (1971)
Principal insulation topics						
Chapter 6	Testing	144 (1987)	—	—	CENELEC HD 345	IEC 894 (1987)
	Classification	—	—	2757 (1986)	—	IEC 85 (1984)
	Co-ordination	—	—	5622 (1979)	—	IEC 71 (1976–1982)
						IEC 664 (1980–1981)
	Thermal endurance	—	—	5691 (1979)	—	IEC 216 (1974–1980)
	Tracking and erosion	—	—	5604 (1986)		IEC 587 (1984)
	Comparative tracking index	—	—	5901 (1980)		IEC 112 (1979)
	Resistivity	—	—	5823 (1985)		IEC 345 (1971)

which is fit for the purpose, i.e. the correct quality (see Section 3.3.4). There is a specific category of drives, which often lead to a highly efficacious solution, when motor and drive are closely combined. These include special-purpose 'submerged' motors for glandless pumps or fans (see Chapter 2), integrated mounting or cooling systems and the like.

3.2.3 Systems approach

3.2.3.1 General

To get the best economical result overall, selection of the motor drive should be carried out as part of the overall system study. This should cover not only the electrical system but also the driven machine and the general environment.

3.2.3.2 The electrical system

The choice of voltage and short-circuit levels was dealt with in some detail in Section 3.1. Translating this to project level, careful consideration has to be given to designing the first preference motor choice, the squirrel cage motor, with a torque–speed characteristic appropriate to the driven machine requirement, utilizing DOL starting. This may result in a reconsideration of the voltage levels, the number of drives deployed, the prevention of simultaneous starting of several motors or the necessity for variable speed to avoid an excessive starting frequency. The stability of the system during fault conditions and the possibility of recovery from faults, particularly where there is a safety connotation, e.g. ships' supplies, is a necessary exercise. Modern computational facilities make this a relatively simple routine matter.

The inevitability of switching overvoltages,[13] and lightning strikes in the rather less likely case of direct connection to an overhead line, needs to be emphasized, and all primary insulation designed and tested accordingly (see Chapter 6).

Proper protection, both against misuse and faults, is also a necessary design exercise under this and the next heading.

3.2.3.3 The load machine

Experience shows that with the wide variety of possible requirements, cases of unsuccessful application of motors usually originate from a lack of communication between user and constructor. This covers such fallacies as, for example, that fan drives cannot stall, remembering the proverbial spanner or less obvious omission of a piece of ducting, and unlimited switching of automatically controlled compressor drives. Two starts in quick succession and three starts per hour does not mean that this regime can be carried on continuously. The undeclared difference between switchgear short-circuit rating and the actual impedance of the supply can cause real trouble. Conversely, requesting starting currents from one motor, which are substantially less than the current required for the totality of installed motors, is not logical.

One of the most common origins of waste is the requirement for noise levels well below those of the driven machine and specifications for vibration performance, which is ultimately negated by lack of attention to the complete rotating assembly, including couplings and bearings and the performance of the supporting structure. Conversely, it behoves the motor constructor to advise the axial as well as the radial forces involved, to allow designs to be carried out properly.

Additionally, where appropriate under reswitching or fault conditions, the inevitable transient torques produced by the motor[14] must be taken into account when designing the driven machine.

3.2.3.4 The environment

Pollution is a serious matter, noise and vibration must be controlled, harmonics and transient voltage variations limited and heat losses properly channelled. To this list one should add that uneconomic use, e.g. lack of reliability and excess use of energy, should be reduced as far as practicable. To do this effectively, the human environment, the skills and deployment of maintenance personnel, spares holdings and overhaul techniques must be considered. The level of maintainability is important, as are various other aspects of quality (see Section 3.4).

Whilst aesthetics are not a usual consideration in motor design, paints and finishes to protect surfaces are important aspects. A general move towards a limited number of detailed specifications with more emphasis on performance requirements is desirable in this respect. Colour coding of equipment has been used with good effect in special situations. Whilst total enclosure with heat exchangers gives maximum protection, the use of so-called NEMA 2 enclosures[15] is of interest with medium-sized squirrel cage motors in areas of modest contamination and humidity. Modern sealed stator windings are entirely suitable when properly designed, manufactured and tested, whilst rotors can be varnished to good effect. This leaves the vulnerable bearings to be sealed to the required level.

Sun canopies and sand precipitators need to be considered in specific, adverse, conditions. In extreme cases, motors may have to be pressurized via chemical (usually activated carbon) filters, to avoid corrosion by such highly corrosive gases as hydrogen sulphide or sulphur dioxide even where there are no collectors involved.

A subject of increasing importance is electromagnetic compatibility. This concerns conducted and radiated interference both from and to the motor and is dealt with in Section 3.3.3.8.

3.2.3.5 Summary

Whilst the list of points to be considered is substantial, it is only necessary to make specific calculations in particular cases. Generally, order of magnitude indications are adequate to point to potential problems. Ultimately, it is a question of engineering judgement to decide on the degree of detailed design required in the more peripheral areas.

3.3 Standards

3.3.1 General

In order to avoid any possible confusion, it is essential to define standards in unambiguous International Standards Organization (ISO) terms:

> A technical specification available to the public, drawn up with the cooperation and consensus or general approval of all interests affected by it, based on the consolidated results of science, technology and experience and approved by a body recognized at the national, regional or international level.[16]

Conversely, a document, however illustrious but not meeting this definition, can only be described as a specification or statement of requirements. The wide application of standards extends over many subjects in the electric motor area, ranging from nomenclature, dimensions, performance aspects, test requirements, to system design and specific prescriptions in connection with hazardous-area motors. Traditionally, electrical engineering has been in the forefront of standards work, and consequently has its own international organization, the International Electrotechnical Committee (IEC), whilst other interests are represented by the ISO.

Most countries subscribe to these organizations; the official body in the UK is the British Standards Institution (BSI), who organize participation in the international work through national committees. The subject is further complicated in Europe by the existence of the Committee Européens de Normalisation Electrique (CENELEC) and the Committee European de Normalisation (CEN). To this is added the work of the EEC which produces 'harmonized standards', which are mandatory for the member states. In principle, all standards should be harmonized with IEC–ISO, but in practice this is not necessarily the case.

The application of standards is generally voluntary, except in situations where legal processes are involved, such as 'approval'. One considerable advantage of standards can be their recognition in law, a practice common in Western Europe generally, but only just beginning to be adopted in the UK.[17] With the increasing stringency of legal obligations with regard to safety and product liability,[18] the approach must be welcomed to aid both constructors and users. Standards have substantial implications in international trade terms, as well as direct commercial effects. The rationalization of supply system voltages and frequencies is an obvious case in point, where the disadvantages of lack of standardization are felt on two counts: (1) electrical, as it affects apparatus performance; and (2) mechanical, as it affects the possibility of installing the apparatus.

3.3.2 Standards making in the UK

The BSI, which has a Royal Charter, is an independent body operating with a permanent staff of over 1000 headed by a director-general, with a management committee structure of considerable complexity, involving industry, in the form of users and constructors; government, in the form of representatives of different departments; and trade unions. As standards making is voluntary and unpaid, the technical and other committees have a formidable task to operate by consensus internally, produce standards in time and take cognisance of the international and European scenes.

The directly relevant industry committees are General Electrical Equipment (GEL) and Power Electrical Equipment (PEL), but clearly there are many cross-connections related to mechanical, material, control and quality-related committees.

It is sometimes alleged that standards incorporate only the lowest common denominator of technology. In reality this is rarely so, if considered against the definition, nor is progress inhibited with a well-drawn-up standard which is consistent, numerate, technically appropriate and especially forward looking. This last point applies, for instance, to dimensional standards, which for obvious manufacturing reasons, take a long time to apply. Similarly, the phasing out of obsolescent and obsolete standards on a definitive time-scale which may be 10 years, is effective in encouraging rather than hindering progress.

In the motor field, user organizations, such as the Engineering Equipment and Material Users' Association (EEMUA), the Electricity Supply Industry (ESI) and trade associations, such as the British Electrical Apparatus and Manufacturers' Association (BEAMA), the Energy Industry Council (EIC), all have a special responsibility for standard making, in so far as they often supply the technical initiatives as a need arises and provide some of the expert manpower to formulate the draft. It is incumbent on all parties to consider economical aspects and commercial pressures, both nationally and internationally.

The effect of government is multi-fold; it is a large user, it affects and effects legislation with special reference to safety, as exemplified in the existence of the Health and Safety Commission and Health and Safety Executive, and it supports the BSI. In recent years, considerable influence has been exercised through the National Economic Development Organization and by various reports, white papers.[19] etc. The input from government research support is also significant. Unfortunately, comparison with other national standards is beyond the scope of this chapter, but some cross-references to European and international standards are included in Table 3.6 relating to apparatus and systems; Table 3.8 covering the principal aspects relevant to motors; Table 3.17 (page 161) dealing with hazardous area (Ex) motors; and Table 3.21 (page 171) for quality assurance.

It must be emphasized that references to standards must always be qualified by the date. The latest amendments should always be checked by reference to the current standards catalogues.[20,21]

3.3.3 Application of British Standards (see Table 3.8, page 145)

3.3.3.1 Dimensional standards and output allocation

The dimensional standardization of induction motor frames has a history going back to the 1920s and illustrates one of the mathematical principles of variety reduction by the use of preferred numbers. There are two basic types of series: (1) arithmetic; and (2) geometric; the latter, known as Renard (R) series, has been standardized and generally applied in 25% (R10), 12.5% (R20) and 6.25% (R40) steps with appropriate rounding for practical purposes. For reference, the basic R20 series is:

100	112	125	140	160	180
200	225	250	280	315	355
400	450	500	560	630	710
800	900	1000			

Whilst slavish conformance with the mathematical principle is counter-productive, a prescribed limited choice obviates unnecessary proliferation without presenting any particular inhibiting features. The leading dimensions of a motor are fixed by the simple arrangement of Figure 3.12, based on: standardized centre height H, which is designated as the 'frame number' or 'frame size'; width, A; length between fixing holes, B; distance between the shaft shoulder and first fixing hole, C; shaft length, E; shaft diameter, D.

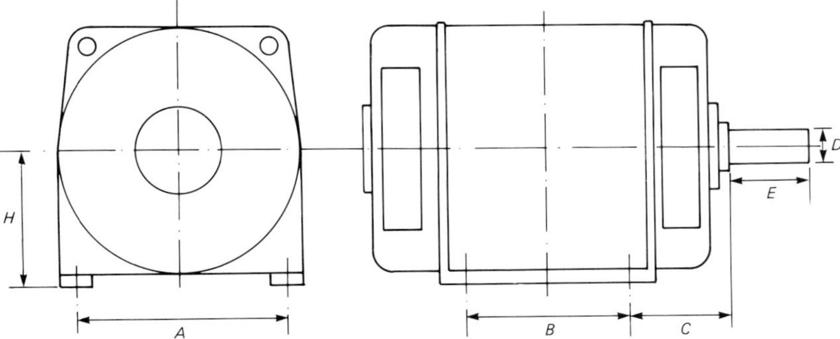

Figure 3.12 Standard symbols for the principal fixing dimensions of horizontal-shaft motors

In order to make a practical choice, there is an IEC grid of dimensions for frame numbers 56–315, a wider range for frame numbers 355–1000, an extension above this following the same series, with an alternative for 'feet-up' mounting. There are also standard flanges for sizes 90–400.

Table 3.9 gives the restricted series 90–400 foot mounted and Table 3.10 gives the wider-series 355–1000 foot mounted preferred range. This can be extended by

Table 3.9 Fixing dimensions of foot-mounted frames all enclosures. (*All dimensions in millimetres*)

Frame No/H	A	B	C		Frame No/H	A	B	C
90S	140	100	56		225S	356	286	149
90L	140	125	56		225M	356	311	149
100S	160	112	63		250S	406	311	168
100L	160	140	63		250M	406	349	168
112S	190	114	70		280S	457	368	190
112M	190	140	70		280M	457	419	190
132S	216	140	89		315S	508	406	216
132M	216	178	89		315M	508	457	216
160M	254	210	108		315L	508	508	216
160L	254	254	108		355S	610	500	254+
180M	279	241	121		355L	610	630	254+
180L	279	279	121		400SL	686	560	280+
200M	318	267	133		400L	686	710	280+
200L	318	305	133					

† Assumes ball and roller bearings.

using the R40 series for the *A* dimension and continuing the R20 series for the *B* dimension at both ends of the range.

The next stage is the allocation of outputs, which is appropriate to small and medium ratings only, in view of the wide variety of types, cooling systems, enclosures, voltages and speeds. The current common range of TEFC squirrel cage motor ratings, two to eight pole, are shown in Table 3.11 in line with European standardization, whilst Table 3.12 shows the more limited range of enclosed ventilated squirrel cage motors. The rating allocations are subject to discussion in the IEC, but so far the advantages of standardization have been allowed to overrule the updating, thereby increasing the ratings of this range, which are practically also used in Eastern European countries. Slipring motor ratings are also available.

The extension of the approach from small- and medium-sized motors to medium- and large-sized motors has been widely accepted as a good discipline, providing simple means of 'order of magnitude' information transfer at the planning and predetailed-design stage. The same approach is used for wires, cables, structural steel and many other materials, allowing the constructor to use preferred sizes, with evident advantages in cost.

3.3.3.2 Ratings and sizes

In order to achieve a manageable overview it is necessary to subdivide the field in some way. It is no use, moreover, to talk vaguely about large or small; numerate criteria have to be applied. Table 3.13 is intended to provide a compromise between physical size and power rating, two very different, though related, concepts. The six categories chosen cover frame numbers 90–1250 and notional outputs 1–20 000 kW. To cater for specific information, kilowatts per revolution per minute is a useful concept to qualify subjective statements in a simple way. The complexity of the situation demands radical simplification in the presentation and, consequently, the information must be considered as indicative only. Apart from voltage levels, there is a particular trend in enclosures and cooling systems. The smallest motors and lowest ratings are generally plain-totally-enclosed changing to totally-enclosed-fan-cooled with an enclosed-ventilated alternative for small and

Table 3.10 Preferred fixing dimensions for large horizontal shaft motors. (*All dimensions in millimetres*)

Frame/H	A	B									C (Ball and roller)	C (Sleeve)
355	610[a]	355	400	450	500[a]	560	630[a]	710	800	900	254[a]	450
400	686[a]	400	450	500	560[a]	630	710[a]	800	900	1000	280[a]	475
450	750	450	500	560	630	710	800	900	1000	1120	315	500
500	850	500	560	630	710	800	900	1000	1120	1250	335	530
560	950	560	630	710	800	900	1000	1120	1250	1400	355	560
630	1060	630	710	800	900	1000	1120	1250	1400	1600	375	600
710	1180	710	800	900	1000	1120	1250	1400	1600	1800	400	630
800	1320	800	900	1000	1120	1250	1400	1600	1800	2000	425	670
900	1500	900	1000	1120	1250	1400	1600	1800	2000	2240	450	710
1000	1700	1000	1120	1250	1400	1600	1800	2000	2240	2500	475	750

Note: (a) See Table 3.9

Table 3.11 Outputs and shaft numbers for totally enclosed fan ventilated or airstream rated cage induction motors. *Single speed, maximum continuous rated, Class E, Class B, or Class F insulation, enclosure IP44, cooling class IC41 or IC48[a]. Suitable for three-phase, 50 Hz, 415 V+ supply*

Frame no.	Output (kW)			
	Synchronous speed (rev./min)			
	3000	*1500*	*1000*	*750*
D90S	1.5	1.1	0.75	0.37
D90L	2.2	1.5	1.1	0.55
D100L	3.0	2.2 and 3.0	1.5	0.75 and 1.1
D112M	4.0	4.0	2.2	1.5
D132S	5.5 and 7.5	5.5	3.0	2.2
D132M	—	7.5	4.0 and 5.5	3.0
D160M	11.0 and 15.0	11.0	7.5	4.0 and 5.5
D160L	18.5	15.0	11.0	7.5
D180M	22.0	18.5	—	—
D180L	—	22.0	15.0	11.0
D200L	30.0 and 37.0	30.0	18.5 and 22	15.0
D225S	—	37.0	—	18.5
D225M	45.0	45.0	30.0	22.0
D250M	55.0	55.0	37.0	30.0
D280S	75.0	75.0	45.0	37.0
D280M	90.0	90.0	55.0	45.0
D315S	110.0	110.0	75.0	55.0
D315M	132.0	132.0	90.0	75.0

Notes: (a) In the case of airstream rated motors these output allocations are applicable only to foot-mounted or flange-mounted machines up to and including frame D250M.

(b) By agreement motors may be wound for the standard output at other voltages not exceeding 660 V.

(c) Previous British rating allocation Appendix A BS 5000–10:1984 still used for 250 size and above.

Table 3.12 Outputs and shaft numbers for enclosed ventilated cage induction motors. *Single speed, maximum continuous rated, Class E, Class B, or Class F insulation, enclosure IP22, cooling Class IC01. Suitable for three-phase, 50 Hz, 415 V[a] supply*

Frame no.	Output (kW)			
	Synchronous speed (rev./min)			
	3000	*1500*	*1000*	*750*
C160M	11.0 and 15.0	11.0	7.5	5.5
C160L	18.5 and 22.0	15.0 and 18.5	11.0	7.5
C180M	30.0	22.0	15.0	11.0
C180L	37.0	30.0	18.5	15.0
C200M	45.0	37.0	22.0	18.5
C200L	55.0	45.0	30.0	22.0
C225M	75.0	55.0	37.0	30.0
C250S	90.0	75.0	45.0	37.0
C250M	110.0	90.0	55.0	45.0
C280S	—	110.0	75.0	55.0
C280M	132.0	132.0	90.0	75.0
C315S	160.0	160.0	110.0	90.0
C315M	200.0	200.0	132.0	110.0

Note: (a) By agreement motors may be wound for the standard outputs at other voltages not exceeding 660 V.

(b) Previous British rating allocation Appendix A BS 5000–10:1984 still used for 250 size and above.

Table 3.13 Overview of categories and principal features of motors

Group	Description	Frame no.[a]	Rating (kW)[b]	Voltage[c]	Type Induction[d]	Type Synchronous[e]	Type Commutator[f]	Enclosure[g]	(Ex)[h]	Bearings[i]
0	Very small	< 90	< 1 (< 0.0075 kW/rev./min)	LT LT	SC/1	PM	U	V, S TE, TEFC	d e	R
I	Small	90–180	1–30	LT	SC/1, (SR)	PM REL	d.c.	V, S TE, TEFC	d e	R
II	Small/medium	200–280	30–200	LT	SC/1, SC/2, SR	PM REL	d.c. (RCM) (SCM)	V, S TEFC	d e	R
III	Medium	315–450	200–2000	LT HT	(SC/1), SC/2, SR	SP	d.c. SCM	V, S TEFC, WP CACW, CACA	d e p	R S
IV	Large	500–1250	1000–10/20 000 (0.5–30 kW/rev./min)	HT	SC/2, SR	SPCG	(d.c.)	V, S (TEFC) CACW, CACA	d e, p	S
V	Very large	> 1250	> 10/20 000 (> 30 kW/rev./min)	HT	SC/3	CG	—	V, CACW	d e, p	S

Notes:

(a) Frame number: shaft height in millimetres for normal horizontal construction

(b) *Rating is indicative only typically 4p ventilated (TEFC<160 frame)*

(c) *Voltage single-phase for Group 0 three-phase (see Table 3.1)*

(d) *Induction motor rotors:* SC/1 *Cast cage*
  SC/2 *Fabricated cage*
  SC/3 *Solid or special rotor*
  SR *Slipring*

(e) *Synchronous motors* PM *Permanent magnet*
  REL *Reluctance*
  SP *Salient pole*
  CG *Cylindrical airgap*

(f) *Commutator motors* U *Universal*
  d.c. *Direct current*
  RCM *Rotor-fed a.c.*
  SCM *Stator-fed a.c.*

(g) *Enclosure* V *Ventilated*
  S *Submerged*
  WP *Weather-protected*
  TE *Plain totally enclosed*
  TEFC *Totally enclosed fan cooled*
  CACA *Closed air circuit air cooled*
  CACW *Closed air circuit water cooled*

(h) *(Ex) hazardous areas* 'd' *Flameproof*
  'e' *Increased safety*
  'p' *Pressurized*

(i) *Bearings* R *Rolling element*
  S *Sliding element*

medium sizes. As sizes increase, cooling becomes more difficult, leading to more sophisticated arrangements with separate air–air (or air–water) heat exchangers. The European continental practice tends towards tube-cooled motors rather than separate heat exchangers.

Types of motors have in practice both technical and economic power and speed limits. These are due to various design aspects such as limitations of the mechanical strength of rotor windings and cores, commutators and sliprings and the limits of current collection and commutation, as well as shaft and bearing design. With the substantial increase in the availability of static frequency converters in terms of rating and frequency, practical limits for the application of variable-speed

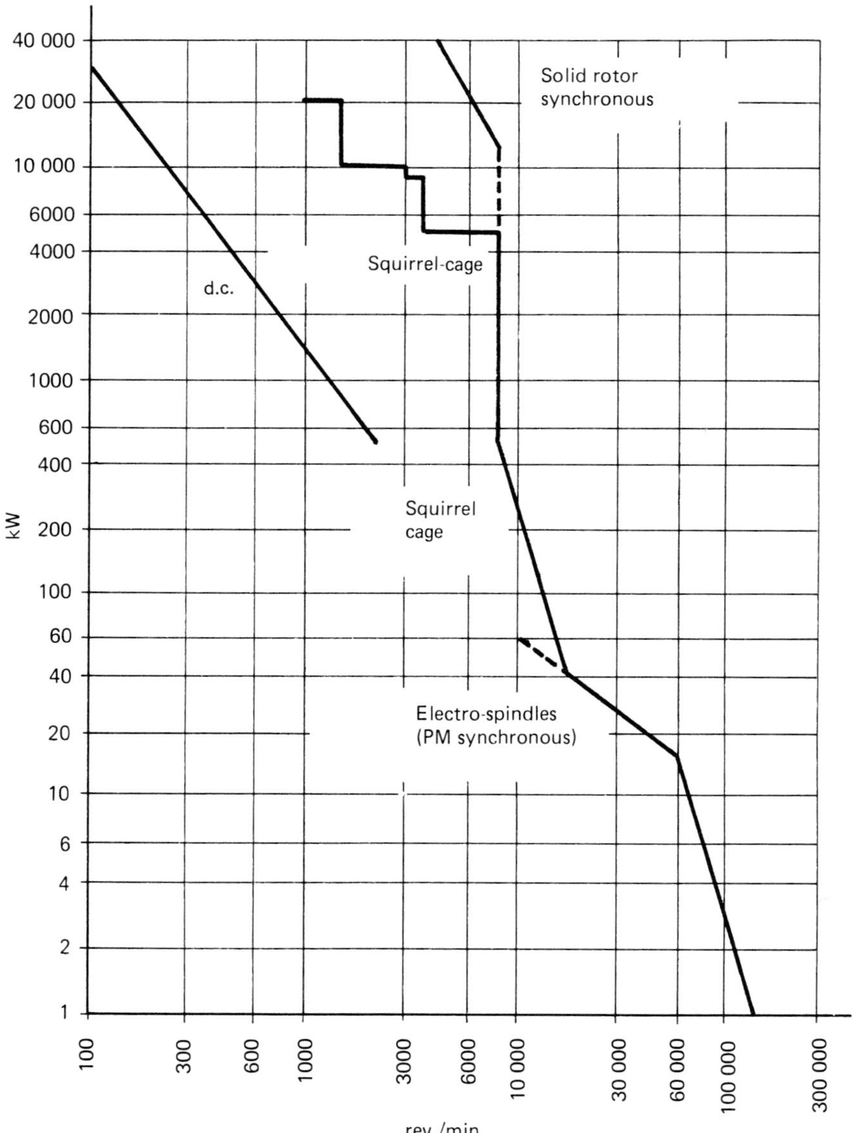

Figure 3.13 Indicative power and speed limitations of motors

Table 3.14 Selected types of construction and mounting arrangement (BS 4999–107)

Symbol	Figure	Bearing	Frame	Shaft extension	Generalities concerning the construction	Attachment or mounting
B3	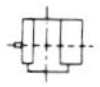	2 end-shields	With feet	Free shaft extension	—	Mounted on substructure
B5	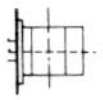	2 end-shields	Without feet	Free shaft extension	End-shield flange access to back. Shaft extension at flange end	Supported by flange
B6		2 end-shields	With feet	Free shaft extension	As B3 but end-shields turned through 90° (if sleeve bearings)	Mounted on a wall. Feet to the left viewing from drive end
B7		2 end-shields	With feet	Free shaft extension	As B3 but end-shields turned through 90° (if sleeve bearings)	Mounted on a wall. Feet to the right viewing from drive end
B8		2 end-shields	With feet	Free shaft extension	As B3 but end-shields turned through 180° (if sleeve bearing)	Mounted on the ceiling. Feet to the right viewing from drive end
B15		1 end-shield	With feet	Free shaft extension	As B3 but without end-shield and bearing at drive end	Mounted on substructure by feet with additional mounting on face of frame at drive end
V1		2 end-shields	Without feet	Free shaft extension at bottom	End-shield flange with access to back. Flange at drive end	Supported by flange at bottom
V2	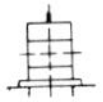	2 end-shields	Without feet	Free shaft extension at top	End-shield flange with access to back. Flange at non-drive end	Supported by flange at bottom
V4		2 end-shields	Without feet	Free shaft extension at bottom	End-shield flange with access to back. Flange at non-drive end	Supported by flange at top
V5		2 end-shields	Without feet	Free shaft extension at bottom	As B3	Wall mounted or on a substructure
V6		2 end-shields	With feet	Free shaft extension at top	—	Wall mounted or on a substructure

synchronous and squirrel cage motors now exceed very substantially the older limits set by d.c. motors, as indicated in Figure 3.13. For a.c. motors, the combined limitations of motor design and frequency conversion equipment apply. Synchronous motors are used for very high ratings at both high and low (below say 375 rev./min) speeds, and there is a wide range of applications in the high rating band where both conventional synchronous and squirrel cage designs are applicable.

3.3.3.3 Enclosures, mountings and cooling systems

The codification for these aspects is laid down internationally in standards which are complicated, complex and often misunderstood. Moreover, there are two distinct aspects to those matters. From the user's point of view, there is a need to arrange the motor to drive a machine in a particular configuration and with a particular speed and power in a specified environment. The constructor has to provide this motor and demonstrate that it meets the specified requirements. There is, consequently, no choice in the mounting, some choice in the enclosure and considerable choice in the cooling system.

For types of construction and mounting, the so-called Code 1 convention, designates machines as either B, horizontal, or V, vertical. A series of numerical suffixes gives further details of the mounting and bearing arrangements. Some of the most common arrangements are summarized in Table 3.14. Code 2 is a very detailed alpha-numeric categorization including additional details of pedestal, gearboxes, etc. The requirement for the enclosure is dictated by the environment and influenced by the cooling system and internal motor design. The basic IP codification for enclosures is based on the first numeral indicating protection of personnel against contact with electrically live or rotating parts and consequently also the degree of protection against ingress of solid bodies. The second number refers to the protection against ingress of harmful water. Table 3.15 lists the most usual enclosures together with the basic test requirements. The actual test details are fully specified in the standard and have to be followed to demonstrate compliance.

The third aspect – cooling systems – is somewhat complicated and, as it represents a range of design solutions, is only of limited interest to the user. There

Table 3.15 Principal types of protection (BS 4999–105)

	Mechanical access	*Water ingress*
IP11	50-mm dia. sphere	Specified drip-proof: test not mandatory
IP21	12-mm dia. test finger	Specified drip-proof: test not mandatory
IP22	12-mm dia. test finger	Specified drip-proof: test not mandatory
IP23	12-mm dia. test finger	Specified spray-proof: test not mandatory
IP44	1-mm dia. steel wire	Specified splash-proof: test not mandatory
IP45	1-mm dia. steel wire	Hose-proof: specified test mandatory
IP54	'Sealed': dust	Splash-proof: specified test not mandatory
IP55	'Sealed': dust	Hose-proof: specified test mandatory
IP56	'Sealed': dust	Deck watertight: specified test mandatory
IP57	'Sealed': dust	Submersible: specified alternative tests required

Notes: (a) The severity of the tests increases with increasing reference numbers.

(b) A suffix S or M means tested with machine stationary or running respectively.

(c) A machine is weather-protected when its design reduces the ingress of rain, snow and airborne particles under specified conditions. This degree of protection is designated by the letter W (placed between IP and the numerals).

Table 3.16 Selected principal methods of cooling (BS 4999–106)

		IC designation
(A)	*Open-air circuit (ventilated)*	
	enclosed ventilated	IC 01 or 06
	weather-protected	IC 01 or 06
	outlet water/air heat exchanger	ICW 37 A01 or 06
(B)	*Closed-air circuit*	
	Totally enclosed	IC 00 40
	Totally enclosed fan-cooled (blow-over)	IC 01 40
	Totally enclosed fan cooled (tube-cooled)	IC 01 51
	Totally enclosed, separately air-cooled	IC 06 40 or 06 51
	(blow-over or tube-cooled with separately driven fan)	
	closed-air circuit air-cooled: integral fans	IC 01 61
	closed-air circuit air-cooled: separately driven fans	IC 06 66
	closed-air circuit water-cooled: integral fans	ICW 31 A61
	closed-air circuit water-cooled: separate fans	ICW 36 A66
(C)	*Submerged*	ICW 01

Note: Above Group II of Table 3.13, motors are often designed with a double-ended cooling system.

are usually various alternatives to be considered for given circumstances, generally environmental, and constructors and national preferences vary widely. Table 3.16 summarizes some of the more usual arrangements, which are illustrated in Figure 3.14 to illustrate some of the executions. There are three basic classifications, the application of which depends on the size of motor:

(1) Through-ventilated.
(2) Closed-air circuit.
(3) Submerged: liquid or gas-cooled.

The impact of the cooling system is considerable in the economic sense, as the limits of temperature, set principally by the stator insulation system, are finite. It is essential to note the difference between temperature in absolute terms and

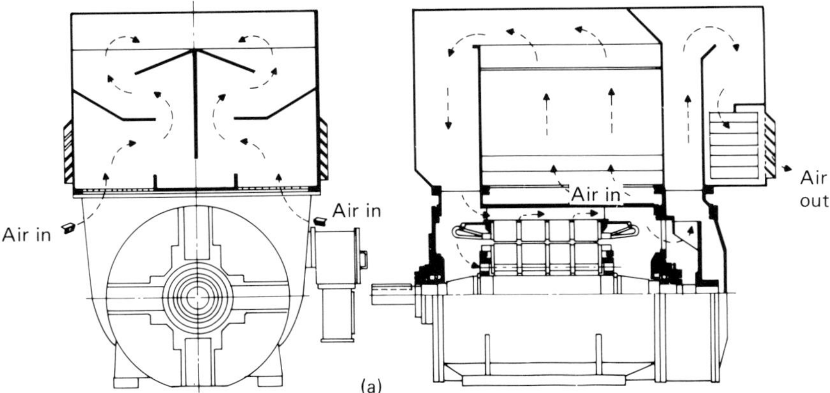

Figure 3.14 Principal methods of cooling. (a) Basic ventilation circuit for weather-protected (NEMA 2 specification) motor design

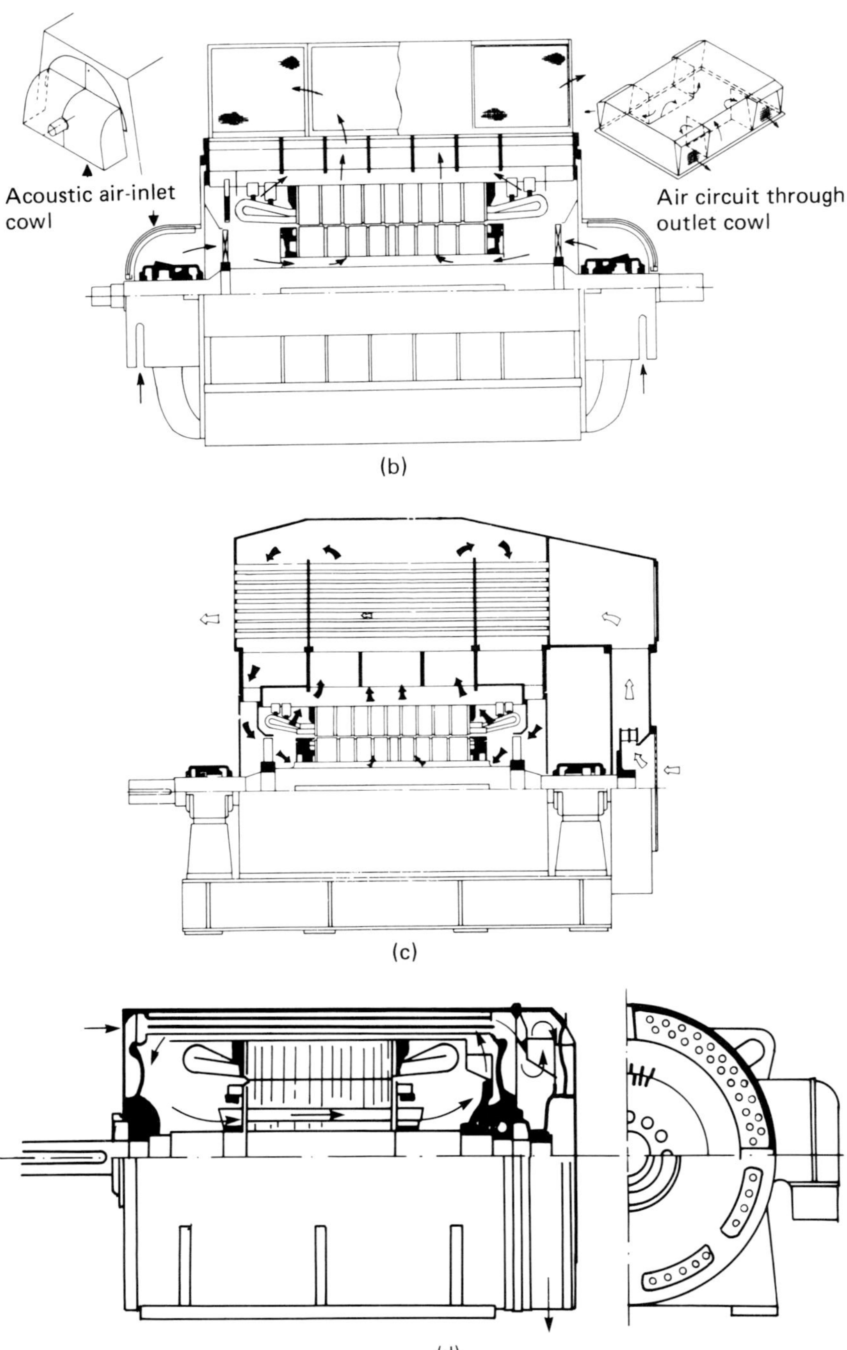

Figure 3.14 (continued) (b) Basic ventilation circuit for ventilated-motor design with inlet- and outlet-noise attenuators. (c) Basic ventilation circuit for closed-air circuit air-to-air heat exchanger motor design. (d) Basic ventilation circuit for totally enclosed fan-cooled machine (tube-cooled)

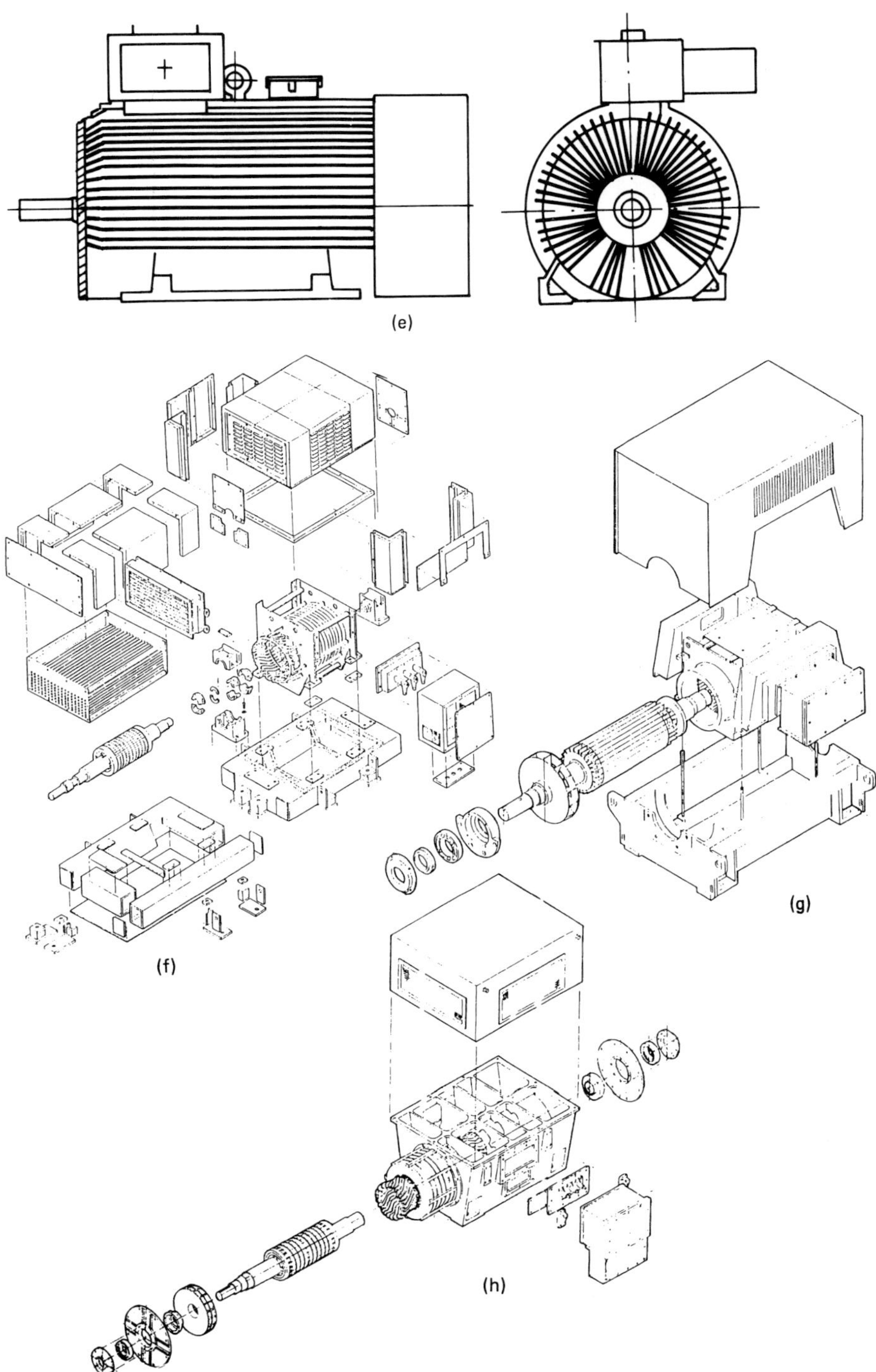

Figure 3.14 (continued) (e) Totally enclosed fan-cooled machine (blowover design). (f) 'Deep bedplate' unit construction design. (g) 'Half box' unit construction design. (h) 'Independent' frame construction

temperature rise. For instance, the use of water–air heat exchangers, provided water is readily available, is generally advantageous above Group II motors. The closed-air circuit has the advantage of environmental protection, both with regard to pollution from outside, water, particles, high ambient temperature, etc., and pollution to the outside, noise, heat, etc. Whilst shaft-driven fans are used for reliability reasons as far as practicable, highly reliable heat exchangers with separate motor-driven fans have been applied very successfully, for reasons of size or speed, particularly variable speed. Motors with collectors, e.g. sliprings, or built-in semiconductors, e.g. brushless synchronous motors, need such protection, which can be enhanced by pressurization with 'clean' air in cases of high levels of chemical contamination, quite apart from (Ex) hazardous area requirements. On the other hand the development of sealed stator windings to NEMA requirements[22] allows reliable application in many situations of weather-protected, i.e. ventilated, squirrel cage motors.

A useful cooling system for motors in groups I and II is the totally enclosed, separate-fan-cooled design, which caters for more arduous variable-speed requirements.

3.3.3.4 Motors for hazardous areas

Table 3.17 gives an overview of the generic features of the four widely used types of protection, which are also cross-referenced in Table 3.13. This inherently complex subject is further confused by the existence of three sets of British Standards, EEC directives and the understandably highly specific approach imposed on users due to the local legal framework in which they have to operate.

To appreciate some of the problems, the technical derivation of different approaches has to be considered.

'Flame-proof' (Ex)d is derived from the requirements of containing a gas explosion within the carcass of the motor. By contrast, 'increased safety' (Ex)e is based on improved electrical and mechanical reliability aspects in combination with an operational regime, which will not ignite the relevant gas, by limitation of the temperatures of the windings for a 'brushless' motor. The (Ex)N concept follows from this at a less sophisticated level, i.e. it is really an enhanced-reliability, brushless motor. The (Ex)p concept has long been accepted as this covers motors with brushes, commutators and sliprings, and has even been extended to pressurizing with an inert gas for additional safety, though this is not practical for brushed motors. It also avoids any technical limitation in the motor design itself.

Normal practice requires certification by a third (independent) party for assurance that the motor meets all the requirements of the relevant standard. This has to be done by a combination of examination of the design, including drawings, and physical testing. This covers gas ignition tests for (Ex)d, detailed electrical tests for (Ex)e and a proof of purging for (Ex)p. Following certification, the operator can obtain 'approval' for use in the particular circumstances from the relevant national regulatory authority. Whilst there is some reciprocity of acceptance of test certification inside the EEC, the subject is fraught with problems, which need early resolution.

3.3.3.5 Terminations and terminal boxes

The increasing use of higher ratings and/or aluminium cables has led to many installation problems, so that considerable efforts have been made to standardize termination systems, both low- and high-voltage. In order to achieve reliability, special care has always to be exercised at joints, both from the electrical (contact) and the dielectric point of view.

Table 3.17 Comparison of the salient aspects of explosion protection methods applicable to rotating electrical machines with selected relevant standards

Method of protection / Symbol	Flameproof (Ex)d	Increased safety (Ex)e	Non-sparking (Ex)N	Pressurized (Ex)p
Approx effect of enclosure requirements on rating of otherwise standard industrial machines	None for temperature classifications T1–T4	Can be considerable depending upon temperature classification and t_e. Also reduced temperature rise in rated service	None for temperature classification T1–T3	None for temperature classifications T1–T4
International Electrotechnical Committee Standard	IEC 79–1:1979	IEC 79–7:1969	—	IEC 79–2:1983
European Standard (CENELEC)	EN 50:018	EN 50:019	—	EN 50:016
Equivalent British Standard	BS 5501:Part 5:1983	BS 5501:Part 6:1986	—	BS 5501:Part 3:1980
Other British Standards	BS 4683:Part 2:1983 BS 5000:Part 17:1981	BS 4683:Part 4:1986 BS 5000:Part 15:1985	BS 4683:Part 3:1972 BS 5000:Part 16:1985	
Restricted surface temperature inside enclosure	No	Yes, including starting, minimum 5s locked rotor	Yes, but not during starting	No
Types of machine to which protection is applicable	All	Cage and brushless synchronous machines	Cage and brushless synchronous machines	All
Reduced temperature rise for given class of insulation	No	Yes	No	No
Hazardous gases for which suitable	Those listed in the standard	All	All	All
Zones for which protection may be acceptable	1,2	1,2	2	1,2

For low-voltage systems, which are generally fuse-protected, reducing the danger from a short-circuit fault to a minimum, adequate space for installation of cables is the main requirement.

For high-voltage motors, it is essential to ensure dielectric reliability, which is best achieved by arranging joints to be covered by insulation after installation whilst, where appropriate, designing clearances and creepages at the joint ignoring insulation. Figure 3.15 illustrates the standard phase-insulated BEAMA 3.3 kV terminal box with stand-off insulator, 6.6 kV insulator type, the standard phase-segregated 11 kV terminal box with connectors and the use of proprietary

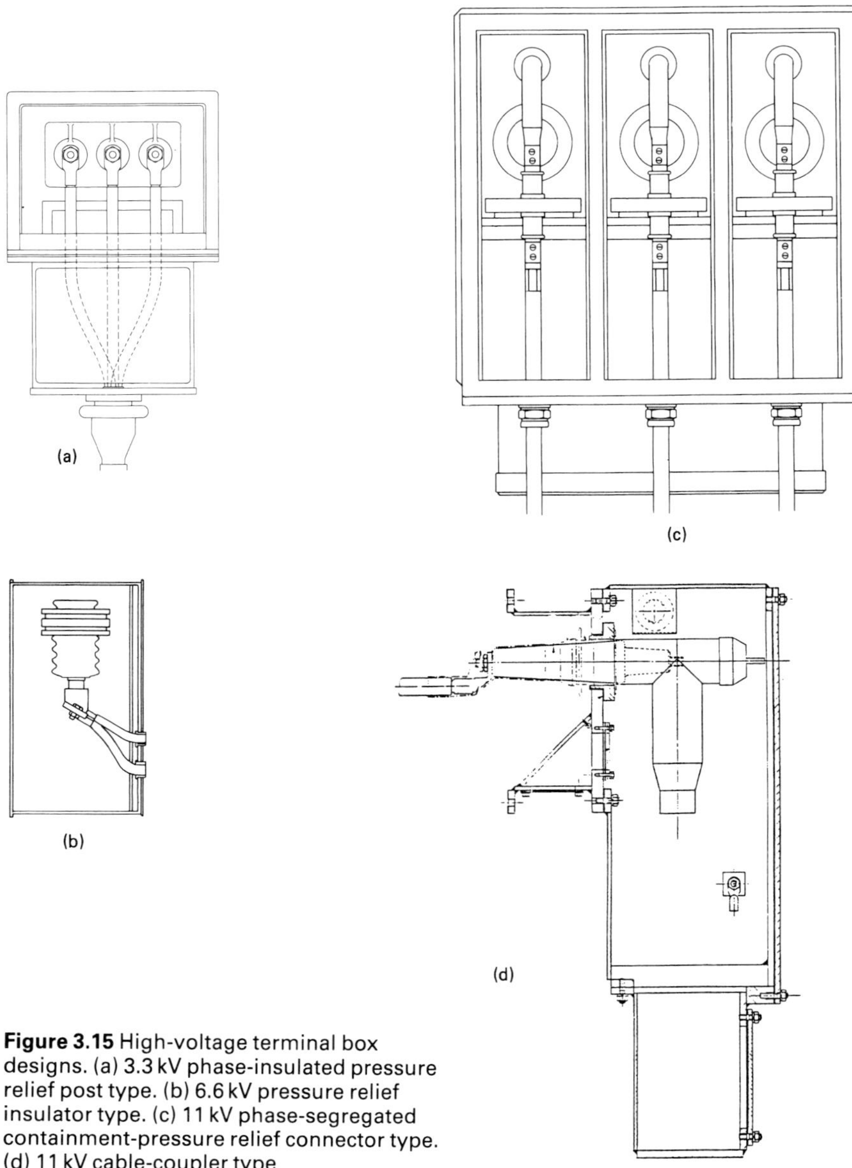

Figure 3.15 High-voltage terminal box designs. (a) 3.3 kV phase-insulated pressure relief post type. (b) 6.6 kV pressure relief insulator type. (c) 11 kV phase-segregated containment-pressure relief connector type. (d) 11 kV cable-coupler type

cable-couplers. In all cases the arrangement is designed and type-tested for: (1) through-fault proofness up to the system level of 26 or 40/44 kA (see Section 3.1); and (2) a 'non-explosive' performance under direct-short-circuit faults, i.e. controlled exhaust of the arcing products and no disruption of the terminal box itself except for the vent. Moreover, single faults must not propagate into multiple faults. The safety aspect[23] is important as, however unlikely a dielectric fault will be with a good modern design, the potential consequences of a high-power fault are too great to be permitted without these extra precautions having been taken to minimize the effect.

3.3.3.6 Vibration, seismic and shock requirements

The simplistic approach to low vibration levels expressed in the standard is only of use in standardized situations. This is due to the considerable and complex influences, which affect the performance *in situ*, where the combined effect of the complete rotating assembly and the support structure becomes evident. However, the prescribed 'free suspension' (uncoupled) vibration measurement on test of the motor is a worthwhile starting point which can be extended to the more general case.

The best practical guides are the well known VDI specifications,[24] which approach the whole problem in terms of size, speed and mounting (see Table 3.18) by assigning realistic vibration levels in terms of velocity (see Figure 3.16) for overall evaluation. There are two major origins of vibration to be considered, mechanical balance of the rotating masses and electrical and mechanical excitation forces. As far as mechanical unbalance is concerned, motor rotors are generally balanced to a prescribed level, which can be checked on test. The driven machine, where testing is obviously difficult, often exhibits higher levels of vibration due to unbalance, and it is therefore common and acceptable practice to unbalance the motor rotor, particularly in the case of pump impellers which are often inaccessible. For vertical and overhung drives, where no standard rules exist, the vibration levels at the free end are inherently amplified, so that levels of, say, twice those for horizontal motors are common and acceptable. Proper alignment, including allowance for thermal growth, has an important effect. Moreover, the coupling must be designed to allow for the normal axial forces where sliding bearings, without thrust capacity, are used on the motor.

The motor produces an almost infinite variety of forces due to the effects of electromagnetic design. Here, as with other phenomena, the problem is three-dimensional and has to be completely integrated into the dynamic design of the shaft and bearing system with its particular stiffness, damping and interactive parameters. Modern computation methods are employed in cases of special importance to obtain a complete picture of the vibration performance.

A particular problem arises with high-speed, generally two-pole motors, in Groups III and IV, the so-called hyper- (as opposed to hypo-) critical shaft designs, i.e. a rotating system having a first major peak in its vibration amplitude characteristic below (as opposed to above) synchronous speed. Whilst it is a normal requirement not to have such a peak within 15–20% of running speed, the unfortunate and misleading terminology of 'critical' speed has connotations of bad performance. In reality, hyper-critical designs have, in general terms, lower vibration levels at running speed than hypo-critical designs, and a correctly designed shaft system will pass through the peak vibration amplitude during acceleration or deceleration without distress. Clearly, the effort involved should not be necessary for standard motors – Groups I and II with rolling bearings.

In view of the many possible excitations, which include the driven machine, the correct way to assess the performance is to compute and of course test, the vibration characteristics (see Figure 3.17 for example) which is easily done by

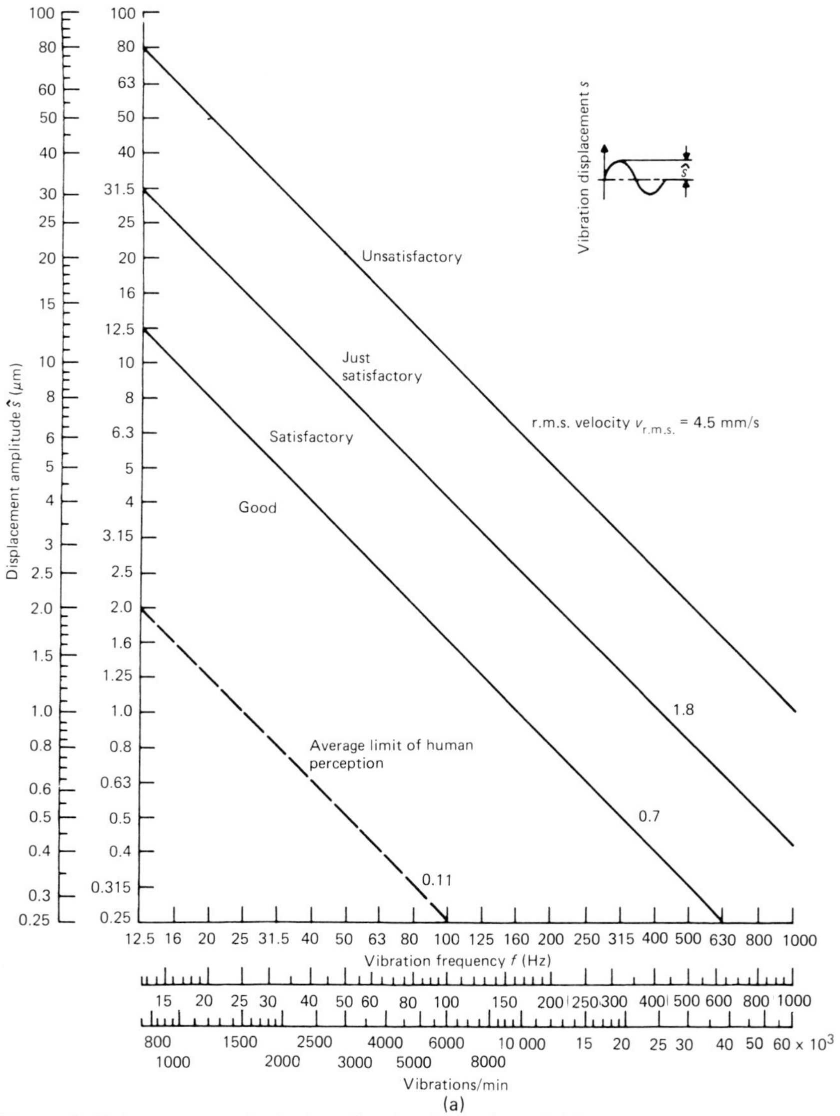

Figure 3.16 Assessment limits for vibration behaviour. (a) Small motors (Group K) according to definition in Table 3.18. (For non-harmonic vibrations, this is valid only for the equivalent displacement amplitude s_{equiv} so that $s_{equiv} = \sqrt{2}v_{r.m.s.}/(2f_B\pi)$ where f_B is the reference frequency.)

variable frequency, including operation up to the prescribed overspeed level. The stability of the design is demonstrated, if the vibration levels before and after the overspeed test remain substantially unchanged, in addition to not exceeding the prescribed levels. It is necessary to distinguish between vibration levels of stationary parts, e.g. bearing housings, and the relative movement between shaft and bearing, when carrying out a critical appraisal.

There are two specialized requirements in the same category: (1) seismic, now normally quoted for nuclear power stations; and (2) 'shock', as normally required for military purposes, where a prescribed low-vibration level may also apply.

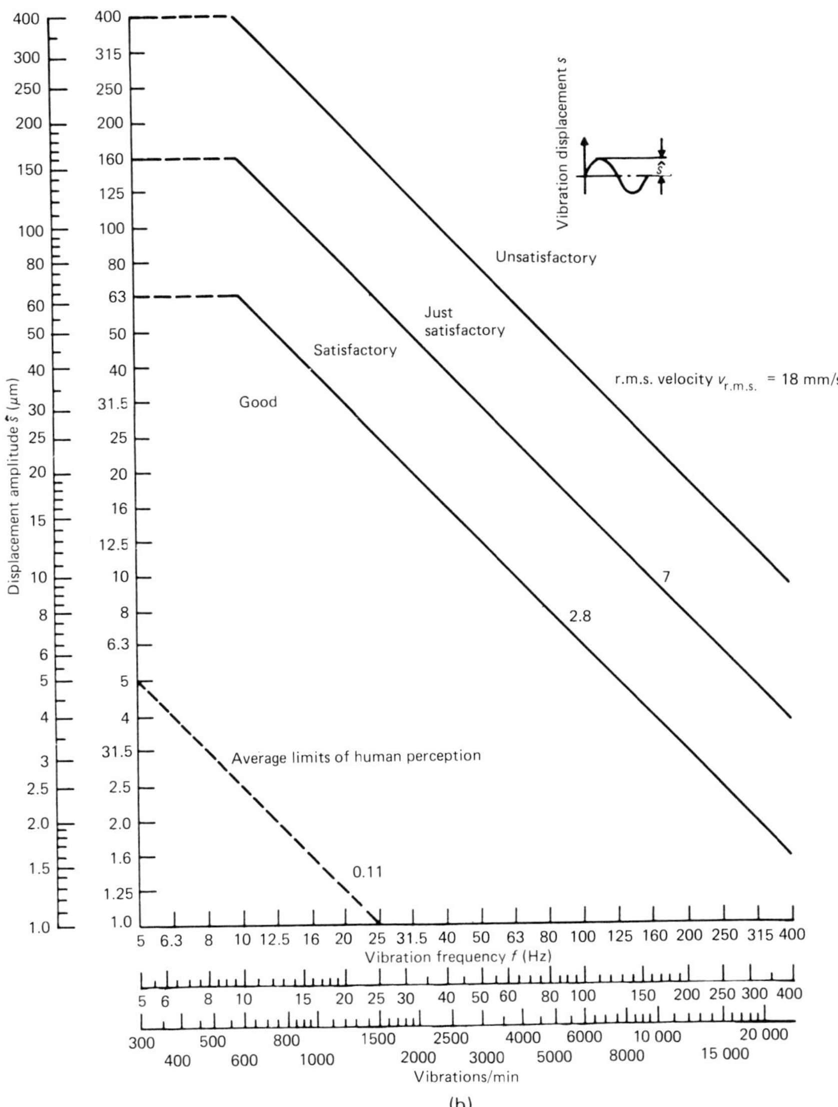

Figure 3.16 (continued) (b) Large high-speed motors (Group T). Machines and turbo-machines are mounted low-tuned on light foundations according to definition in Table 3.18 (for non-harmonic vibrations, valid only for equivalent displacement amplitude)

Whilst seismic design is somewhat specialized in application, it is interesting to note that the level of acceleration forces implied (see Figure 3.18) are not high in absolute terms and relate realistically to the normal behaviour of structures.

3.3.3.7 Acoustic noise

The obvious requirement to limit acoustic noise has long been recognized in legislation, where SPL limits of about 85–90 dB(A) are set for long-term exposure. What is not so well understood is the inherent mechanism of noise generation in

Table 3.18 Vibration intensity levels for different classes of motors

Vibration intensity levels		*Equivalent amplitudes at limits of level*		*Examples of assessment levels for individual machine groups*			
Level designation	r.m.s. velocity at limits of level	Equivalent velocity amplitude	Equivalent displacement amplitude relative 50 Hz	*Group K*	*Group M*	*Group G*	*Group T*
0.28	(mm/s)	(mm/s)	(μm)				
0.45	0.28	0.40	1.25				
0.71	0.45	0.63	2.00	Good			
1.12	0.71	1.00	3.15		Good		
1.80	1.12	1.60	5.00	Satisfactory		Good	
2.80	1.80	2.50	8.00		Satisfactory		Good
4.50	2.80	4.00	12.50	Just satisfactory		Satisfactory	
7.10	4.50	6.30	20.00		Just satisfactory		Satisfactory
11.20	7.10	10.00	31.50			Just satisfactory	
18.00	11.20	16.00	50.00	Unsatisfactory			Just satisfactory
28.00	18.00	25.00	80.00		Unsatisfactory		
45.00	28.00	40.00	125.00			Unsatisfactory	
71.00	45.00	63.00	200.00				Unsatisfactory

Notes: K   Small motors <15 kW.

        M   Medium motors 15–76 kW.

        G   Large motors: rigid foundations.

        T   Large motors: flexible foundations.

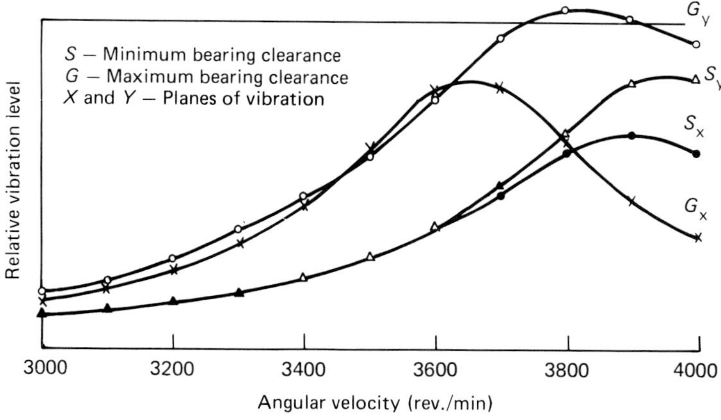

Figure 3.17 Vibration characteristics of 5-MW two-pole motor journal bearing

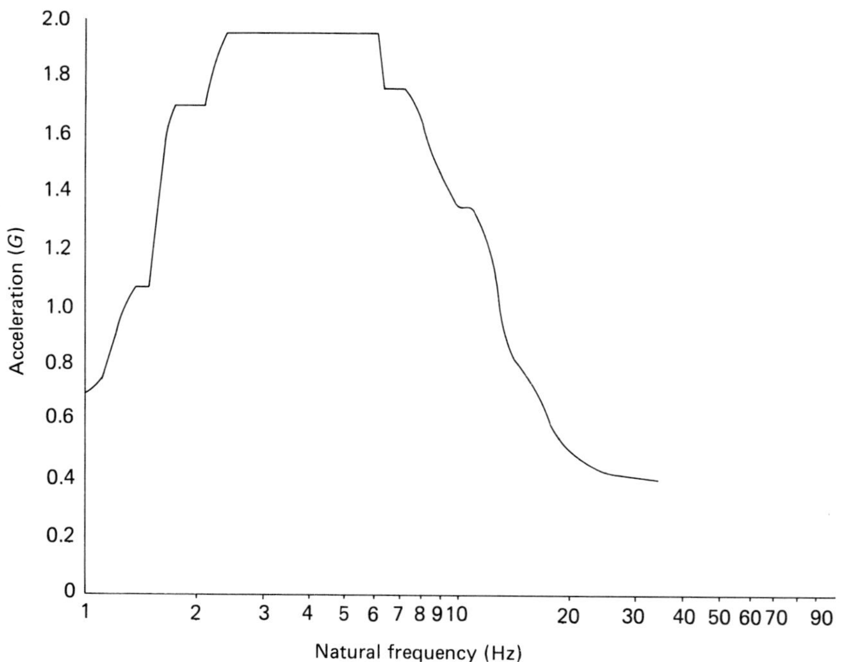

Figure 3.18 Typical undamped seismic requirements (vertical) for British nuclear power stations

motors (see Table 3.19), which has two aspects – aerodynamic and electromagnetic – for noise excitation, and mechanical–structural for noise propagation. The latter subject includes the reverberation characteristics of the location.

Although limited by design, levels inherently increase with power. Discrete frequencies, which may be produced by adverse slot combinations, particularly in squirrel cage motors, should be avoided. On the aerodynamic side, there is also an inherent increase with power. To achieve acceptable performance, the move is

Table 3.19 Limiting sound power levels (BS 4999–109)

Rating kW (or kVA)		*Rated speed (rev./min)*					
Above	*Up to*	*960 and below*	*961–1320*	*1321–1900*	*1901–2360*	*2361–3150*	*3151–3750*
		Sound power level (dB(A)) on no load					
0.0	1.1	76	79	80	83	84	88
1.1	2.2	79	80	83	87	89	91
2.2	5.5	82	84	87	92	93	95
5.5	11	85	88	91	96	97	100
11	22	89	93	96	98	101	103
22	37	91	95	97	100	103	105
37	55	92	97	99	103	105	107
55	110	96	101	104	105	107	109
110	220	100	104	106	108	110	112
220	630	102	106	109	111	112	114
630	1100	104	107	111	111	112	114
1100	2500	107	110	113	113	113	114
2500	6300	108	112	115	115	115	115
6300	1600	110	113	116	116	116	116

Notes: (a) These are based on totally enclosed fan cooled, CACA and similar enclosures. For ventilated motors, levels are 2–5 dB(A) lower.

(b) Corresponding noise-rating numbers, based on pressure and referred to a 3 m hemisphere, will be approximately 18 dB less than the above sound power rating figures.

towards closed-air circuits, with water–air heat exchangers. The use of fan inlet and outlet noise absorbers and acoustic cladding of complete motors is applied where necessary (see Figure 3.14(b)).

3.3.3.8 Electromagnetic compatibility

Electromagnetic compatibility is a subject of increasing importance due to the wide use of sophisticated, and often highly sensitive, instrumentation and control equipment. All electrical motors generate harmonic currents to a lesser or greater degree. There is a series of standards and specifications dealing with this subject, particularly in marine applications, where the closed environment and relatively low generation capacity pose problems, particularly when considering any collector machines, quite apart from static regulators or converters. If there is substantial harmonic pollution in the supply, this can render automatic regulators erratic or even inoperative. It will be understood that the problem arises with both conducted and radiated interference in both directions. Figure 3.19 shows permitted levels in British nuclear power stations.

3.3.3.9 Protection

The myth of the indestructible squirrel cage rotor is illustrative of the problem of motor protection, where there are few standards. An exception relates to thermal protection, with special reference to positive temperature coefficient detectors, 'thermistors'. The requirements of motors can, nevertheless, be described succinctly, and are summarized in Table 3.20, which is by no means exhaustive. (See Section 7.4 for further details.)

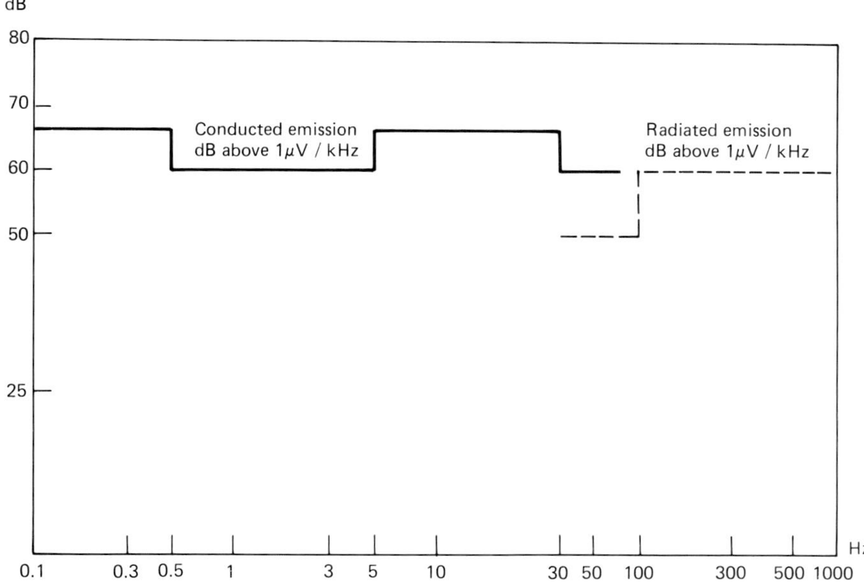

Figure 3.19 Typical electromagnetic compatibility requirements for British nuclear power stations

Electrically, there are three levels of current protection: (1) short-circuit (fault) current; (2) short-time overcurrent in the run-up region; and (3) medium-time overcurrent (overload) in the operating region. However, current cannot be related directly to load, unless processed into its two components, wattful and wattless, nor is it directly related to the thermal endurance of either stator or rotor winding. Furthermore, the load is imposed from outside, and it is highly dangerous to assume that the motor torque reflects the load, which can only be properly measured mechanically.

The stator winding temperature is a prime parameter relating to life expectancy, as would be the rotor winding, including the short-circuit ring for squirrel cage motors, if it could be monitored by simple, reliable means.

The supply voltage needs analysing to ensure that the necessary overload capability is available for the specified duty. The voltage amplitude is not sufficient; it must be measured as the amplitude of the symmetrical components forward, backward and stationary, and its ripple content.

The most critical parameter is not always obvious. Most squirrel cage motors above Group II are rotor-critical and therefore not protected by simple stator current and/or temperature measurement. It is possible to provide sophisticated, usually microprocessor-based, analogue protection systems, which can cope, though this has to be viewed critically in the light of the motor design as applied, for instance, to (Ex)e motors.

The dielectric protection by earth leakage relays is normally satisfactory; for security reasons motors can be operated for limited (always) or unlimited (by special design) periods with one earth on an unearthed system.

The detection of interturn faults, except by voltage injection, is as yet not reliable, whilst vibration monitoring is often included in the protection system, as is over- and under-speed, where appropriate. Reverse running prevention may have to resort to mechanical anti-reverse rotation clutches for real protection, where

Table 3.20 Typical motor protection systems

Reference	Class	Type	Objective
1.1	Current	Instantaneous over-current	Detection of short-circuit fault
2		Excess current 'short' time delay	Detection of abnormal current, e.g. maintained starting current
3		'Overload' current–'long' time delay	Detection of current above rated current, e.g. overload
4		Unbalance current	Detection of single-phasing or other phase asymmetry
5		Differential current	Detection of winding faults by means of phase current unbalance (large machines)
2.1	Voltage	Overvoltage	Detection of abnormal steady state system condition
2		Undervoltage	Detection of abnormal steady state system condition
3		Phase unbalance	Detection of abnormal steady state system condition
4		Over-frequency	Detection of abnormal steady state system condition (self-contained system)
5		Under-frequency	Detection of abnormal steady state system condition (self-contained system)
3.1	Insulation	Earth leakage	Detection of dielectic fault to earth
2		Interturn insulation	Detection of abnormal negative phase sequence fluxes, i.e. includes 1.4, 2.3 and 3.1
4.1	Overheat	Stator winding thermistor (or thermostat)	Single point over-temperature detection
2		Stator winding ETD (thermocouple or resistance thermometer)	Temperature monitoring and over-temperature detection
3		Infra-red squirrel cage rotor temperature detector	Monitoring SC rotor condition, e.g. protection against stalling and excessive starting duty
4		Bearing thermometers or thermocouples	Monitoring of bearing performance
5		Coolant thermometers or thermocouples	Cooling water, lubricating oil, cooling gas monitoring
6		Coolant flow	Indirect overheat monitoring and protection (as 4.5)
5.1	Miscellaneous	Bearing vibration transducers	Monitoring of aggregate mechanical condition
2		Overspeed detection	Prevention of mal-operation due to system and/or motor characteristic
3		Stall protection (particularly SC motors)	Prevention of mal-operation due to seized driven machine or excessive starting demands
4		Reverse rotation (of pump and fan drives)	Detection of abnormal mechanical condition
5		Synchronous motors	Synchronization and asynchronous running protection Monitoring of semi-conductor (rotating) excitation

motors cannot be designed economically for the maximum runaway speed of, for example, hydraulic equipment.

3.3.3.10 Testing

Electric motors are extremely well served by test standards in general, although there are still some gaps in the dielectric field. Almost all reasonable parameters can be found in the relevant standards. For economic reasons, the level of testing ranges from type/simple tests on mass-produced motors to full-routine individual tests for important and/or large motors (see Chapter 8).

There are some areas where, due to technical difficulties, no direct specific tests are available, in addition to the problems of demonstration of life expectancy (see Section 3.3.4).

3.3.4 Quality assurance

No survey of standards can be considered complete without reference to quality assurance. Whilst quality may take on various connotations, for the purpose of selecting electric motors there is only one rational definition, which is 'fitness for purpose'. Like the definition for design and standards, this definition is strongly biased towards economic objectives (see Section 3.4). The need for quality so defined is clear, but the method to be employed to achieve the objective is subject to much discussion and to many standards (see Table 3.21).

Table 3.21 Comparison of quality assurance standards

Grade	International	MOD	BS (industrial)	BS (nuclear)	Canadian (CSA)
(A)				BS 5882 (1980)	Z299.1 (1985)
(B)	ISO 9001 (draft 1987)	AQAP-1 (3rd edn, 1984)	BS 5750:Part 1 (1987)		Z299.2 (1985)
(C)	ISO 9002 (draft 1987)	AQAP-4 (2nd edn, 1976)	BS 5750:Part 2 (1987)		Z299.3 (1985)
(D)	ISO 9003 (draft 1987)	AQAP-9 (2nd edn, 1976)	BS 5750:Part 3 (1987)		Z299.4 (1985)

ISO 9001: Quality system – Model for quality assurance in design/development, production, installation and servicing
ISO 9002: Quality systems – Model for quality assurance in production and installation
ISO 9003: Quality systems – Model for quality assurance in final inspection and test

AQAP-1 MOD Requirement for an industrial quality-control system
AQAP-4 MOD Requirement for an industrial inspection system
AQAP-9 MOD Requirement for an industrial basic inspection system

BS 5750: Part 1 Quality systems: specification for design, development, production, installation and servicing
BS 5750: Part 2 Quality systems: specification for production and installation
BS 5750: Part 3 Quality systems: specification for final inspection and test

BS 5882: A total quality assurance programme for nuclear installation

CSA Z299.1: Quality assurance program requirements
CSA Z299.2: Quality control program requirements
CSA Z299.3: Quality verification program requirements
CSA Z299.4: Inspection program requirements

Normal industrial requirements are laid out in BS 5750 under three basic headings:

Part 1: Design, manufacture and installation.
Part 2: Manufacture and installation.
Part 3: Final inspection and test.

Part 1 is therefore fulfilling the concept of quality assurance, whilst Part 3 covers quality control on important aspects of manufacture. No such grading is applicable to nuclear power stations, where the standard encompasses all aspects of assurance.

However, there are important implications which are not resolved by these documents. It is apparent that the depth and therefore the cost of assurance must in itself be 'fit for the purpose'. Consequently, the concept of quality assurance levels is introduced by a Canadian standard, where the level is determined numerically by a weighted evaluation of six aspects:

(1) Design process complexity.
(2) Design maturity.
(3) Item or service characteristics.
(4) Manufacturing complexity.
(5) Safety.
(6) Economics.

The detailed implications need particular consideration in the light of the application. For offshore,[25] military, nuclear and fossil-fuel power stations, there is heavy emphasis on the design assurance side. Modern computational disciplines, combined with appropriate full-scale and model experimental work, make this a powerful approach to reliability and consequent assurance of total real cost.[26]

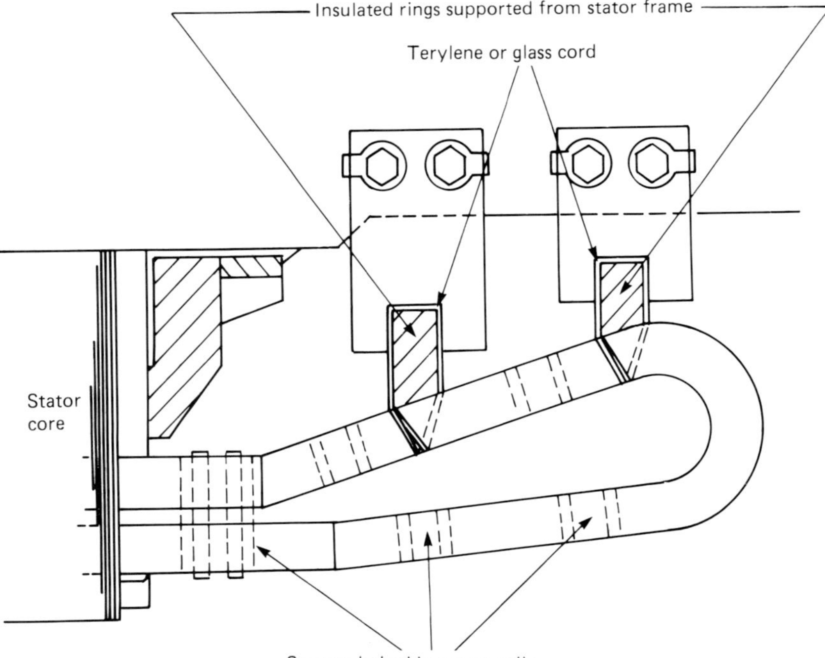

Figure 3.20 Typical high-integrity, high-voltage stator winding overhang support system

Moreover, design is one of the stages where quality can be worked into the product, whilst quality control is simply a numerate method of checking that the design intent has been met, the quality having to be achieved during design and manufacture.

A typical, complex, example is represented by the stator winding mechanical support system (Figure 3.20), which requires detailed analysis of a structure made from relatively weak materials, and requires strict control during manufacture. Finite element analysis, model life testing and experience are combined to achieve the desired performance, reliably and predictably. As a general comment, proliferation of requirements, audits and standards does not in itself produce quality. The combination of a clear numerate statement of the 'purpose', matched by appropriate design and manufacture to achieve 'fitness', produces an economical and reliable motor drive.

3.4 Life-cycle costing

3.4.1 General

3.4.1.1 Introduction

The principle and aim of establishing a realistic economic cost figure, either in absolute or comparative terms, has eluded many generations of engineers, partly due to the inherent difficulties of establishing a viable system, and partly due to the pressures of the 'accountancy' aspects, which reside in the difficulty of defining monetary assets long term. Furthermore, the subject is frequently complicated by the division of responsibility for costs entailed in a large project. Thus, a particular division of the operator/owner may be responsible for the funding of the project, both software (design) and hardware (construction), whereas another division may have to deal with the operation, e.g. direct running costs, spares and maintenance. In some projects, insufficient interplay is found between, for instance, civil and plant engineering. There are consequently many levels of sophistication which may be employed, and in many cases pursued out of context or even ultimately overruled by simple policies such as minimum capital (first) cost. Whatever the motive, there can be no doubt that realistic cost assessment is technically difficult and requires a considerable effort[27] which is, however, worthwhile if the subject is treated rationally and in depth.

3.4.1.2 Identification of costs

In order to facilitate the cost assessment, it is necessary to analyse the component costs of a motor drive. The principle parameters are indicated in Figure 3.21, which assumes that some decision has been made about the requirements. Clearly, it may be necessary to assess alternative solutions within the context of a major project. All aspects have an interactive effect and consequently the diagram is based on a closed-loop approach. Moreover, everything from nuts and bolts to commissioning delays has its real cost. This applies especially to basic design investigations, which are the foundation of cost reduction, if various aspects are considered in depth. The basic execution of a drive in hardware terms is not a straightforward matter viewed in this light. With the possible exception of aesthetic aspects, quite simple things are often overlooked to the detriment of the total cost.

Availability, reliability and maintainability (ARM) is a definitive discipline, which needs numerate attention, just as much as the more obvious operating costs, which are also not as simple as they seem. This is all part of overall quality assurance. Table 3.22 lists the more important parameters, which are always relevant in establishing total costs. In nearly every practical case there is a trade-off

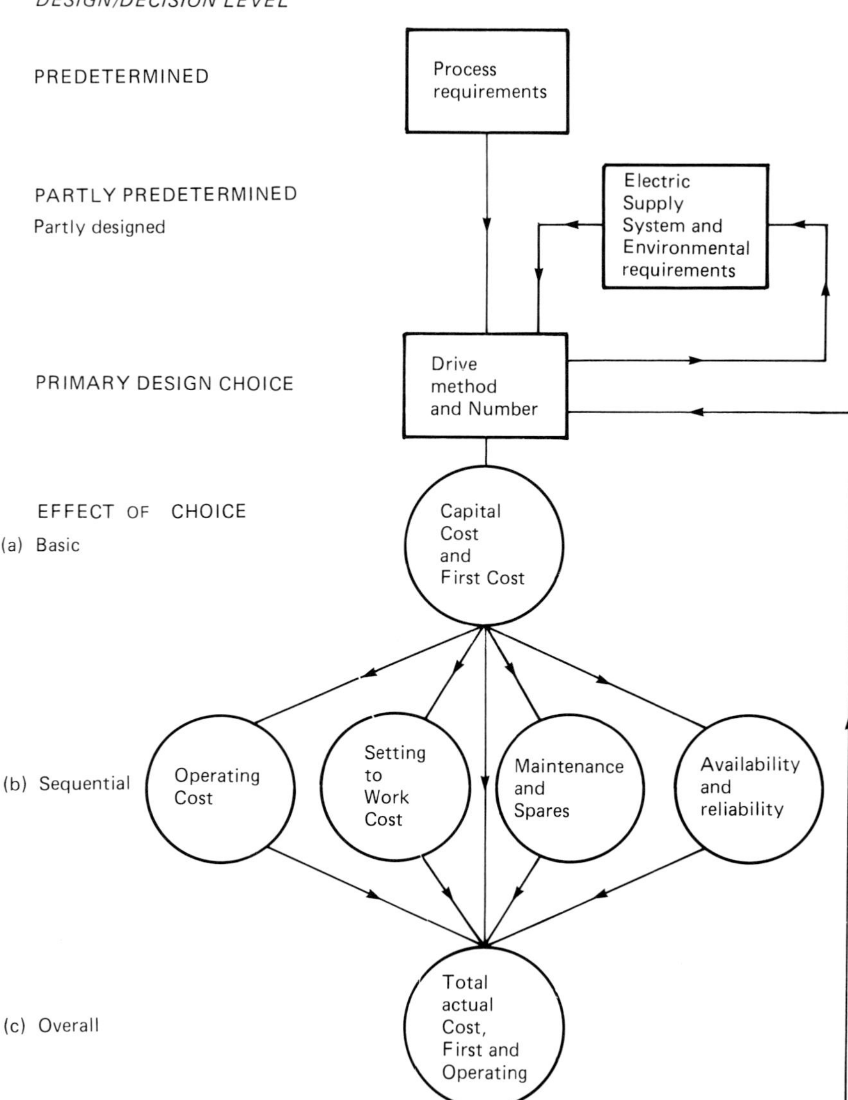

Figure 3.21 Flowchart illustrating the economic aspects of system design

between running and capital costs, not just in the normally recognized area of kilowatt-hours, but particularly maintenance-related aspects, such as spares holding and routine maintenance procedures.

3.4.1.3 Design for reliability and life

Whilst by no means a new concept, a more formalized approach has been adopted recently to demonstrate, on statistical concepts, life expectancies in numerate form. The availability of computation methods for structural and thermal design of windings combined with model test procedures and ultimately feedback from the

Table 3.22 Principal cost factors

(1) Capital cost	Purchase price
(2) Electrical running cost	kWh, kVAr (tariff), kVAr (maximum demand) kVA (capacity of supply: transient current and harmonic distortion limits)
(3) Energy dissipation	Disposal of losses: natural–forced ventilation, water
(4) Setting-to-work	Commissioning
(5) Routine attendance	Spares-holding of wearing parts, lubricants Routine maintenance/supervising personnel Ease of maintenance/supervision: accessibility/monitoring Complexity
(6) Outage costs	Minor faults: *in situ* repairs/spares availability Major faults: Off-site repair/redundancy
(7) Environmental aspects	Noise effect on people, local and/or remote Vibration–effect on equipment w.s.r.t. reliability
(8) Reliability	Design for: mean time between failure : availability : life
(9) Safety	Electrical aspects Mechanical aspects Hazardous area aspects where appropriate

field is an important, though costly, exercise in major projects. It has an incidental advantage, in that it forces user and constructor to formulate properly the requirements or needs and consequential performance. This is demonstrably preferable to arbitrary safety margins (which should not be confused with empirical parameters for aspects too complex to calculate). The great importance of full-scale testing and organized commissioning forms a solid basis for meeting the life expectancy. It should be emphasized that ideas such as 'long' or 'good' can have no part in a cost-based exercise; the requirements and predictions must be numerate and generally fit for the overall purpose.

3.4.2 Establishing 'real' cost

3.4.2.1 Evaluation of losses

As an example of a particular aspect of cost definition, the absolute and/or comparative evaluation of drive losses provides some useful information. To guide the motor designer, it is useful for the user to declare a capitalized loss figure, currently typically £1500/kW for a power station and up to £4000–5000/kW for some particular situations. It must be emphasized that this is only a guide for initial design, but gives an indication when considering various options.

The next step is to consider this problem using the standard annuity calculation:

$$P = \left\{ 1 - \left[\frac{100}{100 + r} \right]^{n} \right\} \frac{100a}{r}$$

where p = annual capitalized cost figure;
a = annual cost of 'power';
r = percentage interest rate; and
n = number of years over which losses are capitalized

'Power' in this context is a complex feature, as electricity tariffs are not a simple concept. As soon as this approach is accepted for a normal life of, say, 20 years, the relatively low significance of first cost often becomes apparent in real long-term costing.

3.4.2.2 Establishing start-up costs

It is now generally accepted that a major cost component lies in the investment in plant during construction, when no revenue is received. The strict programmes exercised for all major projects address this particular aspect, including an allowance for 'setting-to-work'. Efficacious commissioning of motor drives needs the following early expenditure: (1) realistic and direct performance testing at the constructor's works; (2) detailed installation and operating instructions; (3) adequate facilities for minor modifications of control systems; and (4) flexibility of operation to deal with the 'product', e.g. adjustment of flow/pressure characteristic (by mechanical means or variable speed). Additionally, adequate maintenance/ spares facilities are essential. It is inevitable that some unexpected events will occur at this stage. Consequently, a well-prepared software design package will have two effects: (1) most of the more predictable problems will have been obviated; whilst (2) the unexpected problems can be pinpointed with relative ease and speed.

The use of large-scale prototypes or detailed models may well be an economic advantage, when the first cost of the prototypes is offset against the avoidance of lost production.

The availability of trained personnel is an essential part, and this is also the correct time to train the maintenance personnel for the future running of the plant. It is clearly unrealistic to suggest specific rules in this area, but a very modest investment and foresight is always very well worthwhile.

3.4.2.3 Maintenance and spares

Irrespective of the scale of the project or the quality level assigned to it, all equipment has a finite reliability and, consequently, maintenance and repair are inevitable. This is a subject for numerate analysis and must be tackled at the original project specification stage.

Spares fall into several categories, ranging from wearing parts with a limited life, such as bearings, to parts which from experience have the greatest statistical chance of failure and which can be replaced to keep production going.

Maintenance of a routine kind is vital. One of the reasons why bearing failures are still the most common cause of outages is the need for regular lubrication control, i.e. guaranteeing the specified regreasing intervals (which can be very short, in terms of weeks) for rolling bearings and ensuring a proper clean oil supply for sliding bearings.

The need for sliding bearings rather than rolling bearings is one of timing of maintenance, regreasing intervals of less than 3–4 months being generally not recommended, even if the bearing (L10) life is entirely satisfactory.

The general move from motors with brushes, which need regular maintenance, to motors without brushes but utilizing power semiconductors, is basically a switch in the maintenance requirements from skilled, regular attention to fault-finding and a substantial spares holding.

The holding of major spares, stators, rotors or even complete motors or redundancy (see Section 3.4.2.4) is a prime function of the project design. With modern insulation systems, the long-term storage of spare coils for high-tension motors is not practical; holding copper may be appropriate, but with the limited life of insulation materials and the time lags involved in stripping windings, there is usually little incentive for such an investment.

The emphasis should be on maintenance procedures and spares holding to form part of the original project, with personnel and material in place at the commissioning stage.

3.4.2.4 Outage cost

Several complementary concepts provide basic statistical tools to determine the operational performance of equipment. Commonly used terms include:

MTBF   = mean time between failures.
MTTR   = mean time to repair.

$$A = \text{availability} = \frac{\text{MTBF}}{\text{MTBF} + \text{MTTR}}$$

FMEA   = failure mode and effect analysis.
FMECA = failure mode, effect and criticality analysis.
FTA   = fault tree analysis.
ETA   = event tree analysis.

These methods have become an important design discipline for modern, high-integrity equipment of all kinds, not just military or safety-related, in view of the high gearing a minor fault can produce in terms of cost and/or facility, if there is a major plant outage.

The more complex the hardware, which in this sense includes software programmes controlling the hardware, the more important it is to consider from the outset how the plant may be operated under unusual and, particularly fault, conditions. One of the most effective techniques is FMECA. This allows conclusions to be drawn, not only in general, but also in numerical terms. It also emphasizes the inevitable interactive problems, which need proper design attention. For example, where closed-loop automatic control systems are used, a hand (operator) control standby facility should be provided, unless there is no safe way of operating the plant in this mode, which is only rarely the case. A logical consequence of this is to ensure that monitoring and automatic control do not rely on the same transducers, which probably need redundancy, as they are often the weakest link regarding reliability. Most important in the case of electric motor drives is the consideration of the effects of short-time supply failure, even in terms of a few milliseconds, to prevent dangerous conditions arising. There may be a need for deliberately sacrificing equipment, as far as life is concerned, to ensure a safe plant shutdown. The list of possible scenarios is almost endless, but an intelligent event and/or fault tree analysis will give appropriate guidance not only in safety matters, which are routinely considered for nuclear power stations, but for general practical evaluation of effective operational procedures.

These considerations turned into loss of revenue due to process outage will outweigh any reasonable cost associated with prevention at the design stage, quite apart from any safety aspects, which have to be considered in any case. The accent is on 'reasonable', a term which has to be qualified numerically with MTBF figures in particular situations and against a background of 'availability' of 98–99% for a well-designed motor drive, including the necessary supporting services discussed in this section.

3.4.3 Total cost

Whatever jargon is used, total cost is a reality; even if it is not realistic to make a detailed appraisal, the underlying requirements should be borne in mind.[28] There is, as always, a long-term trade-off in early investment, but a superficial view, such as is often taken by ignoring maintenance or spares or concentrating only on one

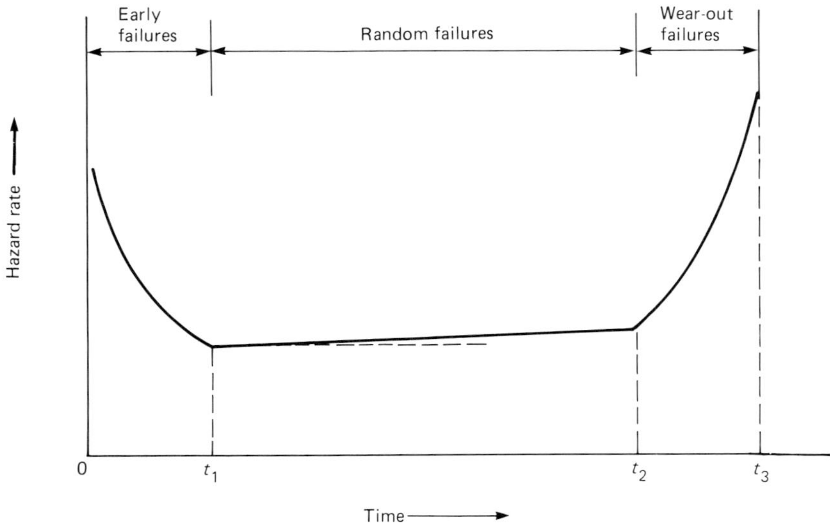

Figure 3.22 Hazard rate as a function of time for a complex plant or equipment

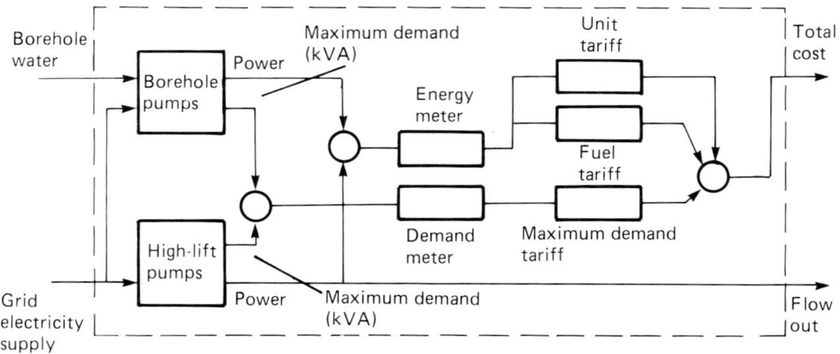

Figure 3.23 Pumping station cost model

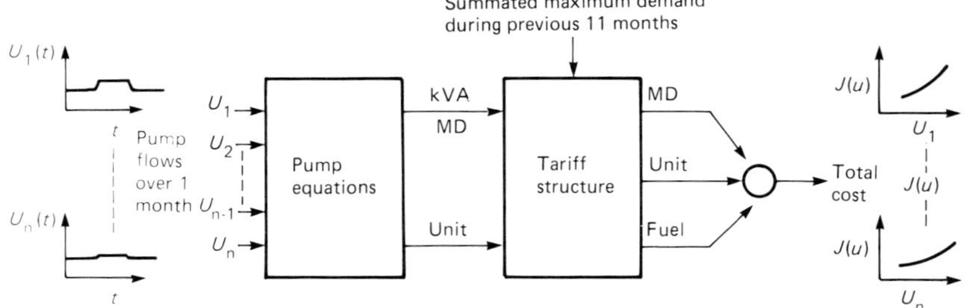

Figure 3.24 Non-linear cost model

feature such as kilowatt-hours saved against capital cost, can be highly misleading. The way forward is undoubtedly by means of a better understanding and consideration of reliability in numerate terms, which generally results in a balance having to be established between complexity and utility. The well-known 'bathtub' effect (Figure 3.22) illustrates the commissioning and early life problems in the context of total life (see Section 10.5 for further discussion).

For a real economic assessment, a much wider system-orientated view must be taken to establish the magnitude of cost of the motor drive, which may often be swamped by quite independent design considerations.

3.4.4 Examples

3.4.4.1 Control of pumping system running costs

East Worcestershire Waterworks[29] introduced an overall system control utilizing mainframe computers in the early 1970s. As a long-term project, the majority of appropriate pump drives had been installed as variable-speed motors; at the time these were NS variable-speed a.c. motors, but the control principles apply to any low-loss, variable-speed system.

The economics of the drives were related to hydraulic requirements, pump performance and tariffs (see Figures 3.23 and 3.24) making up a typical 'real cost' evaluation, which has proved highly successful in reducing real operating costs.

3.4.4.2 Two-speed motor fan drive

A very basic exercise illustrates the use of a two-speed pole amplitude modulation instead of single-speed squirrel cage motor, where the spares and maintenance and setting-to-work costs would be substantially the same, as would the availability.

Table 3.23 Drive particulars, two-speed motor fan drive

Motor: alternative	(a) 430 kW 14-pole single-speed (b) 430/285 kW 14/16-pole PAM two-speed
Supply:	3.3 kV, 50 Hz
Control:	Contactors
Operation:	8760 h Low speed used at weekends in two-speed case
First cost:	(a) 100% (b) 130%

Running cost saving: With two-speed drive: 27%
   The payback period of 1 yr is due to the favourable circumstances of this application. A more usual figure of 2–3 yr can, however, often be achieved

Table 3.24 Capital cost comparison for pumpsets

	Pumpsets	
	2 constant-speed *(%)*	*3 variable-speed* *(%)*
Pumpsets including starters	0.5	2.0
Source work station	59.7	59.7
Reservoir	39.8	19.9
Total	100	81.6

3.4.4.3 Design of pumping system to reduce capital cost

To illustrate the wider aspects of cost savings, a pumping system was considered to meet specific hydraulic requirements. The design compares the use of three variable-speed pumpsets, which include 50% standby capacity, with two fixed-speed pumpsets without standby capacity. In the former case, reservoir capacity of 1 day's supply is needed, whilst in the latter case reservoir capacity of 2 days' supply is required.

The capital cost comparison is as follows as shown in Table 3.24. A 4-fold increase in the cost of pumpsets has in this case resulted in an 18% saving of total capital cost, by good 'system' design. The relatively low cost of the pumping equipment should be noted.

3.5 Applications and special requirements

3.5.1 General

The electric motor is the universal workhorse in virtually all industries because it is:

(1) The most efficacious way of transmitting energy.
(2) The most efficacious method of converting electrical energy to mechanical energy.

All applications and industries have their own special requirements and needs. Constructors endeavour to deal with these by establishing generic solutions, as a basic standardized product, which remains flexible enough to meet the special application needs. It will be clear in this context why it is so important for the user to define his need accurately and numerately, when the constructor has to deal with a wide-ranging variety of solutions at minimum cost. As always, engineering judgements are difficult and complex, especially as the choices available are many. It also has to be said that there are some applications where the electric motor is not the correct solution. When considering specific problems, there are five general aspects which are common:

(1) Process requirements: general, automatic/non-automatic control, variable speed.
(2) Type of motor and type of starter.
(3) Electrical system requirements.
(4) Environmental requirements including safety.
(5) Availability, reliability and maintenance.

The constructor, who has a series of standard solutions available, will propose new solutions where appropriate. Normally, new combinations will have to be 'designed' and consequently a definition of design is quoted here:

> Engineering design is predominantly concerned with wealth creation. It involves the use of scientific principles, numeracy, synthesis, analysis, creativity, decision making, together with the timely consideration of human factors, technical information and market demand in the definition of a product, including systems, to perform pre-specified purposes with the maximum economy and efficiency.[30]

The fourteen application categories discussed here cover a wide range of experience but are not, it must be emphasized, in any way exhaustive. The repeated reference to frequency converters should be explained by the fact that such drives were developed from 1950 onwards using rotating machines, now generally replaced by static converters.

3.5.2 Cement

Cement manufacture represents an important industry, using a great amount of electricity and including motor drives for fans, pumps, conveyors, crushers, kilns and mills. These drives appear under several categories in various combinations. Moreover, as the requirements are for a continuous process, automatic control is widely applied. The very large electrical systems and equipment necessitate high reliability, as any breakdown causes major problems, quite apart from the financial consequences. Energy efficiency is a major requirement. Variable speed is therefore deployed in fan drives, where both the energy saving and controllability of the process are important. The general dust problem is an inherent feature of this application.

Of particular interest is the application of variable frequency (about 1970) to the main mill drive.[31] This consists of a synchronous motor, built as an integral part of the mill, supplied by a cycloconverter at low power frequency, thus avoiding the costly and relatively problematical gear train, as well as achieving energy-efficient variable speed. A typical rating is 6400/5600/(23) kW at 15/13/(0.5) rev./min.

3.5.3 Chemicals, petrochemicals and rubber

These industries use virtually all types and sizes of motor drive, often in hazardous areas, combined with processes which may become hazardous if drive failures occur. Additionally, there are often environmental problems due to chemicals or, as in the case of rubber, penetrating carbon-black dust, particularly harmful to collectors in motors and open contacts in control gear. Many processes are continuous, fully automated with wide deployment of variable speed. The integrated design process is well illustated by compressor drives, particularly high power, where a direct-coupled induction or synchronous motor fed from a frequency converter[32] obviates the need for a gearbox and provides 'loss-free' control.

There are extensive requirements for motors suitable for hazardous areas. These start with non-sparking (Ex)N applications for general refinery use (Zone 2), (Ex)d flameproof or (Ex)e increased safety for Zone 1 together with (Ex)p for collector-type motors (see Section 3.3.3.4).

The use of submerged motors, e.g. stirrers, has found limited application in some chemical processes, where seals are problematical. The choice between mechanical and electrical (variable-speed) control of processes, such as in the manufacture of polyethylene, varies enormously between systems, so that these high-power, low-speed drives often utilize synchronous motors for the constant speed, reciprocating secondary compressors with variable speed primary compressors. The control of current and speed fluctuations and the requirements for additional inertia is an important part of the design process.

By way of contrast, artificial fibre plants with multi-motor drives have a long history of variable frequency, long predating static converters. The use of synchronous motors in a properly space-synchronized manner has to be distinguished from the apparent synchronous performance of such drives. In other words, the stability becomes not only a function of the frequency, but also the load angle, when speed accuracies of 0.1% are invoked, as the rate of change of load angle manifests itself as a speed change. The techniques of dealing with films are similar to those found in the metal (see Section 3.5.8) and paper (see Section 3.5.12) industries, whilst pumps (see Section 3.5.13) are subject to the common restraints of electrical systems.

3.5.4 Materials handling

Machines for moving solid objects, e.g. cranes, hoists and conveyors, have a number of common features. For proper electrical control, regeneration (or dynamic braking) must be available to deal with gravity forces, whilst there are also difficult starting conditions and adverse stall requirements.

Whilst variable speed solves some of these problems, there remain many applications where constant speed as such is acceptable, provided these problems can be obviated. The use of couplings (hydraulic or eddy current) is common, but not always necessary. A squirrel cage motor of the correct rotor design can deal with most conveyor problems, except continuous long-term stalling, where the hydraulic solution, be it a coupling or motor, gives the better service. The use of hydraulics, with its inevitable piping and leakage complications, has only rarely been widely accepted. Cranes and winches of all sorts are generally electrically driven, provided they are designed for the duty. Slipring motors with secondary resistance control are less successful for speed control, as the load-dependent torque–speed characteristic is not generally acceptable. However, drives such as crane travel motors use slipring motors for quasi-synchronized control of several wheels (electrical shaft system).

Design problems arise with the very powerful excavators or draglines where, even if they are supplied by cable, voltage regulation problems arise and the electrical system design needs special attention.

A specialized motor is the 'brake motor', where the motor is axially movable to activate braking, and where multi-speed windings are often employed.

3.5.5 Food and agriculture

This industry ranges from the processing of sugar (beet or cane) to the delicate processes of making chocolate or bottling. It is another typical process orientated application having its own special problems in the need for 'hygienic' motors, i.e. motors of particular cleanliness and designed for a minimum of environmental effect.

Even the technical problems of the processes can be considerable. The fact that a sugar campaign lasts several months and needs to be carried through without interruption is a considerable challenge to design for reliability, both for motors and controlgear. By way of contrast, the manufacture of 'Swiss rolls' represents a complex multi-motor process line, traditionally based on a multi-motor, variable frequency system, which has to cope with the somewhat problematical tensile properties of dough.

Sugar centrifuges representing an exceedingly high inertia load test the design of squirrel cage drive motors, if variable speed is ruled out. The mechanical, mounting and vibration problems of the drive are severe, whilst the squirrel cage duty needs special designs, for single- or multi-speed solutions, such as a two-speed main motor for full speed, and an auxiliary low-speed (high poleage) pony motor, 'spinning' and 'ploughing' respectively. The latter has often been obviated by a low-frequency, typically 5 Hz, auxiliary supply, which can feed a battery of centrifuges. Commutator or cyclo frequency converters are suitable, both being regenerative and therefore able to return the kinetic energy of deceleration to the mains.

3.5.6 Machine tools

The main power drives of most machine tools use fixed-speed motors and gear selection, because the duty is one of constant power. Alternating current and d.c. variable-speed drives are used for sophisticated machines, and they are normal for the NC and CNC associated movements and spindle drives, the latter being commonly special-purpose squirrel cage or synchronous motors with direct drives supplied at the appropriate high power frequency (see Figure 3.13, p. 154). Duty cycle drives such as planers require complex solutions. The drive problems are mostly associated with high-response, high-accuracy servosystems, which are outside the scope of this chapter.

3.5.7 Marine

Ships represent a particular kind of problem (see Section 3.1) owing to their limited generation capacity and critical weight and size requirements. The general use of a.c. has conferred many advantages over the traditional, pre 1940, d.c. systems. Where variable speed is essential, d.c. is still employed, but is being replaced gradually by variable frequency a.c. As a ship requires all services from domestic to propulsion, and includes the hazardous-area problems of tankers and environmental problems of deck-mounted equipment, it provides a wide range of motor applications.

Fans, pumps, compressors and similar machinery form the main loads, the requirements of which are met by suitably designed single- or two-speed squirrel cage motors with DOL starters as described in Section 3.3. The removal of heat poses particular problems in confined spaces, as does the design for good mechanical performance under conditions of externally imposed vibration with a relatively 'live' structure. There is still a background of preference for low voltage, but 6.6 kV has been used. The use of non-sparking controlgear and switchgear (generally vacuum interruptors) has aided progress materially, as has the development of surge proof, air insulated transformers.

The use of isolated neutrals has been dealt with in Section 3.1; the appropriate motor winding dielectric design is of vital importance for reliability, as the failure of power supplies on a ship can have obviously disastrous results.

Propulsion drives with many types of motors are known. The traditional d.c. variable voltage or variable current system is executed with semiconductors instead of rotary conversion equipment. Variable frequency a.c., with proposals for high-power, low-power frequency direct synchronous motor drives, are making steady progress, where the advantage and flexibility of electrical propulsion can be justified economically.

The wide application of bow and stern thrusters uses hydraulic means, either jets or variable flow, to obtain fine control for manoeuvring, including satellite automatic position control for drill ships, where a constant-speed motor of substantial power (Group IV) drives the pumps. The active rudder with variable frequency, submerged squirrel cage motor drive is another specialized requirement.

Naval applications involve further specialities, e.g. shockproofness and low vibration and noise signatures (see Section 3.3.3.6); otherwise, drive technology is in line with normal practice.

3.5.8 Metal industries

The conversion of ores to metal and the subsequent processing into semi-finished goods, ferrous and non-ferrous, strip, plate, rod, wire, beams, etc. is another highly energy-intensive industry. Virtually everything is performed electrically in the drive areas, which range from air- and gas-blowers for coke to the individually driven rolls of a runout table.

There is an internationally recognized standard for d.c. mill motors. Some of the largest d.c. motors are employed in hot steel mills, the ratings decreasing for cold- and non-ferrous mills from Group IV to Group III. Environmental problems loom large, whilst full automation, including overall computer control, has become the norm for modern plant. The widespread use of d.c. variable speed drives is being challenged by variable frequency drives.[31]

The gearless direct squirrel cage motor runout or transport table, allowing satisfactory operation with major distortions of the rolls, which are individually driven by a surface cooled, specially designed high-duty overhung floating motor, is the outcome of long experience in this field and unsurpassed in operational efficacy and reliability. Variable frequency is derived from a cycloconverter, providing a regenerative low frequency. This is clearly more expensive than, for instance, a mechanical linkage with a number of rolls driven by a standard squirrel cage motor, but is an early example of effective life-cycle costing.

3.5.9 Mineral winning

Mineral winning is carried out on a large scale worldwide to obtain the basic raw materials of engineering, e.g. metal ores, uranium, coal, ranging from very deep goldmines to vast open-cast coal pits, with all varieties and sizes of operation. All these plants have various similarities regarding their demand for ventilation, dewatering, main hoists, conveyors and the like, with generally dusty atmospheres and, in the case of underground coalmines, the presence of potentially explosive gases.

The normal underground coal-face drive worldwide is a flameproof (Ex)d Gas Group I squirrel cage motor driving coal-winning machines, conveyors, and pumps. The widespread use of this solution is attested by the dimensional standards as well as special ranges of normalized motors for British Coal. These include conveyor motors with particular electrical, especially torque–speed, characteristics and airflow fan motors for pad suspension.

With the increasing power levels at the pit face, the application of high voltage (above the British mining norm of 1100/550 V) at 3.3 kV, or 2.4/4.1 kV in North America, has become increasingly accepted. Moreover, in terms of good engineering practice, the use of vacuum switchgear and fused vacuum contactors provide two important safety features of non-sparking circuit control and fault energy limitations.

A large amount of power is also consumed in the surface processing plant and the main ventilation equipment. As mines are often laid out for successive extension, and where variable speed is not considered economic, two-speed drives are usually deployed. Direct long conveyor installations from pit to power station are usually variable speed, as are the main hoists. In order to make underground machinery more versatile and eliminate fluid-filled couplings, variable frequency drives are utilized. These require (Ex)d frequency conversion and controlgear, whilst the (Ex)d motors can be filled with inert liquid, e.g. with polyester oils, for minimum size or, more conventionally, TESAC.

3.5.10 Oil and gas industries

3.5.10.1 Offshore and onshore exploration and exploitation

The harsh environment of offshore oilrigs in the North Sea has resulted in increased attention to design assurance to achieve acceptable reliability.[34] This is important not only to safeguard continuous operation but also to minimize the high cost of repair, considering the difficulties involved in removing motors from inaccessible places and transporting them to the mainland for any major repair.

As in other industries, design assurance covers all leading electrical and mechanical aspects. Equally important, however, is the integration of the motor into the overall system, ranging over all the aspects discussed in above sections.

The use of high-power (Group IV) electrical drives may be counter-indicated on pipelines due to the abundance of natural gas. However, it is accepted generally that, where electricity has to be available, the electric motor is a better economic proposition due to its better reliability and lower maintenance costs compared with gas turbines. This is generally the solution adopted on offshore platforms, though on long transcontinental pipe systems, gas turbine drives as well as motors have their place. The speed compatibility of a gas turbine with the pump can be achieved only by a motor without gears by a variable frequency solution.

Downhole pumps with submerged motors are gaining acceptance. These represent a highly complex and special design problem, in view of the long length and small diameter of the motor design involved.

3.5.10.2 Refineries

The problems here are less onerous than offshore, but the generally hazardous atmospheres, large geographical spread and multiplicity of drives presents specific system problems. As a generalization, the common use of (Ex)e motors is a realistic approach to the environment situation, as well as offering reliability at minimum cost.

Large high-speed compressor drives, traditionally valve controlled, are being considered for fitment with frequency converters to achieve a direct gearless motor drive as well as flow and/or pressure control with minimum energy dissipation (see Section 3.3.3.2). Sophisticated process control and condition monitoring are complementary activities needing appropriate design information for the performance of the motor.

3.5.11 Power generation[35]

3.5.11.1 Fossil-fuel power stations

Large- and small-scale power generation represents a process in which the effect of failures of relatively minor pieces of equipment can escalate to a major loss of production. The design of power station auxiliary drives has always received special attention form the point of view of reliability, safety and design assurance. The concomitant increase of unit output, thermal efficiency and availability of generating plant is an excellent example of how attention to detail and early design effort has resulted in a considerable payoff.

Both coal- and oil-fired boilers require substantial auxiliary power, which in turn needs close control to achieve efficient combustion. The backbone of each of these auxiliaries is currently a squirrel cage motor up to substantial sizes (Group IV). After the wide application of variable speed a.c. commutator motors for the 500 MW unit size plants built in the 1960s, the capitalized cost of losses now is unlikely to be sufficiently great to permit the widespread use of the equivalent variable frequency drives.

The concentration is therefore on the squirrel cage motor, single- or two-speed, and its design to meet the specific duties. Generally, the most onerous requirements come from high-inertia drives resulting in difficult starting conditions with supply capacity limited in relation to the motor rating. Starting current of about $4.5 \times$ FLC and starting at 80% voltage against full torque are specified and close matching of motor to the drive and electrical supply system is necessary. The special torque requirements of coalmills have usually resulted in the interposition of hydraulic couplings without speed control facility, although squirrel cage rotors can be designed for direct drive with proper protection, particularly stall protection. For a period, submerged motors of the wet type were used to achieve glandless pump drives for boiler circulation, suction boosters, heater extraction pumps and the like (see Chapter 2).

3.5.11.2 Nuclear power stations

The development of nuclear power is closely linked with improved reliability drives with wide-ranging design assurance and special motor designs.

The current series of advanced gas-cooled reactor (AGR) power stations, with the exception of Dungeness 'B', utilize submerged circulator motors[36] in line with the original concept of the prototype Windscale AGR. This requirement dictates that the motors operate with between 1 atmosphere of air or CO_2 and up to 40 atmospheres of CO_2 directly inside the reactor gas containment. At the expense of the aero and thermodynamic design complications, a glandless circulator is achieved which has major reliability and safety connotations.

The same detailed design analysis will be applied to safety-related motors for the projected pressurized water reactors, including the reactor coolant pump motors, which remains a glanded design in line with normal US practice.

3.5.12 Printing and paper

This technology ranges from the arduous environmental conditions of papermills to the performance requirements of multi-section papermaking machines. The very wet conditions of converting logs to pulp, using high-power Group IV synchronous motors for the grinders, which need careful torque control, requires good transport

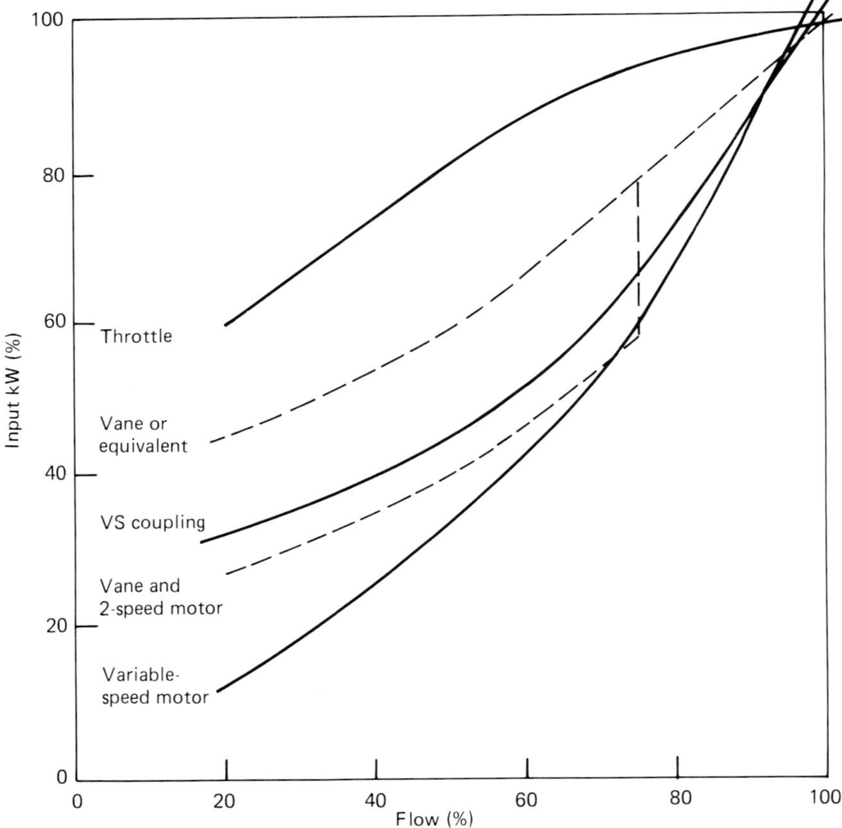

Figure 3.25 Typical inputs for different control systems for centrifugal devices (pump or fan)

technology, first feeding the logs and later pumping the pulp and refining it. The established technology of changing what is virtually a liquid into continuous strip of modest strength, is carried out by automatically controlled multi-motor, usually d.c., drives to obtain the progressive draw (tensioning) during the drying period. Similarly, printing and coating machines use the same basic methods.

The design of the motors, which are required to have fast response, needs to take account of substantial torque margins to accelerate and decelerate the relatively high inertias in addition to the high rates of change of current demanded by the thyristor armature voltage converters. The many auxiliary lines must be of comparable reliability to the machines, to ensure an economic return for these large processes.

3.5.13 Water, gas and ventilation

The transport of liquids and gases is a major electric motor activity in all output categories. The majority of applications use fixed-speed squirrel cage motors, and, where control of flow or pressure is required, hydraulic means, valves, vanes, adjustable blades, etc. There is a high energy loss situation with valves (Figure 3.25) less so with vanes or equivalent control, and the position can be further improved by two-speed motors, particularly in the higher output Groups III and IV (see Section 3.4). The use of slip couplings is a high-loss situation, whilst variable speed motors (see Chapter 4) find wide application in this field from a single controlled drive to fully automated systems (see Section 3.4.4.1).

Motor designs need to take full cognisance of the environmental conditions and the use of outdoor mounting is an example of overall system planning for economic results. The use of lubrication systems common with the driven machine provides economic and reliability advantages.

The use of 'composite' drives, where the motor is an integral part of the driven machine, has already been mentioned in Section 3.5.11 but is not confined to power station applications. Glandless drives for refrigeration compressors, operating in fluids like Freons, water pumps as well as oil pumps, gas circulators – indeed, any application where the elimination of glands has a particular advantage, can be considered.

If this design is coupled with a variable frequency supply, it is possible to eliminate gears as an additional incentive to cost reduction.

The availability of magnetic bearings[37] gives yet another possibility, that of providing a drive requiring no lubricant for the support of the shaft and having a controlled vibration–frequency response, as the automatic control system allows incorporation of variation of stiffness and damping.

3.5.14 Textiles

The processes involved in spinning, treating and weaving have always exercised the motor designer. Many processes require high-speed drives in terms of rotational speed, thus finding early application of variable frequency. The requirements for multi-motor drives to replace long shafts have always been recognized as a potential source of improved reliability, as have high rates of change of speed such as in hosiery knitting machines. A wide range of controlled and uncontrolled small- and medium- (Group I and II) size motors, ranging from commutator to synchronous motors, are employed, designed very much to fit the exact purpose.

3.5.15 Traction

3.5.15.1 Railways

Railway electrification is one of the oldest applications of electric motors and has a specialized technology of its own, e.g. the use of a single-phase high-voltage

16.66 Hz supply to suit a.c. commutator motors. The relatively slow development of mainline, as opposed to metropolitan and suburban, electrification in the UK, has now been reversed and is replacing diesel-electric locomotives. Basically, the use of d.c. overhead supplies for mainline, as opposed to third rail for metropolitan and suburban, is not an effective economic proposition even at 3000 V. It is generally accepted that high-voltage a.c., normally 25 kV, is required and this has led to successive changes. Postwar, locomotives were fitted with d.c. rectifiers, to follow the traditional d.c. techniques of series motors, with reconnections and resistance control. Variable voltage is provided with diesel-electric drives, as is tap-changing. The modern techniques[38] use variable voltage, variable frequency semiconductor converters for phase and frequency conversion and three-phase squirrel cage motors for the axle drives.

Traction motors, d.c. and a.c., require very special design features due to the highly stressed mechanical duty of an axle drive, and are outside the scope of this book.

3.5.15.2 Automotive

The battery-powered road vehicle is basically dependent on the battery performance. It has therefore found only limited practical application in spite of tremendous efforts being made, which have also included many hybrid designs. The basic choice of a d.c. series motor with stepped control has found wide application in, for instance, milk floats and some applications in vans, both not suffering from the limitations of range, as they are by their nature 'homing' vehicles. More recently, a.c. solutions with frequency converters have been proposed and one special design, the switched reluctance motor, has been developed in this connection. To obtain a reasonable steady state and transient performance and range, the minimization of motor losses is all important.

References

1. SCHWARZ, K. K. *Some aspects of reliability and safety*. Institution Electrical Engineers Conference Publication No. 10, pp. 143–149 (1965)
2. SLATER, M. A. 'High-voltage motor control', *LSE Eng. Bull.*, **13**, 2, 9–13 (1975)
3. SLATER (1975) op. cit.
4. SCHWARZ, K. K. *Transient switching overvoltages and compatibility of rotating machine insulation*. IEE *Power Engineering*, Vol. I, pp. 43–48 (1987)
5. LLOYD, M. R. and SCHWARZ, K. K. *Some economical aspects of electrical drive installations with special reference to offshore requirements*. Institution Electrical Engineers Conference Publication No. 254, pp. 237–242 (1985)
6. SCHWARZ (1965) op. cit.
7. SCHWARZ, K. K. 'The design of reliable squirrel cage rotors', *LSE Engrg Bull.*, **9**, 1, 1–10 (1967)
8. SCHWARZ (1967) op. cit.
9. RAWCLIFFE, G. M. and FONG, W. 'Speed-changing induction motors: further developments in pole–amplitude–modulation', *Proc. Inst. Elec. Engrs*, **107A**, 513–528 (1960)
10. SCHOERNER, J. *Speed-controlled industrial drives*. ICEM, pp. 40–49 (1986)
11. BONE, J. C. H. B. and SCHWARZ, K. K. 'Large a.c. motors'. *Proc. Inst. Elec. Engrs*, **120 (-R)**, 1122 (1973)
12. FURSICH, M. 'New converter-fed permanent field motor', ICEM, 1986, pp. 50–56
13. SCHWARZ (1987) op. cit.
14. LLOYD, M. R. *Transient performance of induction motors in electromechanical systems*. Institution Electrical Engineers Conference Publication No. 213, pp. 52–56 (1982)
15. NATIONAL ELECTRICAL MANUFACTURERS' ASSOCIATION. *Motors and generators*. NEMA Publication No. MG1, Section 20.48 (1985)
16. INTERNATIONAL STANDARDS ORGANIZATION ISO, Part 1. (1982)

17. *Standards quality and international competitiveness.* Cmnd 8621, HMSO (1982)
18. DEPARTMENT OF TRADE AND INDUSTRY. *Implementation of EEC directive on product liability.* DTI (1985)
19. HMSO (1982) op. cit.
20. BRITISH STANDARDS INSTITUTION *Catalogue.* BSI (1987)
21. International Electrical Committee *Catalogue of publications* (1987)
22. NATIONAL ELECTRICAL MANUFACTURERS' ASSOCIATION (1985) op. cit.
23. SCHWARZ (1965) op. cit.
24. VEREIN DEUTSCHER INGENIEURE. *Criteria for assessing mechanical vibration of machines.* VDI Publication No. 2056 (1964)
    VDI. *Criteria for assessing the state of balance of rotating rigid bodies.* VDI Publication No. 2060 (1966)
    Assessing the effects of vibration on human beings. VDI Publication No. 2075 (1963)
25. ECONOMIC INTELLIGENCE UNIT. *The North Sea and British industry: the new opportunities.* EIU (1984)
26. LLOYD AND SCHWARZ (1985) op. cit.
27. ECONOMIC INTELLIGENCE UNIT (1984) op. cit.
28. BHADBURY, B. and BASU, S. K. 'Modelling total life-cycle cost'. *Proc. Instn Mech. Engrs,* **200,** 31–40 (1986)
29. BURCH, R. H., FALLFIDE, S., FARMER, A. R., TERRY, P. F. 'An advanced method of water supply and distribution network control'. *LSE Engrg Bull.,* **13,** 3, 1–21 (1976)
30. SCHWARZ, K. K. 'Fitness for purpose – a system design concept'. 10th BPMA Conference, p. 131–139 (1987)
31. MUSIL, R. V. *Survey of large inverter-fed motor designs for variable speed drives.* ICEM, pp. 34–39 (1986)
32. MUSIL (1986) op. cit.
33. MUSIL (1986) op. cit.
34. LLOYD AND SCHWARZ (1985) op. cit.
35. BALCOMBE, R. 'UK practice and experience on motors for major auxiliary drives in power stations', 9th International Conference on Modern Power Stations, AIM, pp. 66.1–66.8 (1985)
36. SCHWARZ, K. K. 'Three generations of submerged circulator motors for AGR power stations'. *Proc. Inst. Elec Engrs,* **132A,** 234–236 (1985)
37. JACKSON, B. and DESTOMBES, Y. 'The potential application of active magnetic bearings to high-power electrical machines'. Institution Electrical Engineers Conference Publication No. 254, pp. 250–254 (1985)
38. APPUN, P. and RUNGE, W. *Three phase a.c. traction drives.* ICEM, pp. 10–20 (1986)

4 Variable-speed drives and motor control

T. A. Lipo and D. W. Novotny

University of Wisconsin, Madison, USA

4.1 Introduction

It might be said that all motoring applications of electric machinery can be considered as drive applications and it is merely a question of whether the application requires a constant speed (or nearly constant speed) or is one in which the motor operates over a continuous or discrete range of speeds. Constant or quasi-constant speed drives have been served by the family of a.c. motors including, most frequently, induction motors but also synchronous, synchronous–reluctance and permanent-magnet motors. These machines inherently operate at a constant or nearly constant speed since their speed of rotation is set by the supply frequency. Discrete numbers of speeds can be achieved by so-called pole-change windings which are obtained by switching the windings of the machine in banks or groups in order to produce different numbers of poles.

Continuously variable-speed drives, on the other hand, have traditionally been the domain of d.c. machines. The capability of variable speed is inherent in a d.c. machine since the field set up by the a.c. on the armature (rotor) remains stationary at an electrical right angle with respect to the field winding, regardless of the speed of rotor rotation, i.e. the armature frequency, owing to the action of the commutator. However, the relatively recent emergence of solid-state frequency changers, i.e. inverters, cycloconverters and the like, have permitted a.c. machinery to be operated with adjustable frequency and consequently with adjustable speed. The switches of the frequency changer effectively act as the solid-state equivalent of the mechanical commutator of the d.c. machine. Alternating current machines operating in co-ordination with adjustable frequency supplies have begun to displace d.c. motor drives in many of their traditional strongholds including electric traction, servos, spindle drives and mill drives. However, a complete transition to a.c. drives is expected to take several more decades, at least. For the present, both families of drives are expected to maintain a strong presence in the market place.

4.2 Direct current motor drives

4.2.1 The d.c. motor drive family

Direct current motor drives can be categorized in essentially two generic types categorized by their type of source as shown in Figure 4.1. The first class contains systems in which the source is d.c. Perhaps the oldest of such configurations utilize the cam controller in which the current to the d.c. motor is adjusted by means of a tapped rheostat actuated, in its traditional form, by a set of cams. When the field winding is connected in series with the armature, such systems comprise the

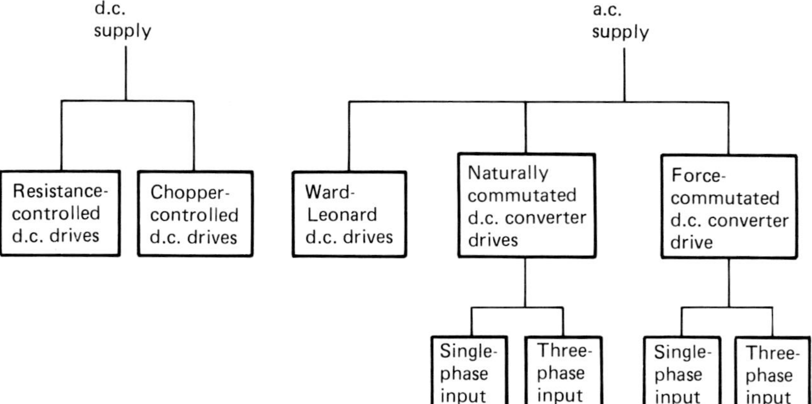

Figure 4.1 Classification of d.c. drive types

traditional traction drive for powering trams, subway cars and so forth. This type of system has been almost entirely replaced by the chopper-controlled d.c. motor in which the voltage adjustment is accomplished by means of one or more solid-state switches.

The second class of d.c. drives contains systems in which the source is a.c. The most common of these is the classical Ward-Leonard system in which the d.c. motor is supplied from a d.c. generator which is, in turn, driven by an induction or synchronous motor. However, this system has also largely been supplanted by modern solid-state systems, primarily line-commutated thyristor converter drives and drives employing a diode rectifier/chopper arrangement. In addition, force-commutated rectifiers employing internal commutating circuits, bipolar junction transistors or gate turn-off thyristors have emerged and will be used increasingly in the future.

4.2.2 Direct current motor equivalent circuit

Figure 4.2 shows the basic circuit representation of the d.c. motor, including armature and field resistances r_a and r_f, and the armature and field inductances L_a and L_f. The armature current is i_a and the field current is designated by i_f. The field current is often made proportional to armature voltage or armature current by parallel (shunt) connection or series connection of the field winding with the armature. For many applications the supply for the armature voltage v_a is separate from that of the field voltage v_f (separately excited machine). However, the effect

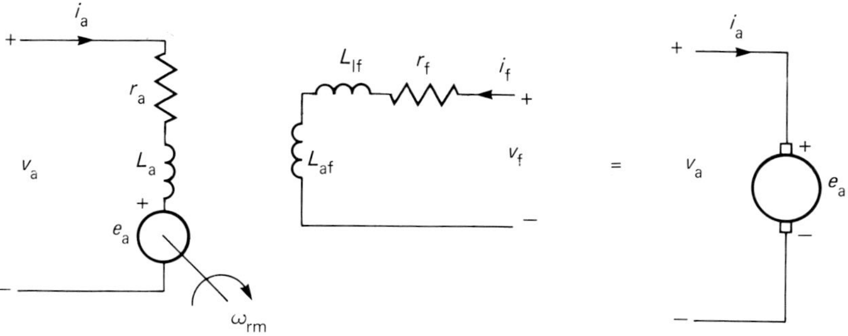

Figure 4.2 Direct current motor circuit

of a shunt connection (field winding in parallel with the armature) or series connection (field winding in series with the armature) can be accomplished by proper programming of the field current. In general, an e.m.f. e_a is induced in the armature winding by its rotation in the flux linkage λ_{af} set up by the field winding. Because this voltage opposes the flow of armature current into the motor when supplied from a voltage source, this voltage is frequently called the back e.m.f. The back e.m.f. is proportional to both the mechanical rotor speed ω_{rm} (rad/s) and field flux linking the armature λ_{af} (weber turns), i.e.:

$$e_a = k_1 \omega_{rm} \lambda_{af} \tag{4.1}$$

A typical saturation curve relating the field current to the armature voltage for several speeds is shown in Figure 4.3.

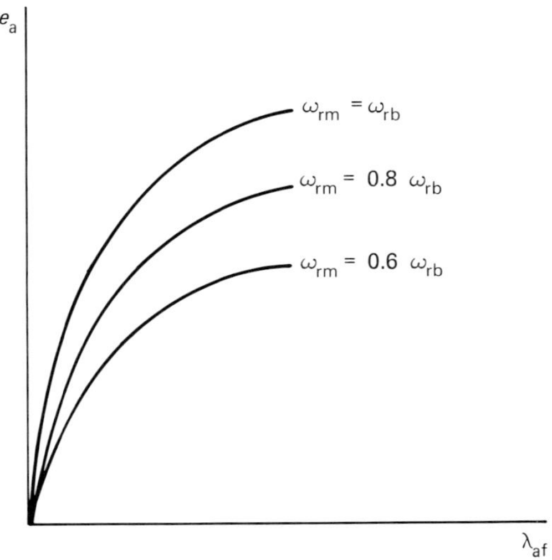

Figure 4.3 Direct current saturation curves as a function of speed

The flux produced by the field current is constrained to be at right angles to the field produced by the armature current. The torque results from the interaction of these two fields and, since they are at right angles, the torque is proportional to the produce of field flux and the armature current, so that:

$$T = k_2 \lambda_{af} i_a \tag{4.2}$$

Finally, from Figure 4.2 it is clear also that in the steady state when $di_a/dt = 0$:

$$v_a = e_a + i_a r_a \tag{4.3}$$

Hence:

$$v_a i_a = e_a i_a + i_a^2 r_a \tag{4.4}$$

Electrical input power = Mechanical output power + power dissipated in armature resistance (4.5)

Since the mechanical output power must equal the developed torque times the rotor speed:

Mechanical output power $= e_a i_a = \omega_{rm} T$ (watts) (4.6)

Note from Equation 4.6 that the constants k_1 and k_2 in Equations (4.1) and (4.2) must be equal when SI units are used.

4.2.3 Principles of speed control

From Equations (4.1) and (4.3):

$$\omega_{rm} = \frac{v_a - i_a r_a}{k_1 \lambda_{af}} \tag{4.7}$$

Two basic modes of speed control are apparent immediately from this equation. If the field flux linkage and the armature current are assumed constant in Equation (4.7) ('constant torque operation' (see Equation (4.2)), then the speed can be adjusted by changing the armature voltage. A plot of speed versus torque for a family of armature voltages is shown in Figure 4.4. Note that for a fixed armature voltage the speed is nearly constant but decreases slightly with load as a result of the armature resistance. For a fixed armature current, the speed increases in direct proportion to v_a. Operation in this mode is somewhat erroneously called the 'constant torque' mode since any value of torque can be obtained at any time and at any speed up to some maximum value, set by the armature resistance losses and/or the current capacity of the supply. This maximum value is often not constant since the maximum permitted torque at a particular speed is set by the cooling capacity of the motor and armature supply.

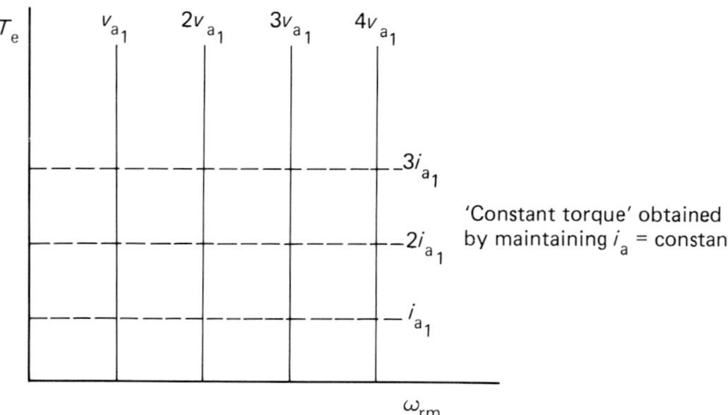

Figure 4.4 Torque vs. speed as a function of armature v_a and armature current i_a

It can be observed from Equation (4.7) that the speed can also be changed by holding the armature voltage and current constant and adjusting the field flux linkage. The speed varies inversely with field flux linkage, so then the product of torque times speed becomes constant ('constant power operation' (see Equation 4.6)). A family of torque–speed curves for changes in field flux linkages, i.e. field current, is shown in Figure 4.5. Since the field flux is reduced to increase speed, this mode of operation is also termed 'field weakening operation'. Again, the output power is not necessarily constant but can take on any value up to a prescribed maximum set by the available armature voltage. The limiting curve is a true constant power characteristic only if armature current is held constant while adjusting the field current. The constant power mode can only be extended over a limited speed range, since the flux provided by the field winding will be limited by saturation at the low end of the speed range and by the possibility of commutation failure (flashover) at the high end.

From Equation (4.7) it is apparent that if the field current is at its maximum, torque can be maintained as speed increases only by a corresponding increase in

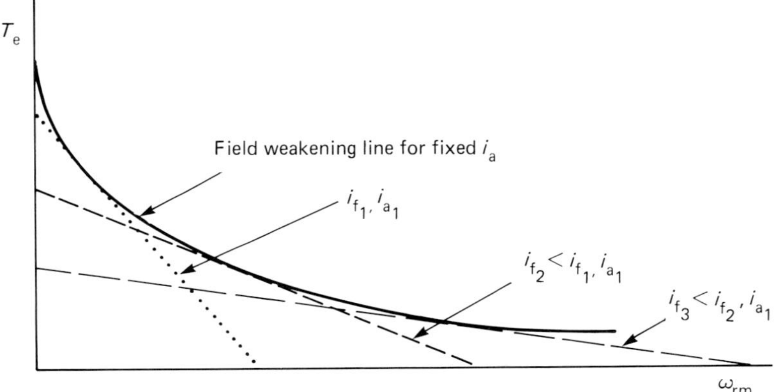

Figure 4.5 Torque–speed curves as a function of field current i_f

armature voltage. The inevitable properties of most sources limit the range of voltage that can be obtained. Hence, the constant torque mode can, in practice, only be extended to some specific speed. However, the speed can continue to increase by switching over to the constant power mode of operation. The torque–speed characteristic of Figure 4.6 results. The switchover point is usually taken as the nominal base speed ω_b. The constant power range can be extended safely by field weakening until speed ω_c. Further increase beyond this speed is possible. However, in order to avoid flashover of the commutator, the field current

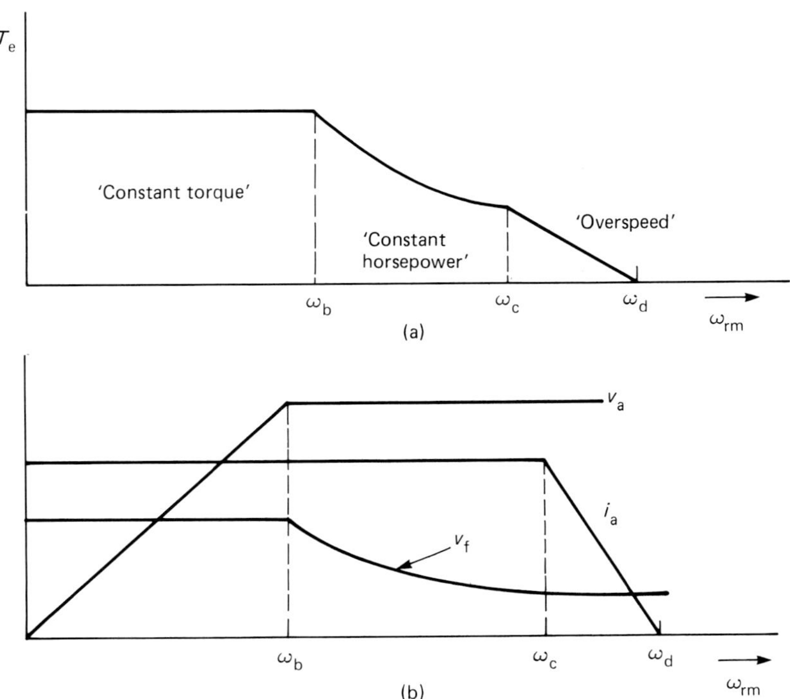

Figure 4.6 (a) Torque–speed curve illustrating constant torque and constant kilowatt regions. (b) Armature voltage v_a, field voltage v_f, and armature current i_a, corresponding to (a)

must again be held constant. The power now drops off with speed. An ultimate speed limit ω_d is reached when the counter e.m.f. $\omega_{rm}\lambda_{af}$ equals the armature voltage v_a. This mode of operation is rarely useful since the power-handling capability of the motor becomes almost nil. The maximum practical speed is limited by the mechanical stresses on the rotor commutator bars and windings or by the inevitable stray losses which effectively load the machine even without an externally loaded shaft.

4.2.4 Direct-current drives with single-phase a.c. input

A more detailed list of the types of drives with single-phase a.c. input is shown in Figure 4.7. Although such single-phase systems normally imply use with power supplies of small rating they are sometimes used for very high-power applications, e.g. with an electric locomotive when the source of power is a single-phase line at the wayside. Each of the drives are shown with an external inductance L_d for the purpose of filtering the d.c. current. Although usually required because of the high harmonic content of the d.c. output voltage, L_d is not essential and can be omitted.

The half-wave converter is the simplest type of converter. Direct-current voltage is adjusted by delaying the triggering point from the positive voltage zero crossing. The angle in electrical degrees between the instant that the thyristor would conduct if replaced by a diode and the actual instant of conduction is called the phase delay

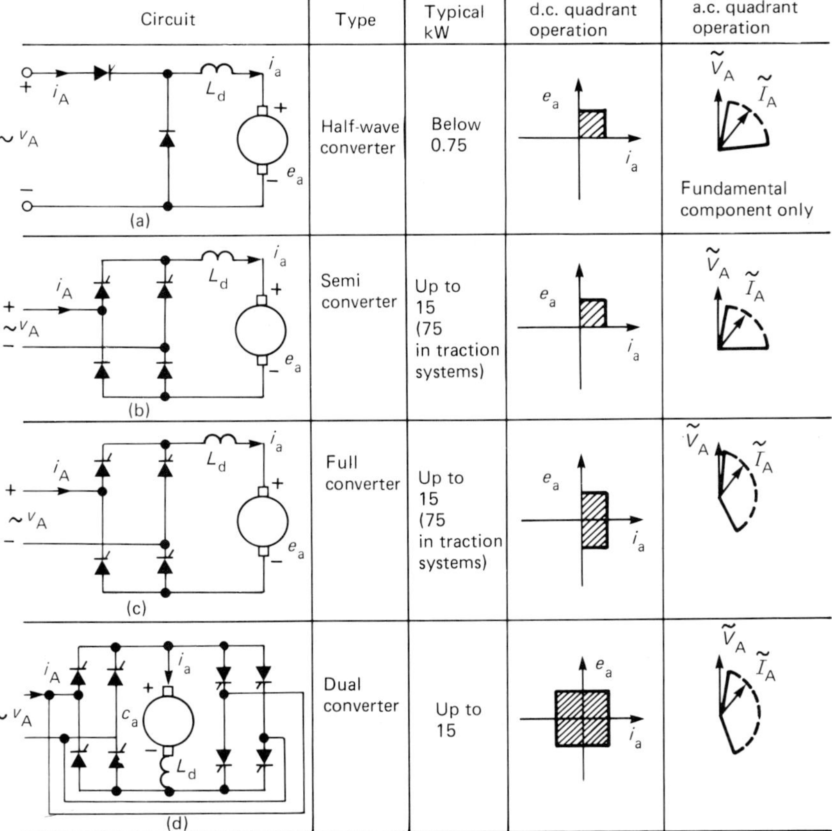

Circuit	Type	Typical kW	d.c. quadrant operation	a.c. quadrant operation
(a)	Half-wave converter	Below 0.75		Fundamental component only
(b)	Semi converter	Up to 15 (75 in traction systems)		
(c)	Full converter	Up to 15 (75 in traction systems)		
(d)	Dual converter	Up to 15		

Figure 4.7 Direct current drive types employing a single-phase a.c. input

angle α. Since only a single thyristor is used, the motor current flows from the supply in unidirectional blocks. This type of behaviour is undesirable since it requires special oversize transformers which can accommodate the unidirectional current flow. The output voltage and current take on positive values only so that power flow must be positive, i.e. into the motor, on both an average and instantaneous basis. Applications are limited to those with ratings below 0.75 kW.

The second type of converter is the single-phase semi-converter, so-called because only the top (or bottom) pair of devices are gate-controlled thyristors. Operation as a semi-converter is also possible by using thyristors in the top and bottom of one leg of the bridge. In each case the circuit behaviour of the converter is the same. Control of the d.c. output voltage is obtained by control of the instants that the two thyristors are allowed to conduct. Because the conducting thyristor together with the diode in the same leg form a short-circuit path, the voltage across the machine can never assume a negative polarity with respect to the plus and minus marks in Figure 4.7. Hence, the current on the d.c. side of the bridge circulates around a 'free-wheel' path whenever the polarity attempts to become negative, and the a.c. current becomes zero during this interval as shown in Figure 4.7. With a sufficiently large phase delay angle, the d.c. current becomes discontinuous. The voltage across the motor is then controlled only partially by the triggering of the two thyristors and the gain associated with the effectiveness of manipulating the angle α to control the d.c. output voltage V_d begins to decrease.

The output voltage, expressed as a function of α for continuous current, is:

$$V_d = \frac{1}{2\pi} \int_{\alpha}^{\pi} \sqrt{2} V_{r.m.s.} \sin \theta \, d\theta$$

$$= \frac{\sqrt{2} V_{r.m.s.}}{2\pi} (1 + \cos \alpha) \tag{4.8}$$

This function is plotted in Figure 4.8. This curve can be considered as the ideal input–output relationship for infinite inductive filtering and zero source impedance.

Figure 4.7(c) shows the fully controlled converter in which thyristors are active in all legs of the converter. Again, the phase control angle α is used to vary the output voltage by delaying the firing of the oncoming thyristors in crosswise complementary pairs. In this manner the freewheel path is avoided and, when α is sufficiently large, the polarity of the d.c. output voltage and, hence, power flow, can be reversed. The d.c. output voltage expressed in terms of α is:

$$V_d = \frac{1}{\pi} \int_{\alpha}^{\pi+\alpha} \sqrt{2} V_{r.m.s.} \sin \theta \, d\theta$$

$$= \frac{\sqrt{2} V_{r.m.s.}}{\pi} \cos \alpha \tag{4.9}$$

A plot of the d.c. output voltage versus α for the case of $L_d \to \infty$ is shown in Figure 4.8. Note that the average d.c. voltage reverses when $\alpha > 90°$. Since the d.c. current is unidirectional, the power flow also reverses at this point and the bridge transits from rectification to inversion. If the bridge operates primarily in the inversion mode, it is useful to define the angle of advance $\gamma = 90° - \alpha$. In practice, the maximum voltage which can be achieved during inversion is somewhat less than the maximum voltage obtainable during rectification. If γ is made sufficiently small the

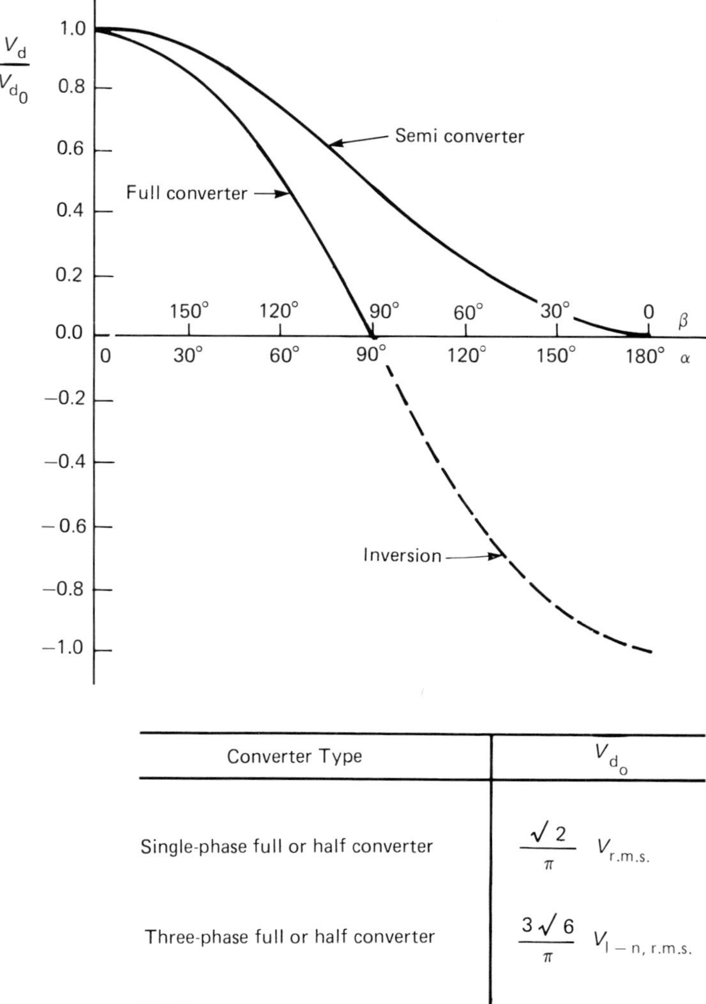

Converter Type	V_{d_o}
Single-phase full or half converter	$\dfrac{\sqrt{2}}{\pi}\, V_{\text{r.m.s.}}$
Three-phase full or half converter	$\dfrac{3\sqrt{6}}{\pi}\, V_{l-n,\,\text{r.m.s.}}$

Figure 4.8 Direct current output voltage of half- and fully controlled bridges as a function of the phase delay α

voltage across the thyristor will not be negative for a sufficiently long period for the device to recover its forward voltage blocking ability. The device will then turn on immediately when forward voltage again appears across the device. In this case a shoot-through is said to occur. Since both thyristors of a given leg of the bridge conduct under such conditions, the d.c. output becomes shorted and large d.c. currents can flow if the d.c. side of the bridge is connected to an e.m.f. source.

The presence of inductance on the d.c. side is generally beneficial to reducing not only currents due to shoot-through but also to limit ripple currents on the d.c. side of the bridge. However, it also tends to reduce the transient response of the drive. Use of inductance on the a.c. side has both positive and negative aspects as well. This inductance consists of mainly that of the input transformer but could also include a separate discrete inductor. Resistance is present also but typically not to an extent to affect behaviour of the bridge.

The process of transferring current from thyristor to thyristor is called commutation. Although the current ideally transfers instantly from offgoing to ongoing thyristor, this process is achieved in finite time in the presence of d.c. line inductance. During the current transfer, the supply is effectively shorted and the d.c. output voltage as well as the a.c. voltage across the bridge input becomes zero.

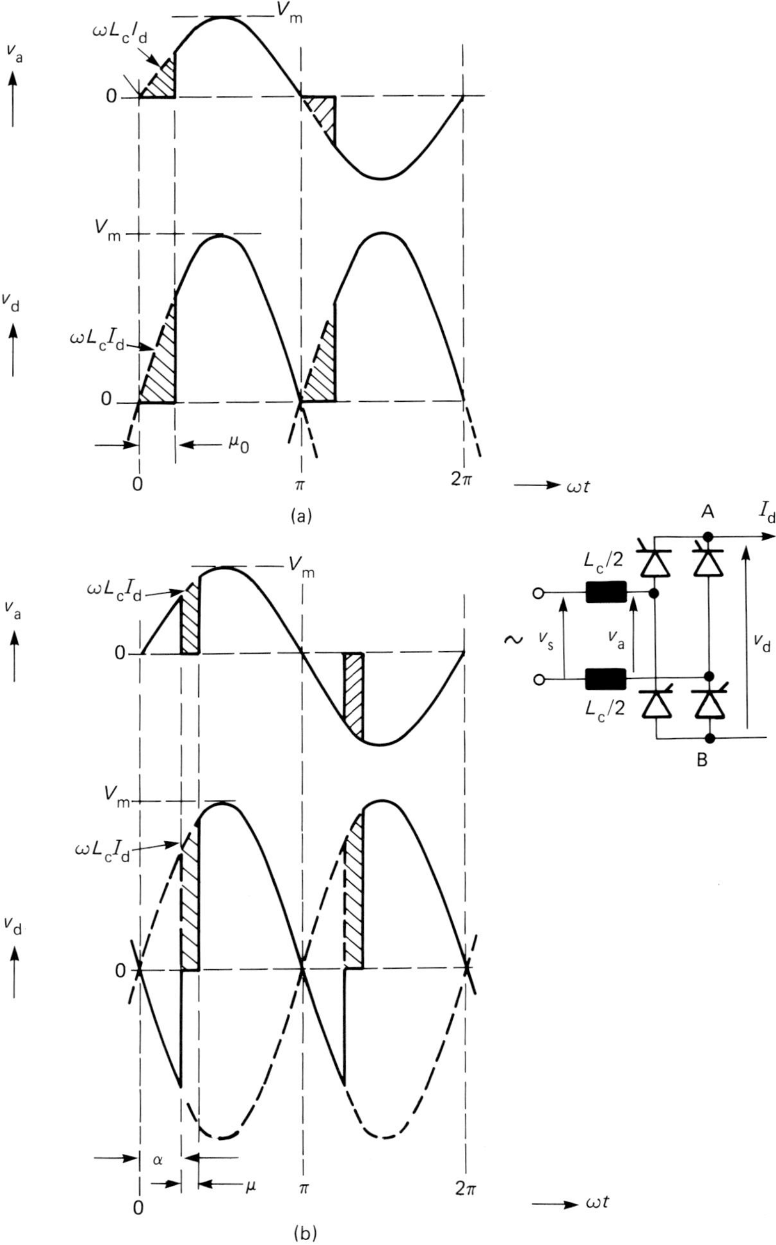

Figure 4.9 Line voltage notching due to commutation overlap. (a) Single-phase diode bridge. (b) Single-phase thyristor bridge

The effect of shorting on the a.c. side is called line notching and is illustrated in Figure 4.9. The effect of shorting on the d.c. side produces an effective drop in the d.c. voltage obtained for a given load current compared to Equation (4.9). The mean voltage drop can be calculated as the 'notch' taken out of the ideal waveform so that:

$$\Delta V_d = \frac{1}{\pi} \int_{\alpha}^{\alpha+\mu} v_d \, d\theta \tag{4.10}$$

where μ is called the overlap angle or commutation angle, Equation (4.10) is evaluated as:

$$\Delta V_d = \frac{1}{\pi} \omega L_c \left[i_a(\alpha+\mu) - i_a(\alpha) \right] \tag{4.11}$$

where $i_a(\alpha)$ is the a.c. line current evaluated at $\omega t = \alpha$ and ω is the a.c. line angular frequency. Since $i_a(\alpha + \mu) = -i_a(\alpha) = I_d$ the d.c. current, Equation (4.11) becomes finally:

$$\Delta V_d = \frac{2}{\pi} \omega L_c I_d \tag{4.12}$$

Note that, while the voltage drop results from inductance, the term appears to have the effect of a resistive drop on the d.c. side. Equation (4.12) suggests the d.c. equivalent circuit of Figure 4.10 which can be used for the analysis of both single- and three-phase fully controlled thyristor bridge circuits. As a result of commutating inductance, the load V–I curves for a bridge have drooping

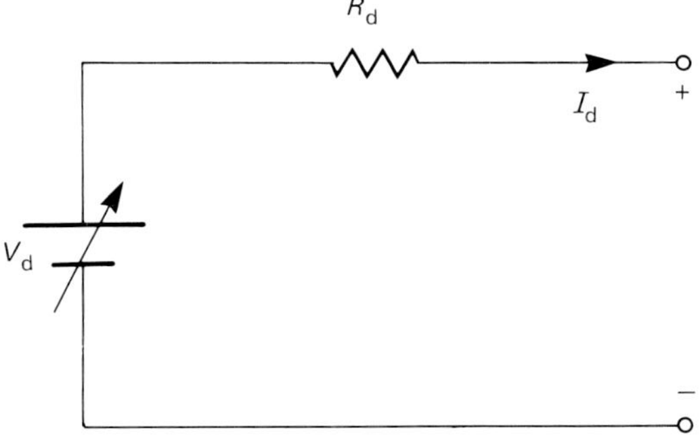

Converter type	R_d	V_d	V_{d0}
Single-phase half bridge	$\dfrac{2}{\pi} \omega L_c$	$\dfrac{V_{d0}}{2}(1 + \cos \alpha)$	$\dfrac{\sqrt{2}\, V_{\text{r.m.s.}}}{\pi}$
Single-phase full bridge	$\dfrac{2}{\pi} \omega L_c$	$V_{d0} \cos \alpha$	$\dfrac{\sqrt{2}\, V_{\text{r.m.s.}}}{\pi}$
Three-phase half bridge	$\dfrac{3}{\pi} \omega L_c$	$\dfrac{V_{d0}}{2}(1 + \cos \alpha)$	$3\sqrt{6}\, V_{\text{l-n r.m.s.}}$
Three-phase full bridge	$\dfrac{3}{\pi} \omega L_c$	$V_{d0} \cos \alpha$	$\dfrac{3\sqrt{6}\, V_{\text{l-n r.m.s.}}}{\pi}$

Figure 4.10 Direct current circuit of half- and fully controlled thyristor bridges

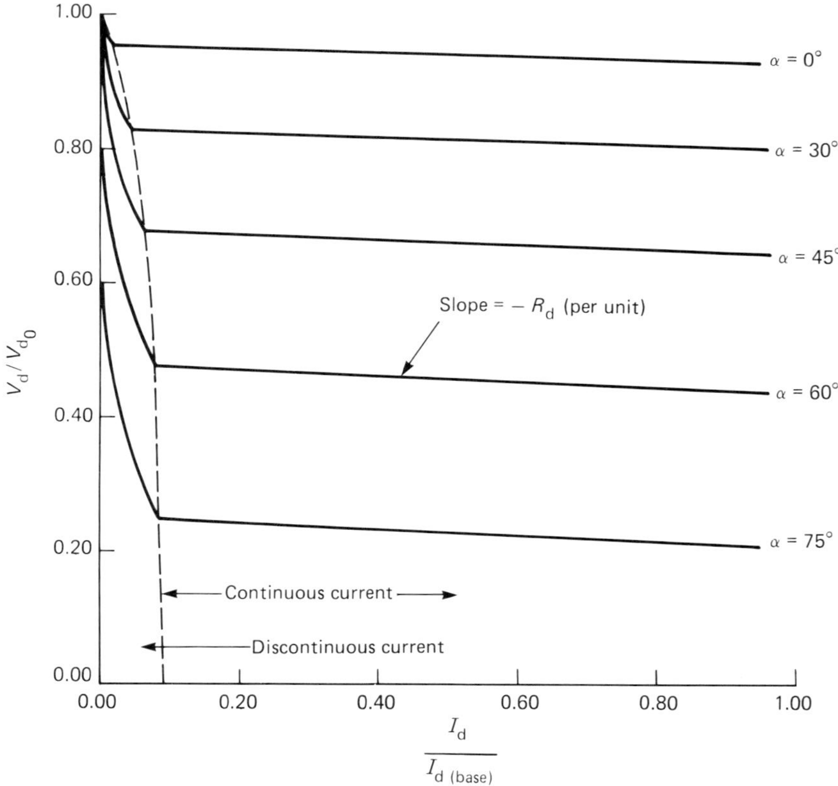

Figure 4.11 Normalized V–I curves resulting from d.c. equivalent circuit showing effects of discontinuous current and equivalent resistance due to commutation overlap

characteristics even though the resistances on the d.c. and a.c. sides are neglected. Figure 4.11 shows normalized V–I curves in which the effective resistance is taken as 0.1 per unit. Figure 4.11 also shows the effect of finite d.c. filter inductance. It is clear that, as filter inductance decreases, the pulsating currents enlarge the region of discontinuity thereby making the control problem even more difficult. A large filter inductance is used typically to make the region sufficiently small for the intended application. Operation in the discontinuous zone calls for a special control to compensate for the decreased gain.

While reversal of d.c. voltage is possible with the fully controlled bridge, power flow cannot be reversed (regenerative or braking operation) unless the polarity of the d.c. motor e.m.f. and, hence, the field current is reversed. Speed reversal is also possible if the field current is reversed. However, the long time constant of the field makes such an approach suitable only for low-cost or moderate-response systems. In order to permit true fast response, four-quadrant operation as a motor or generator in either rotational direction requires that both polarities of the voltage and current be available. Figure 4.7(d) shows the dual converter arrangement in which the reverse polarity is obtained by a reverse-connected bridge. As may be expected, substantial care is required to prevent both bridges from being turned on at the same instant, lest a short-circuit occur. Changeover from one polarity to another requires that the current be brought to zero and remain at zero until all of the previously conducting thyristors acquire their blocking ability. The discontinuous behaviour of the d.c. current near zero d.c. voltage and the associated small current values makes direct sensing of the current somewhat difficult. Determination of the zero current condition is frequently accomplished by sensing the voltages across the offgoing thyristors. If the voltage of all thyristors exceeds some specified amount then the bridge is assumed to be in the offstate and the ongoing bridge is enabled.

A plot showing performance characteristics for the half- and fully controlled bridge is shown in Figure 4.12, assuming perfect filtering by the d.c. link inductor ($L_d \to \infty$). Both bridges contain only odd harmonics, only the first two of which are shown. While the harmonic content of the fully controlled bridge is independent of V_d or control angle α, the half-controlled bridge harmonics vary with output voltage. The harmonic amplitude is always equal to or less than the harmonics of the fully controlled bridge. The half-controlled bridge generally shows better performance over the fully-controlled bridge in terms of displacement factor, harmonic distortion factor and power factor. For this reason, the fully-controlled bridge is frequently operated in a half-controlled mode by continuously gating half of the thyristors (top, bottom of one of the bridges). Since the current circulates

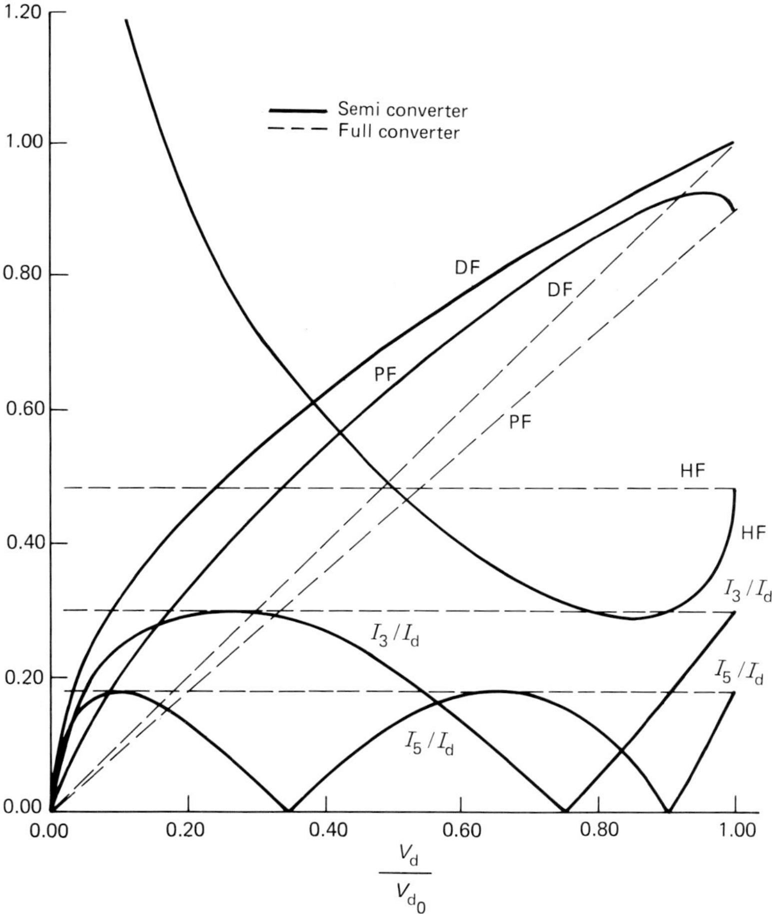

Figure 4.12 Performance characteristics for a single-phase semi-converter and full converter assuming constant d.c. link current (DF = displacement factor; HF = harmonic distortion factor; and PF = DF · HF = power factor)

alternately through the two fully-gated thyristors, control of the current can be regained by simply retarding their firing pulses to be in step with the normally gated thyristors. Reversal of d.c. voltage polarity, if so desired, can then commence.

4.2.5 Sequence control using a semiconverter

Figure 4.13 shows a scheme which has been used extensively in electric locomotive applications employing a high-voltage, single-phase catenary. In this case a number of simple diode bridges are connected in series with a single half-controlled bridge. Although two such bridges are shown in Figure 4.13, any number is possible.

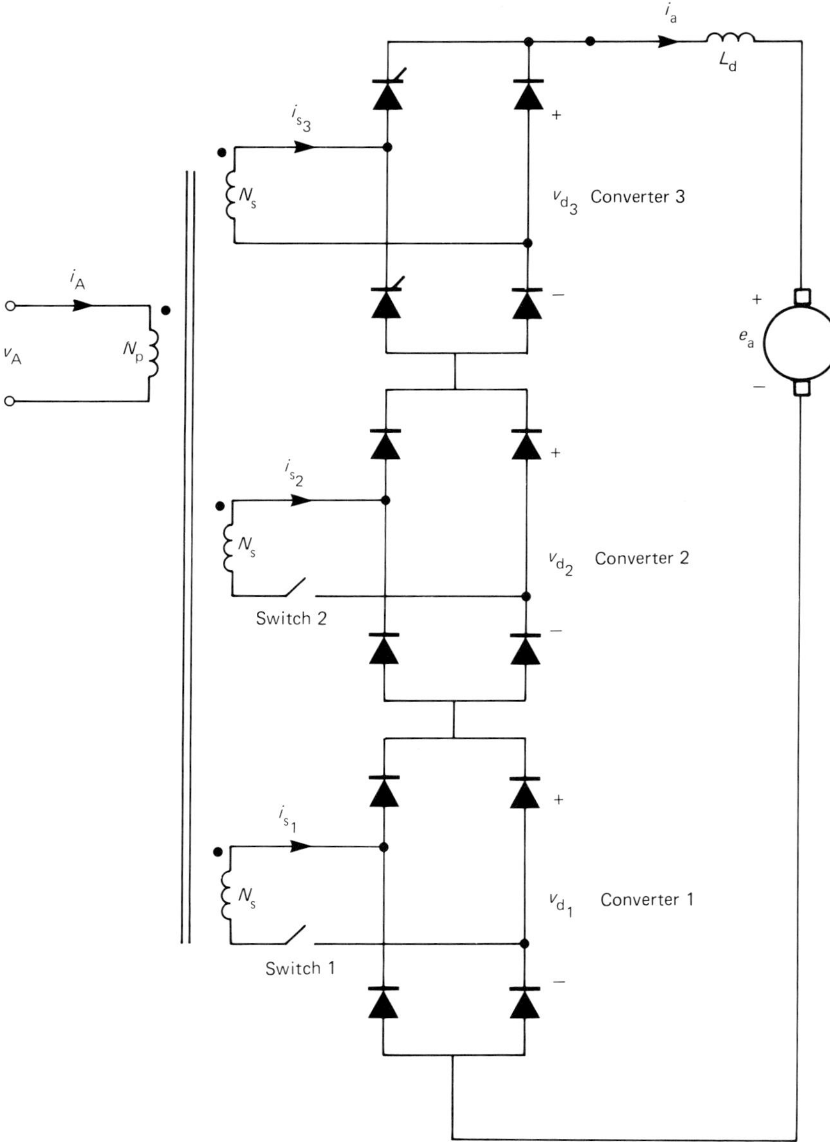

Figure 4.13 Sequence control arrangement using multiple diode bridges and a single half-controlled bridge

Initially, at standstill the d.c. motor e.m.f. is zero and the voltage requirement from the bridge network is small. Switches 1 and 2 are opened and the d.c. output voltage appears only across the half-controlled bridge. As the phase control angle α decreases from π radians, the voltage across the motor begins to increase. The diode bridges act simply as free-wheeling short-circuit paths for the d.c. current. When α approaches zero and the fully-controlled bridge reaches its maximum, switch 1 is closed and the control angle of the half-bridge simultaneously switched back to $\alpha = \pi$. Voltage can now again be increased by reducing the control angle α.

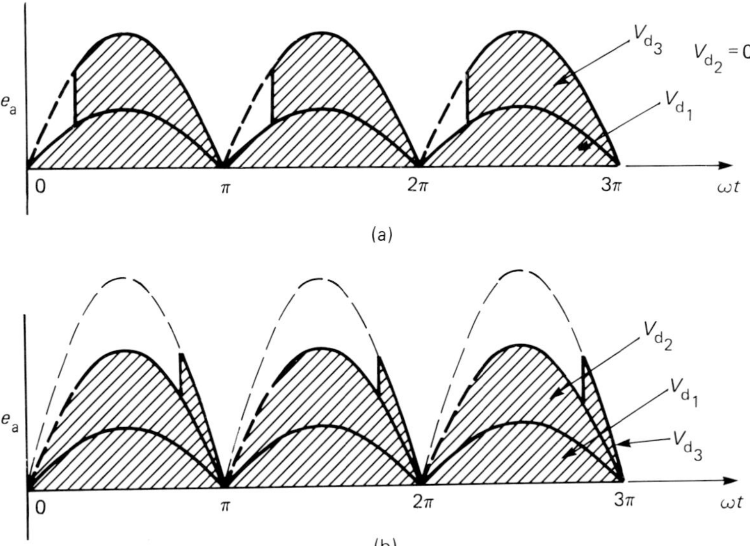

Figure 4.14 Waveforms of sequence controller showing output voltage. (a) Just before switching-in a diode bridge. (b) Just after switching-in a diode bridge

Figure 4.14 shows the two conditions: (1) when switch 1 is closed, switch 2 is opened and α is approaching zero (Figure 4.14(a)); and (2) at a slightly higher output voltage when switches 1 and 2 have been closed and α again retarded to a large value (Figure 4.14(b)). Because the two diode bridges operate at near unity displacement factor the overall power factor of the composite system at high voltages is improved. Figure 4.15 shows the displacement factor and power factor for sequence controllers employing one- and two-diode bridges.

4.2.6 Direct-current drives with three-phase a.c. input

Figure 4.16 gives a tabulation of the most popular varieties of drive configurations which utilize a three-phase supply. The half-bridge configuration of Figure 4.16(a) is rarely used since, again, the current on the a.c. side of the converter is pulsating unidirectional current rather than a.c. The half-controlled bridge of Figure 4.16(b) provides a path for reverse current flow in the a.c. lines and is therefore more practical. Again, the d.c. voltage is always positive or zero since the freewheeling action of the three diodes prevents negative voltage from appearing across the output terminals. The circuit is limited to motoring operation.

The reference point for the angle α is again chosen as the point of earliest possible conduction, e.g. in this case when line-to-line voltages v_{ca} and v_{ab} intersect for thyristor 1. Although the initial reference time is different for all three

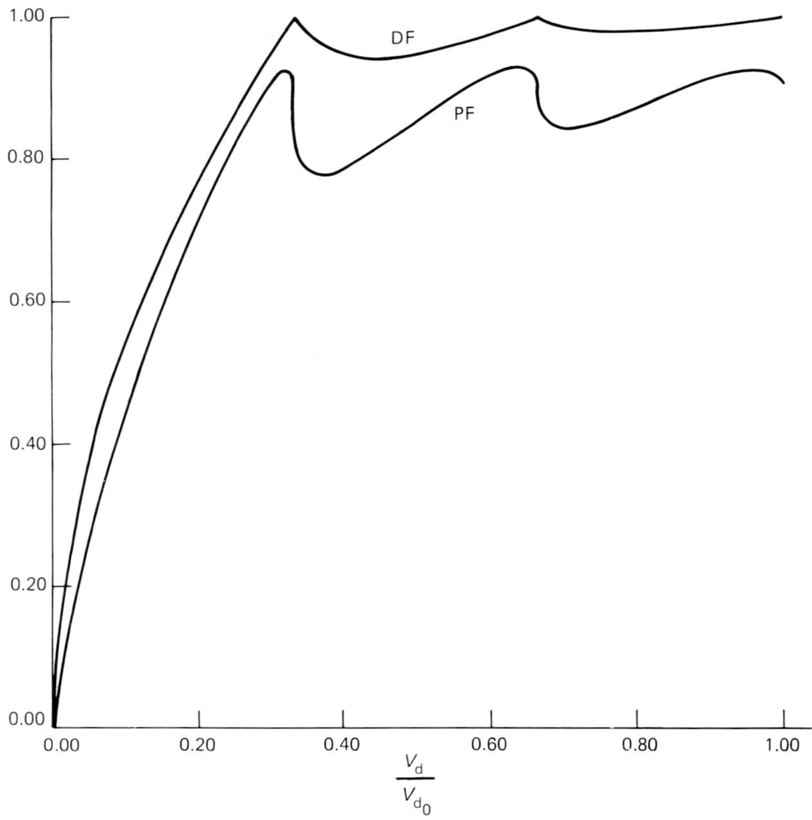

Figure 4.15 Displacement factor and power factor of sequence controller as a function of output voltage employing two diode bridges

thyristors, symmetrical gating results in the same α for each thyristor. Since a given thyristor conducts for 120°, the d.c. average voltage can be calculated as:

$$V_d = \frac{3}{2\pi} \int_{\alpha+\pi/3}^{\alpha+\pi} \sqrt{2}V_{ll-r.m.s.} \sin\theta\, d\theta$$

$$= \frac{3\sqrt{2}V_{ll-r.m.s.}}{2\pi}(1 + \cos\alpha) \tag{4.13}$$

$$= \frac{3\sqrt{6}V_{ln-r.m.s.}}{2\pi}(1 + \cos\alpha) \tag{4.14}$$

This result is very similar to the half-controlled single-phase bridge. When the single-phase bridge is connected across the same line as the three-phase bridge, the ratio of the maximum open-circuit d.c. voltage of the three-phase bridge to that of the single-phase bridge is:

$$\frac{\dfrac{3\sqrt{2}}{2\pi}}{\dfrac{\sqrt{2}}{\pi}} = 3/2$$

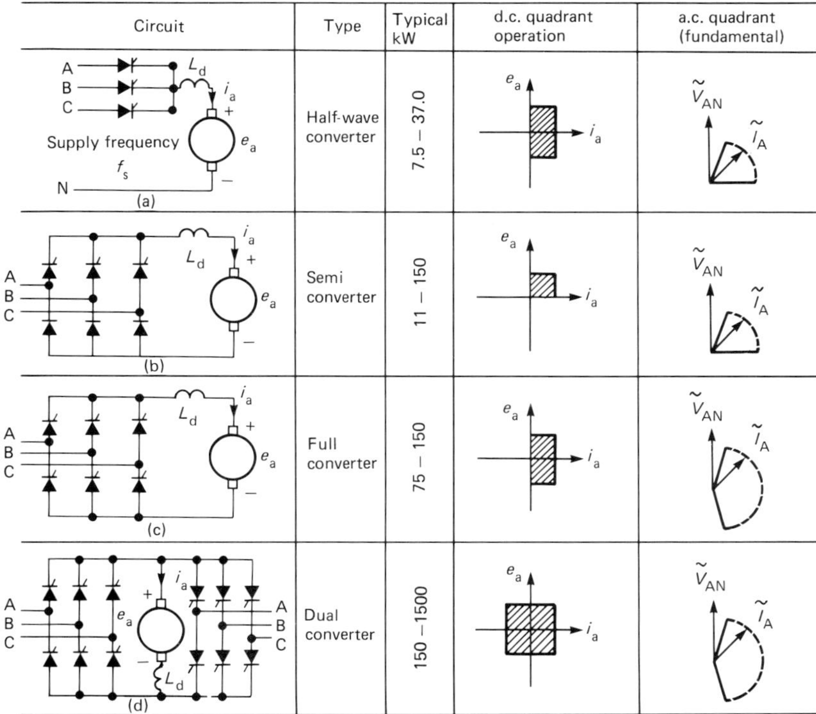

Circuit	Type	Typical kW	d.c. quadrant operation	a.c. quadrant (fundamental)
(a)	Half-wave converter	7.5 – 37.0		
(b)	Semi converter	11 – 150		
(c)	Full converter	75 – 150		
(d)	Dual converter	150 –1500		

Figure 4.16 Direct current types employing a three-phase a.c. input

Hence, while the three-phase bridge uses 50% more devices than the single-phase bridge, it is also capable of 50% more voltage.

The positive portions of the a.c. current are controlled by the angle α, while the negative portions are uncontrolled owing to the three diodes, resulting in an unsymmetric waveform (even harmonics). Plots of the displacement factor, harmonic distortion factor and power factor, as well as the first few harmonics of the half-controlled bridge, are given in Figure 4.17.

In Figure 4.16(c) is shown the full-wave three-phase equivalent of the single-phase fully-controlled bridge. Since both positive and negative voltages can be obtained, the system is capable of both motoring and regenerating with reversal of the field current. The thyristors of the full bridge are delayed symmetrically with respect to both positive and negative line-to-line voltage intersections. Since a given pair of thyristors conducts for only 60°, the average d.c. voltage is calculated as:

$$V_\mathrm{d} = \frac{3}{\pi} \int_{\alpha+\pi/3}^{\alpha+2\pi/3} \sqrt{2}V_{\mathrm{ll-r.m.s.}} \sin\theta\, d\theta$$

$$= \frac{3\sqrt{2}V_{\mathrm{ln-r.m.s.}}}{\pi} \cos\alpha \tag{4.15}$$

$$= \frac{3\sqrt{6}V_{\mathrm{ln-r.m.s.}}}{\pi} \cos\alpha \tag{4.16}$$

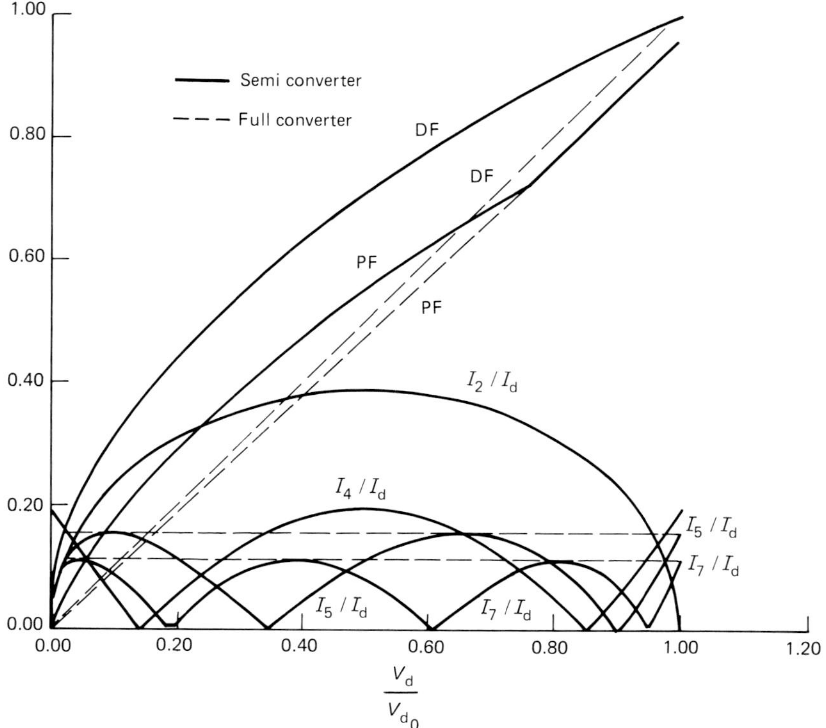

Figure 4.17 Performance curves of three-phase half-controlled and fully controlled three-phase bridges assuming constant d.c. output current (DF = displacement factor; PF = power factor)

A plot of the output voltage as a function of α for both the full- and semi-controlled converter is shown in Figure 4.8.

The line notching discussed in Section 4.4 also occurs with the three-phase connection. Typical a.c. and d.c. voltage waveforms for $\alpha = 15°$ and with 15° overlap are illustrated in Figure 4.18. It can be seen that voltage zeros occur again twice per cycle and additional voltage jumps appear in the waveform due to commutations between thyristor pairs in other lines. While the two a.c. lines become shorted during commutation, the bridge output is not shorted until commutation overlap reaches 60°. This condition occurs only with abnormally large values of commutating inductance or during severe current overloads. However, a 'notch' is taken out of the waveform which again reduces the effective output voltage.

In the case of the three-phase full bridge, it can be shown that the 'effective resistance' due to commutation is:

$$\Delta V_d = \frac{3}{\pi}\, \omega L_c I_d \tag{4.17}$$

Hence, the equivalent circuit of Figure 4.10 also applies for the three-phase full bridge with proper interpretation of the open-circuit voltage V_{d_0} and resistance R_d. Commutation overlap against causes a droop in the $V\text{–}I$ characteristic, as shown in Figure 4.19. However, the simple linear droop and the equivalent circuit of Figure 4.10 becomes modified when the commutation delay reaches 60°. A third mode occurs when the overlap angle μ reaches 120°. The reader is referred to more

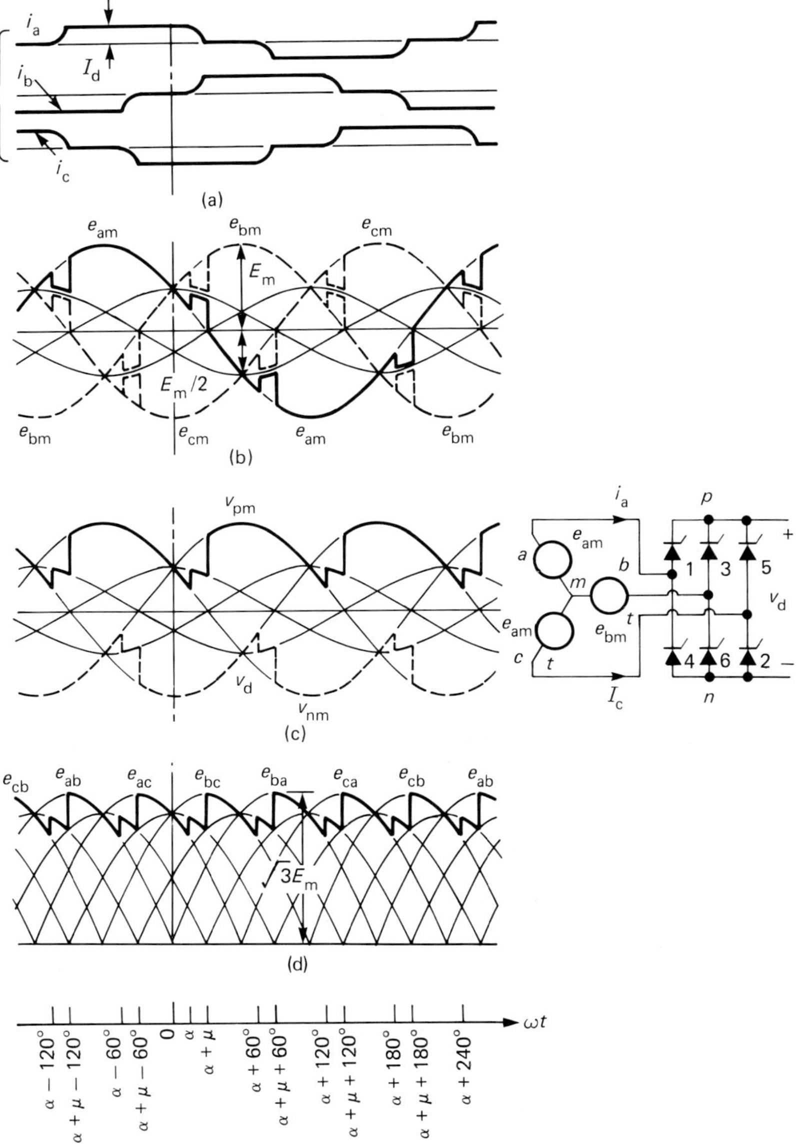

Figure 4.18 Illustrating line notching with a three-phase fully controlled bridge for the case $\alpha = 15°$, $\mu = 15°$

advanced texts for further discussion of these modes. Plots of displacement factor, power factor and dominant current harmonics are given in Figure 4.17.

Polarity reversal of both d.c. output current and voltage are achieved with a pair of fully controlled converters connected in inverse parallel (Figure 4.16(d)). This connection is popular for fast-response two-quadrant drives in which the speed is unidirectional, but the machine must make rapid transitions between motoring and braking (regeneration).

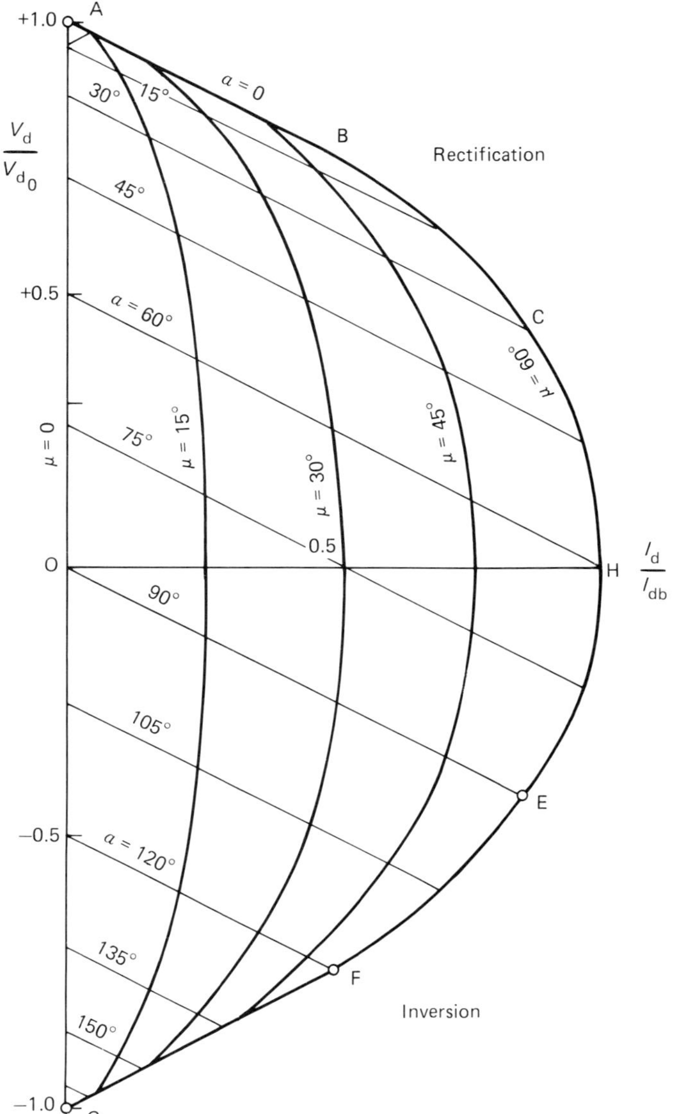

Figure 4.19 Average d.c. output voltage as a function of average d.c. current for the three-phase bridge connection

4.2.7 Useful three-phase converter variants

The number of converter variants for the simple single- and three-phase connections is regrettably enormous and no comprehensive compilation can be attempted here. However, mounting interest in decreasing harmonic distortion in the line and increasing power factor has resulted in two variations of the basic three-phase bridge becoming increasingly popular. These two schemes are shown in Figure 4.20. The first converter (Figure 4.20(a)) uses a free-wheeling thyristor in parallel with the bridge. The extra thyristor is fully gated-on during the rectifying

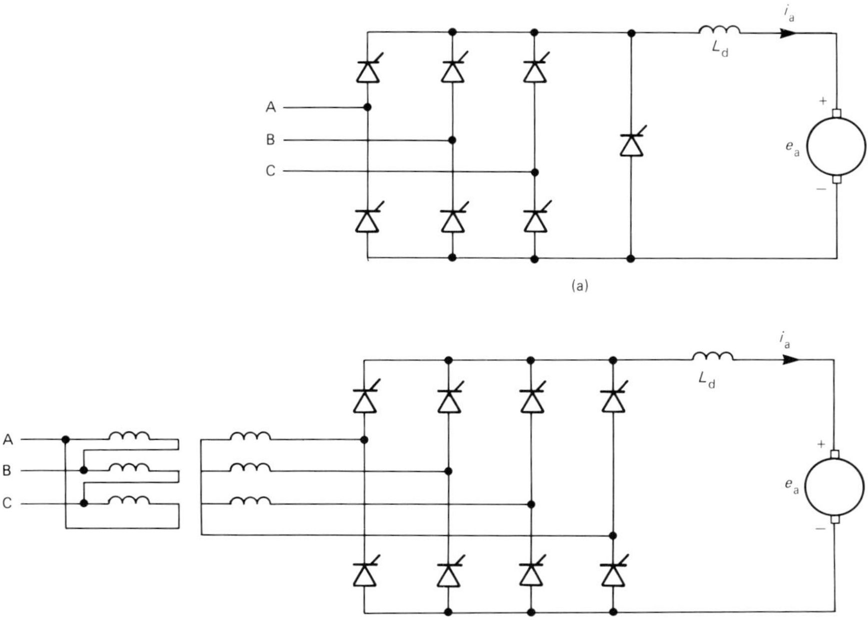

Figure 4.20 Two useful three-phase bridge variants

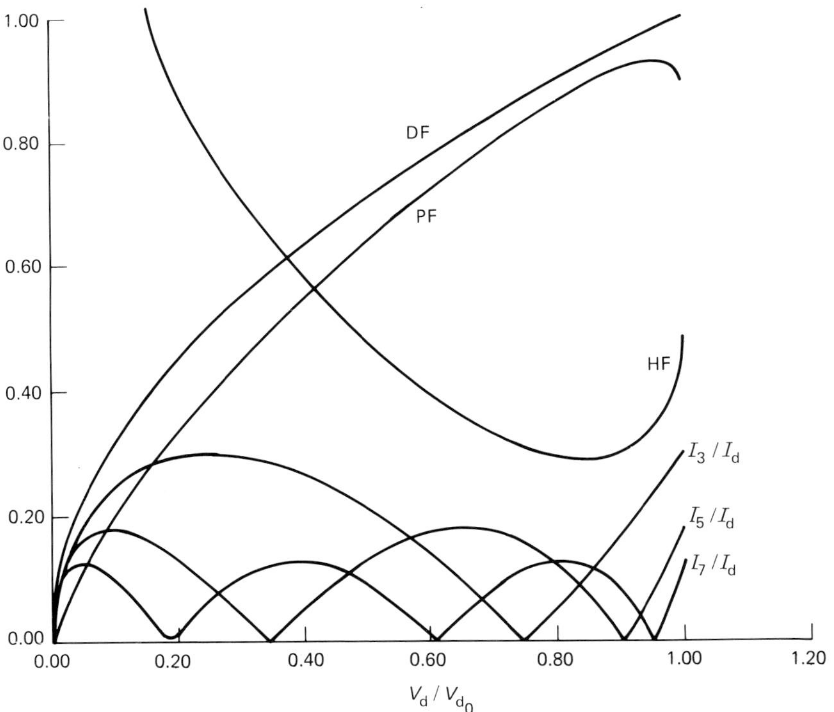

Figure 4.21 Performance characteristics of a four-legged bridge (see Figure 4.12 for symbols)

operation and operates as a diode. This thyristor provides a free-wheeling path for motor currents when the control angle α exceeds 60°. The a.c. currents remain symmetrical (no even harmonics) and the power factor is improved compared to the standard fully controlled bridge. When a command to invert is issued, complete control of the bridge is regained by simply inhibiting the gate pulses to the free-wheeling thyristor. Plots of displacement factor, harmonic distortion factor, power factor and the fifth and seventh harmonic components are shown in Figure 4.21 and can be compared with Figure 4.17. As an alternative to this concept the conventional full bridge can be utilized in a free-wheeling mode by proper programming of the thyristors.

The second connection (Figure 4.20(b)) uses a 'four-leg' bridge so that current is permitted to flow from the neutral of the associated transformer. The four-leg converter again provides a free-wheeling path which can now be controlled during the inverting as well as rectifying mode. A set of performance characteristics are also included in Figure 4.21.

4.2.8 Speed-reversing configurations

In a reversing drive, control of speed is required in both the forward and reverse directions. In some applications, speed reversal is required only occasionally, as with a tram or subway car. In other situations, continuously reversing speed is desired, e.g. in a hot strip mill. The Ward-Leonard system has, until recently, been the favoured scheme since it provides inherent regeneration capability. However, the expense associated with the induction motor and d.c. generator and overall low efficiency has provided the impetus for faster, more efficient, thyristor converters.

Figure 4.22 shows a family of three-phase converter configurations suitable for regenerative reversing drives. The basic principle is either to reverse the polarity of the motor e.m.f. with respect to the inverter voltage (field current reversal) or to reverse the polarity of the motor current (armature current reversal) as speed reverses.

Since the power supplied to the field is only a fraction of the armature power rating, the field reversal method is the system of choice if cost is a major factor. The field current can be reversed either with a set of contactors or with a dual-converter (Figures 4.22(a) and (b)). Single-phase converters are generally used since the large field time constant provides more than adequate filtering to avoid discontinuous currents. The current is brought to zero as quickly as possible by inversion and the contactors switched or the reverse polarity bridge enabled. Because of the large field time constant (frequently several seconds) the converters are designed to be operated with a much higher voltage than that required for normal steady-state operation. Direct-current voltages of the order of 2–5 times normal can then be applied to rapidly reverse the field current by the technique of field forcing. The approach remains relatively slow, even with field forcing, since the process of reducing the current to zero, detecting the condition, operating the contactors (or enable pulses) and then increasing the current in the other direction takes in the order of a half second or more.

With the armature current reversal method, the field current is kept constant during the speed reversal and armature current is reversed. This can be accomplished much faster, since the time constant associated with the armature is almost 2 orders of magnitude smaller than the field. For relatively low-cost systems, armature current reversal is obtained by pairs of forward and reverse contactors (Figure 4.22(c)). However, armature current must again be brought to zero, the contacts opened, the bridge set into the invert mode to match reversed motor e.m.f., the contacts closed and the current finally increased in the opposite direction. Hence, the system is only capable of moderately fast response.

The problems concerned with contactors can be avoided by use of the

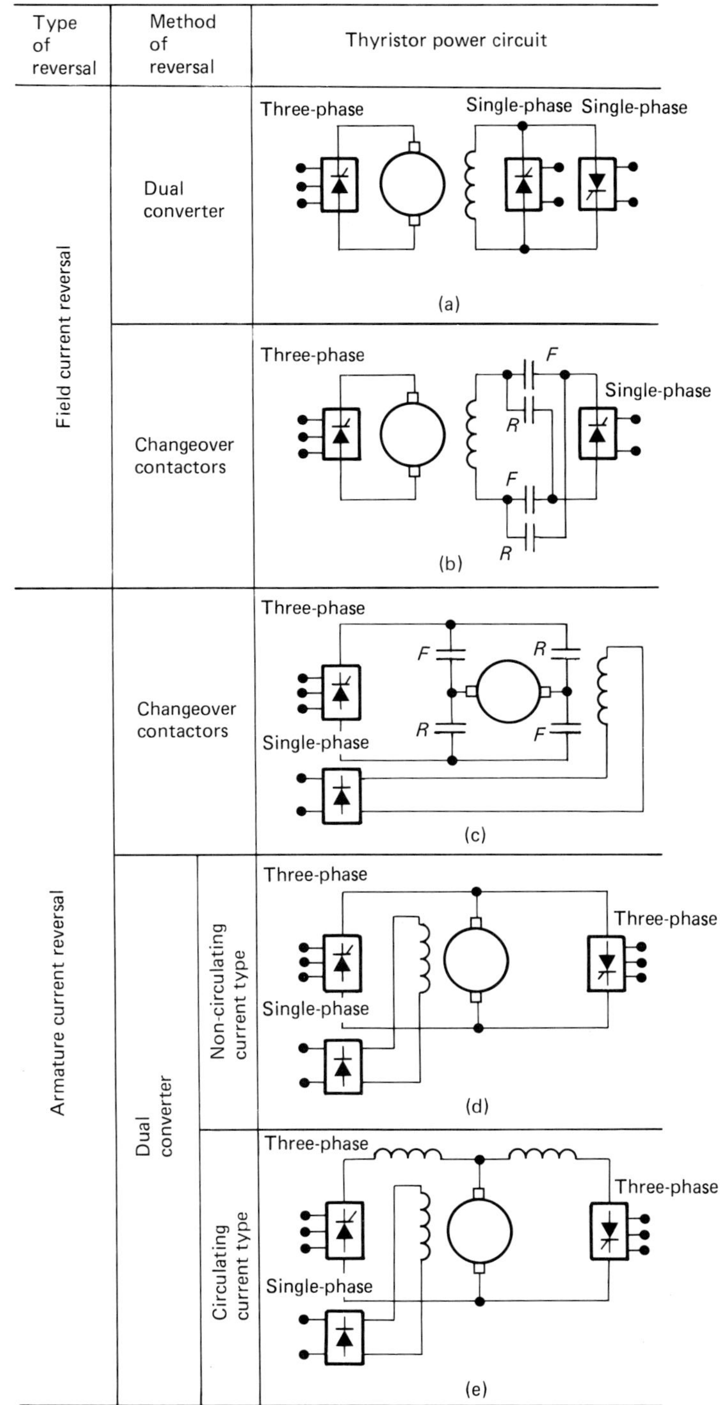

Figure 4.22 Speed-reversing configurations

double-bridge arrangement (Figure 4.22(d)). In principle, it would be useful to have both converters enabled and let the armature current naturally find a path through the positive to the negative converter when the command to reverse (or decrease) the speed is issued. However, while the mean voltages of the bridges may be set equal, the ripple components do not cancel and these components of the d.c. voltage allow large currents to circulate around the two bridges. One solution is to control the bridges so that only one bridge is active at a time. Zero current detection is again used to determine when the reverse polarity bridge should be enabled. The time involved in this process is of the order of 10 ms, so a very rapid transition is possible.

An alternative solution to the circulating current problem is to insert a d.c. reactor in series with the circulating current to limit the ripple current to a reasonable value (Figure 4.22(e)). Because one converter operates continuously in the rectifying mode and the other in the inverting mode, reversal of armature current is very rapid. Although this system is inherently faster than the double bridge without the reactor, the difference is slight and the added cost of the coupled reactor usually makes this system uneconomical.

4.2.9 Chopper-controlled d.c. drives

A number of industrial applications take power from d.c. rather than from a.c. voltage sources. Subway cars, battery-operated vehicles, trolleybuses, forklift trucks, and minehaulers are but a few examples. In such cases the power is derived from a fixed d.c. bus. However, to satisfy the requirements of speed control, the d.c. voltage must be adjusted with the speed of the motor. The traditional method of achieving voltage control was the cam controller which is, in effect, a resistor which can be varied in steps. The solid state equivalent of the cam controller is the d.c./d.c. chopper. Various chopper configurations and their quadrants of operation are shown in Figure 4.23. As for naturally commutated bridge converters, the inductance shown in series with the motor is to limit the ripple current. The inductor is most often omitted since the ripple frequency of the chopper is typically at least an order of magnitude greater than that of the bridge converter.

An important difference between rectifier bridge connections and chopper connections is the need for a so-called force-commutated switch, i.e. the switch must be capable of turning off without the aid of external e.m.f.s from the supply or the load. This turn-off requirement is satisfied either with a device such as a transistor or gate-turn-off thyristor or with a conventional thyristor utilizing a special commutating circuit.

The principle of operation of a chopper is relatively simple, e.g. the operation of the first-quadrant,step-down chopper of Figure 4.23(a) is shown in Figure 4.24. The force-commutated switch periodically connects the d.c. supply to the load and then disconnects it for a predetermined period. Since the motor load is inductive, when the switch is disconnected the motor armature current finds a path through the free-wheeling diode. Hence, the voltage across the motor alternates between the d.c. voltage $V_{\text{d.c.}}$ and zero. The average voltage across the motor is simply:

$$< v_{\text{d}} > = \frac{t_{\text{on}}}{t_{\text{on}} + t_{\text{off}}} V_{\text{d.c.}} \tag{4.18}$$

$$= \frac{t_{\text{on}}}{T} V_{\text{d.c.}} = d\, V_{\text{d.c.}} \tag{4.19}$$

where T is the total switching period. The quantity d is called the duty cycle of the chopper. The average d.c. output voltage can be varied in two ways: (1) constant frequency modulation; and (2) variable frequency modulation. In the case of

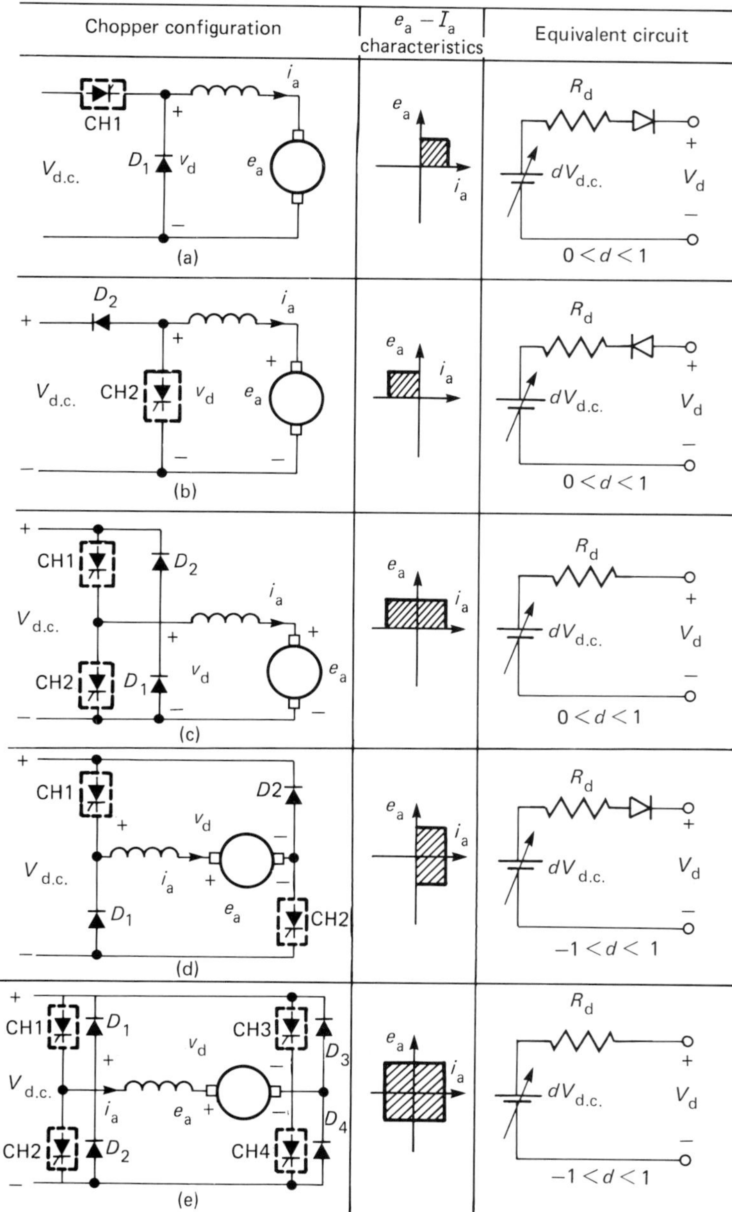

Figure 4.23 Direct current chopper configurations and their quadrants of operation. (a) First-quadrant chopper. (b) Second-quadrant chopper. (c) Two-quadrant type 1 chopper. (d) Two-quadrant type 2 chopper. (e) Four-quadrant chopper

constant frequency modulation, the chopping frequency and, therefore, the period T is kept constant and the on-time t_{on} is varied. This type of modulation is frequently called pulse width modulation. When variable frequency modulation is used, the pulse width t_{on} is kept constant, and the off-time t_{off} and therefore also the period T are varied. Since a chopper working at a continuous maximum frequency

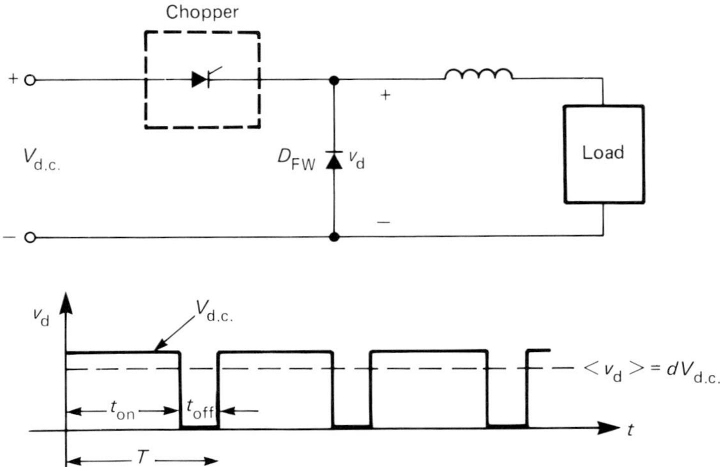

Figure 4.24 Principle of operation of a step-down chopper

set by limitations of the switch generally produces the best overall efficiency, the constant frequency system is usually the preferred scheme for chopper drives. Since neither the voltage nor the current of the step-down chopper can reverse polarity, the equivalent circuit of the step-down chopper (neglecting harmonics) is a variable voltage source in series with an ideal diode (Figure 4.23(a)).

In the second-quadrant regenerative chopper (Figure 4.23(b)), the roles of the diode and switch are reversed. When the motor is shorted by closing the solid state switch, the current rises in a negative sense. Since the current in the armature inductance cannot change instantly, the current finds a path through the diode when the switch is opened. It can be noted that the voltage across the inductor plus motor must rise to a value equal to the d.c. supply voltage $V_{d.c.}$ in order for the diode to conduct. The e.m.f. E_a is fixed for a given operating condition so that the quantity di/dt must be sufficiently large to allow the diode to conduct. Since the armature inductance is relatively small this implies that a very large pulse of current will occur if an external filter inductor is not used. The inductance must be chosen to ensure continuous braking at rated current down to zero speed. The equivalent circuit of the regenerative chopper is again a controllable voltage source but this time with a reverse-connected diode.

The two-quadrant type 1 chopper of Figure 4.23(c) is essentially the first two circuits superimposed. Motoring is obtained by modulating switch 1 while braking is obtained by modulating switch 2. Very fast response in motoring or braking is possible with this circuit, but speed reversal requires reversing the armature polarity with two pairs of contactors (Figure 4.22(a)) or reversing the field. The equivalent circuit of this chopper is simply a controllable voltage source proportional to the duty cycle d of switch 1 without a series diode, as shown in Figure 4.23(c).

In the two-quadrant type chopper of Figure 4.23(d), switches 1 and 2 are turned on at the same instant, which applies voltage pulses to the motor in much the same manner as the step-down chopper. However, when both switches are turned off, the motor voltage is reversed and the current returned to the supply rather than allowed to circulate in the free-wheeling diode. An average voltage of either polarity can be obtained by adjusting the duty cycle d of the switches. A duty cycle of 0.50 corresponds to zero average output voltage. Motoring in one direction is possible but braking or speed reversal requires field or armature voltage reversal.

The average value equivalent circuit is a reversible voltage source in series with an ideal diode.

In the four-quadrant chopper of Figure 4.23(e), both the motor voltage and current are reversible. When switch 3 is turned off, switch 4 is turned on and the resulting circuit reverts to the two-quadrant type 1 chopper of Figure 4.23(c). (Diode 3 cannot conduct when E_a has the polarity shown.) Motoring and braking in the forward direction are obtained by modulating switches 1 and 2. For reversal, switch 1 is turned off, switch 2 on and the polarity of the current reversed by modulating switches 3 and 4. The equivalent circuit of the four-quadrant chopper is a simple reversible voltage source proportional to the duty cycle of switch 1 (forward direction) or switch 3 (reverse direction), as illustrated in Figure 4.23(e). Hence, the duty cycle effectively varies from -1 to $+1$. The four-quadrant chopper is clearly the most versatile configuration and provides the fastest response.

4.2.10 Small-signal equivalent circuit converter models

While the 'average value' equivalent circuits shown previously are useful for steady state or quasi steady state calculations, control studies require a more detailed representation of the relation between the control input and the converter output. Figure 4.25 shows a simple circuit used to control the gating of a fully controlled single-phase bridge. Although a rudimentary analogue circuit, it is similar to many digital firing schemes which count clock pulses up to a specified (but variable) value set by an input control signal. Zero crossings of the input voltage initiate a ramp wave which is compared with the control voltage e_c. When the ramp wave exceeds the control voltage, gate pulses are sent to the appropriate thyristors.

If e_{rm} is the maximum value of the ramp, then the d.c. output voltage as a function of e_c can be written as:

$$V_d = V_{d_0} \cos \alpha \tag{4.20}$$

$$= V_{d_0} \cos \left(\frac{e_c}{e_{rm}} \pi \right) \tag{4.21}$$

where

$$V_{d_0} = \frac{2\sqrt{2}V}{\pi} \tag{4.22}$$

Hence, the transfer characterized is non-linear.

It is useful to linearize the gain for the purpose of control analysis. The incremental gain is obtained by taking the partial derivative of V_d with respect to e_c

$$\frac{\Delta V_d}{\Delta e_c} = \frac{\pi V_{d_0}}{e_{rm}} \sin \left(\frac{\pi e_c}{e_{rm}} \right) \tag{4.23}$$

In most cases the control angle α is restricted to values near 90° since the full voltage V_{d_0} is only needed under severe conditions. Let α_0 denote the value which produces rated voltage. The average gain which occurs between the control angles α_0 and $\pi - \alpha_0$ is:

$$\frac{\Delta V_d}{\Delta e_{c \, (ave)}} = \frac{\pi V_{d_0}}{e_{rm}} \frac{1}{\pi - 2\alpha_0} \int_{\alpha_0}^{u - \alpha_0} \sin \alpha \, d\alpha \tag{4.24}$$

$$= \frac{V_{d_0}}{e_{rm}} \frac{2\pi}{\pi - 2\alpha} \cos \alpha_0 \tag{4.25}$$

In addition to the gain change with α, the converter bridge is also a sampling device which converts the desired voltage e_c to an amplified equivalent $V_{d_0} \cos \alpha$ and every π electrical radians. The sampling process is unfortunately a non-linear sampler, since the sample is not taken at precise instants of time but occurs only somewhere during each half cycle. Figure 4.25 illustrates the problem. Although the control voltage e_c changes at θ_0, the change is not recognized until θ_1. The effective delay

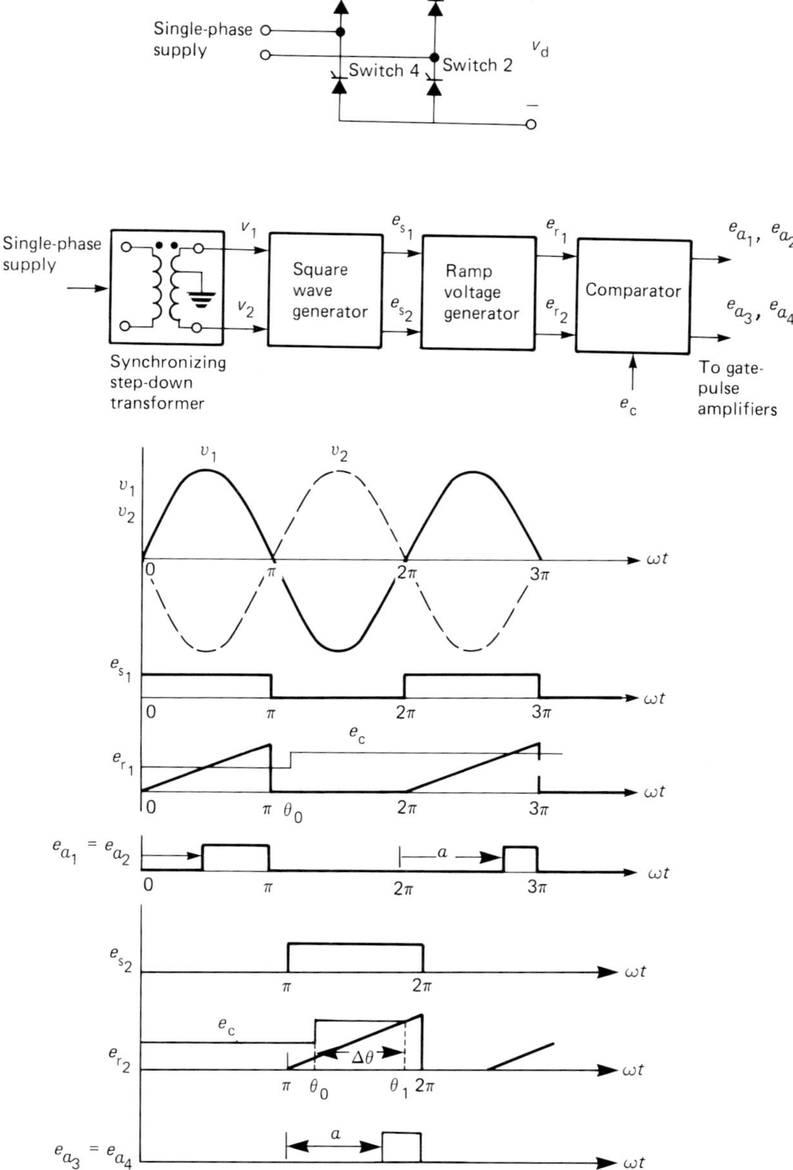

Figure 4.25 Simple control circuit for gating a single-phase bridge

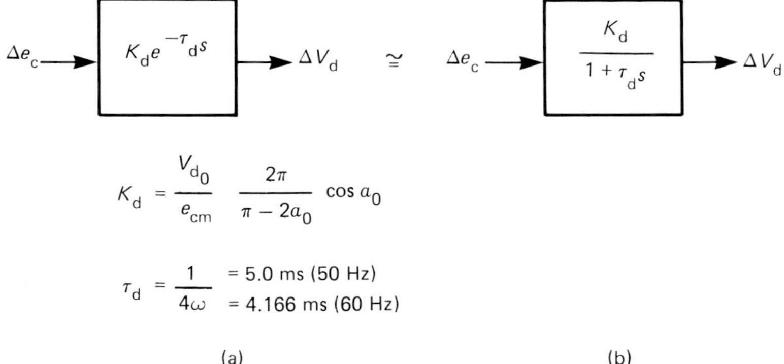

$$K_d = \frac{V_{d_0}}{e_{cm}} \; \frac{2\pi}{\pi - 2a_0} \; \cos a_0$$

$$\tau_d = \frac{1}{4\omega} \quad \begin{array}{l} = 5.0 \text{ ms (50 Hz)} \\ = 4.166 \text{ ms (60 Hz)} \end{array}$$

(a)   (b)

Figure 4.26 Transfer function of a fully controlled single-phase bridge with ramp comparison

$\Delta\theta$ in the control depends upon when the change occurs and can vary anywhere from 0 to π radians. It is conventional to assume that, on average, the delay corresponds to $\pi/2$ electrical radians which corresponds to 5 ms and 4.166 ms for 50 and 60 Hz systems respectively.

The overall transfer function of the single-phase fully controlled bridge with a ramp comparison controller is shown in Figure 4.26(a). In practice, for purposes of analysis, the sample delay function block is replaced by a first-order lag having the same time constant as the delay (Figure 4.26(b)).

Since the gain of the transfer characteristic is essentially the slope of the function $\cos \alpha$, it becomes small for conditions near maximum output voltage. Hence, the response deteriorates as the speed (and, hence, e.m.f.) increases. This tendency toward poor response may, in turn, limit the maximum speed range in the 'constant torque' mode. The problem is overcome partly by the biased cosine method of gate control shown in Figure 4.27. In this case the control signal e_c is compared with a cosine wave which is derived from the line voltage by integration. The phase angle α is given by:

$$\alpha = \cos^{-1}\left(\frac{e_c}{e_{rm}}\right) \tag{4.26}$$

The output voltage of the bridge is therefore:

$$V_d = V_{d_0} \cos \alpha \tag{4.27}$$

$$= \frac{V_{d_0}}{e_{cm}} e_c \text{ for } - e_{rm} \leqslant e_c \leqslant e_{rm} \tag{4.28}$$

which has a constant gain independent of α. The transfer functions of Figure 4.26 are also valid for the biased cosine method with only modification of the gain constant. The biased cosine approach also linearizes the transfer characteristic of the half-controlled bridge.

In practice, the delay angle is not permitted to reach a value less than 165° or so, in order to allow the offgoing thyristor time to regain its forward blocking ability. The method is probably the most popular method used in industry when analogue

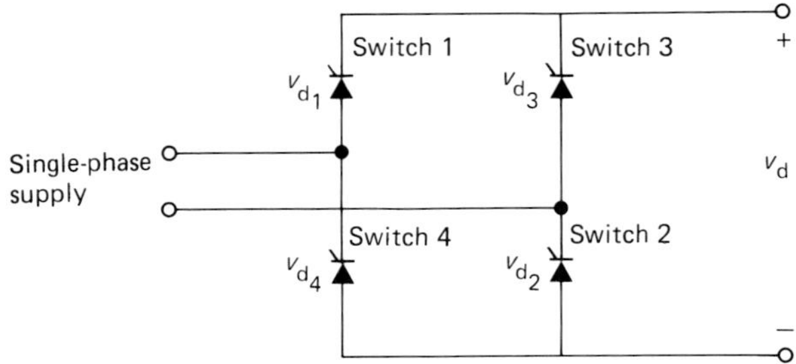

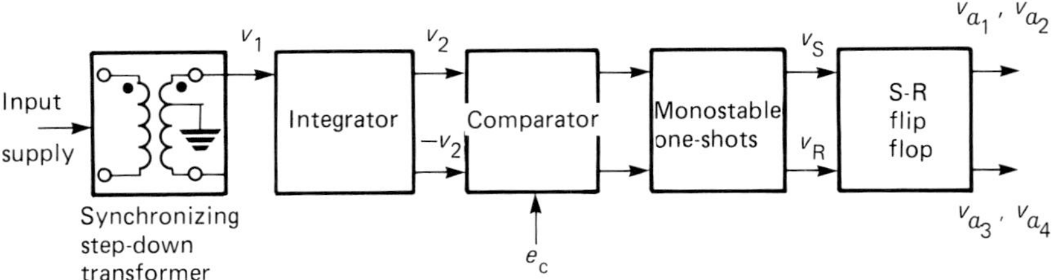

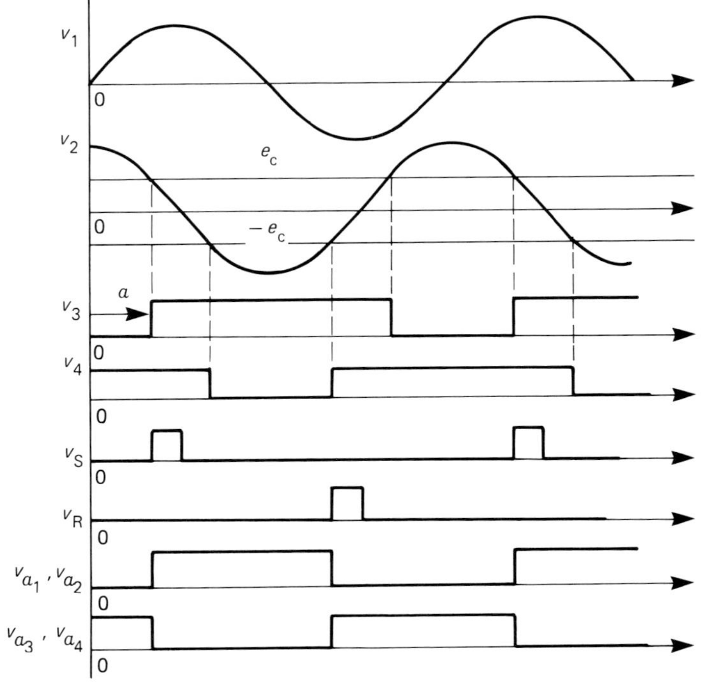

Figure 4.27 The biased cosine method of gate control

circuitry is used. It is less common in digital implementations since the generation of the cosine function is more difficult.

The problems involved in calculating the transfer function of three-phase converters is a relatively straightforward extension of the single-phase case. The results are also given in Figure 4.26.

4.2.11 Block diagram representation of a d.c. motor

If p denotes the operator d/dt then the differential equations describing transient behaviour of the machine are:

$$V_d = (r_a + r_d)i_a + (L_a + L_d)pi_a + e_a \tag{4.29}$$

$$v_f = r_f i_f + p\lambda_f \tag{4.30}$$

$$\lambda_f = L_{lf} i_f + \lambda_{af} \tag{4.31}$$

$$\lambda_{af} = f(i_f) \tag{4.32}$$

$$e_a = k_1 \omega_r \lambda_{af} \tag{4.33}$$

$$T_e = k_1 i_a \lambda_{af} \tag{4.34}$$

$$T_e = T_l + B\omega_r + Jp\omega_r \tag{4.35}$$

In these equations r_a and L_a denote the internal resistance and inductance of the motor armature, r_d and L_d the external resistance and inductance of the d.c. link filter, and L_{lf} the leakage inductance of the field. Equation (4.32) represents the open-circuit saturation characteristic of the main field flux linking the armature λ_{af}.

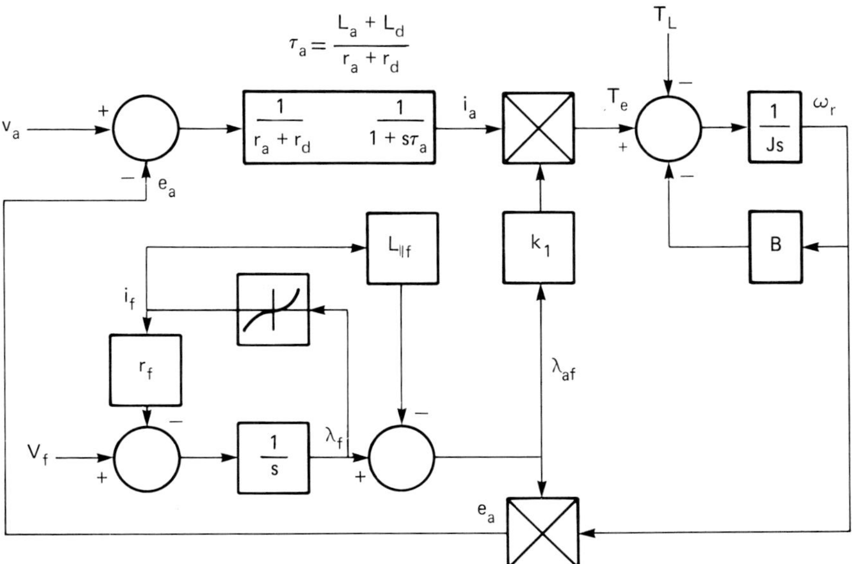

Figure 4.28 Block diagram representation of a d.c. machine

The quantity J is the moment of inertia of the rotor armature including connected load and B is the viscous damping due to windage and friction.

These seven equations generate the block diagram representation of a d.c. motor shown in Figure 4.28. Note that two multipliers (non-linearities) are involved making the response of the motor a function of speed as well as torque.

4.2.12 Small signal behaviour of a d.c. motor with constant flux

In order to obtain a clearer view of the transient behaviour of a d.c. machine, the block diagram of Figure 4.28 has been simplified by assuming that the field flux remains constant during a transient. Also, it is assumed that the transients of concern are sufficiently small that the saturation curve (Equation (4.32)) can be replaced by a linear approximation. Figure 4.29 results. Upon transforming

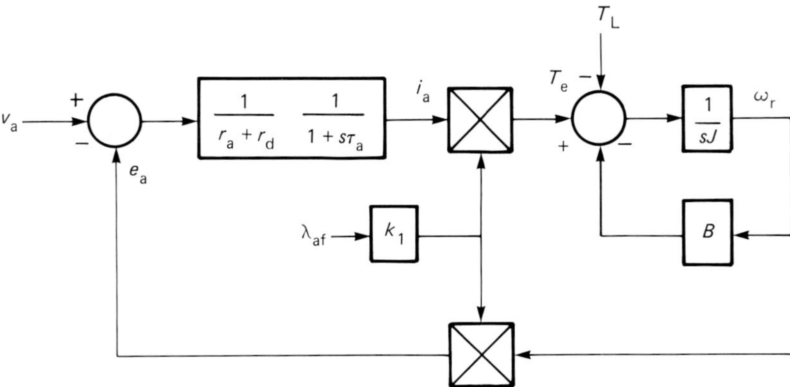

Figure 4.29 Simplified block diagram assuming constant field flux linkage

Equations (4.29–4.34) to Laplace variables by means of the Laplace transformation, and assuming that the field flux is constant, the following transfer function for $I_d(s)$ as a function of $V_d(s)$, is obtained:

$$I_a(s) = \frac{V_a(s)}{r_a + r_d} \frac{s\tau_m}{1 + (1 + s\tau_a)\,s\tau_m} \tag{4.36}$$

where τ_a is the armature time constant,

$$\tau_a = \frac{L_a + L_d}{r_a + r_d} \tag{4.37}$$

and τ_m is the so-called mechanical time constant, and

$$\tau_m = \frac{(r_a + r_d)J}{k_1^2 \lambda_{af}^2} \tag{4.38}$$

The quantity $k_1^2 \lambda_{af}^2/(r_a + r_d)$ corresponds to the slope of the motor torque–speed curve, and thus provides a damping effect. The quantity τ_m is equivalent to inertia divided by a damping term and, hence, corresponds to a time constant.

The roots of the denominator polynomial contain the eigenvalues or characteristic roots of the system. They are:

$$s_{1,2} = \frac{1}{2\tau_a} \left(-1 \pm \sqrt{1 - \frac{4\tau_a}{\tau_m}} \right) \tag{4.39}$$

It is useful to define the ratio:

$$\alpha = \frac{\lambda_{af}}{\lambda_{af_0}} \tag{4.40}$$

where λ_{af_0} corresponds to maximum (rated) flux linkage. Equation (4.39) can be written:

$$s_{1,2} = \frac{1}{2\tau_\alpha} \left[-1 \pm \sqrt{1 - (2\alpha)^2 \frac{\tau_a}{\tau_{m_0}}} \right] \tag{4.41}$$

where $\tau_{m_0} = \dfrac{(r_a + r_d)J}{k_1^2 \lambda_{af_0}^2}$

The ratio α varies from unity down to some minimum value set by the commutation limit of the machine. As α varies, the poles move as shown in Figure 4.30. The points shown as α_{max} correspond to maximum flux condition, while the points α_{min} correspond to the minimum flux condition. It is readily seen that at a given speed, field weakening tends to make the motor mechanical time constant decrease since reduced flux makes the slope of the motor torque–speed curve greater. When complex, the frequency of the roots is:

$$\omega = \left[\frac{\alpha}{2\tau_\alpha} \right] \sqrt{\frac{4\tau_a}{\tau_{m_0}} - \frac{1}{\alpha^2}} \tag{4.42}$$

The damping ratio is:

$$\zeta = \frac{-Re\,(s_1)}{|s_1|}$$

$$= \frac{1}{2\alpha} \sqrt{\frac{\tau_{m_0}}{\tau_a}} \tag{4.43}$$

and the natural frequency is:

$$\omega_n = \sqrt{s_1 s_2} \tag{4.44}$$

Figure 4.30 shows the case where α_{max} results in complex conjugate roots. However, the end points for α_{max} may be either complex or real depending on the size of the machine. The critical value is:

$$\tau_m = 2\alpha^2 \tau_\alpha \tag{4.45}$$

Most small machines have real roots for rated excitation, resulting in overdamped response. However, the trend is to have faster armature time constants but underdamped response, i.e. complex roots, as the machines become larger. Oscillatory response is also the case for small servomotors which have elongated slender shafts to reduce inertia. This conclusion can also be verified by observing

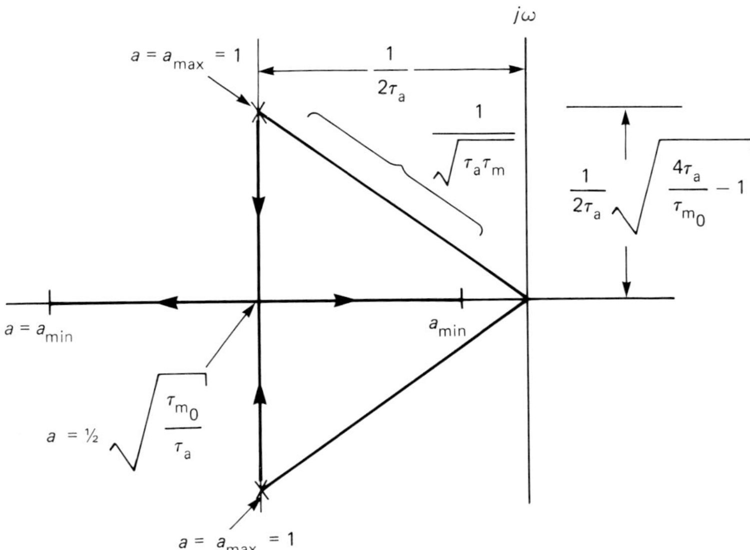

Figure 4.30 Motion of the characteristic roots of a d.c. motor as a function of the field flux linkage

Table 4.1 Typical armature and mechanical time constants for a range of machine sizes rated at 1150 rev./min. $\tau_e = L_a/r_a$, $\tau_m = J/B$

NEMA frame	τ_{rated} (J)	τ_e (s)	τ_m (s)
283	20.34	0.0073	0.036
284	40.67	0.0074	0.039
286	59.66	0.0099	0.022
288	81.35	0.016	0.020
365	101.69	0.026	0.018
366	162.70	0.015	0.018
367	203.37	0.026	0.013
368	244.05	0.021	0.014
503	406.75	0.028	0.018
504	508.43	0.084	0.013
505	610.12	0.083	0.022

the effect of varying J in the time constant τ_m. A table of typical armature and mechanical time constants for a range of machine sizes is shown in Table 4.1.

Examination of Equation (4.38) indicates that the motion of the characteristic roots (Equations (4.41)) can also be observed for any value of the flux linkage ratio α. In particular, when the field flux is zero:

$$s_{1,2\,\omega=0} = 0, \; -\frac{1}{\tau_a} \tag{4.46}$$

Hence, when the field is not excited, the mechanical pole moves to the origin and the other pole to the reciprocal of the open-circuit armature time constant. It can thus be said that the field flux introduces coupling between the electrical and mechanical time constants of the d.c. machine.

4.2.13 Speed control of d.c. motors

The speed-controlled d.c. drive with a converter supply is, at present, the most widely employed form of high-response d.c. drive. The speed signal is normally obtained from a small permanent-magnet generator. For higher accuracy the speed signal can be taken from a magnetic or optical encoder generating several thousand pulses per revolution of the shaft. The frequency of the pulse train is proportional to speed which is determined by counting repetitively the number of pulses over a fixed time or, alternatively, determining the elapsed time for a fixed number of pulses. The controller is most often equipped with an inner current loop which serves to protect the converter from excessive current as well as provide good dynamic system response. The normal practice is to obtain the current feedback signal from the a.c. side of the converter. The measured a.c. quantities are then rectified to convert it into the required armature current variable. The motor generally has constant field excitation.

A typical speed-controlled system diagram is shown in Figure 4.31. Note that the e.m.f. feedback internal to the machine 'overlaps' the external current feedback loop, so that block diagram manipulation is required before conventional control loop analysis can proceed. Because of the problems involved, it is often sufficient to assume that the speed e.m.f. is zero. Divergences in the gain characteristics with and without speed feedback first occur at frequencies corresponding to the reciprocal of the armature time constant which is generally sufficiently high that they may safely be neglected. Conventional proportional plus-integral controllers are most often used to regulate both armature current and speed.

Cost considerations, or reliability questions associated with the speed sensor, often preclude use of explicit speed feedback. Since the e.m.f. is proportional to speed, the rotor speed can be varied implicitly by feedback of the e.m.f. Figure 4.32 shows a typical block diagram of such a voltage-controlled system. Again the system is typically equipped with an internal current loop. 'Overlapping' of the inherent motor e.m.f. loop and the internal current loop again causes some difficulty in maintaining fast response over a wide speed range. The inner current loop is frequently a simple integral controller. The block in the voltage feedback loop is inserted to compensate for the armature resistance and inductance, but is frequently omitted.

For good utilization of the machine, field current is held at its maximum value while the armature voltage is controlled as speed increases. When maximum armature voltage is reached the control must smoothly transit to field weakening. A representative control scheme which accomplishes this purpose is shown in Figure 4.33. The purpose of the field voltage controller is to limit the induced armature current beyond base speed. The input to the field voltage controller is the maximum armature voltage V_{d_0}. During armature control, the field voltage controller is saturated so that maximum field voltage is obtained. However, when the armature voltage begins to reach V_{d_0} the regulator becomes active and starts reducing the field voltage so as to maintain the armature voltage constant. The block diagram shows that the system is now highly non-linear so small signal linearization techniques must be used to set the controller gains. The reader should refer to advanced textbooks on this subject for further information.

4.3 Fixed-frequency a.c. drives

4.3.1 Introduction

Variable-speed a.c. drives utilizing fixed-frequency, variable-voltage controls are severely limited in size and efficiency by the fundamental properties of a.c. machines. With the exception of brush-type 'universal motors', which operate essentially the same as d.c. series motors, all a.c. machines produce exactly or very

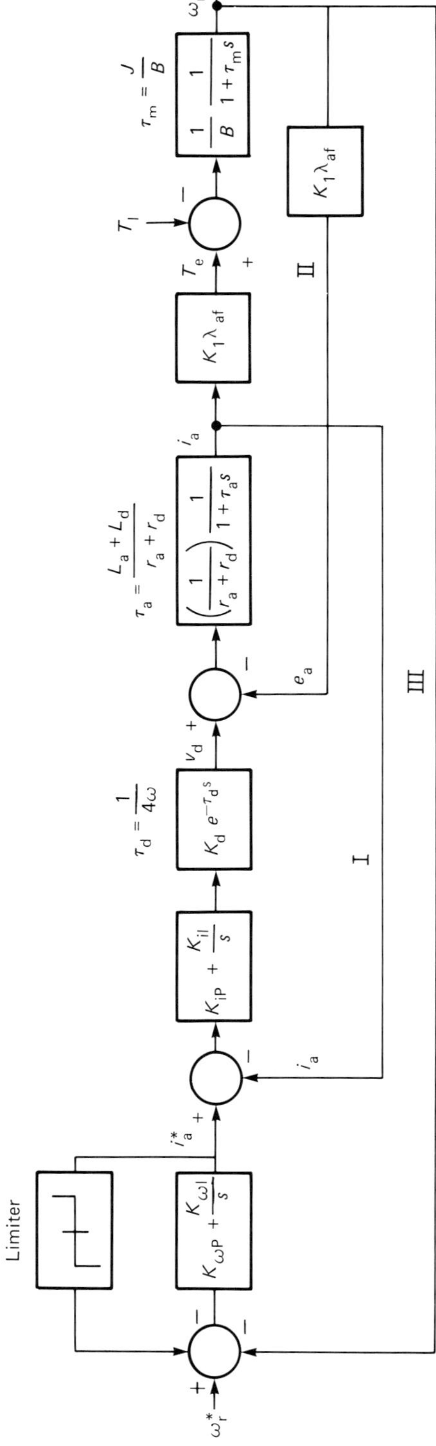

Figure 4.31 Speed control system with explicit speed feedback and an inner current loop

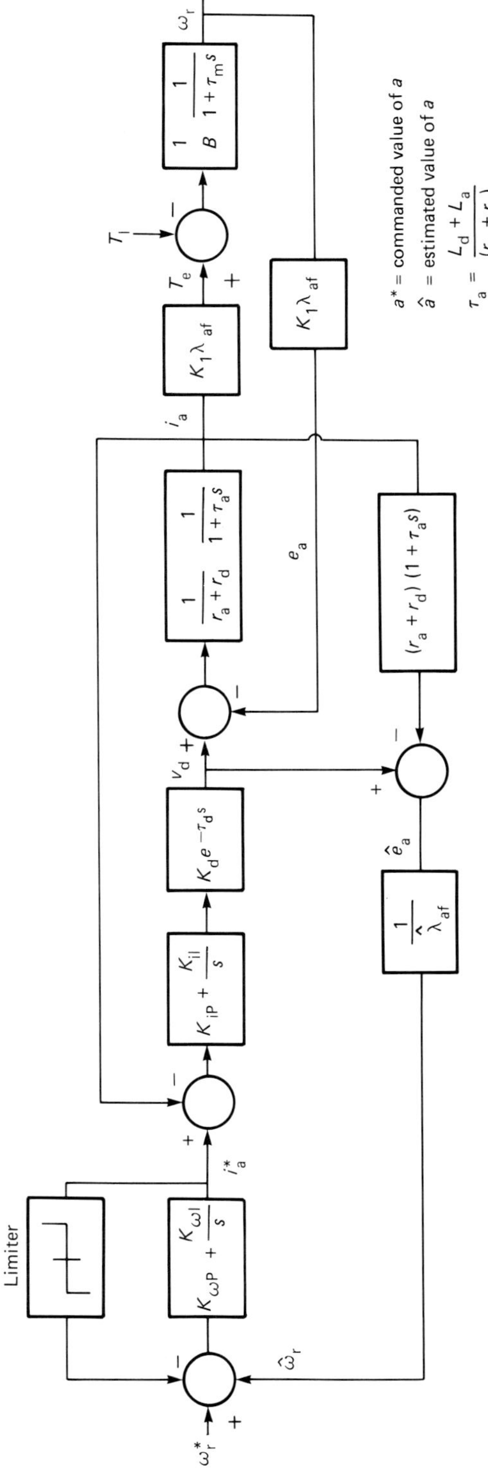

Figure 4.32 Speed control system with e.m.f. feedback with an inner current loop

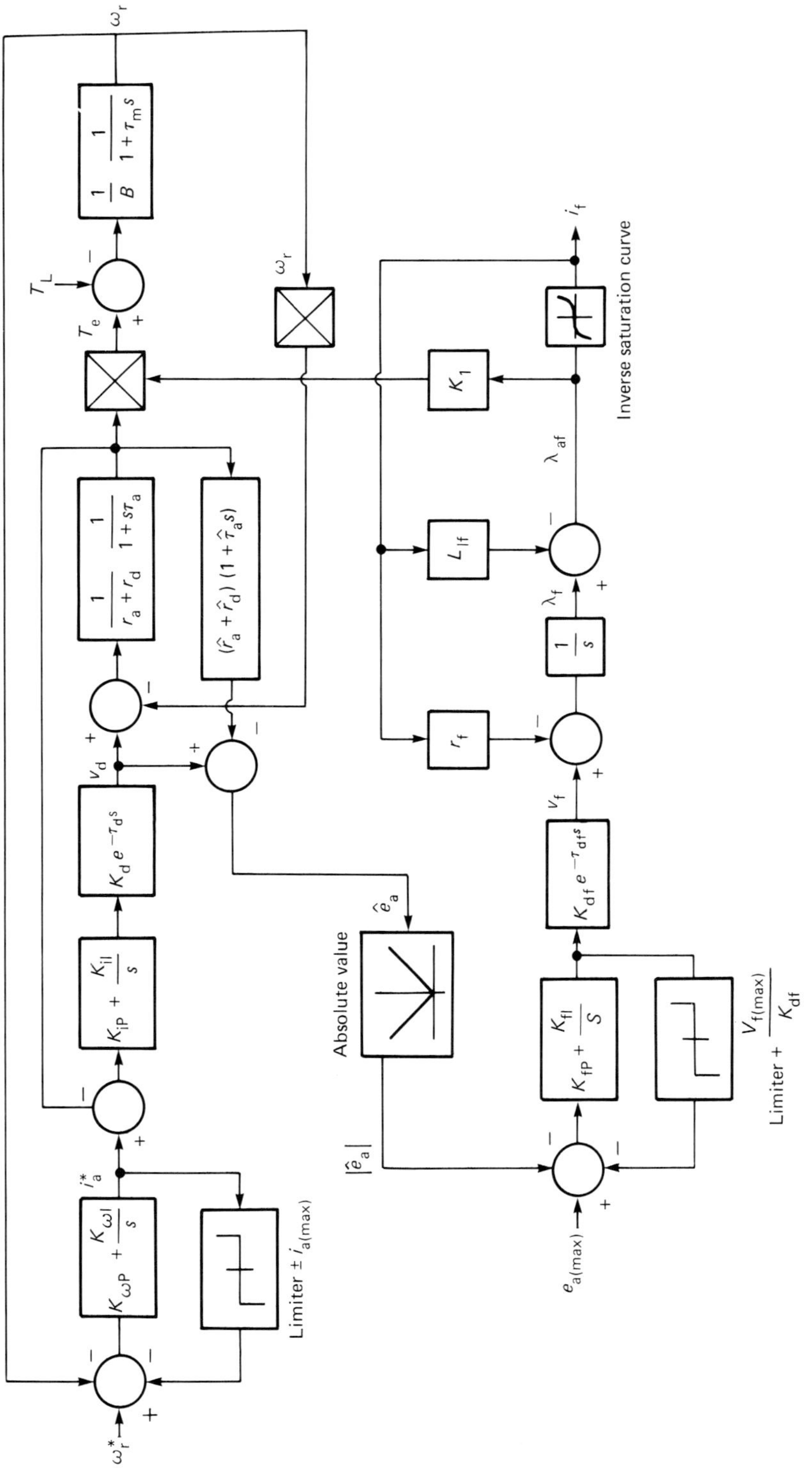

Figure 4.33 Speed control system for a d.c. machine operated in the field weakening range

nearly constant speed when operated from a fixed-frequency supply. Reduced-voltage operation of an induction machine will result in lower speed but this requires increased slip and the rotor I^2R losses are accordingly increased. This type of high slip drive is therefore limited in application to situations where the high losses and low efficiency are acceptable and, generally, where the speed range is not large. Such drives are generally limited to relatively low power ratings because of cooling problems.

4.3.2 Eddy-current drives

One variation of constant-frequency control which has seen wide use even at substantial power ratings is the eddy current drive. This system consists of an eddy current clutch placed between a constant-speed drive motor and the variable-speed load. By controlling the d.c. excitation to the clutch magnetic circuit, the slip of the clutch is controlled to obtain speed control.

An eddy current clutch is identical in principle to an induction machine in which both the rotor and stator are allowed to rotate. With d.c. excitation of the 'stator' winding, a magnetic field rotating at the speed of the 'stator' is created. This rotating field interacts with the rotor (eddy current) winding exactly as in a normal induction motor and a coupling torque is produced. The torque–slip characteristic is the same as in an induction motor. In this case, however, the slip is given by:

$$S = \frac{\omega_{in} - \omega_o}{\omega_{in}} \tag{4.47}$$

where ω_{in} = input shaft speed; and ω_o = output shaft speed

Since the input and output torque must be the same, the relation between the input and output is:

$$\frac{P_o}{P_{in}} = \frac{\omega_o}{\omega_{in}} = 1 - S \tag{4.48}$$

Thus, the efficiency is directly related to the speed ratio of the coupling (excitation power neglected). The power loss equal to SP_{in} is dissipated as heat in the eddy current armature of the drive. For reasons of efficiency and cooling, eddy current drives are usually limited to small speed ratios (small slip); typically not more than a 10–30% speed reduction is employed.

The physical construction of an eddy current clutch can take a variety of forms including both radial and axial flux types. Also both slipring-brush type and brushless designs are available, the brushless systems having the disadvantage of multiple airgaps. In all cases, careful design consideration must be given to provide adequate cooling of both the field coils and the eddy current armature. Typically, the torque rating for small slip is determined by field coil heating and for high slip by armature heating.

4.3.3 Voltage-controlled induction machine drives

Stator voltage control of induction machines is used for current limiting during starting, to reduce speed below rated speed for speed control and, most recently, to reduce losses at light load and thus improve overall efficiency in variable-load applications. The methods of control include:

(1) *Autotransformers or motor winding taps* – primarily for motor starting but also for speed control in small, usually single-phase, motors.

(2) *Series reactors* – again, primarily for motor starting but sometimes applied to speed control in small motors.
(3) *Phase control of series-connected Triacs or back-to-back-connected inverse parallel thyristor pairs* – widely used for all three types of applications.

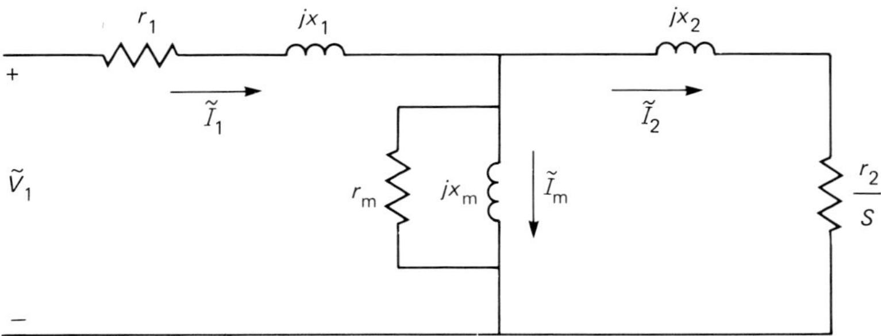

Figure 4.34 Induction motor equivalent circuit

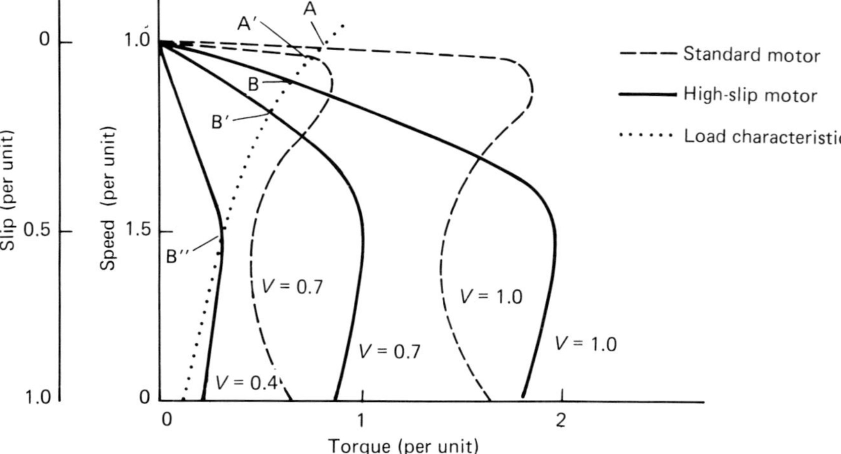

Figure 4.35 Constant voltage torque–speed curves

4.3.3.1 Basic principles

In all cases, the basic principles of voltage control can be obtained readily from the conventional induction motor equivalent circuit shown in Figure 4.34 and the associated constant voltage speed–torque curves illustrated in Figure 4.35. The torque produced by the machine is equal to the power across the airgap divided by synchronous speed:

$$T = \frac{P}{2}\frac{P_\mathrm{r}}{\omega_\mathrm{e}} = 3\frac{P}{2}\frac{I_2^2 r_2}{S\omega_\mathrm{e}} \tag{4.49}$$

where P = number of poles

The peak torque points on the curves in Figure 4.35 occur when maximum power is transferred across the airgap and are easily shown to take place at a slip:

$$S_\mathrm{MT} \approx r_2/(x_1 + x_2) \tag{4.50}$$

From these results and the equivalent circuit, the following principles of voltage control are evident:

(1) For any fixed slip or speed, the current varies directly with voltage and the torque and power with voltage squared.
(2) As a result of (1) the torque–speed curve for a reduced voltage maintains its shape exactly but has reduced torque at all speeds (Figure 4.35).
(3) For a given load characteristic, a reduction in voltage will produce an increase in slip (from A to A′ for the conventional machine in Figure 4.35, for example).
(4) A high-slip machine has relatively higher rotor resistance and results in a larger speed change for a given voltage reduction and load characteristic (compare A to A′ with B to B′ in Figure 4.35).
(5) At small values of torque, the slip is small and the major power loss is the core loss in r_m. Reducing the voltage will reduce the core loss at the expense of higher slip and increased rotor and stator I^2R loss. There is an optimal slip which maximizes the efficiency and varying the voltage can maintain high efficiency even at low torque loads.

4.3.3.2 Inverse parallel thyristor or Triac phase control

When inverse parallel thyristors or Triacs are used to obtain voltage control, the motor voltage and current are no longer sinusoidal. Figure 4.36 illustrates a typical three-phase arrangement and the associated current waveform. Although the current and voltage harmonics produce additional losses and torque ripple, the fundamental component determines the average torque and the largest portion of the power and power losses.

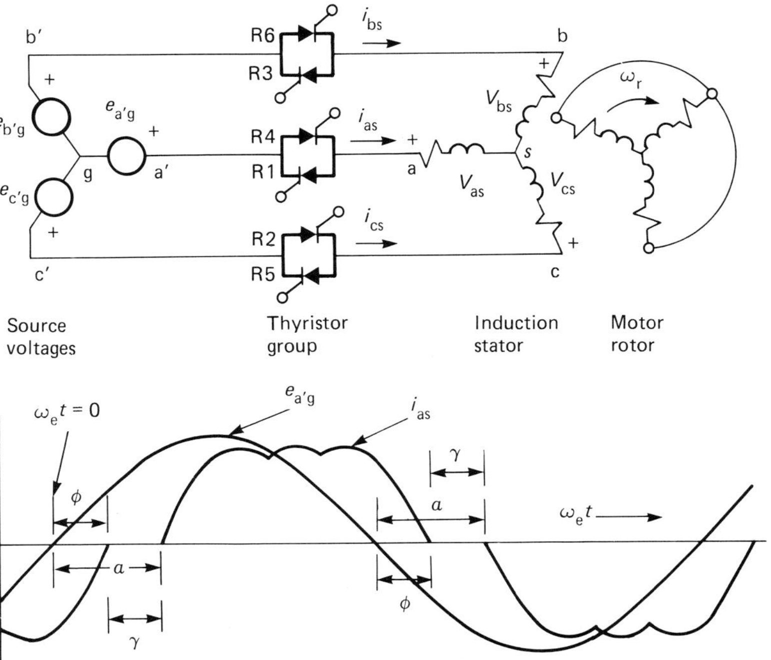

Figure 4.36 Induction motor voltage controller employing inverse parallel thyristors and resulting current waveform

It has been shown that a very accurate fundamental component model for the effect of phase-controlled inverse parallel thyristors or Triacs is a series reactance given by:

$$x_{eq} = x_s' f(\gamma) \tag{4.51}$$

where $x_s' = x_1 + x_2 x_m/(x_2 + x_m)$ (stator transient reactance);

$\gamma = $ hold-off angle;

and
$$f(\gamma) = \frac{\dfrac{1}{\pi}(\gamma + \sin \gamma)}{1 - \dfrac{1}{\pi}(\gamma + \sin \gamma)} \quad \text{(single-phase)} \tag{4.52}$$

or
$$f(\gamma) = \frac{\dfrac{3}{\pi}(\gamma + \sin \gamma)}{1 - \dfrac{3}{\pi}(\gamma + \sin \gamma)} \quad \text{(three-phase)} \tag{4.53}$$

This reactance can be added in series with the motor equivalent circuit to model a phase-controlled system. For typical machines the accuracy is well within acceptable limits; the accuracy decreases when the harmonic power delivered to the machine becomes significant compared to the fundamental volt-amperes. Thus, the approximation is better in larger machines and for smaller values of γ. In most cases of interest, the error is quite small; however, the harmonic power losses and torque ripple are entirely neglected.

4.3.3.3 Motor starters

Motor starters are intended to provide a reduction in starting current and also an unavoidable reduction in starting torque. Autotransformers have the advantage of reducing the input current as the square of the voltage ratio because of the transformer action. Thus, both input current and torque are reduced by the square of the voltage ratio. In contrast, reactance or inverse parallel thyristor starters reduce the current by the voltage ratio and the torque by the square of the ratio.

Unlike autotransformer or reactance starters which have only one or two steps available, an inverse parallel thyristor starter can provide stepless and continuous 'reactance' control. These electronic starters are often fitted with feedback controllers which allow starting at a preset constant current, although simple timed starts are also available. Some electronic starters are equipped to short-out the inverse parallel thyristor at the end of the starting period to eliminate the losses due to forward voltage drop during running. Other options include 'energy savers' which vary the voltage during variable-load running conditions to improve efficiency. Such options are described in Section 4.3.3.5.

4.3.3.4 Speed control

Variable-voltage speed controllers must contend with the problem of greatly increased slip losses and the resulting low efficiency. In addition, only speeds below synchronous speed are attainable and speed stability may be a problem unless some form of feedback is employed.

An appreciation of the efficiency and motor heating problem is available from Equation (4.49) rewritten to focus on the rotor I^2R loss:

$$3I_2^2 r_2 = SP_{\mathrm{g}} = S\frac{2}{P}\,\omega_{\mathrm{e}}T \tag{4.54}$$

The inherent problem of slip variation for speed control is clearly indicated. For example, for a constant torque load the rotor loss is linearly dependent on the slip. This relation leads to the upper bound on induction motor efficiency of $(1 - S)$, as previously stated for eddy-current drives.

As a result of the large rotor losses to be expected at high slip, voltage control is only applicable to loads in which the torque drops off rapidly as the speed is reduced. The most important practical case is fan speed control in which the torque required varies as the speed squared. For this case, equating the motor torque to the load torque results in:

$$3\frac{P}{2}\,\frac{I_2^2 r_2}{S\omega_{\mathrm{e}}} = K_{\mathrm{L}}\omega_{\mathrm{e}}^2 (1 - S)^2 \tag{4.55}$$

Solving for I_2^2 as a function of S and differentiating yields the result that the maximum value of I_2^2 (and hence $3I_2^2 r_2$) occurs at:

$$S = 1/3 \tag{4.56}$$

or at a speed of two-thirds of synchronous speed. If this worst case value is substituted back to find the maximum required value of I_2^2 and the result used to relate the maximum rotor loss to the rotor loss at rated slip, the result is:

$$\frac{\text{Maximum rotor } I^2R \text{ loss}}{\text{Rated rotor } I^2R \text{ loss}} = \frac{0.148}{S_{\mathrm{R}}(1 - S_{\mathrm{R}})^2} \tag{4.57}$$

where S_{R} = rated slip. Figure 4.37 illustrates this result and from the curve it is clear that, to avoid excessive rotor heating at reduced speed with a fan load, it is essential that a high-slip machine with a rated slip in the range of 0.2–0.25 be

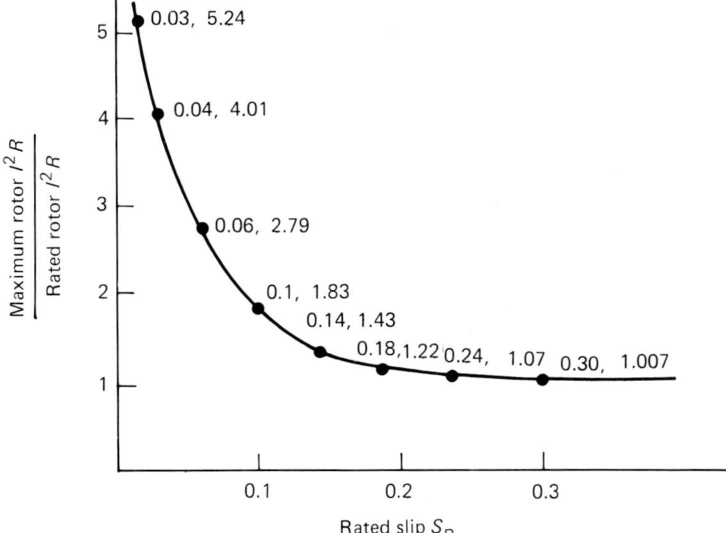

Figure 4.37 Worst case rotor heating for induction motor and fan load – worst case speed = 0.667 per unit

employed. While the use of such high slip machines will avoid rotor overheating, it does not improve the efficiency; the low efficiency associated with high slip operation is inherent in all induction machines.

As noted previously, speed stability is an inherent problem in voltage-controlled induction motor drives at low speeds. This is a result of the near coincidence of the motor torque characteristic and the load characteristic at low speed. The problem occurs primarily when the intersection of the motor torque characteristic and the load characteristic occurs near or below the speed of maximum motor torque (see point B″ in Figure 4.35).

As a result of the poor efficiency, need for high-slip motor designs and poor speed stability, voltage-controlled induction motor drives are restricted to small machines and almost exclusively to fan drives. The most common application is to small fans where winding taps are used to achieve voltage control. These systems usually use single-phase motors.

4.3.3.5 Efficiency improvement in variable-load systems

Simply stated, efficiency improvement by voltage control is achieved by reducing the applied voltage whenever the torque requirement of the load can be met with less than full motor flux. The reduced motor flux results in reduced core loss and also in reduced stator copper loss, since the magnetizing component of stator current is reduced. The reduced airgap flux, however, requires a larger slip to produce the required torque compared to operation at full-rated flux. Hence, the slip-dependent rotor loss and load component of the stator copper loss are increased. By proper adjustment of these two loss components, the fundamental component losses can be minimized for most part-load operating conditions. Since continuous variation of the voltage is required, the only practical controller is a Triac or thyristor phase-back system. Harmonic losses in the motor as well as thyristor losses are introduced, however, so that the net energy saving is decreased. These additional losses could, in fact, outweigh the savings gained by reducing the fundamental loss components.

The nature of the efficiency improvement attainable can be illustrated by considering an idealized situation wherein the motor parameter variations are neglected and the voltage controller is assumed to be modelled by an ideal adjustable amplitude sinewave source in series with a resistance. Under these assumptions the equivalent circuit of Figure 4.34 applies. As is well-known, conventional constant-voltage operation of an induction machine results in a motor slip which varies in proportion to the required load torque. In general, the load torque is considered as the independent variable. Clearly, from Figure 4.34, the only variable quantity in the circuit is the slip, and hence the input impedance, power factor, input power, and efficiency may be considered as functions of slip. For fixed input voltage, load changes are accommodated by changes in the slip S. When slip varies, so do the power factor, current, input power and efficiency. If, as an alternative, the voltage to the machine is adjusted to supply the required load while the slip is held constant, then the efficiency and power factor would remain fixed. It follows that if the optimal slip frequency can be located and the power factor held constant at its corresponding value, then this optimal value of motor efficiency could be maintained over a range of part-load conditions. Regrettably, however, the non-linearities in the motor and thyristor characteristics prevent this ideal value from being attained.

The type of behaviour which can be attained is illustrated in Figure 4.38. In this figure the dot–dash curve represents conventional constant voltage operation. The efficiency peaks at about 92% at 80% of rated torque and falls precipitously below about 40% of rated torque. The dashed curve represents operation from an ideal, adjustable amplitude, lossless sinusoidal voltage supply. Although a constant

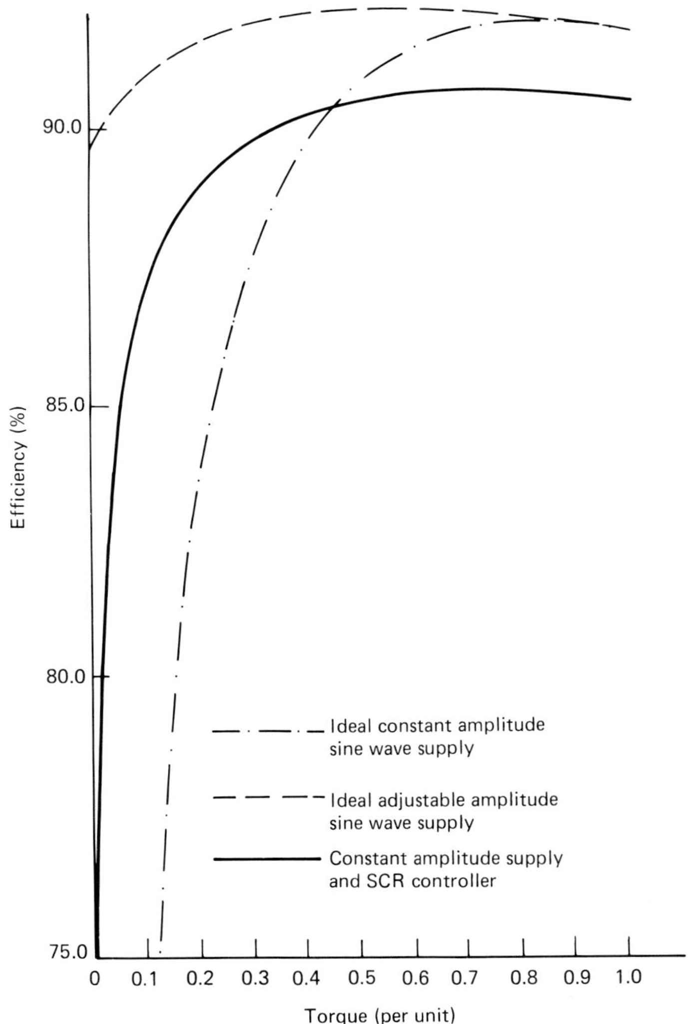

Figure 4.38 Calculated frequency for three modes of operation for a 5.5 kW, 230 V three-phase induction motor

parameter analysis indicated that the motor should operate at a constant efficiency, the efficiency actually increased somewhat and then declined. This behaviour is caused by variations in the magnetizing reactance and core loss resistance, resulting from changing flux levels in the motor. Figure 4.39 illustrates these parameter changes which have a significant influence on the efficiency improvement which is actually attainable with voltage control.

The solid curve in Figure 4.38 represents the best efficiency attainable with an inverse parallel thyristor or Triac controller. At the higher torques (above 0.9 per unit) there is no phase-back and the reduced efficiency is entirely caused by the thyristor forward voltage drop. At lower torques the harmonics produced by the phase-back create additional losses. The point where the solid and dot–dash curves cross (at about 0.45 per unit) represents operation where the additional

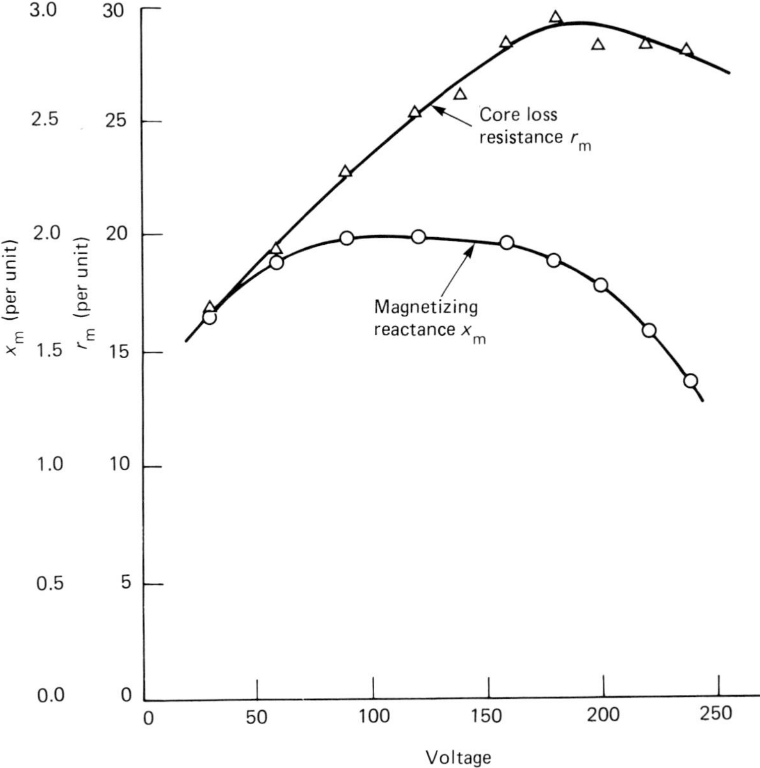

Figure 4.39 Change in magnetizing reactance and core loss resistance for the 5.5 kW, 230 V machine of Figure 4.38

controller-related losses are equal to the reduction in fundamental component losses. Only below this point is there a net energy saving and clearly a substantial portion of the load cycle must lie in this region to offset the higher losses at larger loads. The actual load where this occurs will be different for every motor.

Figure 4.40 illustrates the importance of the duty cycle on net energy savings for the same motor as that in Figure 4.38. The numbers on each curve represent the two extremes of the duty cycle. For example, if the duty cycle consists of alternate periods of 0.01 and 1.0 per unit of torque, the uppermost curve applies and a net saving occurs only when the time spent at the larger load is less than about 70%. Note that for part loads greater than about 0.5 per unit, no savings are possible.

It is quite clear from Figure 4.40 that relatively long periods at light load (low duty cycle) are necessary to obtain any net savings. Other factors which enhance the potential for savings are lower peak efficiency in the motor, very little load during the light-load interval, overvoltage operation of the motor and motors with a high ratio of core loss:copper loss. Figure 4.41 illustrates the effect of overvoltage and an oversized motor. The curves for duty cycles where the larger load is only 0.5 per unit clearly show substantial energy savings even for 100% duty cycle.

Application of voltage controllers for efficiency improvement must be examined carefully to establish that savings are possible and that the payback period is short enough to justify the installation. Since a phase-controlled thyristor motor starter already contains the control hardware to allow efficiency optimization, such installations offer some of the best candidates for application of the concept.

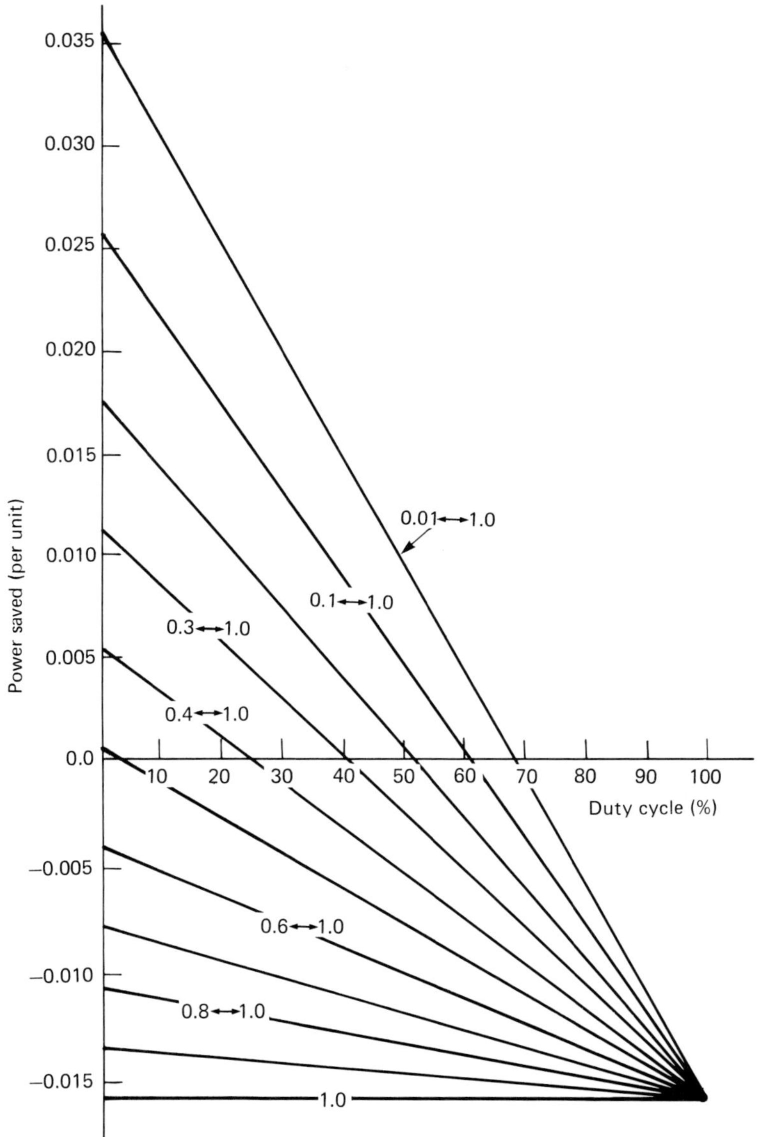

Figure 4.40 Power saved vs. duty cycle for the machine of Figure 4.38. (Figures on curves are extremes of the duty cycle)

4.4 Naturally commutated variable-frequency drives

4.4.1 Introduction

In high-power applications, such as turbo-compressors, induced and forced draft fans, and boiler feed pumps, very high mechanical efficiencies can be achieved through the use of adjustable-speed motor drives resulting in substantial energy savings. Such high-power applications are typically beyond the range of both

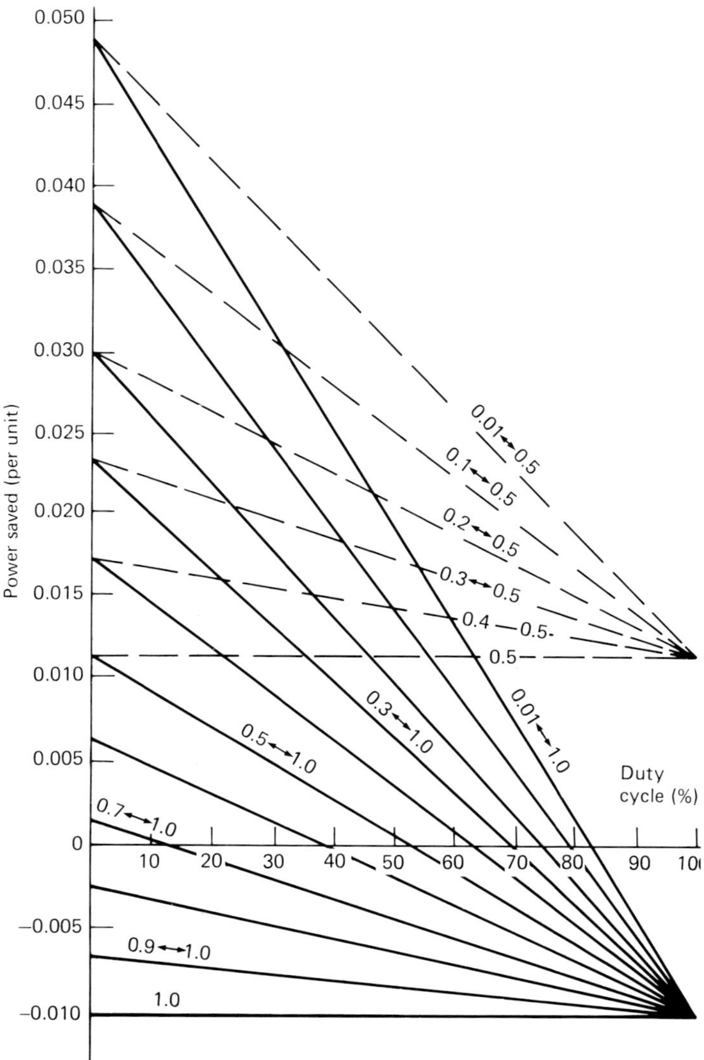

Figure 4.41 Power saved for the machine of Figure 4.38 for operation at 10% overvoltage: solid lines, 1.0 per unit for nominal load; dashed lines, 0.5 per unit nominal load

conventional d.c. motor drives and force-commutated induction motor drives. The load-commutated inverter-fed synchronous motor drive has emerged as a leading choice for such applications, particularly for those with fan-load-type characteristics. However, line commutated cycloconverter drives have also been used for low-speed applications. Load commutation is made possible by the induced e.m.f. of the synchronous machine which eliminates the need for the expensive forced commutation circuitry required in a conventional current source inverter drive. Load-commutated inverter-fed induction motor drives are also encountered, particularly in retrofit applications, employing capacitor banks at the motor terminals to achieve the leading power factor needed for load commutation.

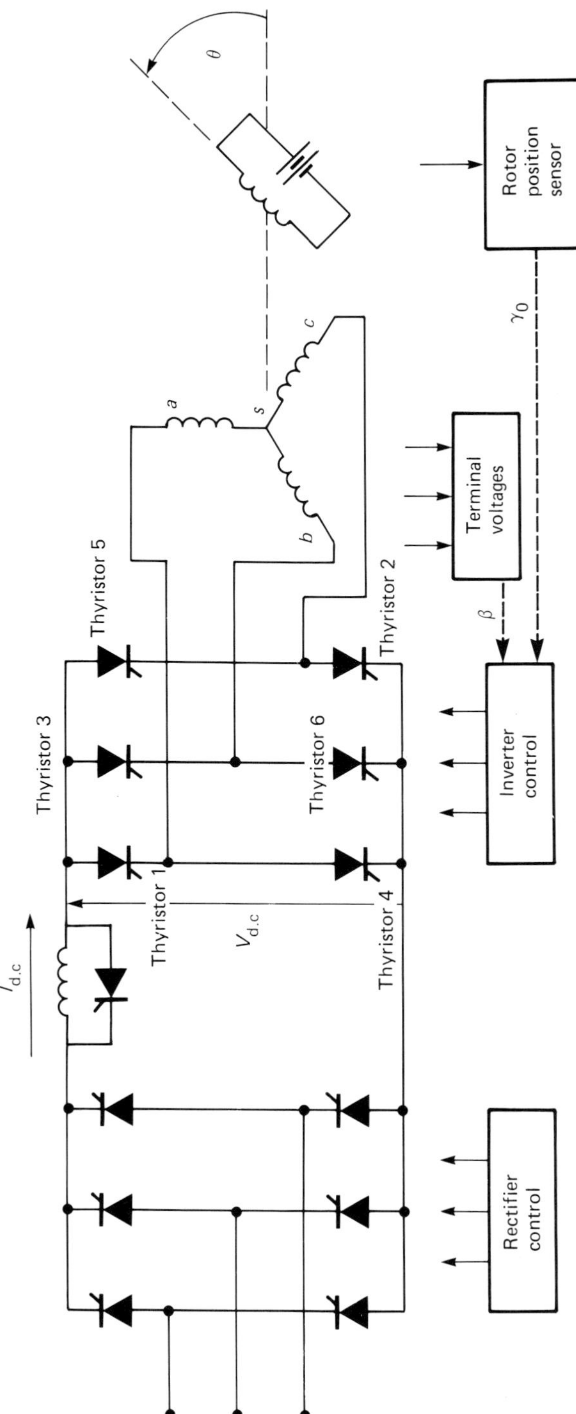

Figure 4.42 Basic load-commutated inverter synchronous motor drive system

4.4.2 Load-commutated inverter synchronous motor drives

The basic load-commutated inverter synchronous motor drive system is shown in Figure 4.42. In this drive, two static converter bridges are connected on their d.c. side by means of a so-called d.c. link having only a link inductor on the d.c. side. The line side converter ordinarily takes power from a constant frequency bus and produces a controlled d.c. voltage at its end of the d.c. link inductor. The d.c. link inductor effectively turns the line side converter into a current source as seen by the machine side converter. Current flow in the line side converter is controlled by adjusting the firing angle of the bridge and by natural commutation of the a.c. line.

The machine side converter normally operates in the inversion mode. Since the polarity of the machine voltage must be instantaneously positive as the current flows into the motor to commutate the bridge thyristors, the synchronous machine must operate at a sufficiently leading power factor to provide the volt-seconds necessary to overcome the internal reactance opposing the transfer of current from phase to phase (commutating reactance). Such load e.m.f.-dependent commutation is called load commutation. As a result of the action of the link inductor, such an inverter is frequently termed a naturally commutated current source inverter.

Figure 4.43 illustrates typical circuit operation. Inverter thyristors 1–6 fire in sequence, one every 60 electrical degrees of operation, and the motor currents

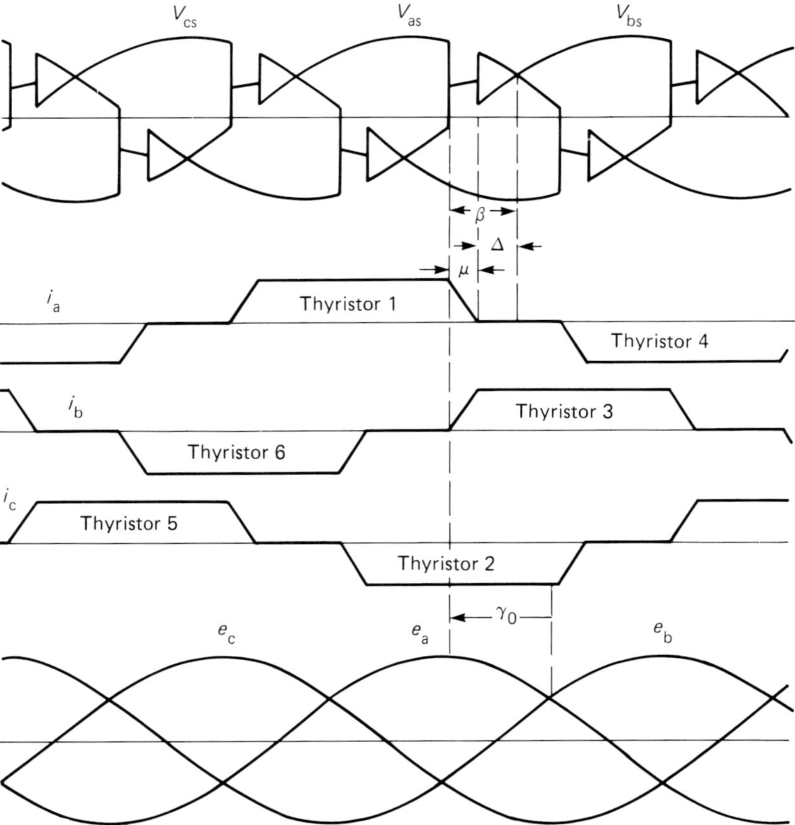

Figure 4.43 Idealized load-commutated inverter synchronous motor waveforms

form balanced three-phase rectangular waves. The electrical angles shown in Figure 4.43 pertain to commutation from thyristor 1 to 3. The instant of commutation of this thyristor pair is defined by the phase advance angle β relative to the machine terminal voltage V_{ab}. Once thyristor 3 is switched on, the machine voltage V_{ab} forces current from phase a to phase b. The rate of rise of current in thyristor 3 is limited by the commutating reactance, which is approximately equal to the subtransient reactance of the machine.

During the interval defined by the commutation overlap angle μ, the current in thyristor 3 rises to the d.c. link current $I_{d.c.}$ while the current in thyristor 1 falls to zero. At this instant, V_{ab} appears as a negative voltage across thyristor 1 for a period defined as the commutation margin angle Δ. The angle Δ defines, in effect, the time available to the thyristor to recover its blocking ability before it must again support forward voltage. This time $T_r = \Delta/\omega$ is called the recovery time of the thyristor. The phase advance angle β is equal to the sum of μ plus Δ. The angle β is defined with respect to the motor terminal voltage. In practice it is useful to define a different angle γ_0 measured with respect to the internal e.m.f. of the machine.

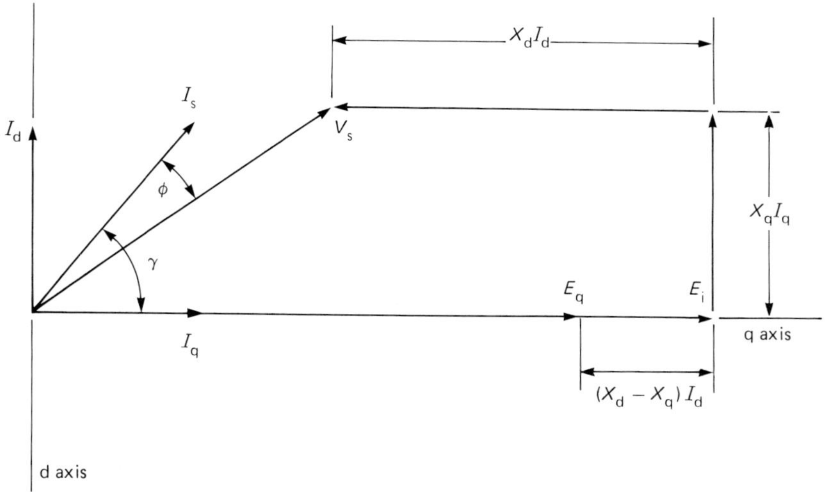

Figure 4.44 Per phase phasor diagram of load-commutated inverter synchronous motor drive operation

This angle is called the firing angle. Since the internal e.m.f.s are simply equal to the time rate of change of the rotor flux linking the stator windings, the firing angle γ_0 can be located physically as the instantaneous position of the salient poles of the machine, i.e. the d axis of the machine. Hence, in general, the system is typically operated in a self-synchronous mode where the output shaft position (or a derived position-dependent signal) is used to determine the applied stator frequency and phase angle of current. The fundamental component per-phase phasor diagram of Figure 4.44 illustrates this requirement. In this figure the electrical angle γ is the equivalent of γ_0 but corresponds to the phase displacement of the fundamental component of stator current with respect to the e.m.f. Spatially, γ corresponds to 90° minus the angle between the stator and rotor m.m.f.s and may be called the m.m.f. angle. A large leading m.m.f. angle γ is clearly necessary to obtain a leading terminal power factor angle ϕ.

4.4.3 Torque production in a load commutator inverter synchronous motor drive

The average torque developed by the machine is related to the power delivered to the internal e.m.f. E_i and, from Figure 4.44, can be written as:

$$T_e = \frac{3E_i \cos \gamma}{\omega_{rm}} \tag{4.58}$$

where ω_{rm} is the synchronous mechanical speed. Since E is speed-dependent, the apparent speed dependence vanishes and Equation (4.58) takes the form:

$$T_e = K\lambda_{af}I_s \cos \gamma \tag{4.59}$$

where λ_{af} is the field flux linking the stator phase winding. Thus, for a fixed value of the internal angle γ, the system behaves like a d.c. machine (as for field-oriented drives) and direct steady state torque control is possible.

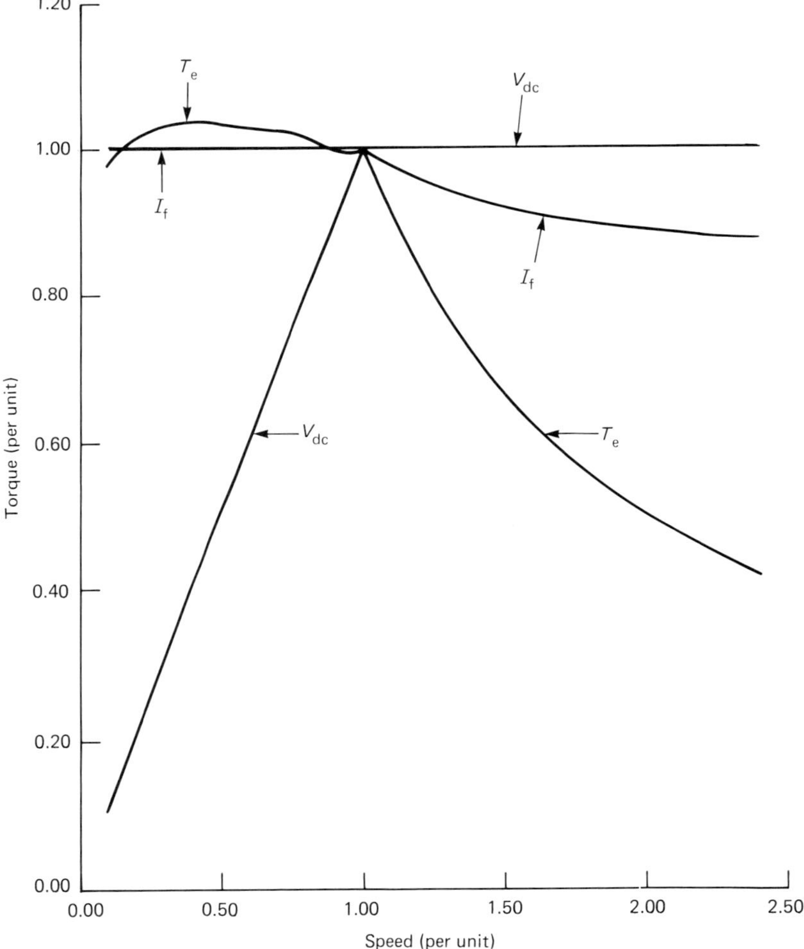

Figure 4.45 Capability curve of a load-commutated inverter synchronous motor drive with constant d.c. link current and constant commutation margin time. Field-weakening operation assumes operation at constant d.c. link voltage

4.4.4 Torque capability curves

One useful measure of drive performance is a curve showing the maximum torque available over its entire speed range. A synchronous motor supplied from a variable-voltage, variable-frequency supply will exhibit a torque–speed characteristic similar to that of a d.c. shunt motor, as previously shown in Figure 4.6(a) (page 194). If field excitation control is provided, operation above base speed in a field-weakened mode is possible and is used widely. The upper speed limit is dictated by the required commutation margin time of the inverter thyristors.

Figure 4.45 is a typical capability curve assuming operation at constant-rated d.c. link current, at rated (maximum) converter d.c. voltage above rated speed and with a commutation margin time Δ/ω of 26.5 ms corresponding to $\Delta = 12°$ at 50 Hz. At very low speeds, where the commutation time is of the order of the motor transient time constants, the machine resistances make up a significant part of the commutation impedance. The firing angle must subsequently be increased to provide sufficient volt-seconds for commutation. The resulting increase in internal power factor angle reduces the torque capability. At intermediate speeds the margin angle can be reduced to values less than 12° to maintain 26.5 ms margin time and slightly greater than rated torque can be produced.

Above rated speed the inverter voltage is maintained constant and the drive, in effect, operates in the constant kilovolt-ampere mode. The d.c. inverter voltage

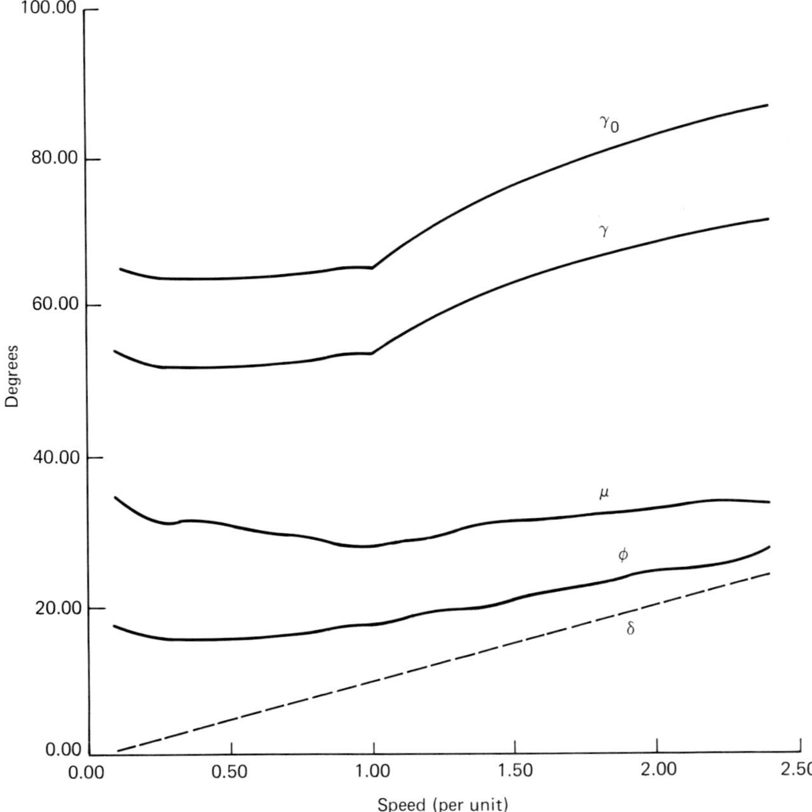

Figure 4.46 Characteristic electrical angles of a load-commutated inverter synchronous motor drive corresponding to Figure 4.45

reaches the maximum value allowed by the device ratings and the maximum output of the rectifier. Although Figure 4.45 shows a weakening of the field in the high-speed condition, the reduction is not as great as the inverse speed relationship indicated in Figure 4.6. This again is a consequence of the constant commutation margin angle control. Since the margin angle increases with speed, i.e. frequency, to maintain the same margin time, the corresponding increase in power factor angle results in a greater demagnetizing component of stator m.m.f. This offsets partly the need to weaken the field in the high-speed region. A plot of the machine angles corresponding to Figure 4.45 is shown in Figure 4.46.

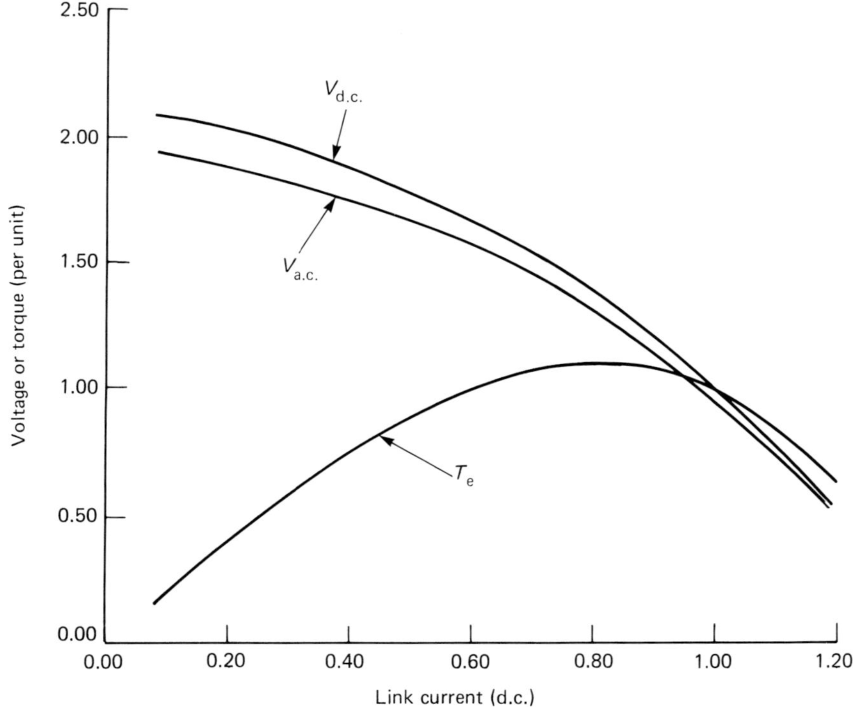

Figure 4.47 Motor torque, a.c. and d.c. voltages for operation at rated speed and rated field current as a function of d.c. link current. Margin angle $\Delta = 10°$

4.4.5 Constant speed performance

When the d.c. link current is limited to its rated value, the maximum torque can be obtained from the capability curve (Figure 4.45). However, operation below maximum torque requires a reduction in the d.c. link current. When field current is held constant, the torque, voltage, m.m.f. angle γ and power factor angle φ vary as shown in Figures 4.47 and 4.48 as a function of the d.c. link current. It can be noted that the inverter d.c. voltage increases at small values of torque so that it is typically necessary to control the field current and d.c. link current simultaneously as a function of load. When the field current is adjusted to keep the angle Δ at its limiting value, the curves of Figure 4.49 result. It can be noted that the torque is now essentially a linear function of d.c. link current so that the d.c. link current command becomes, in effect, the torque command.

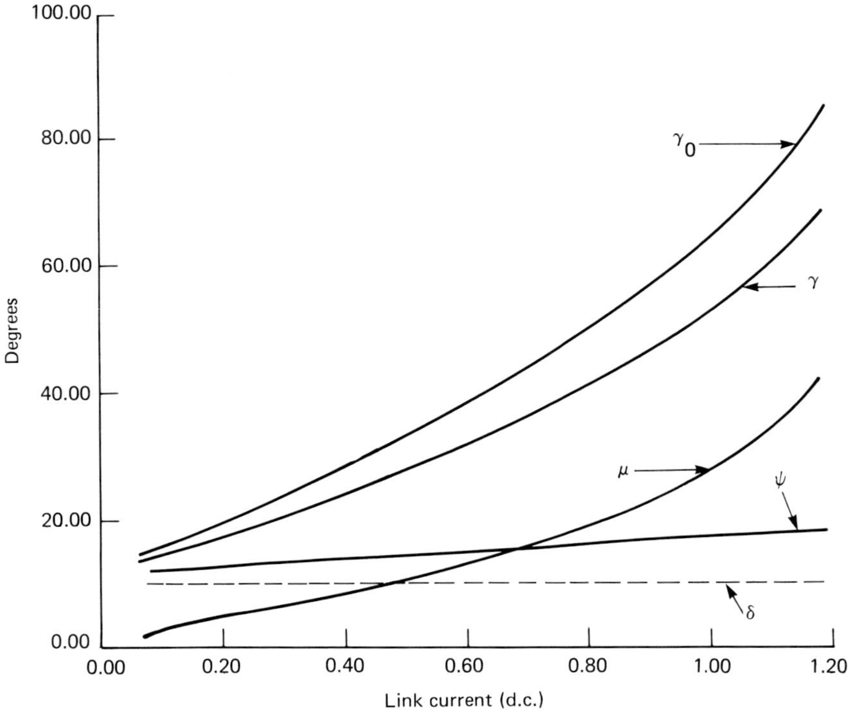

Figure 4.48 Electrical angles corresponding to Figure 4.47 for operation at rated speed and rated field current as a function of d.c. link current

4.4.6 Control considerations

Direct control of γ by use of a rotor position sensor has traditionally been applied in load commutator invertet drives but has largely been replaced by schemes using terminal voltage and current sensing to indirectly control γ. Direct control of the commutation margin angle Δ (more correctly, the margin time Δ/ω where ω is the motor angular frequency) illustrated in Figure 4.43 has the advantage of causing operation at the highest possible power factor and hence gives the best utilization of the machine windings. The waveforms in Figure 4.43 also demonstrate that changes in the commutation overlap angle μ resulting from current or speed changes produce significant differences between the actual value of γ and the ideal value γ_0. For this reason, compensators are required in direct γ controllers. This compensation is automatic in systems based on controlling the margin angle Δ.

4.4.7 Starting considerations

Load commutation of a synchronous machine is accomplished by adjusting the field current such that the e.m.f. is sufficiently large (and lagging the current) to ensure transfer of current from phase to phase. The e.m.f. is proportional to rotor speed so operation at or near zero speed is not possible, as the e.m.f. approaches zero. Commutation of the machine side bridge, however, can be maintained if the line side bridge, which always has sufficient commutating kilovars, is controlled to produce zero current intervals on the d.c. link, as shown in Figure 4.50. Since the d.c. current becomes zero, the currents in all thyristors are also zero. The zero

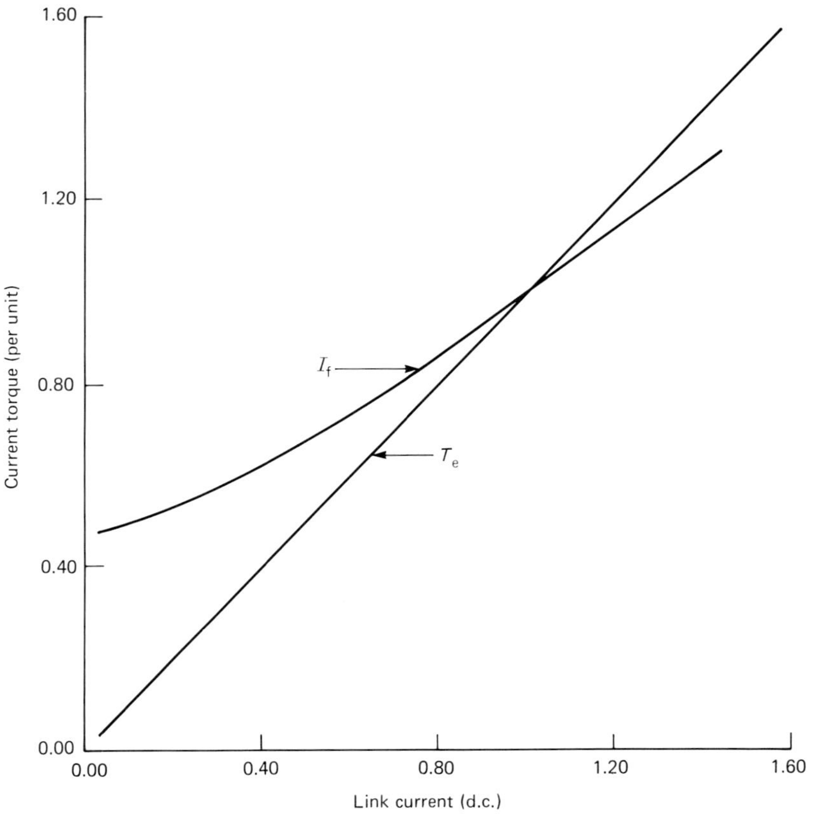

Figure 4.49 Motor torque and field current for operation at rated speed with rated inverter a.c. voltage. Margin angle $\Delta = 10°$

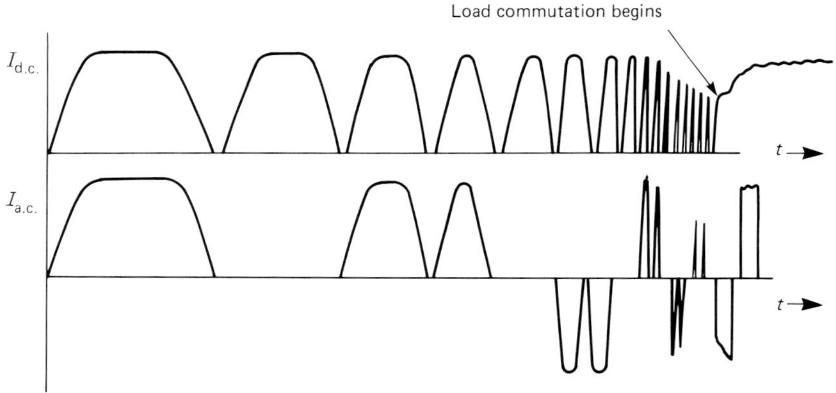

Figure 4.50 Direct current link current and motor phase current illustrating artificial commutation at low speed

current intervals allow the previously conducting thyristors comprising the load side converter to recover their blocking ability. The bridge thyristors are sequentially fired in pairs to direct the current through the two appropriate phases and, hence, synchronize the stator m.m.f. to the rotor as it begins to rotate. Since six commutations occur per cycle, the frequency of the d.c. link pulses must be 6 times the motor line frequency. Because of the inductance afforded by the d.c. link inductor, the pulsing procedure can only be continued to 2–3 Hz, at which time normal load commutation is initiated. In most cases rated motor torque cannot be supported by artificial commutation even to 2–3 Hz. Fortunately, however, such drives find their greatest application in fan-type loads in which the torque requirements at low speed are relatively modest.

4.4.8 Load-commutated induction motor drives

Because induction machines require reactive energy for excitation, they are inherently incapable of supplying the reactive power needed to load-commutate the machine side bridge. Load commutation can be preserved, however, if capacitors are added to help provide self-excitation of the machine as well as to commutate the bridge. Such a motor drive configuration is shown in Figure 4.51. The system is most widely used in large retrofit applications in which the energy savings are substantial but do not warrant replacing the installed induction motor by a synchronous machine.

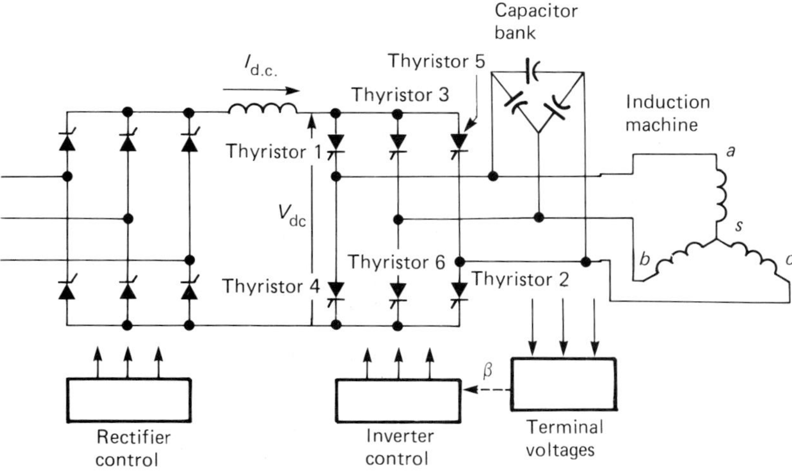

Figure 4.51 Load-commutated induction motor drive

4.4.9 Cycloconverter-fed synchronous and induction motor drives

Certain applications, e.g. ball mill drives, inherently require low-speed operation. The low-speed shaft is typically obtained by using a conventional four-pole machine together with a gearset having a large gear ratio. The cost and maintenance of these large gears has prompted recent application of variable-speed drives for such applications. These high-power applications are naturally served by cycloconverter drives since the switching requirements of these converters inherently make them suitable for low-frequency applications. Figure 4.52 shows a schematic of a typical cycloconverter drive. Since commutation is obtained from the supply line, the

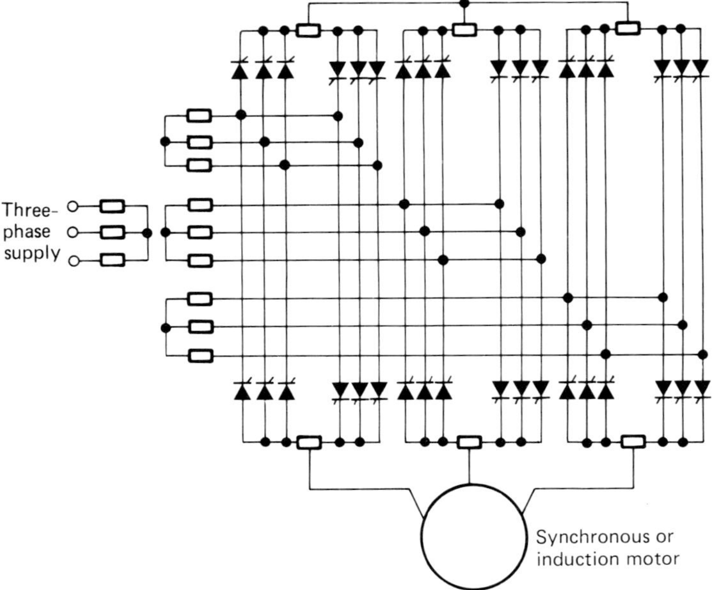

Figure 4.52 Cycloconverter-fed a.c. motor drive

machine can be either an induction or synchronous machine. However, since such applications typically require a very large torque capability at low speed, the most frequently used machine is the synchronous type.

4.5 Forced commutation variable-frequency drives

4.5.1 Introduction

Variable-frequency a.c. drives are now available from fractional kilowattage to very large sizes, e.g. to 15 000 kW for use in electric generating stations. In large sizes, naturally commutated converters are more common, usually driving synchronous motors. However, in low to medium sizes (up to approximately 750 kW) forced commutated converters driving induction motors are almost exclusively used.

4.5.2 Forced commutation inverters

At present, essentially all forced commutation converters are d.c. link inverters. There are two basic power circuit configurations in use: (1) the voltage source inverter; and (2) the current source inverter. An important variation of the voltage source inverter configuration is the pulse width modulated inverter which incorporates voltage control as well as variable frequency operation in the basic inverter power circuit.

4.5.2.1 Inverter topology and power switches

Figure 4.53 illustrates the basic power circuit topology of voltage source inverters and current source inverters. Only the main power-handling devices are shown; auxiliary circuitry such as snubbers or commutation elements are excluded. The circuits are illustrated with thyristors as the switching elements. In the voltage source inverter, bipolar transistors, gate turnoff thyristors or field effect transistors are commonly employed up to several hundred kilowatts. Thyristors are universally employed in current source inverter systems and in the high-power range for voltage source inverter systems.

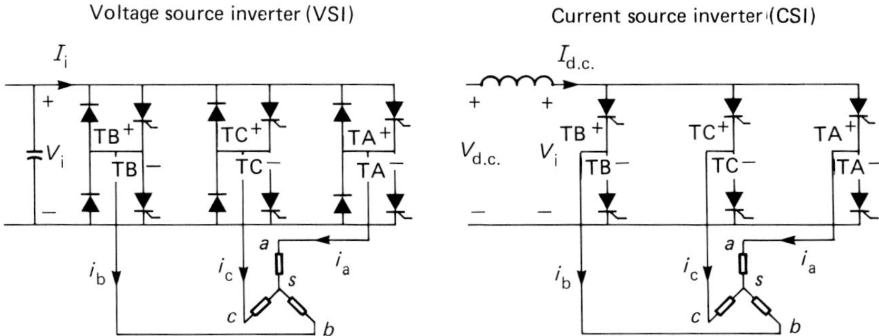

Figure 4.53 Power circuit configuration of voltage source and current source inverters

As shown in Figure 4.53, the voltage source inverter can be distinguished from the current source inverter by the presence of reverse-connected diodes across each main switch in the voltage source inverter. These 'feedback' or 'reactive' diodes provide reverse-current paths such that when a particular power switch is turned on, one input d.c. terminal and one output a.c. terminal are connected regardless of the direction of the load current. The voltage source inverter is also characterized by the d.c. link capacitor shown in Figure 4.53. This capacitor is required to allow momentary reverse currents in the d.c. link and can be omitted only if the d.c. source is fully bilateral (battery supply, for example). The series d.c. link inductor in the current source inverter serves to allow momentary voltage swings on the d.c. bus.

4.5.2.2 Basic conduction modes and waveforms

Figures 4.54 and 4.55 illustrate the six circuit connections and the resulting steady-state output waveforms associated with the basic inverter circuits. In the voltage source inverter, each power switch is turned on for a full 180° of the output cycle resulting in three 'on' switches at all instants of output. All three output terminals are thus connected to the two input terminals at all instants and there is always one line-to-line short circuit at the output. As a result, all output voltages are directly related to the input voltage (leading to the term voltage source inverter) but there is a 'free' output current circulating in the short-circuit path. The resulting output waveforms are shown in Figure 4.54 and follow directly from the conduction mode diagrams.

In contrast, the power switches in a current source inverter are only 'on' for a 120-segment of the output cycle resulting in only two 'on' switches at every instant. There is, therefore, always one open-circuited output terminal and the three-phase output currents are directly related to the d.c. current. There is a degree of freedom in the output voltage because of the open circuit; the system is the dual of the voltage source inverter. The waveforms are identical to those in a voltage source inverter except the current is the controlled quantity, as shown in Figure 4.55.

For an inductive load, such as a motor, the current cannot follow the ideal rectangular pattern of Figure 4.55 and the switching is slowed down deliberately, to

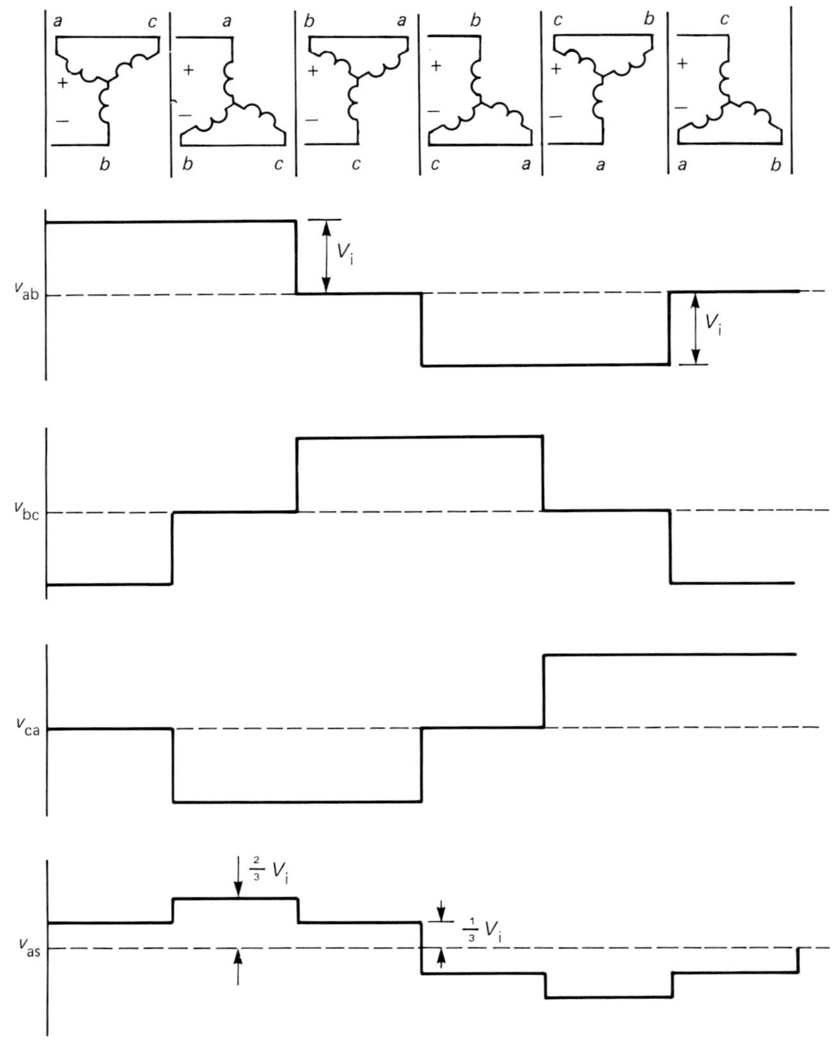

Figure 4.54 Conduction modes and output voltage waveforms in a six-step voltage source inverter

form a more trapezoidal pattern. Unlike the voltage source inverter, where a wide variety of switching devices and commutation arrangements are in use, there is essentially a single preferred commutation circuit for the current source inverter, as shown in Figure 4.56. This circuit automatically and sequentially (automatic and sequential current inverter) commutates the current from one thyristor to the next following a simple gate signal to the oncoming thyristor. The capacitors shown in Figure 4.56 must be co-ordinated with the load inductance to ensure commutation.

4.5.2.3 Pulse width modulated operation

The six-step operation described in the preceding section provides a variable-frequency output wave with the amplitude controlled by the d.c. link voltage or

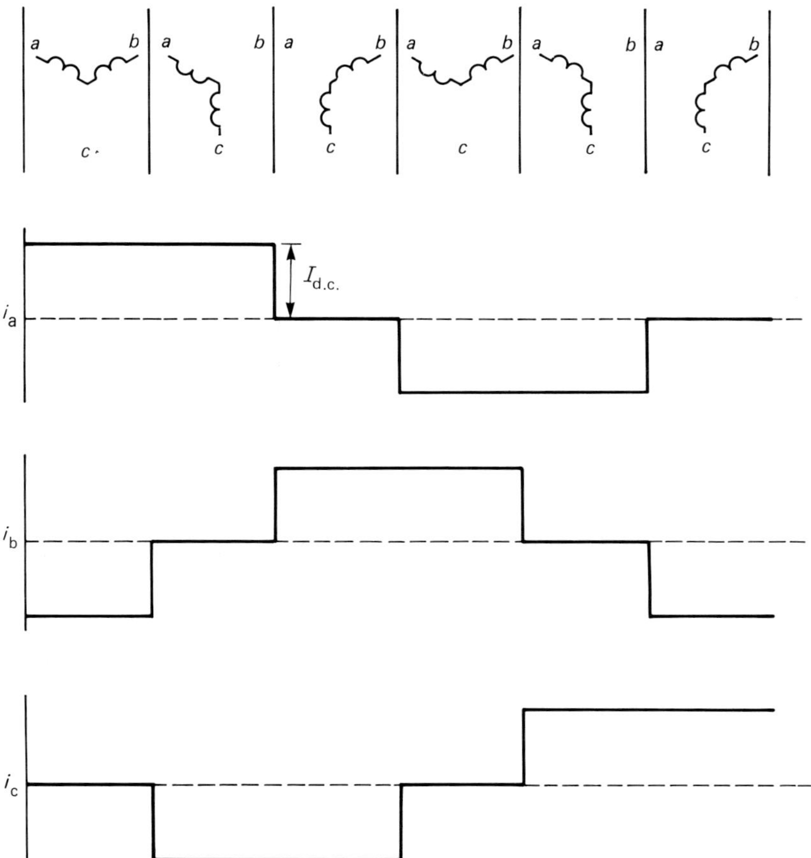

Figure 4.55 Conduction modes and output voltage waveforms in a six-step current source inverter

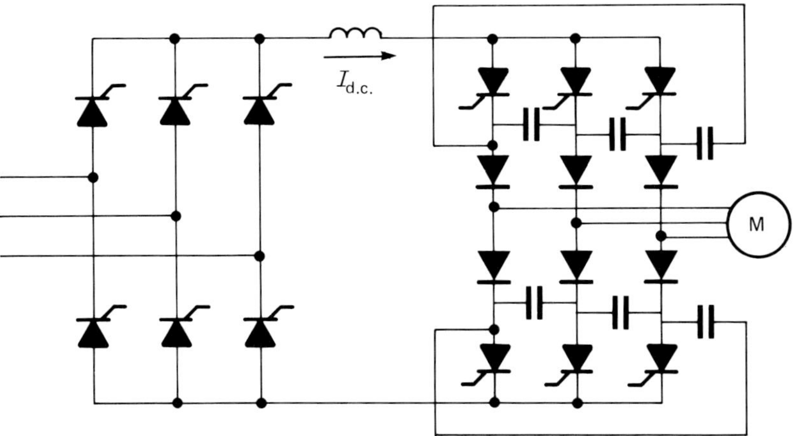

Figure 4.56 Schematic of an auto-sequentially commutated current source inverter

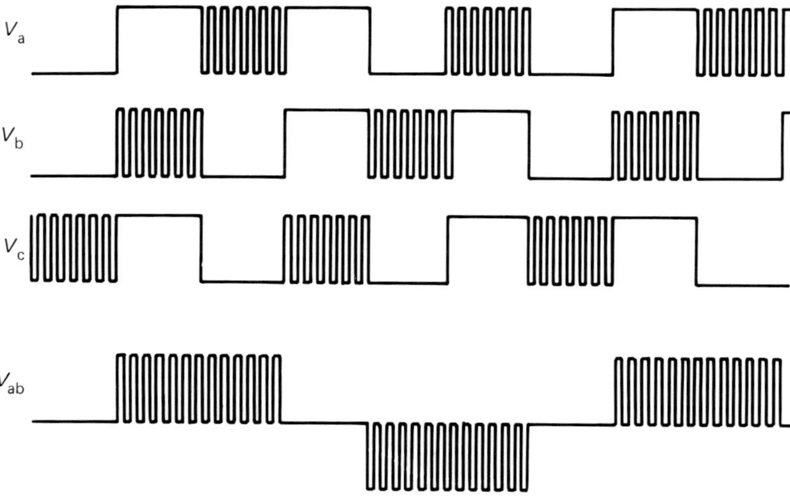

Figure 4.57 Typical block modulation waveforms showing voltages from line-to-negative d.c. bus and line-to-line voltage V_{ab}

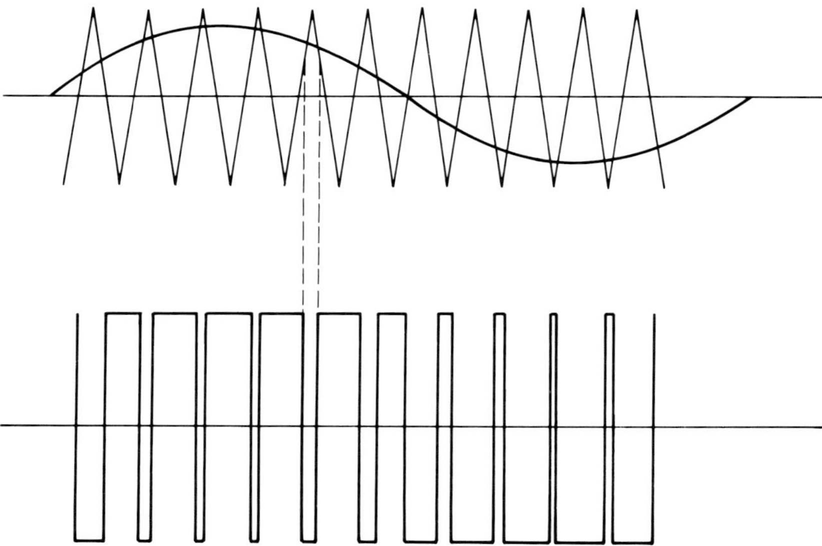

Figure 4.58 Illustration of sinusoidal pulse width modulation using sine-triangle comparison

current. Variable-amplitude operation thus requires a controlled front end converter; a phase-controlled rectifier or d.c. chopper is commonly used (Sections 4.2.4, 4.2.6, 4.2.9). An alternative is pulse width modulation of the main power switches to cause them to operate as choppers. Except at very low frequencies, this alternative is limited to the voltage source inverter because of limitations of the automatic and sequential current inversion converter.

Pulse width modulation–voltage source inverter operation results in an additional conduction mode in which all three output terminals are connected to one input terminal resulting in a three-phase short-circuit at the output.

Modulation techniques fall into two classes: (1) those which operate at a fixed ratio to the fundamental switching frequency (block modulation); and (2) those in which the switching ratio is continuously changing, usually sinusoidally, to synthesize a more nearly sinusoidal output (sinusoidal pulse width modulation).

A simple form of block modulation is shown in Figure 4.57 where the modulation is limited to the middle 60° of each conduction period. Block modulation is the simplest type of modulation and has for the most part been replaced by various forms of sinusoidal pulse width modulation. This has the advantage of reducing greatly the low-order harmonics and this provides much smoother low-speed operation. Figure 4.58 illustrates one means of generating sinusoidal pulse width modulation by comparing a high-frequency triangular wave with a fundamental-frequency sinewave. The intersections of the two waves provide the on–off signals for the modulated wave. Digital versions of this and other pulse width modulation strategies are rapidly replacing older analogue schemes.

Pulse width modulation operation offers the advantages of smoother torque production, reduced motor losses and elimination of a separate voltage control stage. As offsetting features, pulse width modulation switching losses are high, inverter control circuitry is more complex and interactive, and motor acoustic noise is often higher. In general, pulse width modulation inverters all revert to six-step operation above base frequency.

4.5.2.4 Steady state performance

A Fourier series representation of the rectangular and six-step waveforms of the voltage source inverter (Figure 4.54) yields the following relations between the d.c. link inverter voltage V_i and the r.m.s. line and phase fundamental voltages:

$$V_{\text{ll(r.m.s.)}} = \frac{\sqrt{6}}{\pi} V_i$$

$$V_{\text{ln(r.m.s.)}} = \frac{\sqrt{2}}{\pi} V_i \qquad\qquad (4.60)$$

The harmonic spectra vary inversely as the harmonic order h:

$$V_{\text{ll}(h)} = \frac{1}{h} V_{\text{ll(r.m.s.)}}$$

$$V_{\text{ln}(h)} = \frac{1}{h} V_{\text{ln(r.m.s.)}} \qquad\qquad (4.61)$$

with all even and triple harmonics ($h = 3,9,15\dots$) absent. The duality of the current source inverter (Figure 4.55) results in the expressions:

$$I_{\text{ll(r.m.s.)}} = \frac{\sqrt{6}}{\pi} I_{\text{d.c.}} \qquad \text{(delta connected machine)}$$

$$I_{\text{ln(r.m.s.)}} = \frac{\sqrt{2}}{\pi} I_{\text{d.c.}} \qquad \text{(wye or delta connected machine)} \qquad (4.62)$$

with the same harmonic spectra as in the voltage source inverter.

In these polyphase inverters, the input d.c. link power has a sixth harmonic pulsation but no second harmonic. The fundamental frequency reactive power of the a.c. motor load is thus not reflected to the d.c. link but is supplied locally by the inverter switching itself (the circulating short-circuit current in the voltage source inverter, for example). A steady state real power balance thus yields the relation between input and output fundamental current in a voltage source inverter (and voltage in a current source inverter):

$$I_i = \frac{3\sqrt{2}}{\pi} I_{ll(\text{r.m.s.})} \cos \theta \qquad \text{(voltage source inverter)} \qquad (4.63)$$

$$V_i = \frac{3\sqrt{6}}{\pi} V_{ln(\text{r.m.s.})} \cos \theta \qquad \text{(current source inverter)} \qquad (4.64)$$

The fundamental frequency–voltage current relationships for the basic voltage source inverter and current source inverter circuits are conveniently summarized by the per-phase fundamental component equivalent circuits shown in Figure 4.59. The inherent reactive supply capabilities of the basic inverter systems are clearly illustrated by the adjustable reactive elements (which are always in resonance with the load) used in these circuit models. In the pulse width modulated equivalent circuit, the variable-ratio transformer represents pulse width modulated voltage control in terms of the voltage reduction factor a_1 resulting from pulse width modulation operation.

4.5.2.5 Controlled-current operation

In torque-controlled drives, the inverter is typically operated in a controlled current mode in which the amplitude and phase of the motor stator current are both regulated. The current source inverter is operated readily in this mode by regulating the d.c. link current (amplitude of a.c.) and varying the firing instants of the inverter thyristors to regulate the phase of the output a.c. At low output frequencies, a form of pulse width modulation is sometimes employed to give smaller steps in the phase of the output current waveform. Compensation for the commutation time of the inverter is usually required at high output frequencies.

Pulse width modulation inverters with external regulating loops are employed where more precise and rapid current control is needed (servo drives, for example). Figure 4.60 illustrates the basic scheme. Current feedback signals are compared with the reference input current waves, and the error signals are employed to develop firing signals for the pulse width modulation inverter. The output current is a reproduction of the reference wave with a superimposed ripple at the pulse width modulation carrier frequency. A variety of current regulation schemes are employed, the simplest being an on–off hysteresis-type controller, and the most commonly used being some variation of sine-triangle pulse width modulation where the reference wave replaces the sinusoidal signal. The current sensors must have wide bandwidth, including d.c., and little or no offset to achieve acceptable performance. The method is called current-regulated pulse width modulation.

4.5.3 Forced commutation adjustable-speed a.c. drives

Adjustable-speed a.c. drives employing forced commutation inverters and induction motors are used very widely in applications which do not require precise control of motor torque (fans, centrifugal pumps, etc.). These drives operate 'open loop' with respect to motor speed and, while there are many variations, all essentially provide a programmed voltage–frequency profile at variable frequency and utilize the normal speed–torque characteristics of the induction motor. Synchronous motors (permanent magnet or reluctance type) are also used where precise speed control is required.

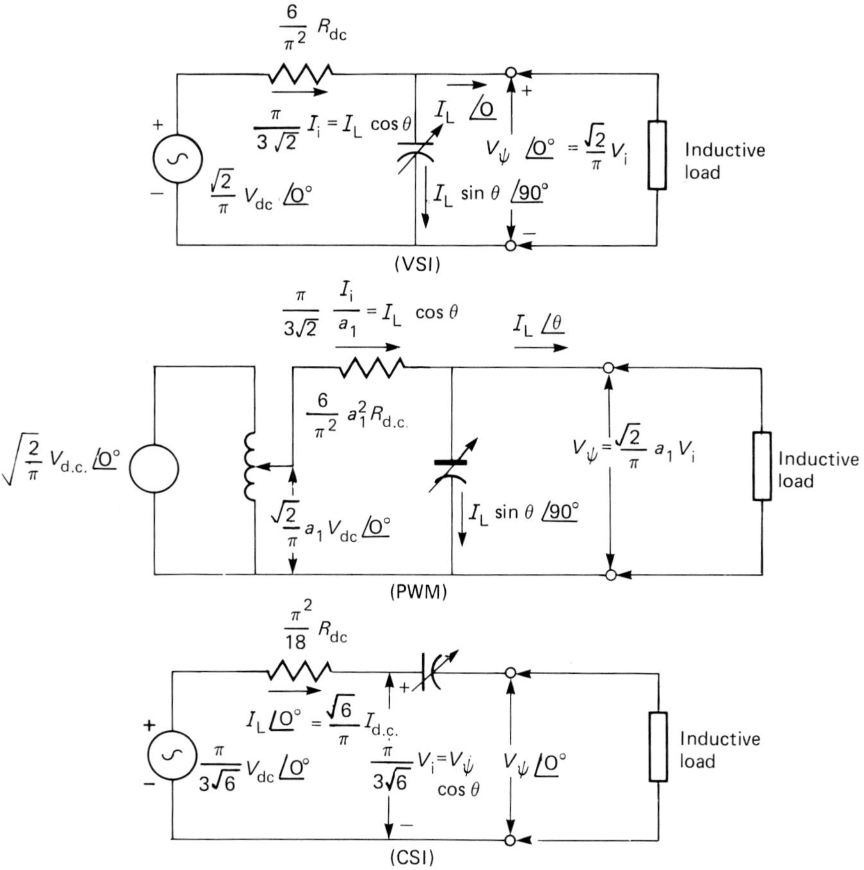

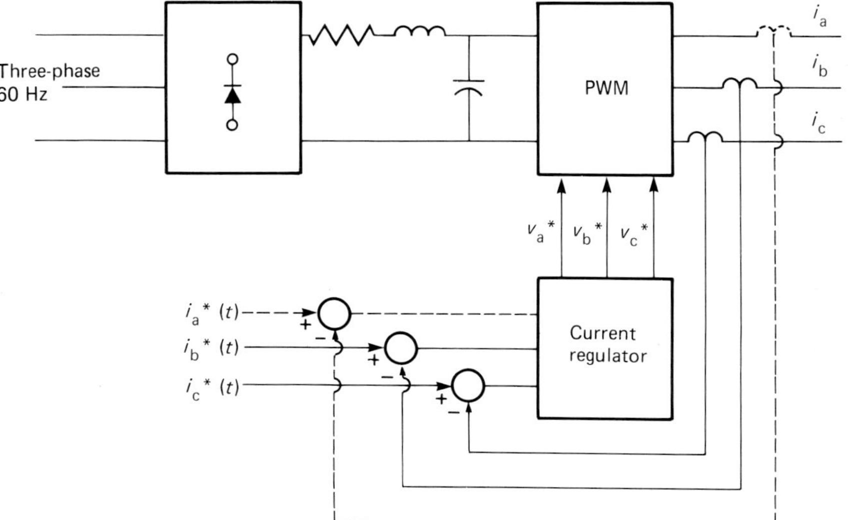

Figure 4.59 Per-phase fundamental component equivalent circuits of inverters.
(a) Voltage source inverter. (b) Pulse width modulated inverter. (c) Current source
inverter

Figure 4.60 Pulse width modulated inverter with current regulator to produce a
controlled three-phase current source

4.5.3.1 Induction machine capability curves

When operated with a variable-frequency supply, the induction machine has a capability curve very much like that of a d.c. machine. Below base speed the machine is operated at constant volts/hertz and provides constant torque capability at constant current and constant slip frequency. Increased voltage/hertz (voltage boost) is required at low frequencies to compensate the stator IR drop. The reduced cooling at low speed will typically require some derating. Above base speed, field-weakened operation is attained by increasing the frequency with a fixed terminal voltage (reduced voltage/hertz) as dictated by the maximum output voltage available from the power converter. To utilize fully the converter and motor current ratings, the slip frequency can be increased with the result that, for a limited range, the motor power output is nearly constant. Unlike the d.c. machine, the upper speed for constant power is limited since, at some speed, the slip frequency reaches the breakdown value. Beyond this speed, the slip frequency is best held constant and the power falls off inversely as the speed (torque falling as speed squared). These properties are summarized in Figure 4.61.

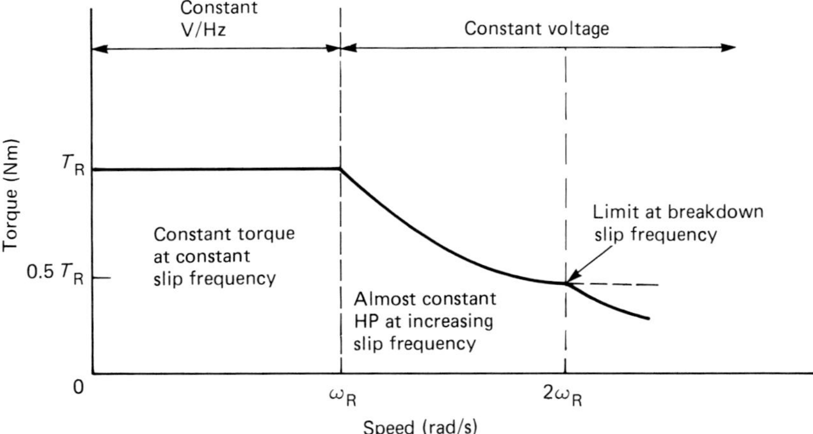

Figure 4.61 Capability curves for an induction machine operated from a controlled variable frequency supply, with 2:1 field-weakening range

4.5.3.2 Six-step voltage source inverter and pulse width modulation drives

Six-step voltage source inverters and pulse width modulated inverters are employed as adjustable speed drives by providing a d.c. bus supply. For battery-driven systems, the pulse width modulator is appropriate because of its inherent voltage control capability. In a.c.-supplied systems, a diode rectifier is used in pulse width modulated drives and either a phase-controlled rectifier or diode bridge-chopper system for d.c. bus voltage control is employed for six-step voltage source inverter systems. A d.c. link filter is commonly used in either case, although the link inductor is sometimes omitted in very small drives. The diode rectifier systems offer better line-side power factor and reduced line notching for low-speed operation.

In the a.c.-supplied systems, the maximum line-to-line fundamental r.m.s. inverter output voltage $V_{0(ll-r.m.s.)}$ is determined by the full-wave rectified input voltage ($3\sqrt{2}/\pi$ times the source r.m.s. line-to-line voltage) and is equal to:

$$V_{0(ll-r.m.s.)} = \frac{\sqrt{6}}{\pi}\left(\frac{3\sqrt{2}}{\pi}\right) V_{S(ll-r.m.s.)} = 1.05\, V_{S(ll-r.m.s.)} \tag{4.65}$$

Most systems provide rated frequency at nominal supply voltage to match standard motors. Adjustments for low-frequency voltage boost and base speed voltage/hertz trim are normally provided. A voltage regulator to handle fluctuations in the supply voltage is typical and extended output frequency to about twice rated frequency at constant voltage for constant power operation is often provided. Because of the very limited overcurrent capability of the power switches, an instantaneous overcurrent trip is always provided, and many (but not all) systems have an internal current limit loop which reduces the voltage and frequency when necessary to avoid unnecessary instantaneous overcurrent trip shutdowns.

The primary differences between six-step and pulse width modulated systems relate to motorside harmonics. The voltage harmonics present in the output wave create harmonic motor currents, additional losses and torque pulsations. The harmonic currents are limited primarily by the machine leakage reactance and, hence, higher leakage machines are preferred. For this reason, oversized machines may cause excessive harmonics and high peak currents even when operated at light load.

In the six-step drive, the harmonics and low-speed torque pulsations are unavoidable. Pulse width modulated drives offer potentially much reduced low-order harmonics and small pulsation torques at the expense of generally increased inverter switching losses. A major problem with pulse width modulation is increased motor acoustic noise.

Regenerative operation requires a reversal in d.c. link current and thus a bidirectional d.c.–a.c. converter, dual bridge, for example. A controlled resistor across the link is often provided as an option for limited dynamic braking.

4.5.3.3 Six-step current source inverter drive

Unlike the voltage source inverter and pulse width modulator, which have natural voltage source characteristics, the current source inverter requires feedback control to allow operation as an adjustable speed drive. The most common control scheme is illustrated in Figure 4.62. The outer control loop is a machine terminal voltage-regulating loop. Typically, the output of this loop serves as a reference for a

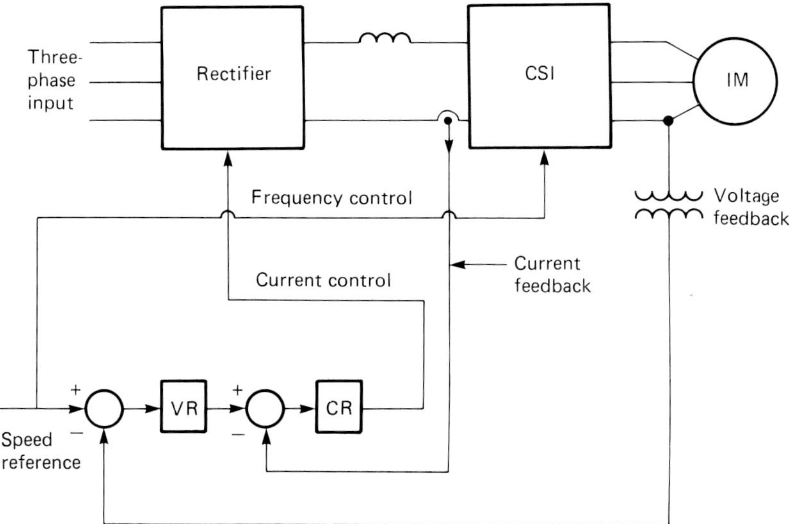

Figure 4.62 Block diagram of a current source inverted induction motor drive with motor voltage feedback loop and open-loop frequency control

current regulator which controls the d.c. bus supply, either a phase-controlled rectifier or diode rectifier–chopper. In effect, the scheme of Figure 4.62 converts the current source inverter to an equivalent voltage source and open-loop motor operation like that on a voltage source inverter or pulse width modulation system results.

Essentially the same type of voltage/hertz schedule as used in the voltage source inverter and pulse width modulator is employed, except that the upper frequency is typically much less than twice-rated frequency because of commutation time requirements. Utilizing the full-wave rectified source voltage as the maximum available Equation (4.64) results in:

$$V_{0(\text{ll–r.m.s.})} \cos \theta = V_{S(\text{ll–r.m.s.})} \tag{4.66}$$

which indicates that the maximum available motor voltage again allows operation at rated frequency and nominal voltage. The d.c. bus current is directly proportional to motor current (Equation (4.62)).

The most significant differences between current source inverter and voltage source inverter or pulse width modulator drives concern motor harmonics, regeneration and fault currents. In the current source inverter, the harmonic currents are determined by the inverter and always are proportional to the fundamental motor current (varying inversely as the harmonic order). The motor leakage reactance determines the harmonic voltage and, hence, lower leakage machines are preferred; a lower-power motor can thus cause excessive harmonic (spike) voltage because of its higher leakage inductance.

Regeneration is much simpler in current source inverter drives because the link voltage, rather than link current, reverses in regeneration. There is thus no need for a dual converter or other bidirectional input converter. The large link inductor also slows down any tendency for rapid link current changes, making fault protection much simpler than in voltage source inversion or pulse width modulation systems with their large link capacitors.

4.5.3.4 Starting characteristics

The very limited overcurrent capability of power electronic converters requires that motor starting be achieved always by operating the motor at a low slip frequency, where torque per ampere is high. This requires starting with a very low stator frequency (1–3 Hz) and slowly ramping the frequency up as the motor accelerates. The problem in adjustable speed drives is that motor speed is not known and, hence, the proper frequency ramp is not known. All drives provide for adjustment of the ramp time to meet various starting requirements. A variety of other special functions are employed in adjustable speed drives to assist in starting; among the most effective are current limit loops which reduce frequency and voltage automatically if the current reaches the limit value. Such current loops are also effective for load variations which could otherwise cause overcurrent trips.

Even with current-limit control, the starting torque of adjustable speed drives is limited and this often becomes a determining factor in drive selection. Use of low-speed machines operated at higher than nominal frequency (equivalent to a gear reduction) is often a more effective solution than a larger drive or one of the more expensive, torque-controlled drives.

4.5.4 Forced commutation adjustable-torque drives

Torque-controlled a.c. drives are the a.c. equivalent of a separately excited d.c. machine operated with a current-regulated armature supply. In such machines, a true torque command input port is possible and the output torque follows this command as rapidly as the current regulator supplying the armature can change the armature current. The same type of behaviour can be produced in a.c. machines, but the current regulator must now control both amplitude and phase of the current (which has led to the name 'vector control'). The orthogonal space angle between the armature m.m.f. and field flux which is forced by the d.c. machine commutator must also be attained in the a.c. machine. This spatial orientation requirement is referred to as 'field orientation' in a.c. machine control.

4.5.4.1 Synchronous machine field orientation

Field orientation of a synchronous machine is simple conceptually because the location of the rotor field axis (axis of field winding) is determined readily by a rotor position transducer. Use of optical encoders or resolvers is common.

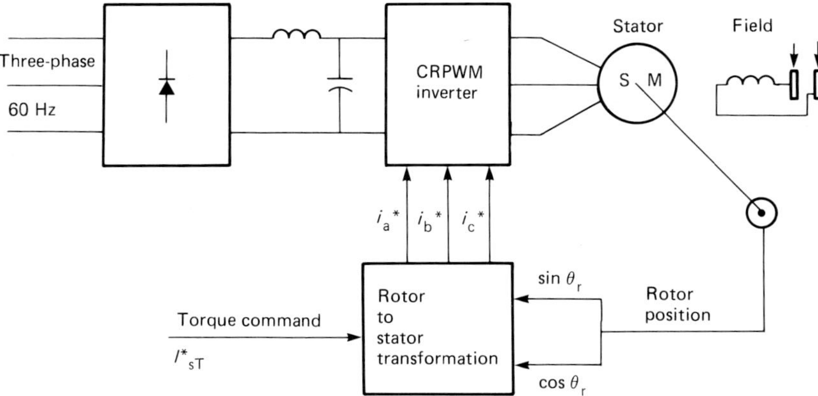

Figure 4.63 Synchronous machine torque control (field orientation) using a current-regulated pulse width modulated inverter

Orthogonal orientation of the rotor field axis and the armature m.m.f. axis is attained by using the rotor position information to control the phase of the armature current using a current-regulated pulse width modulated inverter. Figure 4.63 illustrates the basic concept in block diagram form. The 'rotor-to-stator transformation' is described by the relations:

$$i_a^* = i_{sT}^* \cos \theta_r$$

$$i_b^* = i_{sT}^* \left(-\frac{1}{2}\cos \theta_r + \frac{\sqrt{3}}{2}\sin \theta_r \right) \tag{4.67}$$

$$i_c^* = i_{sT}^* \left(-\frac{1}{2}\cos \theta_r - \frac{\sqrt{3}}{2}\sin \theta_r \right)$$

and ensures the proper orientation of the fields in the machine.

With a controlled 90° angle between the armature m.m.f. and field flux, the torque is given by:

$$T_e = K\lambda_{af} i_{sT} \tag{4.68}$$

where λ_{af} is the field flux linking the stator and i_{sT} is the instantaneous amplitude of the stator current. The system frequency is always matched to shaft speed, and the system is a true a.c. equivalent of the d.c. machine (Equation 4.2).

Systems of the type illustrated in Figure 4.63, utilizing permanent magnet machines are available as servo drives for use in machine tool control, for example. A closely-related type employs trapezoidal flux distribution in the machine and a simple six-step encoder to achieve similar performance (brushless d.c. machine). Control of torque pulsations is a problem in these trapezoidal flux machines.

An important variation of the system in Figure 4.63 is a scheme which replaces the shaft encoder with counter e.m.f. measurement to obtain rotor field position information. Such systems employ typically 120° gating of the inverter switches in a pulse width modulated inverter and utilize the 60° non-conduction interval to sample the counter e.m.f. These systems fail at high speed, when the motor currents persist in the inverter diodes and prevent measurement of the counter e.m.f. Typically, these systems are intended for speed control as opposed to servo applications.

4.5.4.2 Induction machine field orientation

Although the basic field orientation concept is essentially the same as for load commutator inverter synchronous motor drives, implementation in an induction machine is more involved because the rotor field position is not directly measurable and both the excitation and torque-producing currents must co-exist in the stator winding. The control principle is illustrated in the equivalent circuit shown in Figure 4.64. The rotor flux and torque-producing components of the stator current are explicitly shown as $I_{s\phi}$ and I_{sT} and the flux and torque equations in terms of these variables are also given in the figure. Field orientation is accomplished by controlling independently these two current components.

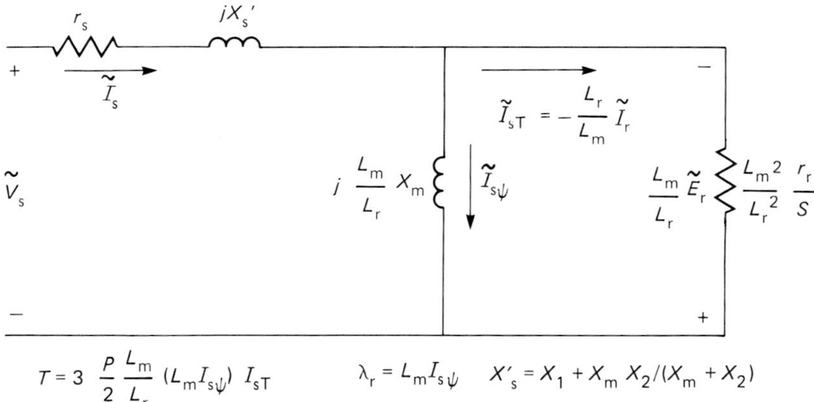

$$T = 3\,\frac{P}{2}\,\frac{L_m}{L_r}\,(L_m I_{s\psi})\,I_{sT} \qquad \lambda_r = L_m I_{s\psi} \qquad X'_s = X_1 + X_m X_2/(X_m + X_2)$$

Figure 4.64 Equivalent circuit showing torque component I_{sT} and rotor flux component $I_{s\phi}$ of stator current

Direct control is accomplished by detecting stator or airgap flux and calculating the rotor flux position using measured stator currents and machine constants. The flux component $I_{s\phi}$ is then placed in phase with the rotor flux and the torque component in quadrature to the flux. Figure 4.65 illustrates the basic concept in block diagram form; the transformation labelled T^{-1} in the figure is an implementation of:

$$i_a^* = i_{sT}\cos\theta_{rf} + i_{s\psi}^*\sin\theta_{rf}$$

$$i_b^* = \left(-\frac{1}{2}i_{sT}^* - \frac{\sqrt{3}}{2}i_{s\psi}^*\right)\cos\theta_{rf} - \left(-\frac{\sqrt{3}}{2}i_{sT}^* - \frac{1}{2}i_\psi^*\right)\sin\theta_{rf} \qquad (4.69)$$

$$i_c^* = \left(-\frac{1}{2}i_{sT}^* - \frac{\sqrt{3}}{2}i_{s\psi}^*\right)\cos\theta_{rf} - \left(-\frac{\sqrt{3}}{2}i_{sT}^* + \frac{1}{2}i_\psi^*\right)\sin\theta_{rf}$$

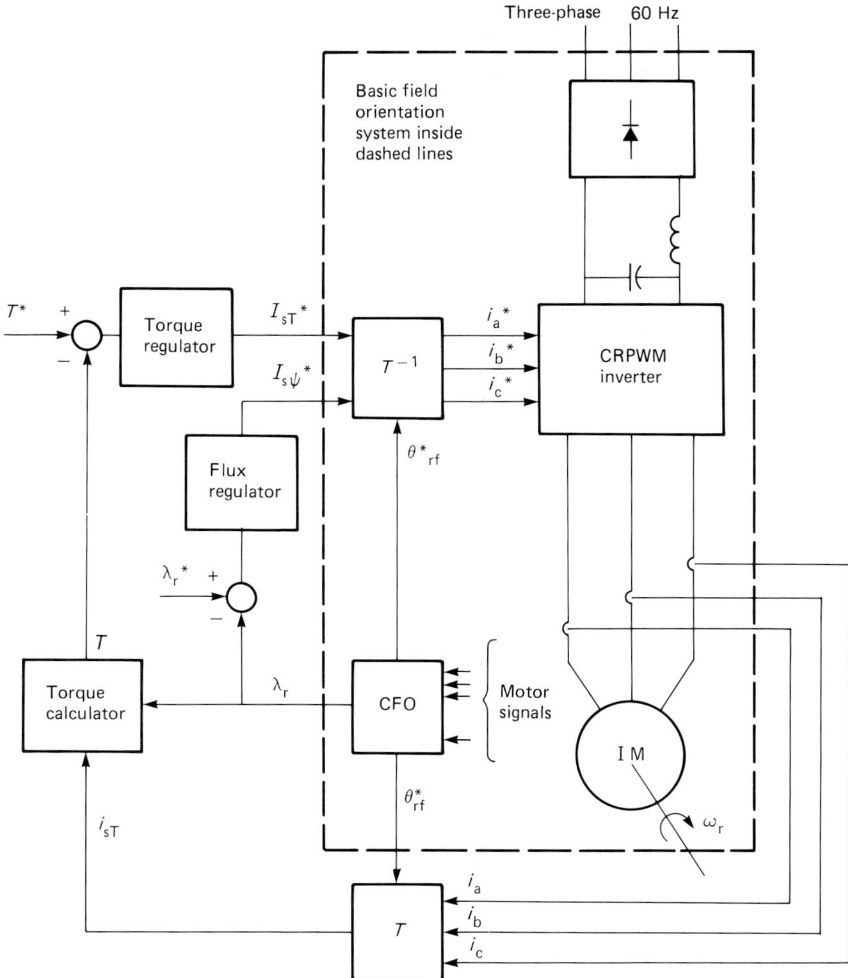

Figure 4.65 Direct field orientation using a current-regulated pulse width modulator. Torque and flux-regulating loops shown as options

Flux and torque regulators can be added to the basic system as shown in Figure 4.65.

A more common implementation which determines indirectly the rotor flux position is illustrated in Figure 4.66. This approach is based on the slip relation:

$$S\omega_e = \frac{r_r}{L_r}\,\frac{I_{sT}}{I_{s\psi}} \qquad \text{or, if } \lambda_r \text{ is a variable, } = \frac{r_r}{L_r}\,L_m\frac{I_{sT}}{\lambda_r} \tag{4.70}$$

which is a statement of the fact that the division of the stator current into the two components I_{sT} and $I_{s\varphi}$ is controlled by the slip frequency $S\omega_e$ (see equivalent circuit in Figure 4.64). In this implementation, the two current components are used to determine the required slip frequency and this result is added (preserving phase information) to rotor position information to obtain the rotor field angle. The transformation T^{-1} is the same as in direct control. This scheme can be viewed as a special form of co-ordination of current and slip frequency which results in decoupled flux and torque control.

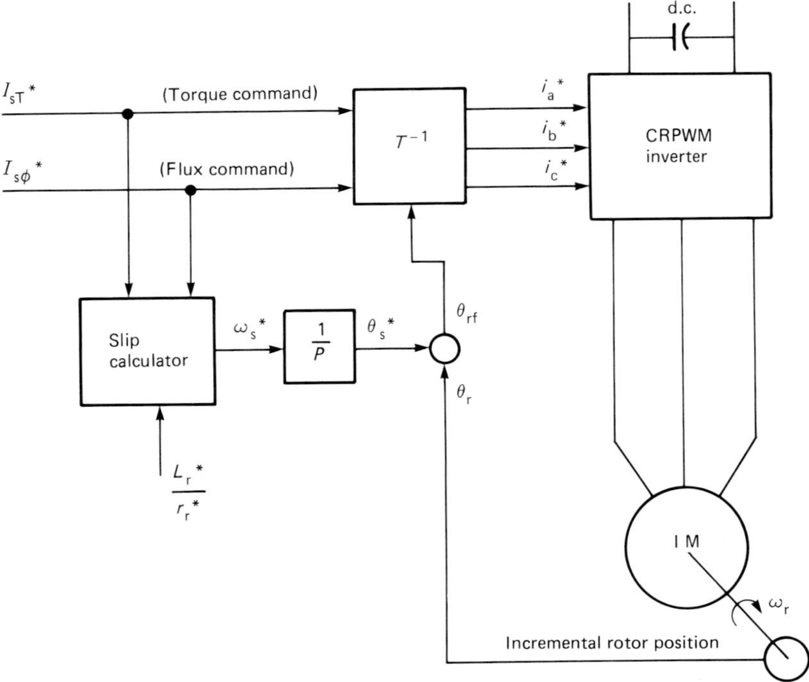

Figure 4.66 Induction machine indirect field orientation using a current-regulated pulse width modulated inverter

The indirect method of Figure 4.66 is the more commonly used form of field orientation control. It is available for servo- and speed-control applications in smaller sizes (to 75 kW) using pulse width modulated inverters and in larger sizes (to several thousand kilowatts) using current source inverter systems. Figure 4.67 illustrates a current source inverter field-orientation system. The major difference compared with a synchronous machine implementation is that current control is directly in terms of amplitude and phase, as shown in the figure. The compensation path shown in the figure, providing phase change information, is essential to correct field orientation for transient conditions.

In all cases, the implementation is by digital means. This is especially important in indirect control, where the summation of slip frequency and rotor speed signals must be very accurate and must preserve phase information. Both the direct and indirect schemes are machine-parameter-dependent. In direct control, the dependence occurs in the rotor flux position computation and in indirect control in the slip frequency calculator. A number of schemes for correcting adaptively for parameter variations have been utilized, but improved systems are in development. Generally, in low-power systems, the parameter dependence is weak and acceptable performance is attainable without adaptation.

4.6 Slip power controlled systems

4.6.1 Introduction

A class of drives which are particularly useful for high-power applications requiring a relatively narrow speed range utilizes the principle of slip power control. This type of drive is suitable, for example, for driving large centrifugal pumps and fans.

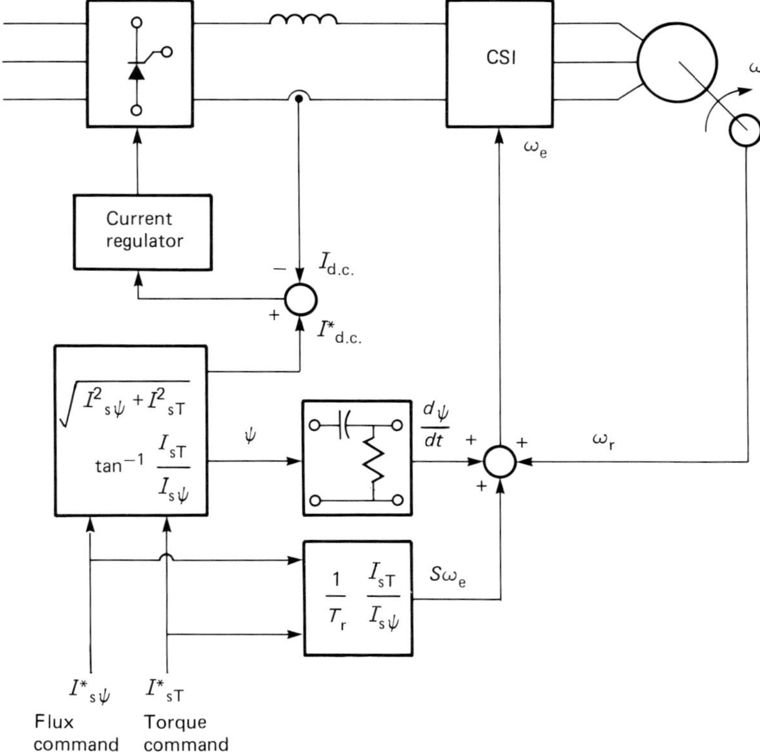

Figure 4.67 Induction machine indirect field orientation using a current-regulated current source inverter

Other recent applications involve balancing turbine generator rotors and flywheel energy storage for large pulsed-power supplies. These systems all employ a wound rotor induction machine in which the stator circuit is connected directly to the supply while the wound rotor circuits are connected to a power converter. The energy is either fed back to the supply, converted to additional useful mechanical power or dissipated in a resistor bank. All three possibilities are simply modern equivalents of traditional methods of speed control which formerly required special rotating machines, usually commutator machines, to accomplish the necessary conversion of slip frequency rotor power to d.c. or line frequency.

4.6.2 Speed control by rotor power dissipation

The simplest type of speed-control scheme for a wound rotor induction machine is achieved by adjusting the rotor current and therefore the torque by use of rotor resistance control. Conventionally, the rotor resistance is altered manually in discrete steps by sequentially shorting resistors or in relatively smooth fashion by liquid rheostats. While both methods of rotor resistance variation are still in wide use they have the disadvantage of high maintenance and poor speed of response. The modern equivalent of rheostatic control of speed is shown in Figure 4.68. In this system, the wound-rotor induction motor is connected to a three-phase diode bridge. The bridge, in turn, feeds a chopper which is placed in parallel with a load

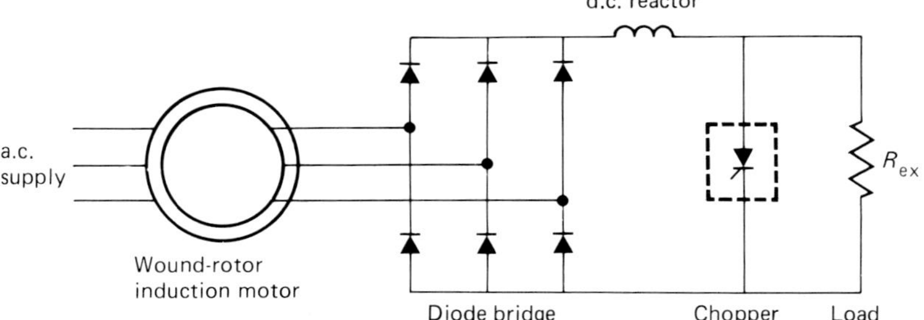

Figure 4.68 Speed controller employing solid state control of rotor resistance

resistor. The current in the resistor is controlled by the duty cycle (see Section 4.2.9):

$$d = \frac{t_{on}}{t_{on} + t_{off}} \tag{4.71}$$

It is apparent that the external resistance can be varied smoothly from zero up to the value set by R_{ex}. Speed of response is much improved with feedback control of the chopper input current. Since the controller is contactless, reliability is also clearly enhanced.

A set of typical torque–speed curves is shown in Figure 4.69. A set of performance characteristics for a typical case of $d = 0.7$ compared with the same parameters for $d = 1$ (external resistor completely shorted) is shown in Figure 4.70. It is noted that while the power factor with chopper control is improved compared to no control ($d = 1$), the efficiency η has been decreased. This general behaviour can be shown to be an inherent feature of the rotor resistance control technique.

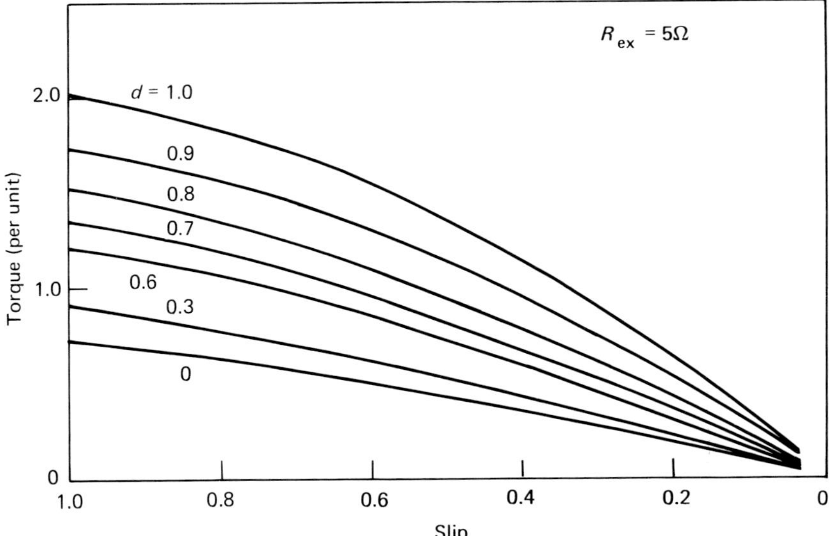

Figure 4.69 Torque–speed curves as a function of the mark:space ratio d

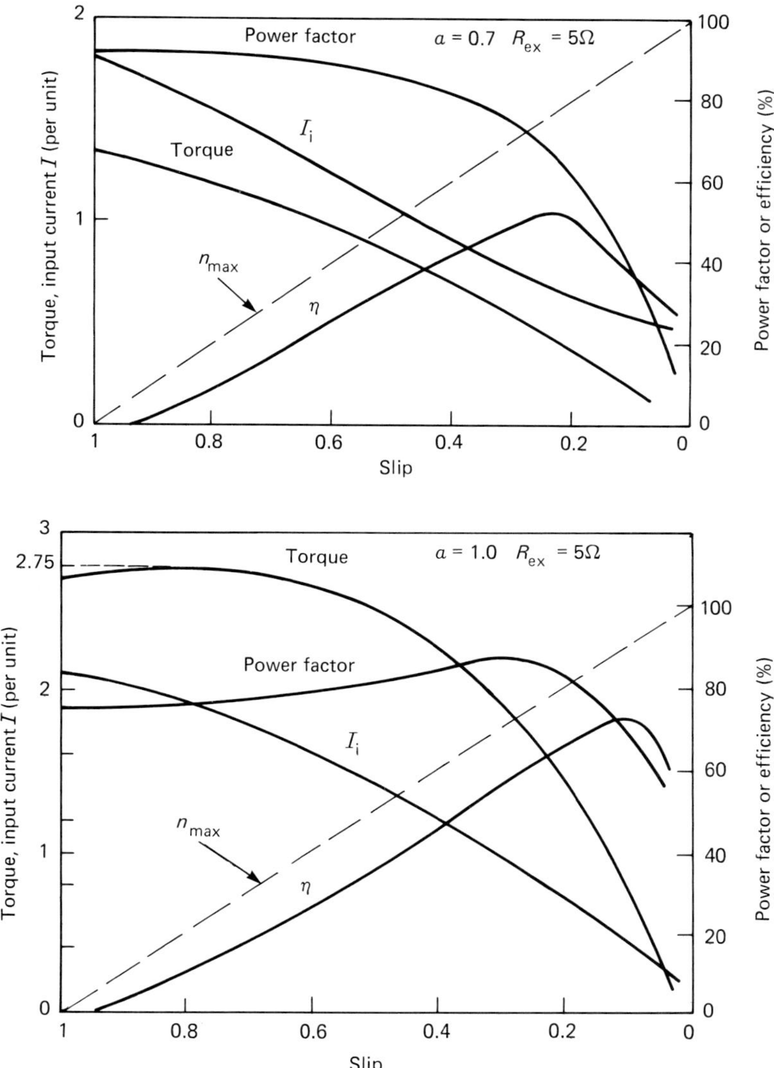

Figure 4.70 Performance curves of controller for $d = 0.7$ and 1.0

4.6.3 Need for slip power recovery

In general, the total power per phase crossing the airgap in an induction motor may be divided into two portions: (1) that portion dissipated as rotor copper loss; and (2) that portion converted to mechanical power to drive the load. The copper loss may be considered as dissipated by the rotor resistance $R_{2(tot)}$ which represents the sum of the rotor resistance r_2 and the equivalent resistance representing the chopper load $R_{2(ext)}$ as shown in Figure 4.71. The remaining power developed by the rotor and represented by the resistance $R_{2(tot)}(1-S)/S$ corresponds to the mechanical power developed. The mechanical power is therefore:

$$P_{em} = 3I_2^2 R_{2(tot)} \frac{(1-S)}{S}$$

$$(4.72)$$

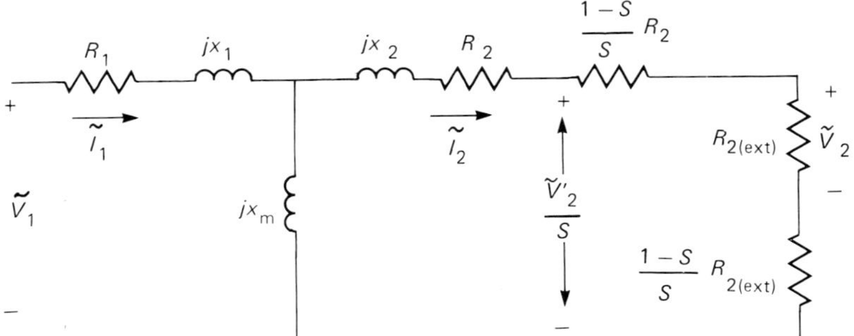

Figure 4.71 Per-phase equivalent circuit of wound rotor induction motor

However, the power delivered to the airgap from the stator is the sum of the power dissipated in both $R_{2(tot)}$ and $R_{2(tot)}\,(1 - S)/S$ so that:

$$P_{ag} = 3I_2^2 \left(R_{2(tot)} + R_{2(tot)}\,\frac{(1 - S)}{S} \right) \tag{4.73}$$

$$= 3I_2^2\,\frac{R_{2(tot)}}{S} \tag{4.74}$$

The ratio of useful rotor power to total rotor power is therefore:

$$\frac{P_{em}}{P_{ag}} = 1 - S \tag{4.75}$$

Hence, at half speed, for example, where $S = 0.5$, the efficiency of energy conversion with slip power dissipation is never greater than 50% even if the rotor resistance of the motor itself were zero. The efficiency clearly becomes progressively worse as the speed range of the drive increases. In practice, the efficiency is always somewhat poorer than Equation (4.74) due to stator copper and iron losses. The theoretical maximum efficiency is shown in Figure 4.70. This result clearly demonstrates the need to recover the energy dissipated in the external resistance when the speed range becomes relatively large.

If other means for speed control are used rather than dissipation techniques, Equation (4.74) remains valid. The amount of power which must be removed from the rotor P_{em} is that portion of the airgap power which is not converted to mechanical energy or, from Equation (4.74):

$$P_{ag} - P_{em} = SP_{ag} \tag{4.76}$$

Hence, the rating of the power recovery device varies directly with the difference between synchronous and the minimum rotor speed, i.e. with the maximum expected slip frequency.

4.6.4 Speed control by slip power utilization

One method for avoiding the heavy penalties in losses arising from slip power dissipation is to convert this power to useful electromechanical output power. The traditional approach to this type of speed controller is to utilize a rotating commutator machine (synchronous converter) to convert slip frequency power to d.c. power. The d.c. power is then supplied to a conventional d.c. motor which is

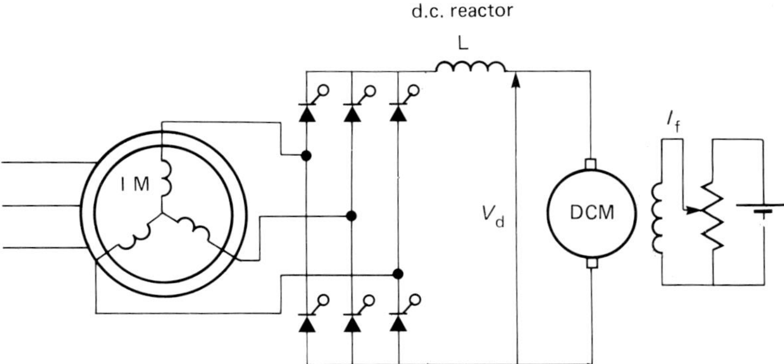

Figure 4.72 Static Kramer system

coupled to the same shaft as the wound rotor induction machine. The torque supplied to the load is the sum of the torque produced by the induction and d.c. machines. Systems of this type have traditionally been called Kramer drives.

The solid state equivalent of the Kramer drive is shown in Figure 4.72. The principle of speed control can be illustrated by considering an ideal case where the winding resistance and leakage inductance of the induction motor as well as the armature resistance, armature reaction and magnetic saturation are neglected. In this case the overlap angle of commutation of the six-pulse bridge is zero so that for a firing angle α we obtain for the bridge output voltage:

$$V_{\rm d} = \frac{3\sqrt{2}}{\pi} SV_2 \cos \alpha \tag{4.77}$$

where $S > 0$. The e.m.f. induced in the d.c. motor, say $E_{\rm d}$, is given by:

$$E_{\rm d} = k_1 I_{\rm f} (1 - S) \omega_{\rm e} \tag{4.78}$$

where k_1 is the e.m.f. constant of the machine. Combining these two equations we obtain:

$$S = \cfrac{1}{1 + \cfrac{3\sqrt{2} E_{\rm d} \cos \alpha}{\pi k_1 I_{\rm f} \omega_{\rm e}}} \tag{4.79}$$

It is apparent that speed can be controlled by varying either α or $I_{\rm f}$. In practice it is more convenient to keep α constant and to vary $I_{\rm f}$. In particular, if $\alpha = 0$ the thyristor bridge of Figure 4.72 can be replaced by a simple diode bridge. Figure 4.73 shows a typical speed-control characteristic for such a system.

Like the Kramer drive, the static Kramer drive is little used due to the relatively high cost of the d.c. machine together with the necessity for field control which limits the speed of response of the system. In addition, speed can be varied smoothly only to approximately 90–95% of synchronous speed. At this point the voltage at the rotor sliprings SV_2 is only about 5–10% of its maximum zero speed value. The losses in the thyristor bridge and d.c. machine contribute an effective resistive IR drop which equals SV_2 and therefore prevents a further increase in speed. Operation can be continued above synchronous speed by reversing the field current, in which case the d.c. machine acts as a generator, feeding power to the

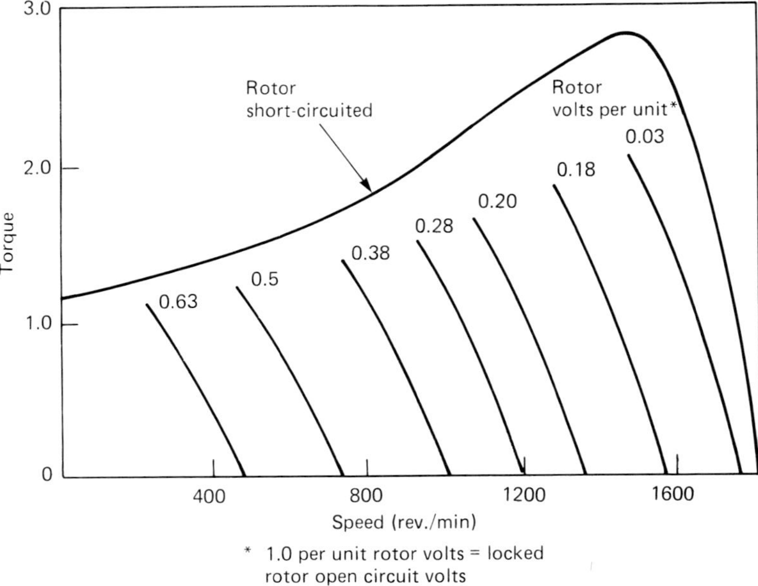

Figure 4.73 Speed control characteristics of static Kramer system

rotor through the bridge which now operates in the inverting mode. However, the transition from sub- to super-synchronous operation is only possible with a mechanical prime mover torque to carry the machine through synchronous speed ($S = 0$). This is, of course, impossible with typical loads so that this motor drive is normally restricted to speeds below synchronous speed.

4.6.5 Speed control by slip power recovery

The third alternative for managing slip power is simply to feed it back to the supply. Such drives are called Scherbius drives. The original Scherbius drive required two types of a.c. commutator machines, one operating as a frequency changer to convert slip frequency power to line frequency power (or vice versa) and the other to control the flow of the slip frequency power extracted from the rotor of the wound rotor induction machine. The solid state equivalent is called a static Scherbius drive and is shown in Figure 4.74. In this case the a.c. commutator machine is replaced by a cycloconverter. The cycloconverter shown in Figure 4.74 is only one of many which could be employed and is called a suppressed d.c. link cycloconverter. Currents which flow on both line and load sides of the cycloconverter take on quasi-rectangular waveshapes which are much the same as the currents in a d.c. current link system (see Section 4.4.2). Another type of cycloconverter is shown in Figure 4.52 (page 246).

It is interesting to note that both the frequency conversion as well as control of power flow is accomplished in a single device rather than with cascaded machines as was the case for the original Scherbius drive. The system retains the same essential features of the Scherbius drive, i.e. the system can operate as a motor both above and below synchronous speed as well as synchronous speed. A power flow diagram for the four possible modes of operation is shown in Figure 4.75. It can be noted that slip power is returned to the line for rotor speeds below synchronous speed ($S = 0$) and taken from the line at speeds above synchronous, so that the

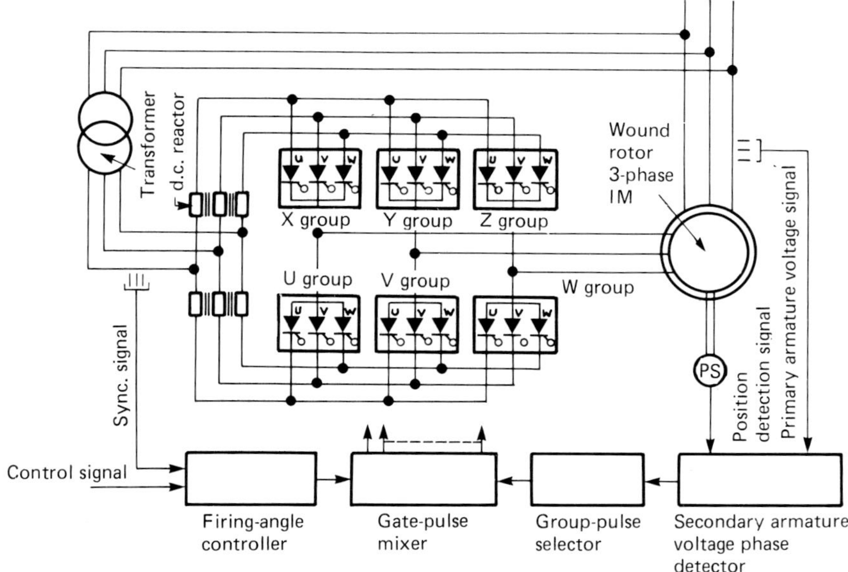

Figure 4.74 Cycloconverter-type static Scherbius system

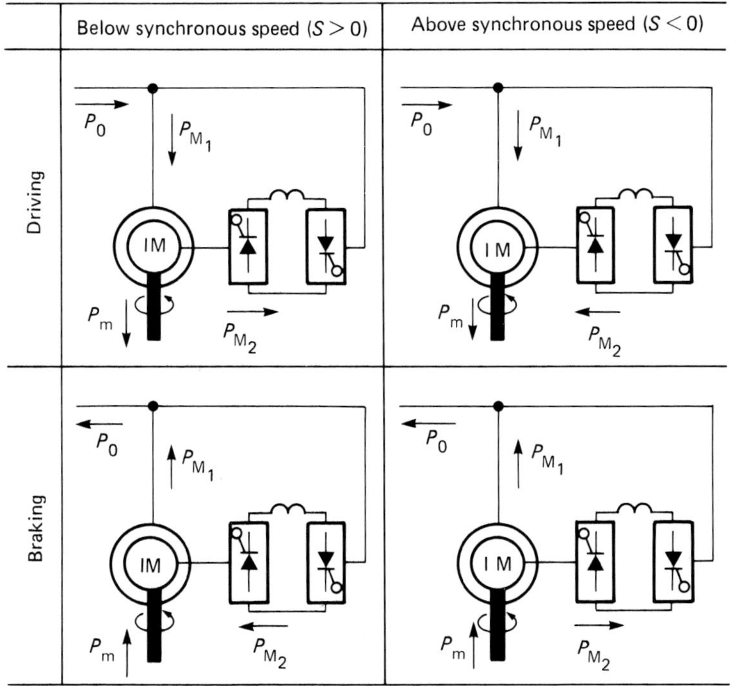

Figure 4.75 Power flow diagram

cycloconverter must therefore be capable of bidirectional power flow even when operating only in the motoring mode.

It is clear that the speed range of this drive is essentially doubled for the same rating of converter, compared to the static Kramer drive. However, the speed range of the system is typically limited to about 50–33% above and below synchronous speed due to the switching limitations of the cycloconverter which is limited to a maximum frequency ratio of about 2–3:1. If necessary, the speed range can be increased by increasing the number of phases on the high-frequency side of the cycloconverter (line side). This can be accomplished by equipping the transformer between the converter and the line with the required number of secondary phases. The disadvantage of this approach is a proportionate increase in the number of thyristors with the phase number.

The presence of the transformer in Figure 4.74 provides a number of additional important benefits. First, since the range of motor slip is only ± 0.33, the maximum voltage required on the slipring side of the cycloconverter is small, i.e. $S_{max}V_2$ where $V_2 \approx V_1/\alpha$ and α is the effective stator:rotor turns ratio. This rotor voltage ranges from 100 V to perhaps 2000 V and is typically only a fraction of the rated stator voltage V_1. The rating of the cycloconverter is optimized when the voltage rating on both sides of the converter is equal. Hence, the turns ratio of the transformer can be selected such that $N_1/N_2 \approx \alpha/S_{max}$ to minimize the voltage rating of the cycloconverter. Secondly, the transformer also provides the isolation required for proper operation of the cycloconverter in order to avoid the rotating 'short circuits' which would occur without such isolation. (The alternative to transformer isolation is to isolate each of the three rotor phases which, in turn, requires six brushes and sliprings.) Thirdly, the transformer leakage inductance limits 'shoot-through' currents which occur when the cycloconverter misfires.

4.6.6 Variants of the static Scherbius drive

The emergence of power electronic circuits as a replacement for cumbersome electromechanical converters has inevitably resulted in numerous approaches to accomplish a single purpose. In the case of the Scherbius drive, considerable confusion reigns since the definition of what constitutes a Scherbius drive has never been defined precisely. In general, the original Scherbius drive was capable of: (1) returning rotor power to the line below synchronous speed; (2) taking rotor power from the line above synchronous speed; and (3) taking power from the line to excite the rotor when operating exactly at or very near synchronous speed. In case (3) the machine operates, in effect, as a synchronous machine. This latter case is important if motoring above and below is required since a motoring torque capability must be preserved near synchronous speed in order to accelerate the load through synchronous speed.

Figure 4.76 shows a variant of the static Scherbius drive which employs a d.c. current link with a six-pulse bridge-type thyristor rectifier and thyristor inverter. Below synchronous speed, rotor power at slip frequency is rectified to d.c. power and then subsequently inverted to supply frequency. The roles of the rectifier and inverter reverse above synchronous speed and requirements (1) and (2) above are satisfied. However, when the rotor approaches synchronous speed the voltage available to commutate the bridge connected to the rotor becomes progressively smaller until commutating ability of the rotor side bridge is lost. In practice, the d.c. link current can be 'pulsed' at 6 times slip frequency in much the same manner as during starting of a synchronous motor drive (see Section 4.4) and some torque-producing capability maintained. However, this results in only a fraction of rated torque so that requirement (3) cannot be satisfied entirely.

A modified configuration which meets the requirements of full torque-producing capability in the region near synchronous speed is shown in Figure 4.77. In this

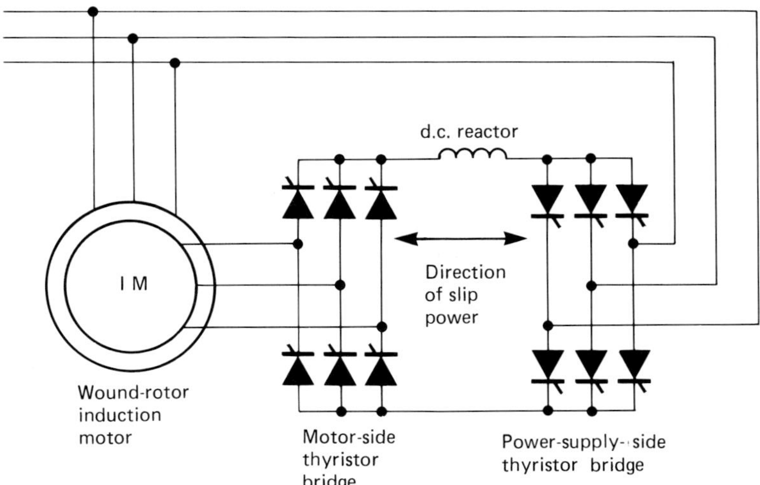

Figure 4.76 Static Scherbius system employing d.c. current link with thyristor rectifier and thyristor inverter

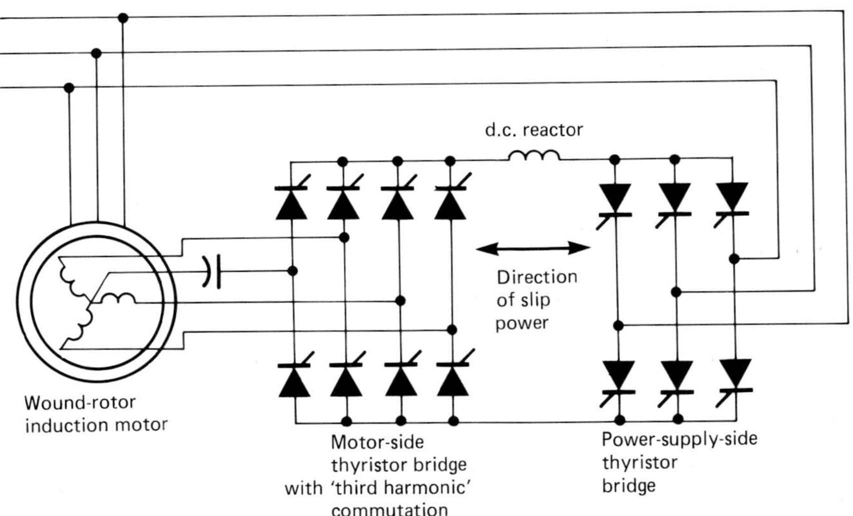

Figure 4.77 A third implementation of a static Scherbius system utilizing a d.c. current link with force-commutated rotor side bridge

system the rotor-side bridge is force-commutated near synchronous speed just prior to the point where natural commutation ceases to be effective. Since the commutation circuit is brought into play only near synchronous speed, the volt-ampere rating of the commutating capacitor is relatively modest. However, the method does require an extra brush and slipring. Other types of force-commutated bridges would also clearly satisfy requirement (3), e.g. the six-step current source inverter of Figure 4.56 (page 249). However, the increased cost generally does not warrant these more complicated converters.

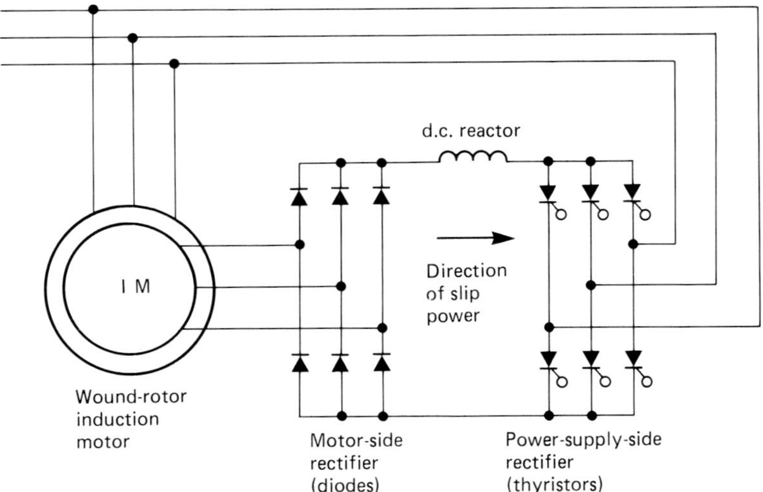

Figure 4.78 The sub-synchronous static Scherbius system. The d.c. current link with diode rectifier and thyristor inverter

Another variation of the d.c. current link arrangement is shown in Figure 4.78. In this case the rotor-side thyristor bridge is replaced by a simple diode bridge. Since the diode bridge is incapable of inversion (reverse polarity of the d.c. link voltage) this system satisfies only requirement (1) and is termed a subsynchronous static Scherbius drive. The motivation for resorting to a diode bridge is primarily one of cost, since a simple diode bridge is much cheaper than a thyristor bridge of the same rating. The maximum speed for this configuration is about 95% of rated speed.

The subsynchronous static Scherbius drive operates over the same speed range as the rotor dissipation scheme of Section 4.6.2 but offers a substantial increase in efficiency at lower speeds. The efficiency improvement is accomplished, however, with the penalty of lower power factor which is even lower than that of the motor itself, owing to the commutation requirements of the converters.

The efficiency deteriorates near its maximum speed condition due to the inability of the rotor-side bridge to effect commutation. In general, the line-side inverter provides an average d.c. voltage which, when impressed on the d.c. link inductor, gives rise to the desired motor current. However, the inverter voltage also contains harmonics related to the line frequency. At low inverter voltages, the inverter is nearly sawtooth in waveshape and, at a sufficiently small inverter voltage near synchronous speed, the instantaneous polarity of the inverter voltage reverses sign even though the average value of the voltage remains the same, i.e. negative polarity relative to rectification. This instantaneous voltage finds a short-circuit path through the rotor-side diode bridge which creates substantial extra losses in the d.c. link inductor. The onset of 'circulating d.c. current' is frequently chosen as the limiting value for the maximum speed of the drive. If the losses in the d.c. link inductor are excessive at the maximum desired speed point, the problem is best solved by utilizing the 'four-legged bridge' shown in Figure 4.20(b) (page 209). This connection permits average voltages very near zero to be obtained with very low ripple as well as an improvement in power factor.

The subsynchronous static Scherbius drive is probably the most widely used of the doubly fed induction motor family and is sometimes called a subsynchronous converter cascade. It is also erroneously called a static Kramer drive or Kramerstat by associating its inability to operate above synchronous speed with the classical Kramer drive. However, it more correctly corresponds to certain traditional-type Scherbius drives, so-called single-range equipment, which recover slip power but are only capable of subsynchronous operation.

4.6.7 Starting considerations

Doubly fed induction motor drives are ideally suited to operation over a limited speed range since the converters connected to the rotor need only be rated for slip power. However, this restriction clearly results in a starting problem since the converters would clearly encounter voltages greater than their rated value if the motors were accelerated from rest. Fortunately, most applications involve either a fan-type load in which the power varies with the cube of rotor speed or a load which appears only when the motor has reached the desired speed condition. In such cases the motor can be started with reduced voltage or with a small pony motor.

If the motor accelerates an appreciable load or if the overall system has a high inertia, then a finer control is required. A controlled start is frequently accomplished by using the rotor power dissipation method of speed control, using a set of resistors which are sequentially shorted as the motor comes up to speed. A more elegant scheme is shown in Figure 4.79. In this method the stator is shorted and the motor runs 'inside out' with the cycloconverter operating as a variable-frequency supply for the rotor which now plays the role of the 'stator' or primary. When the machine reaches about half speed, the output frequency of the cycloconverter has reached one-half supply frequency. At this point the contactors are thrown and the stator is connected to the line. The rotor power flow suddenly reverses. The rotor frequency remains at one-half supply frequency and the motor can now continue to accelerate with close control of the motor torque, the cycloconverter output frequency becoming smaller as speed increases.

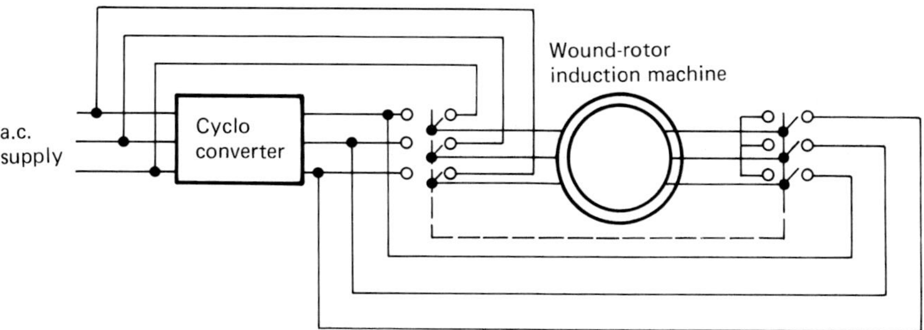

Figure 4.79 Starting scheme for static Scherbius drive

4.6.8 Other considerations

Doubly-fed induction motor drives are frequently the drive of choice, particularly for large applications. However, they differ from more conventional voltage source inverter and current source inverter induction motor drives in several important aspects.

4.6.8.1 Power factor

Without external correction of power factor, conventional squirrel cage induction motor drives as well as synchronous motor drives always operate at a lagging power factor. This is because the line-side converter requires VARs from the line to accomplish commutation. With the static Scherbius drive, however, this is not necessarily the case. In many respects the doubly fed induction machine operates exactly as a synchronous machine. In fact, if the rotor of the doubly fed machine operates synchronously then the power fed to the rotor is d.c. and the machine is essentially identical in behaviour to a cylindrical-rotor synchronous machine. With proper thermal design of the rotor windings, the rotor current can be increased to raise the power factor of the stator windings to a leading condition which will cancel the VARs required by the cycloconverter itself and thereby operate at unity power factor. This component of rotor current continues to be supplied to the rotor for conditions away from synchronous speed. It adds vectorially to the torque-producing component and therefore increases the thermal loading on the rotor, affecting the size of the machine, as well as having an impact on the current rating of the cycloconverter. In principle, the input power factor can be improved to unity over the full speed range provided the motor and converter are rated adequately.

The drives of Figures 4.76 and 4.77 also have the potential for power factor control. However, the commutation requirements of the two converter bridges must now be overcome. Again, the kilovolt-ampere ratings of the converters are increased significantly so that only that portion of power factor correction which can be obtained at light load without appreciable derating of the converter is frequently sought. An important disadvantage of the subsynchronous static Scherbius drive compared to the various other types of Scherbius drives is its inability to influence power factor.

4.6.8.2 Torque pulsations

The presence of a converter on the rotor circuit results in the generation of rotor current harmonics and subsequently torque pulsations in much the same manner as with the stator-side force-commutated converter configurations. The frequency spectrum of the voltage source-type cycloconverter (Figure 4.49) is very difficult to describe due to the modulation process inherent in cycloconverter operation. However, in general, the output spectrum contains a high-frequency component at 6 times the line frequency (for a six-pulse bridge) and also integral multiples of the output frequency plus or minus the line frequency due to the modulation process. The amplitude of these sub-harmonic frequencies becomes progressively larger and, hence, torque pulsations increase as the difference between line frequency and output frequency increases, i.e. as the rotor speed departs from synchronous speed. A commonly accepted upper limit for the ratio of output fundamental frequency to input (line) frequency is $pf_{line}/15$ where p is the pulse number of the basic converter bridge making up the cycloconverter (typically 3, 6 or 12).

The harmonic spectrum of the basic d.c. current link-type of system (either suppressed or explicit) is defined by the basic switching properties of the converter bridge. If the filtering action of the d.c. link inductor is good then the principal harmonics impressed on the rotor circuit are simply odd multiples of slip frequency, excluding the triple harmonics. Torque pulsations then occur at multiples of 6 times slip frequency. Since the frequency spectrum swept during changes from minimum to maximum speed is very large, resonances with the natural frequency of mechanical system inevitably occur. The problem is more difficult than in voltage source inverter or current source inverter squirrel cage induction motor drives since doubly fed induction motor drives tend to operate continuously at potential mechanically resonant conditions whereas the area of concern for single-fed drives is at very low speed which is generally not a normal operating point for most applications. If the drive is large, as is frequently the case with these systems, shaft breakage can occur. The problem is best alleviated by the use of a resilient coupling which introduces sufficient added damping at the resonant conditions so as to limit

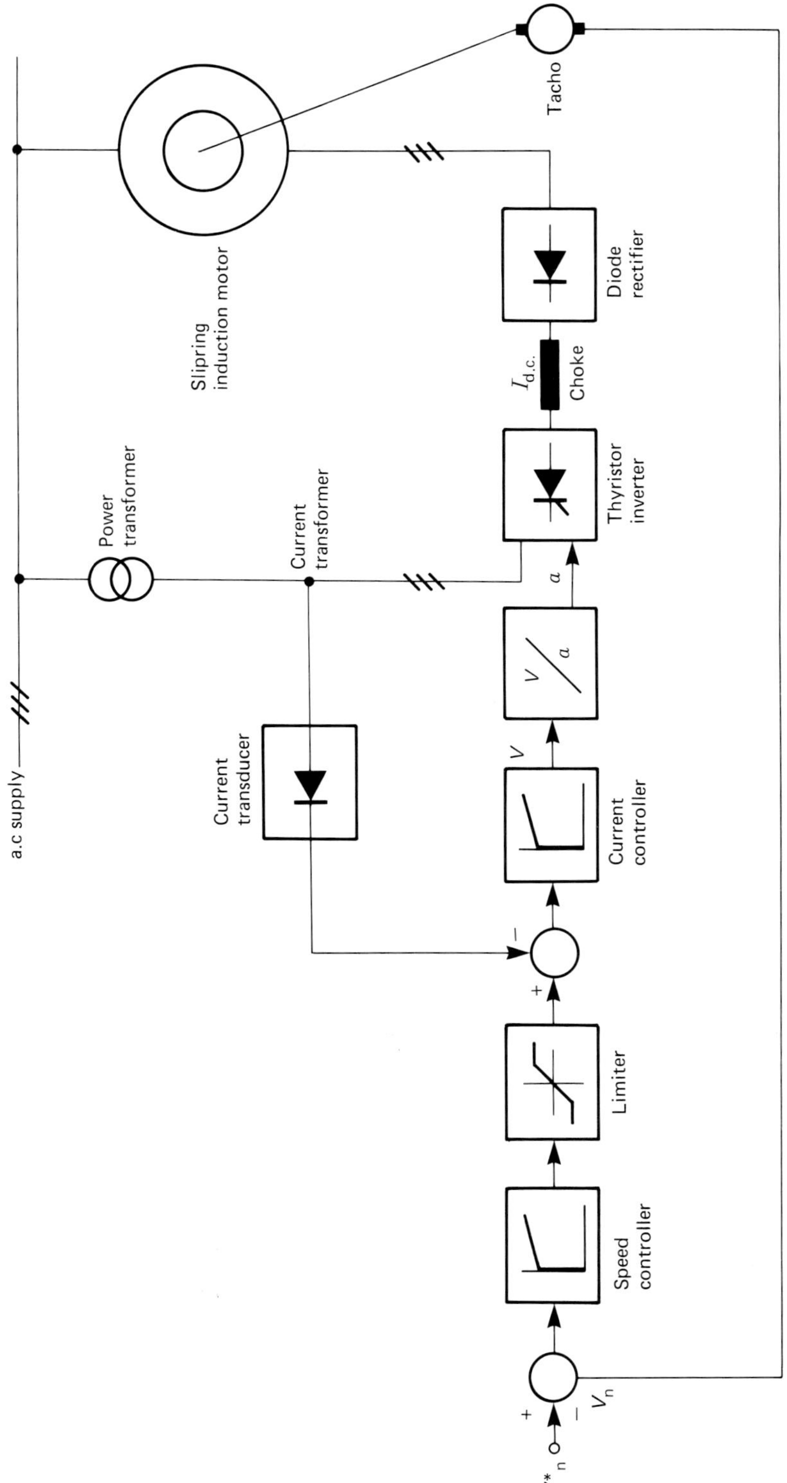

Figure 4.80 Speed control system for sub-synchronous Scherbius drive

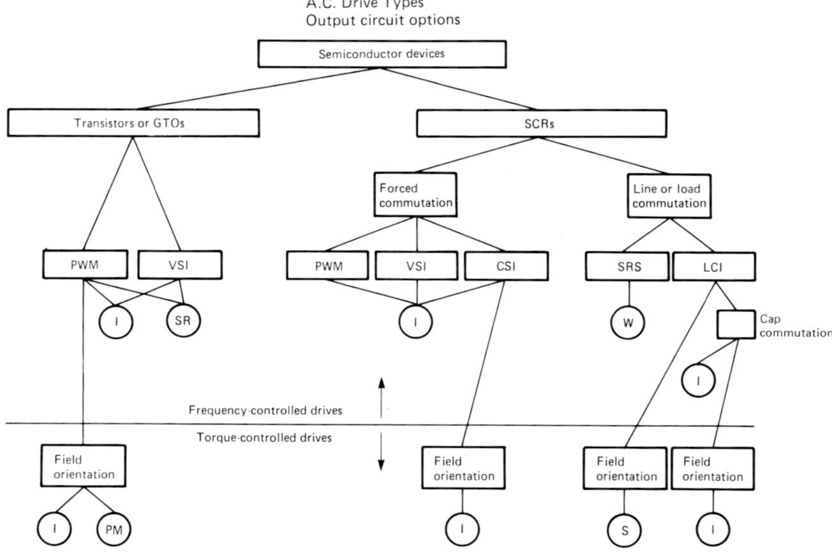

Figure 4.81 Basic output circuit classification of a.c. drive types

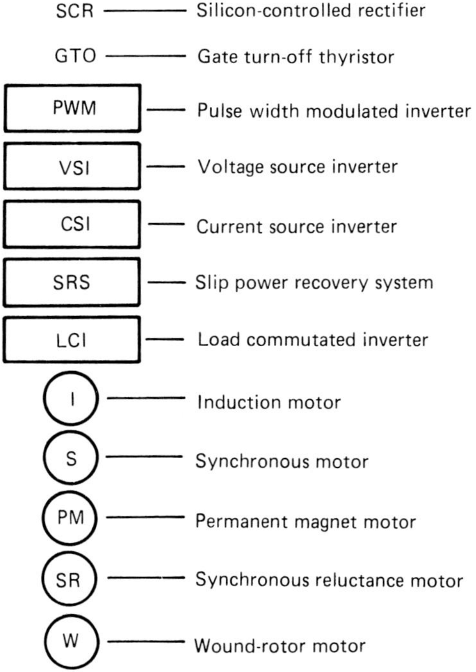

Figure 4.82 Symbols used in Figure 4.81

the amplitude of the mechanical resonances to stress values that can be accommodated in the shaft design.

4.6.8.3 Speed control

In the vast majority of cases, doubly fed induction motor drives involve speed-controlled systems such as fan or pump loads in which the speed of response is not of great concern. A typical closed-loop speed control scheme for such applications is shown in Figure 4.80. The internal current control loop provides good current-limiting action as well as continuous regulation to afford protection of the converter devices. The outer speed control loop is the major loop and maintains the speed of the motor constant at the set value independent of load variations. The d.c. I_d is sensed indirectly by means of current transformer–rectifier configuration connected on the a.c. side of the thyristor bridge. The speed is sensed with a conventional tachogenerator.

It can be noted that the subsynchronous static Scherbius drive has inherently the poorest performance of the doubly fed induction motor drive family since the diode bridge allows only the amplitude of the rotor current to be affected by the line-side converter. If required, faster response is afforded by other configurations in which the phase as well as the amplitude of the rotor current can be controlled rapidly. The static Scherbius drive with a d.c. current link (Figure 4.76) is capable of fast control of rotor current amplitude and phase for conditions sufficiently far from synchronous speed. However, the number of opportunities to influence the phase of the rotor current is dependent on slip frequency, so that phase control of the rotor current is lost as the speed of the drive approaches synchronous speed ($S \rightarrow 0$). On the other hand, since both amplitude and phase of the rotor current are adjusted by line-connected thyristors, the static Scherbius drive affords the opportunity for excellent control capability near synchronous speed. However, in this case response slowly deteriorates as the rotor speed moves away from synchronous speed (slip frequency increases). Applications requiring a relatively narrow speed range but fast response have been slow in appearing. Nonetheless,

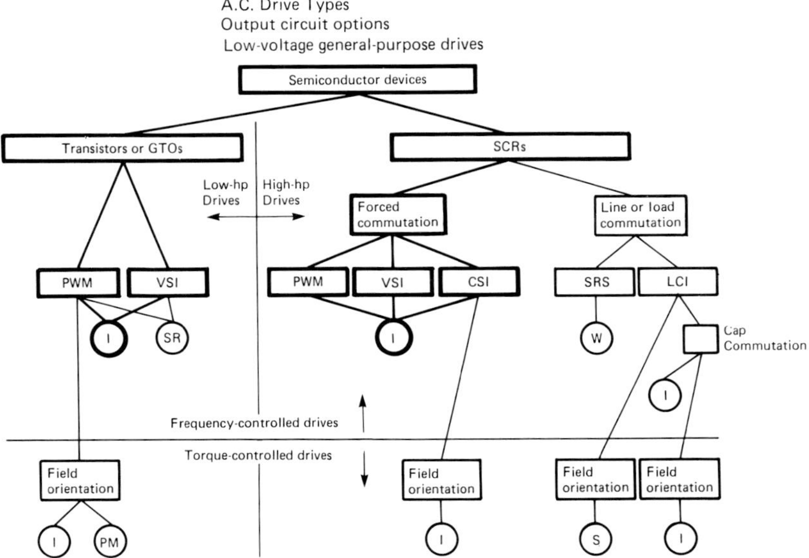

Figure 4.83 Output circuit classification – low-voltage general-purpose a.c. drives

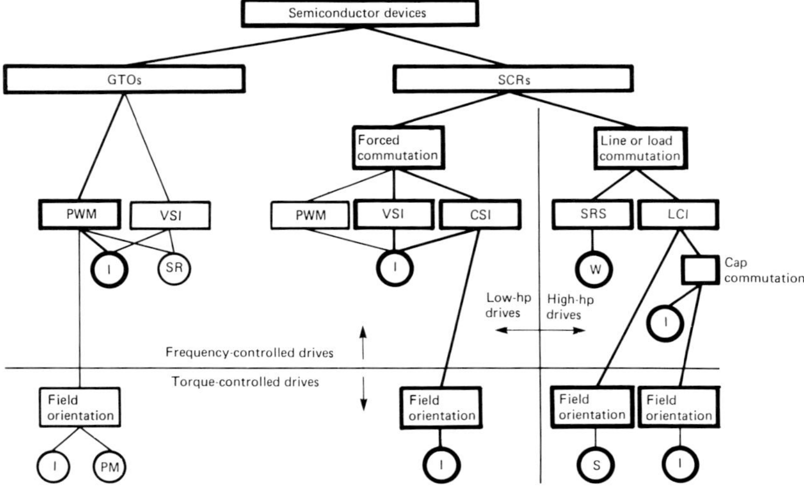

Figure 4.84 Output circuit classification – medium-voltage a.c. drives

fast response 'field-oriented' control of the type discussed in Section 4.5.4 is practical and will almost certainly move out of the laboratory as time progresses.

4.7 Summary of a.c. drive types

Figure 4.81 provides a summary and classification of the various a.c. drives discussed in the preceding sections. A list of symbols used in the figure is given in Figure 4.82. The diagram in Figure 4.81 is based on a number of classification items, the most important of which is the type of semiconductor switching device employed in the drive. Secondary classification items include the type of inverter circuit, whether forced or natural commutation is employed, the type of motor and the type of control strategy used. In general, control complexity increases from top to bottom in the figure and drive power level from left to right.

For low-voltage (460 V or less) general-purpose drives, only certain types of drives are commonly used. This fact is indicated in Figure 4.83 where the commonly employed drive types are shown by highlighting those types on the general diagram. The power dividing line on this figure is currently somewhere in the 150–225 kW range and is expected to move up to higher power as device ratings increase.

Figure 4.84 illustrates the commonly used drive types in medium-voltage, higher power applications. The power dividing line in this figure is currently in the 750–1500 kW range, although forced commutation current source inverter drives up to 3750 kW are available.

Bibliography

SEN, P. C. *Thyristor d.c. drives*. Wiley Interscience, New York (1981)

SEN, P. C. and MA, K. H. J. 'Rotor chopper control for induction motor drive: TRC strategy', *IEEE/IAS Annual Meeting Conference Proceedings* (1974)

OWEN, E. L. and WEISS, H. W. 'Efficient and reliable synchronous motors for large a.c. adjustable-speed drives', *Proceedings American Power Conference* (1981)

ROSA, J. 'Utilization and rating of machine commutated inverter-synchronous motor drives', *IEEE Transactions on Industry Applications*, **IA–15**, 2, 155–164 (1979)

LEONHARD, W. *Control of electrical drives*. Springer-Verlag, Berlin (1985)

TSUJI, Y., TAKEDA, M. and TSUCHIYA, T. 'Super synchronous Scherbius', *Fuji Elec. Rev*, **20**, 4, 164–170 (1974)

OLIVIER, G., STEFANOVIC, V. R. and APRIL, G. E. 'Evaluation of phase-commutated converters for slip-power control in induction drives', *IEEE IAS Annual Meeting Conference Proceedings*, pp.536–542 (1981)

KRISHNAN, T. and RAMASWAMI, B. 'Slipring induction motor speed control using a thyristor inverter', *Automatica*, **11**, 419–424

KAZUNO, H. 'A wide-range speed control of an induction motor with static Scherbius and Kramer systems', *Elec. Engrg. Japan*, **89**, 2, 10–19 (1969)

LAVI, A. and POLGE, R. J. 'Induction motor speed control with static inverter in the rotor', *IEEE Transactions on Power Apparatus and Systems*, **PAS-85**, 1, 76–84 (1966)

SHEPHERD, W. and STANWAY, J. 'Slip power recovery in an induction motor by the use of a thyristor inverter', *IEEE Transactions on Industry and General Applications*, **IGA-5**, 1, 74–82 (1969)

COLBY, R. S. 'Steady-state performance of load-commutated inverter-fed synchronous motor drives', MSc. thesis, University of Wisconsin, Madison (1983)

DEWAN, S. B. and STRAUGHEN, *Power Semiconductor Circuits*, John Wiley & Sons, New York (1975)

BOSE, B. K. *Power Electronics and AC Drives*, Prentice Hall, New Jersey (1987)

SHEPHERD, W. *Thyristor Control of AC Circuits*, Crosby Lockwood Staples, St. Albans, UK (1975)

NOVOTRY, D. W. and KING, T. L., 'Equivalent circuit representation of current inverter driven synchronous machines', *IEEE PAS Trans.*, **100**, 6, 2920–2926 (1981)

CORNELL, E. P., LIPO, T. A., 'Modelling and design of controlled current inverter induction motor drive systems', *IEEE-IAS Trans.* **IA-13**, 6, 321–330 (1977)

JARC, D. A., NOVOTRY, D. W., 'A graphical approach to a.c. drive classification', *IEEE-IAS Annual Meeting Conference Proceedings* (1986)

BLASCHKE, F. 'A new method for the structural decoupling of a.c. induction machines', *IFAC Symposium*, Dusseldorf, Germany, 1–15 (6.3.1), Oct. (1971)

HASSE, K. 'Zum dynamischen, Verhalten der Asynchronmaschine bie Betrieb mit variabler Standerfrequenz und Standerspannung', **ETZ-A**, 89, H.4, 77–81 (1968)

ROWAN, T. M. and LIPO, T. A. 'A quantitative analysis of induction motor performance improvement by SCR voltage control', *IEEE-IAS Trans.*, **IA-19**, 545–553 (1983)

5 Materials and motor components

I. W. Grayer

Chief Engineer, Motor Department, Brown Boveri, South Africa

5.1 Introduction

The materials and components used to construct an electric motor are the result of technical and commercial development. Before describing these components and their manufacture, we should look briefly into the history of the electric motor to see how these circumstances have influenced the choice and design of materials and components.

As a young engineer at Bruce Peebles (now Parsons Peebles) of Edinburgh, Scotland, it was suggested to me by the general manager W.F.D. Hart that I should plot a graph of motor output per unit volume of active material against year of manufacture. Since this graph yielded surprising results, I have since made a frequent habit of asking designers their opinion as to the shape of this curve. Invariably their expectation is that, no matter what was the early gradient, the graph will now be nearly horizontal. Examination of such a graph,[1] (Figure 5.1) shows a rapid rise in the early 1960s, at a time when hand calculation methods were being replaced by computer-aided calculations.

Commerical pressures in Europe in the early 1970s, which resulted in the amalgamation of some manufacturing companies, caused design engineers to

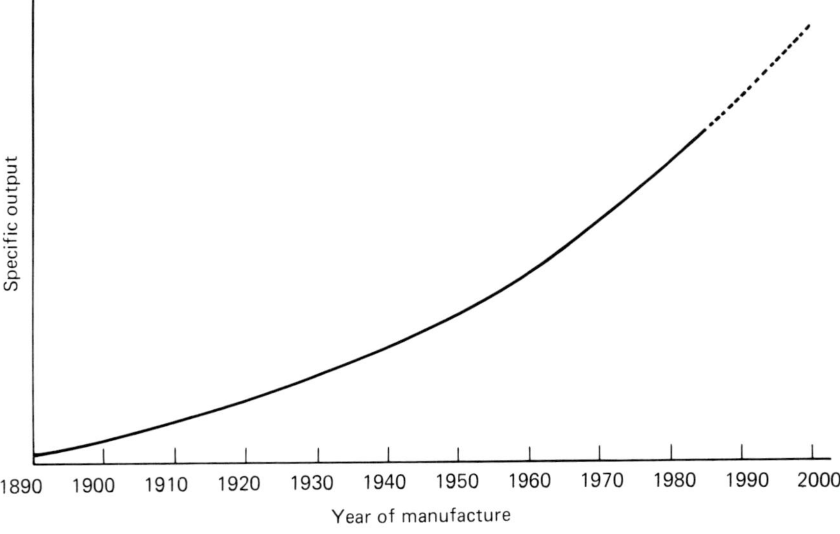

Figure 5.1 Specific output (output from a fixed amount of active material) against year

further squeeze increased outputs from frames in the competition to stay in business. This competition has probably been the main reason for the extensive mechanical and electrical design analysis of electric motors.

Figure 5.1 is projected to the year 2000 which shows a 30% output increase over 1985. When it is considered that an immediate increase of 11% is possible by using full Class F temperature levels (the majority of motors presently being offered with Class B temperature rise but using Class F insulation), it can be understood that the graph is probably realistic.

5.1.1 Development

The early motors tended to have short core lengths of large diameter for any particular speed, and the shaft height above the fixing feet was relatively large. After the Second World War industry, being the major user of electric motors, was being rebuilt and some 25 years later this industry was still expanding. Electric-motor manufacturers were often asked to replace a motor of another make, and a convenient way of doing this was to use a lower centre height motor capable of the output and to add an adaptor plate under the new motor. Whether for this reason or because manufacturers' costing systems tended to a lower cost for the low centre height motors, the new range of modular constructed motors available in the late 1970s and early 1980s incorporated as much electrical steel as possible within the centre height. Once this philosophy had been followed by some manufacturers, the others could not ignore the trend. It is always possible to raise a motor centre height, but difficult to lower it. Costing also showed that the long core of small diameter was less expensive to manufacture than the short core of large diameter.[2]

Consider two motors ((1) and (2) below) having the same active rotor volume and, hence, nearly the same output.

(1) Diameter = D, length = L Airgap area = πDL
(2) Diameter = $2D$, length = $L/4$ Airgap area = $\pi DL/2$

The airgap flux densities will be approximately the same, so motor (2) will have half the flux of motor (1). Motor (2) must therefore have twice the stator turns per phase compared with motor (1), and this will increase the manufacturing cost.

The resulting trend was towards machines with smaller diameters and longer core lengths and led to development of the two-pole motor running above its first critical speed. Extensive theoretical work was undertaken to ensure that the rotor always ran stably under the influence of unbalanced magnetic pull during starting and running, the shaft and frame stiffness being carefully considered as well as the size of the radial airgap between stator and rotor.

5.1.2 Motor shape and construction

From the foregoing discussion it is seen that the modern motor will tend towards a longer core of smaller diameter and will have a low centre height for a given output.

The need to manufacture components in parallel with each other to save manufacturing time resulted in the modular construction concept. This involves manufacture of frames, cores, coolers and bearings all proceeding together and the finished components being assembled to form a complete motor.

Coolers are usually top-mounted on the motor, rather than being built into the motor frame. Stator core units are wound separately from the frame for ease of access and to occupy as little space as possible in the insulating resin tank, particularly where vacuum pressure impregnation systems are used.

The present state of the art (1987) is an extension of the foregoing principles, together with the introduction of modern manufacturing methods such as the 'just in time' principle, and with computer-aided draughting and computer-aided manufacture having strong influences.

Two basic motor types are being produced in standard ranges such as shown in Figures 5.2–5.4. Figure 5.2 shows a large synchronous motor using pedestal bearings located outside the main enclosure. Figures 5.3 and 5.4 show the bearings

Figure 5.2 Pedestal bearing motor. (*Courtesy:* Brown Boveri Company)

Figure 5.3 End-shield bearing motor. (*Courtesy:* Brown Boveri Company)

Figure 5.4 Small end-shield bearing motor. (*Courtesy:* GEC Machines, Benoni)

located in the end bracket of the motor. This end-shield type of construction is normal for small motors, but both end-shield and pedestal types are used in the larger sizes.

Any manufacturer will have a standard range of motors established, or in course of development, with sizes denoted by the centre height.

These will range from 100 to 315 mm and 355 mm for the low-voltage motors (up to 550 V), and from 355 to 1000 mm for the high-voltage motors (up to 13 800 V). The numbers owe their origin to the Renard series of equal percentage steps, and in the development of standard ranges of motors this series of numbers has been used extensively. Copper sizes and motor frames both commonly follow the Renard series.

Referring again to Figure 5.1, specific outputs will continue to increase as new materials become available and as knowledge of losses improves. For an aircooled motor, the losses for a given temperature rise are substantially constant, so that any increase of output will tend to be associated with an increase of efficiency of conversion from electrical to mechanical energy.

Reduced to its simplest form, the electric motor must be constructed in a cost-effective way which will locate the rotor centrally to the stator bore at all times. The shaft must transmit the transient and steady torques, and the stator must transmit the counter torque to the foundations.

The electrical supply will be brought in to the stator windings via bushings or terminals, according to the practice of the customer and the supply authority. The motor enclosure and cooling will be chosen to suit the operating conditions, and safety devices monitoring temperatures and flow of oil or water will be incorporated to ensure warning is given of abnormal operation.

In view of the requirement for consistent quality, every component, whether base material such as steel or specialized manufactured item such as a bearing, will be specified fully by the motor manufacturer and inspected and tested to the specification before being accepted for manufacture of the motor.

5.2 Stator cores

The usual a.c. motor is connected to a three-phase supply on its stationary part, the stator. The stator therefore carries a magnetic field which, at each point, alternates at supply frequency and must be constructed of flux-carrying material, suitably laminated in planes to carry the flux but to minimize eddy current losses.

5.2.1 Iron losses

Two types of loss occur in the laminated steel:

(1) Hysteresis loss due to the cyclic magnetization of the steel firstly in one direction and then the other. This component is proportional to the supply frequency.
(2) Eddy current loss, which is approximately proportional to the square of the supply frequency and varying with the square of the lamination thickness.

Guaranteed maximum total losses will be obtained from the steelmaker's curves for different grades and thicknesses of steel used, and these losses are then usually increased some 15% for estimation purposes to allow for the additional losses expected in operation, after punching the diameters and slots.

5.2.2 Steel grades and thicknesses

Manufacturers typically keep in stock a limited number of steel types for the range of motors manufactured. Small motors will use a steel of some 0.65 mm thickness and losses of 6 W/kg at 1.5 T. Medium-sized high-voltage motors will use 0.5 mm thickness and 4 W/kg, and large high-voltage motors 0.35 mm thickness and 3.3 W/kg.

Insulation between the laminations for small motors may be accomplished by steam-bluing the steel surfaces, whereas the larger laminations will employ a phenolic or synthetic resin insulation on one or both sides of the lamination. Insulation on both sides has the advantage of protecting the steel against corrosion during storage or shipping, whereas single-sided insulation can ensure the laminations are always punched and stacked in one direction, thus avoiding contact between punching burrs.

5.2.3 Manufacture

5.2.3.1 Manufacture of laminations

Figure 5.5 shows a single stator lamination from an open-slot high-voltage motor, and Figure 5.6 a single segment from a large high-voltage motor of slow speed. Steel can be manufactured in rolls of up to 1220 mm width and if the required output can be obtained within this back-of-core dimension then single-piece laminations will be used. For motors requiring a larger back-of-core lamination diameter, the circle will be formed from a number of steel segments. Six segments per circle is normal for the standard range of motors and each segment will then be the same if the number of slots is divisible by 6. It is not essential to have a whole number of slots per segment and, on the larger motors, nine or twelve segments per circle may be used with whole or fractional slots per segment. It is the responsibility of the designer to assess the cost implication of the choice of a suitable slot number to give the desired performance, against the choice of a slot number for ease of manufacture.

While laser technology is being used to produce special shapes in thin materials of limited number, stator core laminations are normally produced by a punching process. As each slot is produced by the same tool they will all be identical, but the action of the punch and die causes a burr to be formed on the underside of the steel. With pre-insulated steel, it is important to control the size of this burr such that it will not short to the next lamination and thus form a short-circuited turn in the core.

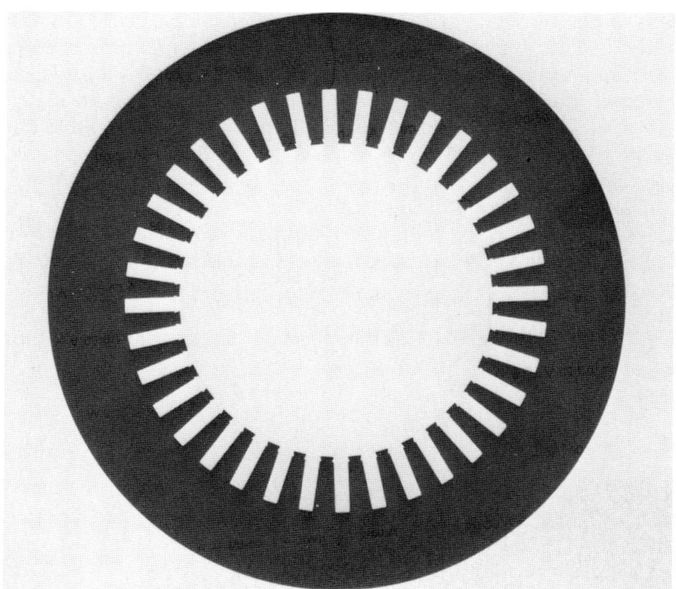

Figure 5.5 Circular stator lamination. (*Courtesy:* Brown Boveri Company)

Figure 5.6 Segmental stator lamination. (*Courtesy:* Brown Boveri Company)

Figure 5.7 Lamination magnified 100 times. Insulation is shown at A, 0.012 mm thick. The burr is shown at B, which must be less than 0.012 mm

Figure 5.7 shows a lamination magnified 100 times. The insulation is 12 μm thick on one side only and the allowable burr must therefore be less than 12 μm. The burr size will be kept to a minimum by keeping the punch and die sharp, and will be controlled by inspection after a set number of operations. If half this insulation thickness is applied on each side of the lamination, the maximum permissible burr size will be unaffected.

The larger motors will employ a cooling system where the air passes through radial ducts formed between packets of laminations. These ducts are formed by welding either angles or flats to a thick punched lamination to give air spaces of between 8 and 12 mm. The ducts will be spaced about 60 mm from each other so a pack of 1 m core length will comprise some 14 duct spacer plates and 1720 0.5-mm thick main core plates.

Figure 5.8 shows a typical spacer plate with angles spot-welded to form cooling ducts 10 mm wide.

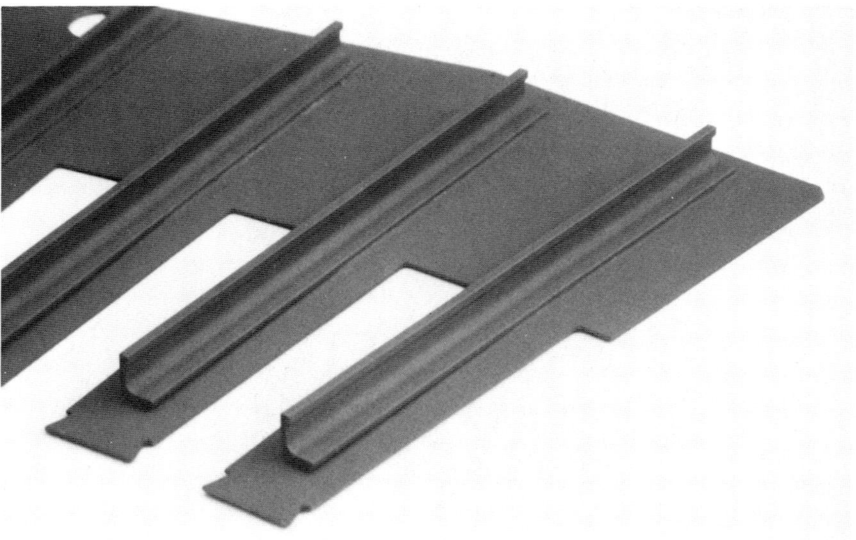

Figure 5.8 Duct spacer plate. (*Courtesy:* Brown Boveri Company)

5.2.3.2 Manufacture of stator core

The complete set of stator laminations and duct spacer plates will be assembled together to form a core unit. The laminations may be assembled directly into the stator frame or, more probably, a core pack will be created round a stacking device. Lining bars located in the slots will be used to ensure careful alignment of the diameters and slots. When the stack is complete it will be pressed between heavy-section end-plates and the laminations welded together on the back of core. For small motors the weld will be directly to the laminations only, possibly in a specially punched semicircular groove. On the larger motors, angular or rectangular core bars will be welded to the outside diameter to provide a rigid steel cage to support the lamination pack. Figure 5.9 shows a medium-size core pack after stacking. The heavy-section core bars are welded to alternate lamination packets and also to the compression rings. A light-section steel ring is shown to support the temperature detector leads after winding. Depending on manufacturing methods, a moderate degree of dressing with a file may be employed to remove

Figure 5.9 Assembled stator core. (*Courtesy:* Brown Boveri Company)

any high spots on the laminations which would otherwise damage the winding insulation. Dressing must, however, be kept to an absolute minimum or short-circuits will result between laminations which will show as localized hotspots when the stator is excited.

Some manufacturers check for any damage by carrying out a ring flux test on the stator core before winding. This test is also carried out on any motors which have had a damaged winding removed.

The number of turns for the flux test may be calculated as follows. For 1.5 T and a 240 V supply at frequency f, the number of turns is equal to:

$$240/(f \times 1.5 \times 4.44 \times L_{fe} \times D_{fe}$$

where L_{fe} = net iron length, in metres; D_{fe} = depth of core behind the slot, in metres.

After the stator has been excited in this manner for 10 min, examination is made by hand or thermometer for any hotspots on the core and teeth. The core should remain cool, with not more than 15°C rise above ambient temperature. Hotspots will quickly be observable and too hot to touch.

If hotspots are found this may necessitate a complete core rebuild, or etching of the area where the lamination insulation has been destroyed.

Etching can be carried out using a 12 V battery connected to a felt pad soaked in a solution of 20% phosphoric acid. The pad should be applied to the area for 1–2 min and, after the hotspot is cleared, the area must be cleaned carefully to remove any traces of acid.

Removal of hotspots in this manner is usually confined to stators being repaired after damage has been caused when removing a winding or when arcing from a damaged winding has occurred in service. The process is unlikely to be economical for small motors, when it is better to provide a new core unit.

Figure 5.10 Core unit undergoing flux test. (*Courtesy:* Brown Boveri Company)

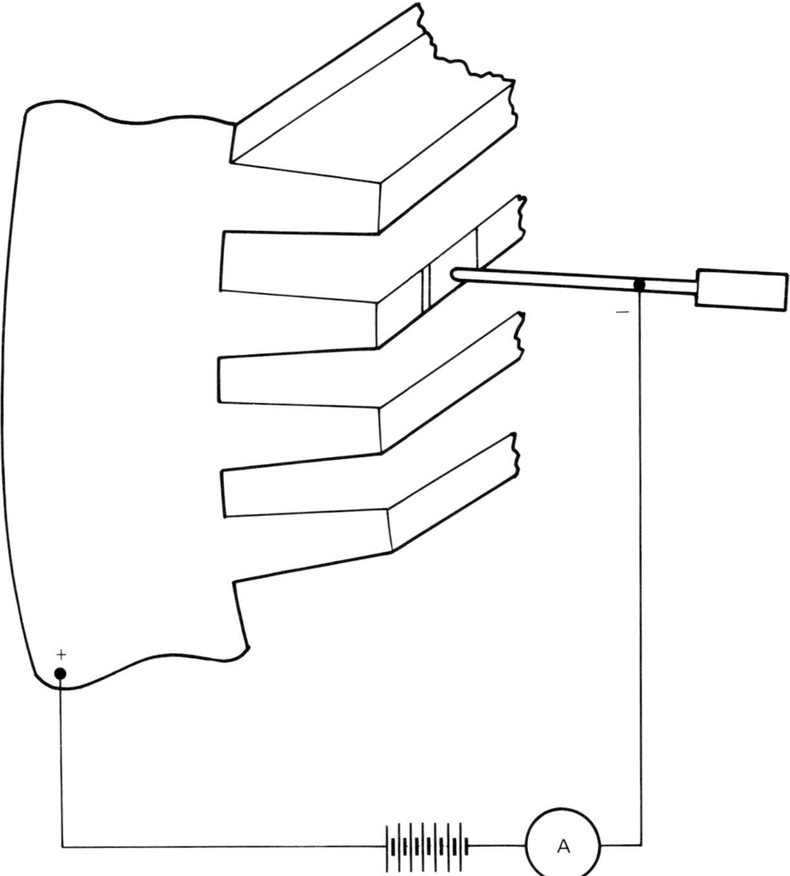

Figure 5.11 Circuit for etching damaged areas

Figure 5.10 shows a medium-sized stator being checked for hotspots by a ring flux test. Figure 5.11 shows the circuit for etching damaged areas. The current passed may be up to 50 A.

5.3 Winding types

Many different winding types have been used in the manufacture of electric motors and various descriptions have been used by designers and repairers to refer to them in conversation and correspondence. Such terms as 'hairpin', 'concentric', 'butterfly', and 'pulled diamond' have been used to describe mainly high-voltage windings. The principal winding types presently in use for the range of a.c. motors from about 1 kW to 20 000 kW are as follows.

5.3.1 Low voltage

Low-voltage small motors (up to 525 V, 50 kW) will use windings formed of round wire covered with an insulating varnish. The wires will not be held in a rigid mass before winding into the slot, and one term used to describe such a winding is 'random mush'. The nature of this winding is such that it can be inserted into the semi-closed stator slot wire by wire.

For motors of a size capable of being wound by machine, each slot will contain one coil-side and the series phase groups will be wound in a concentric manner.

When hand insertion of the coils is used, each slot will contain two coil-sides and the winding can now be short-pitched to save copper and reduce the m.m.f. harmonic content of the winding.

Figure 5.12 shows a semi-closed stator slot, the slot opening being large enough for the insulated wire to be inserted without damage.

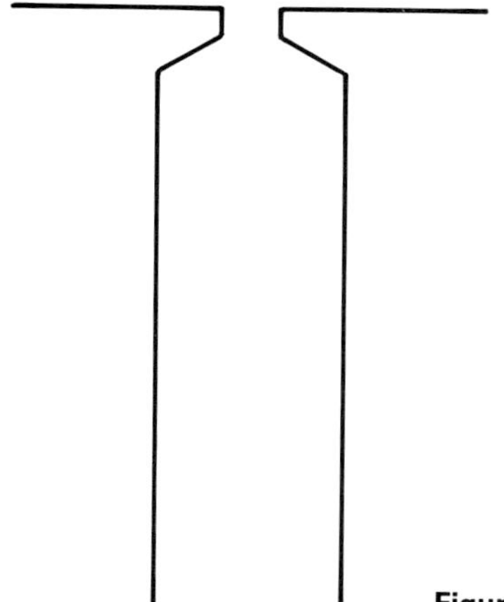

Figure 5.12 Semi-closed stator slot

5.3.2 High voltage

A high-voltage winding will have the conductors arranged in a definite relationship to each other, and the slot opening usually will be as large as the slot width, for radial insertion of the coil-side. The hairpin coil, which was quite common some years ago, was inserted axially from the slot end into a semi-closed slot, but this necessitated each turn being hand-connected at the open coil end, with considerable added labour cost.

The pulled diamond coil was therefore introduced and at present is the most popular choice of coil for winding high-voltage electric motors. Motors which are very large with respect to their voltage use a single turn per coil, each coil-side being a transposed bar with Roebel transposition to minimize eddy and circulating current losses.

The stator winding is the first component in the conversion of electrical to mechanical energy in a motor. A winding will be designed so that when it is connected to a three-phase voltage source, a rotating fundamental field is produced with adequately low harmonic content. Additionally, differences in flux linkage between individual coil-sides and individual conductors may cause circulating and eddy current losses, if not carefully controlled by the winding design.[3]

5.3.3 Pulled diamond winding

This term is used to describe the manufacture of both low- and high-voltage windings with two coil-sides per slot. The ease with which this winding can be manufactured and designed to control losses has made it a most popular winding for all electric motors. All the coils are identical and, in the case of multi-turn high-voltage coils, a natural transposition of the conductors takes place in winding the coil loops. Each coil will have only two connection points to be made to adjacent coils.

5.3.3.1 Coil design

The design engineer will determine the core diameter and length and the number and size of the turns of copper to be used in the coils. Insulation thickness will be determined from standards set by the insulation engineer (see Chapter 6).

To calculate the coil resistance and reactance it is necessary to know the coil dimensions accurately, and consequently the determination of the dimensions of the coil will be an early stage of the basic design calculation programme. This programme starts with the actual coil shape in the wound machine, with correct clearances between coil legs and to earth material, and from this is determined the dimensions for the stages of coil manufacture, and in particular the distance between the pins to wind the coil loops.

5.3.3.2 Coil manufacture

A high-voltage coil is described below, though a similar process is used for a low-voltage diamond coil, the low-voltage coil usually being wound into the stator after pulling to shape.

Depending on the voltage to earth and between turns, the basic conductor insulation is chosen. Typically this will comprise either a layer of polyester imide enamel with a protective layer of glass covering, or layers of epoxy-impregnated polyester film on bare copper. For high-voltage, large, motors layers of mica tapes may be used. This covering will often be carried out at the copper supplier's works and the copper supplied on drums for the first part of the coil-manufacturing process.

Looping. From the design programme the loop size will have been predetermined, as will the number of wires in parallel and the number of turns. The looping machine is set to the correct dimensions and the required number of loops wound with the specified turns. Figure 5.13 shows looping in progress.

The loops will be protected by a layer of sacrifice tape and transferred to the pulling machine.

Figure 5.13 Winding coil loops. (*Courtesy:* Brown Boveri Company)

Pulled diamond.   The loops are positioned between the pins in the pulling machine and the straight parts of the coils are gripped by four clamps. Each loop is now pulled into a diamond shape with the correct included angle between the coil legs and with the winding overhang shaped. To ensure conformity in the shape of each coil, these are finally formed in a wooden coil-former specially made for the coil set. Figures 5.14 and 5.15 show the pulling machine and a coil-former made of hard wood. The above process will be used where resin-rich or vacuum impregnated coils are being manufactured, though the remaining process of applying the main insulation will differ and is described in Chapter 6.

5.3.4 Winding

Individual coils will be wound into the stator core units and a careful process control will be followed to ensure that any fault in insulation is discovered at a stage where it can be rectified. Coil-to-coil joints will be made by brazing and the group connections completed.

Finally, the impregnation or varnish treatment will be completed and the wound core tested at a higher voltage than the final flash test of twice line voltage plus 1000 V for 1 min.[4]

Figure 5.16 shows a wound core unit, with outer bracing of the winding overhang to prevent movement due to the high end-winding forces which are experienced during switch-on.

5.3.5 Winding design

The low-order phase-belt harmonic fields produced owing to the distribution of the winding in discrete phase groups cause losses in the rotor bars of a cage motor and

Figure 5.14 Pulling the diamond shape. (*Courtesy:* Brown Boveri Company)

Figure 5.15 Forming the pulled coil. (*Courtesy:* Brown Boveri Company)

it is usual to control these losses by short-pitching the stator winding.[5] A pitch of 83% gives harmonic pitch factors for fifth and seventh harmonics of 25.88%, and 96.6% for the fundamental field. A coil pitch of 83%, 5/6 times full pitch, is therefore normal, or as close to this as the number of slots per pole and other considerations allow.

In the case of two-pole motors, a winding of 83% pitch is difficult to wind as the dimension across the coil legs will probably be greater than the bore of the stator. In this case a pitch of around 52% will often be used; if considered desirable, the low-order belt harmonics can then be controlled by interspersing the coil groups.[6]

Figure 5.16 Wound and partly braced stator. (*Courtesy:* AEG)

In practice, the control of the low-order belt harmonics only affects the temperature rise of the motor to a small degree, but can be effective in controlling the heating of the rotor bars during normal operation, particularly if the ventilation of this part of the two-pole motor is limited.

5.3.6 Pole-change windings

The Dahlander winding has been popular for many years as a means of obtaining a 2:1 speed ratio from an induction motor. The winding will typically be connected in two circuits star for the high-speed, and single-circuit delta for the low-speed. By so doing, it can be seen that the current in half the winding is reversed in the delta connection, and poles of only one polarity are formed by the winding. Consequent poles of opposite polarity are formed automatically between these poles and, hence, the motor runs at half its original speed. The distribution factor for the winding can be calculated in the usual way but the average number of phase belts per pole is now 1.5, not 3, which reduces the winding factor by about 14%.

The stator coils are all identical and the coil pitch may be chosen to give the necessary flux for the performance required at each speed. Figure 5.17 shows the connections for the two speeds.

In the same manner, i.e. reversing the current in half of each phase winding, Rawcliffe, Burbidge and Fong[7] showed that two discrete speeds could be obtained of ratio much closer to unity than 2:1. Again, the coils are all identical and of the same coil pitch, but the connection of the coil groups is specially arranged.

This type of motor has been built successfully by many European and American manufacturers for driving Banbury mixers and induced- and forced-draught fans in sizes up to 7500 kW. The theory of the arrangement of the coil grouping and their connection for optimum performance is complex, and reference may be made to the many papers, e.g. those by Rawcliffe and Fong from Bristol University, on this subject.[8–14]

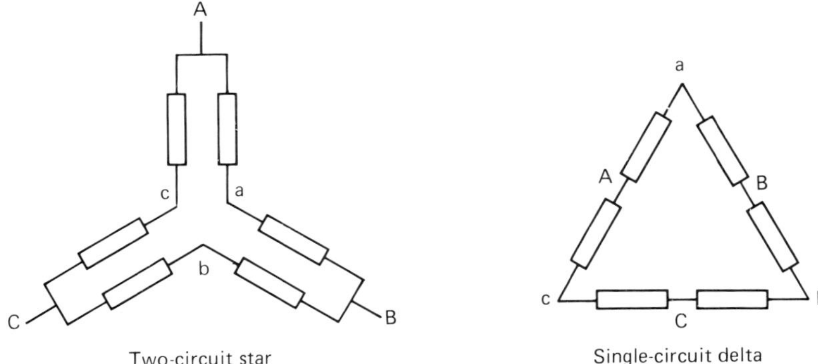

Figure 5.17 High speed: connect supply to A–B–C, join a–b–c. Low speed: connect supply to a–b–c. If the winding factors are equal in both modes of connection, the flux of the delta connection is 0.866 times the flux of the star connection

Theory of pole amplitude modulation

The close-ratio two-speed motors referred to above are more generally known as pole amplitude modulated motors as the principle is similar to amplitude modulation used in radio communication.

The rotating m.m.f. produced by a conventional winding of $2p$ poles may be written as $M_1 = A \sin(p\theta - \omega t)$. If this m.m.f. is modulated by a wave of $(2p \pm 2q)$ poles, the resulting m.m.f. may be written as:

$$M_2 = A \sin(p\theta - \omega t) \cdot \sin(p \pm q) \tag{5.1}$$

which may be expanded to give:

$$M_2 = A/2 \left[\cos(\pm q\theta - \omega t) - \cos(2p \pm q)(\theta - \omega t)\right] \tag{5.2}$$

The positive sign gives:

$$M_2 = A/2 \left[\cos(-q\theta - \omega t) - \cos(2p + q)(\theta - \omega t)\right] \tag{5.3}$$

The negative sign gives:

$$M_2 = A/2 \left[\cos(q\theta - \omega t) - \cos(2p - q)(\theta - \omega t)\right] \tag{5.4}$$

The first terms of Equations (5.3) and (5.4) represent a rotating $2q$-pole field; the second terms are unwanted harmonics. If the modulating wave is of $(2p + 2q)$ poles, this is known as sum modulation. Similarly, if the wave is $(2p - 2q)$ poles, the term used is 'difference modulation'. The choice of sum or difference modulation and also the choice of coil pitch is made to control the second term or unwanted harmonic. In many cases, these yield zero resultant in a polyphase machine.

Despite the complexity of the pole amplitude modulation theory, the actual winding comprises identical coils having the same pitch. The winding diagram presents little difficulty to an experienced armature winder who, by numbering the coil group ends, can readily make the unusual group connections. The pole amplitude modulation winding well illustrates the versatility of the pulled diamond coil.

5.3.8 Slot wedges

All windings, whether low- or high-voltage, will have some form of wedge at the mouth of the slot to keep the winding firmly in the stator slot. The wedges prevent vibration of the slot portion of the coil during operation.

Since the days of hairpin windings inserted into semi-closed slots, designers have been aware of the reduction of no-load losses achieved with slots having small slot openings. The use of open stator slots to insert the diamond coil increased the no-load losses and, until designers understood the mechanism of this loss production, rules were created as a guide to avoid excessive losses. One such rule was that the slot opening should not exceed 8 times the radial airgap dimension. The stator slot openings cause harmonic losses in the rotor laminations and cage winding.

One way of controlling these losses is to close effectively the stator slot with a magnetic wedge. Wedges have been made from small pieces of lamination steel or of a mixture of iron filings and epoxy resin. Unfortunately, the early wedges were prone to failure, as the airgap forces were now partly produced upon the wedges. Wedges vibrated loose and fouled the rotors of motors, sometimes causing sufficient damage to require a complete stator rewind. For reasons better known in the individual companies, some companies ceased using magnetic wedges whilst others sought to calculate the forces involved and devised methods of ensuring the wedges stayed in place. The bad experiences of the early wedges also caused some users to state that magnetic wedges should not be used in their motors.

Currently, the manufacture and fitting of reliable magnetic wedges containing either iron particles or wires is well established and more manufacturers are using this type of wedge in open-slot induction motors. The avoidance of magnetic wedges by some manufacturers has led to an interesting theoretical development in that these manufacturers have probably spent more time on the analysis of the harmonic losses. Users of magnetic wedges have, on the other hand, developed a design philosophy of fewer stator coils and higher airgap flux density in their motors. There is little doubt that magnetic wedges do reduce the losses of motors of small pole pitch, but that the introduction of a saturable component in the stator slot can be disadvantageous on large motors of limited starting current.

The fibre wedges will be continuous in length with bevels cut in the region of the airducts to facilitate the flow of cooling air. Magnetic wedges are not easily machined and so will be confined to the core packet lengths only.

5.4 Rotor shafts and couplings

The rotor shaft is that part of the electric motor which transmits the output torque to the driven equipment, usually via a coupling, though in some cases the shaft will be directly coupled to the driven load. Any shaft must be capable of transmitting the worst torsional stresses and have sufficient stiffness to minimize deflection under conditions of unbalanced magnetic pull.

Small motors will employ round shafts of adequate steel quality to transmit the torques and have good machining properties. Larger motors, where welding to the basic shaft material is required, will be constructed of a low-carbon, high-tensile steel. Carbon content will be less than 0.25% to ensure ease of sound welding. Figure 5.18 shows the shaft of a 10 MW, 1500 rev./min motor with shallow welded-on ribs, and Figure 5.19 a shaft with deep ribs to facilitate airflow and improve shaft stiffness on a slow-speed motor.

In the development of motor design, there has been a trend to increase the airgap forces and reduce diameters for a given output. This has necessitated careful consideration of the extent to which flux will pass through the shaft structure so that sufficient area will be present under the rotor core for air to enter the radial ducts.

Figure 5.18 Machined high-speed shaft. (*Courtesy:* Brown Boveri Company)

Figure 5.19 Machined low-speed shaft. (*Courtesy:* Brown Boveri Company)

5.4.1 The two-pole motor

The shaft of a two-pole motor will be either circular or have very shallow ribs, probably machined from the shaft material. According to the manufacturer's practice or the customer's specification, the shaft will be designed to run either in excess of or below its first critical speed. In both cases the predicted critical will be at least 15% away from the running speed.

Figures 5.20 and 5.21 relate to two different shaft designs for the same 3600 kW motor. That for Figure 5.20 has a smaller diameter under the rotor core and passes through its first critical speed to reach running speed. The shaft analysed in Figure 5.21 has a larger diameter under the rotor core and is intended to run below its first critical speed.

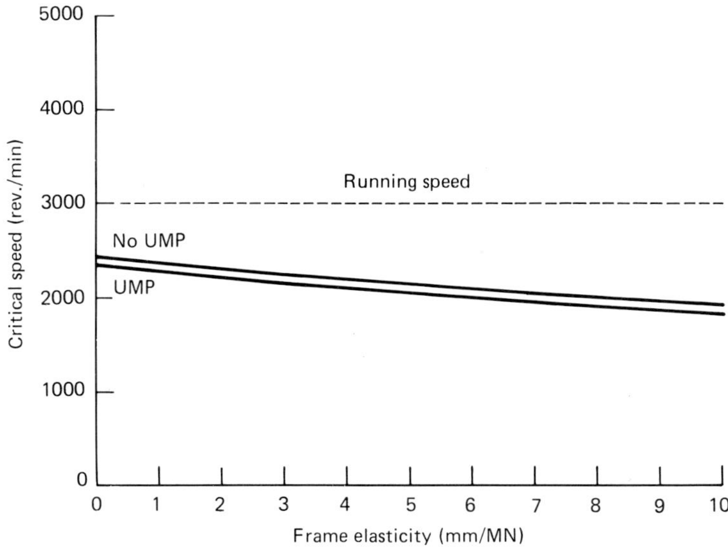

Figure 5.20 Critical speed of smaller diameter shaft

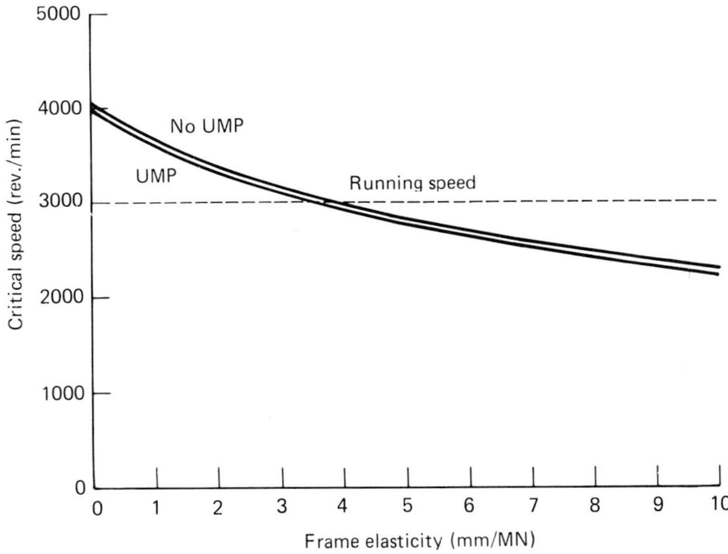

Figure 5.21 Critical speed of larger diameter shaft

Examination of the calculated curves of critical speed plotted against variable frame elasticity show that the small-diameter shaft design is much less sensitive to change in frame elasticity and, hence, its critical speed is more predictable.

5.4.2 Shaft end

The coupling will either transmit the torque by a heavy shrink fit, or will use a shaft key and a slight interference fit of k6 or m6 and H7 (International Standards

Radial misalignment

a_1, b_1, c_1 and d_1 are readings from the dial indicator R at the locations a = top, b = bottom, c = right, d = left. The readings are entered in the formulas to obtain the values of radial misalignment (Table A)

Table (A) Radial misalignment

Measuring location	First Measurement	Second Measurement	Example	
Vertical				
top	a_1	a_2	2.5	2.8
bottom	b_1	b_2	3.1	2.8
difference	$a_1 - b_1$	$a_2 - b_2$	−0.6	0
half difference	$\dfrac{a_1 - b_1}{2}$	$\dfrac{a_2 - b_2}{2}$	−0.3	0
$\hat{=}$ misalignment				
+ Left-hand coupling is higher than the right-hand one			Left-hand coupling raised by	
− Left-hand coupling is lower than the right-hand one			0.3 mm	
Lateral				
right	c_1	c_2	3.8	2.8
left	d_1	d_2	1.8	2.8
difference	$c_1 - d_1$	$c_2 - d_2$	2.0	0
half difference	$\dfrac{c_1 - d_1}{2}$	$\dfrac{c_2 - d_2}{2}$	1.0	0
$\hat{=}$ misalignment				
+ Left-hand coupling displaced to right of right-hand coupling			Left-hand coupling moved	
− Left-hand coupling displaced to left of right-hand coupling			1 mm to left	
Check for measuring error	$\dfrac{a_1 + b_1}{c_1 + d_1} = 1 = \dfrac{a_2 + b_2}{c_2 + d_2}$		$\dfrac{2.5 + 3.1}{3.8 + 1.8} = 1 = \dfrac{2.8 + 2.8}{2.8 + 2.8}$	

Alignment report sheet for radial misalignment – 4 turns, each of 90°

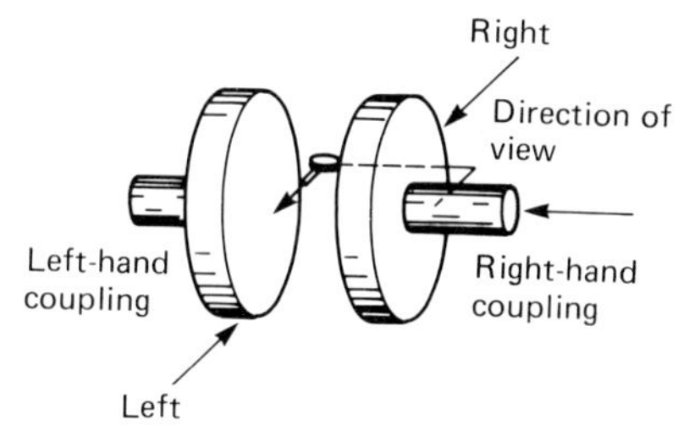

Figure A Schematic for Table A

Axial gap and misalignment. The axial gap is determined by taking readings from the two dial indicators AI and AII, whereby the first reading from the top indicator AI is designated by e_1 and that from the bottom indicator AII by h_1. The values for vertical and lateral misalignment can be determined as shown in Table B; axial displacement (in example 0.2 mm) during the measurement does not affect the results (Figure B)

Table (B) Angular misalignment

Measuring location	Dial AI	Dial AII	Example
Vertical			
top	e_1	g_1	5.0 4.2
bottom	f_1	h_1	6.2 5.0
gap	$\dfrac{(f_1 - e_1) - (g_1 - h_1)}{2}$		$\dfrac{(6.2 - 5.0) - (4.2 - 5.0)}{2} = 1\,\text{mm}$
+ the gap is greater at the top			The gap is greater by 1 mm at the top
− The gap is greater at the bottom			
Lateral			
left	i_1	l_1	4.0 3.6
right	k_1	m_1	4.8 4.0
gap	$\dfrac{(k_1 - i_1) - (l_1 - m_1)}{2}$		$\dfrac{(4.8 - 4.0) - (3.6 - 4.0)}{2} = 0.6\,\text{mm}$
+ the gap is greater at the left			The gap is greater by 0.6 mm at the left.
− The gap is greater at the right			

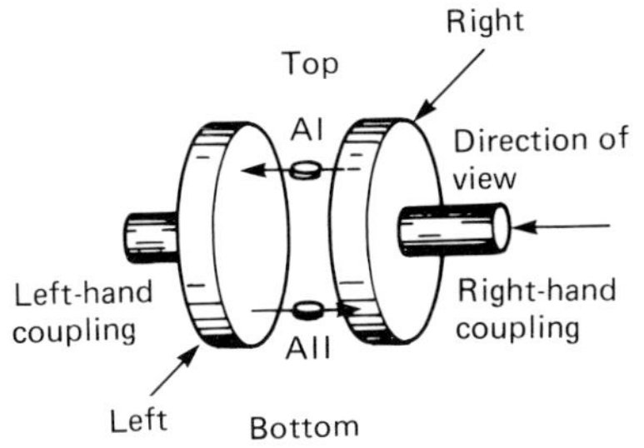

Alignment report sheet for axial gap – 4 turns, each of 90°.
In the example it is assumed that an axial movement of the shaft of 0.2 mm took place during the measurement.

Figure 5.22 Coupling alignment. (*Courtesy:* Brown Boveri Company)

Organization Tolerance System). Key sizes are standardized according to the shaft diameter and are listed in BS 4999:Part 10 for shaft sizes up to 150 mm diameter.

The shaft-end diameter will be designed to transmit the worst case of transient torque the motor can produce, which will be the two-phase, short-circuit torque. In the case of a welded shaft this often means that the shaft end may be larger than the driven shaft and a larger coupling may be required than is necessary solely for the driven equipment.

5.4.3 Couplings

The purpose of the coupling is to connect the two shaft ends together mechanically to transmit the torque, without disturbing the normal running of the motor in its bearings or of the driven equipment. Couplings therefore have the capability of a limited amount of misalignment, though if it is appreciated that any flexible membrane of a coupling will become quite rigid when transmitting full-load torque it is clear that the allowable misalignment must be very small.

5.4.3.1 Coupling alignment

With the coupling halves mounted on the shafts, they must be aligned axially, angularly and radially. To do this it is necessary for the rotor to be turned.

Vertical motors will require the rotor to be suspended from a lifting device if the motor has white-metalled bearings, and in the case of horizontal motors oil must be added to the bearing shells. The shafts of motors with rolling bearings can be turned without difficulty.

5.4.3.2 Axial alignment

The coupling halves will be set with the required gap between their faces, according to the coupling manufacturer's catalogue.

5.4.3.3 Angular alignment

It is important that the coupling faces be parallel to each other within the stated tolerance and, as the shaft may move axially as it is turned, two readings are taken at any time (Table (B) in Figure 5.22). A clock gauge can be used if the flanges are far apart, otherwise feelers will be used.

5.4.3.4 Radial alignment (see Table (A) of Figure 5.22)

With the magnetic chuck on the motor coupling and the clock pointer on the driven coupling, the motor shaft is rotated and a series of readings taken. The total indicated reading of the clock should not exceed the limits set out in the coupling-maker's catalogue for the size of coupling being used. For a Bibby resilient coupling as shown in Figure 5.23, this will be from 0.001 × coupling outside diameter for a 120 mm coupling to 0.0005 × diameter for a 1000 mm coupling.

5.4.3.5 Limited end-float

When the motor is fitted with white-metalled bearings and the thrust faces are designed only for intermittent or light thrust, it is important to ensure that the journal shoulders will not come into contact with these faces during operation. If contact is made, shaft expansion may impose a continuous and large thrust, up to a value where the coupling teeth or pins will slide under load.

Figure 5.23 Bibby resilient coupling. (*Courtesy:* Wellman Engineering, South Africa)

This is accomplished by ensuring the coupling halves have restricted movement and setting the shaft and journal positions so that the motor journal shoulders will always run clear of the bearing thrust faces. It should be noted that motors often have white-metalled thrust faces to take the light intermittent thrust of the shaft as the rotor oscillates axially on uncoupled start-up or switch-off.

5.4.4 Cage bars and end-rings

The reliability of the induction motor owes much to the construction of its cage rotor. Small motors up to around 355 mm centre height will use cast aluminium rotor bars and short-circuiting rings. Not only are these rotors easily cast to reduce costs, but the area of the bars and rings gives a high heat capacity.

Large motors will use copper bars and rings, sometimes using a copper alloy to increase the resistivity of the cage to give the required run-up torque and minimize heating of the windings.

All rotor bars will exhibit to some extent the Boucherot effect. The lower part of the bar, being embedded deeper in the rotor iron and linked with more cross-slot leakage flux, will have a higher reactance than the upper part when the slip frequency is high, i.e. at start. This causes the current to be concentrated in each bar towards the slot mouth, and this displacement of the current produces an effective increase of the rotor bar resistance.

Rotor bar breakages have been troublesome in the past, but as the stress mechanism is now understood and as there is a tendency to use rectangular bars of low reactance on modern machines, bar breakages are rare.

Rotor bars having reduced section near the slot mouth are sometimes described as T- or L-bars and are used when a high starting torque is required, e.g. for a coal mill. Alternatively, a double-cage design may be used which has a high-resistance outer cage separate from the normal inner cage. When using either a double cage or a reduced-section bar, care must be taken at the design stage to ensure the heat capacity is adequate.

5.4.4.1 Endrings

To complete the cage winding, the individual bars must be connected together. This is accomplished by brazing a heavy-section copper or alloy ring to the ends or underside of the rotor bars. In the case of small motors this ring will be cast at the same time as the bars. Figure 5.24 shows the brazing process on a large rotor where the rings are brazed to the bar ends. The brazing jig is constructed especially to

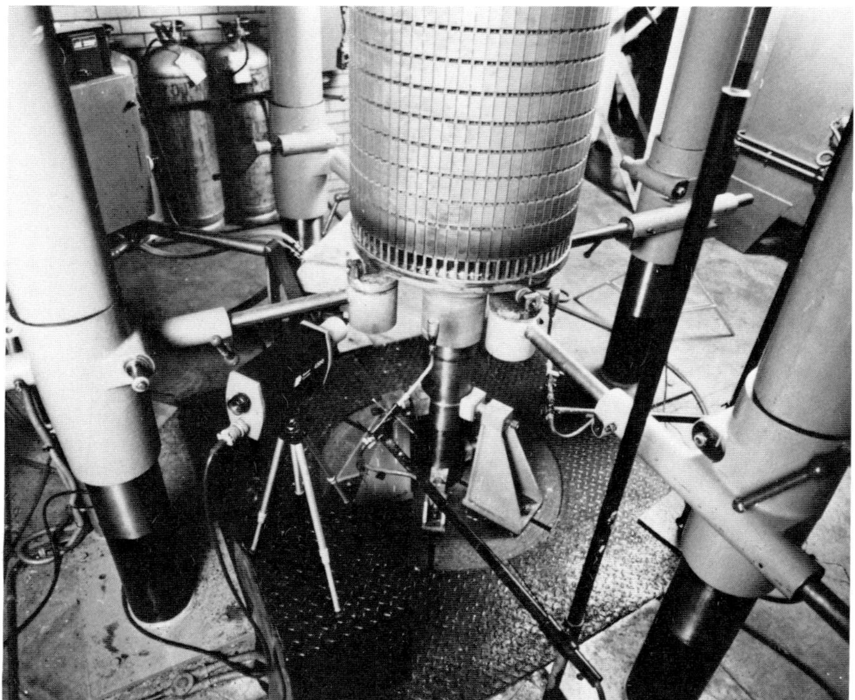

Figure 5.24 Brazing the short-circuit ring to the rotor bar ends. (*Courtesy:* Brown Boveri Company)

allow the rotor to turn slowly, the heating being from a medium-frequency generator. Gas jets maintain the ring at a moderate temperature during the process. This method allows ring materials to be used which have a high tensile strength and which would otherwise become annealed during a prolonged heating process. Figure 5.25 shows a ring being hand-brazed to the underside of the rotor bars. This process is less expensive and is used when the ring stresses are low. The brazing process is controlled to give a joint which is sound both mechanically and electrically, the latter being accomplished with a good brazing fillet.

5.5 Bearings, lubrication and bearing insulation

Electric motors will use either rolling or sleeve bearings.

5.5.1 Rolling bearings

'Rolling' is the term used to describe ball- and roller-bearings, and 'sleeve' the term used to describe white-metalled bearings where the journal runs on an oil film directly against the bearing surface.

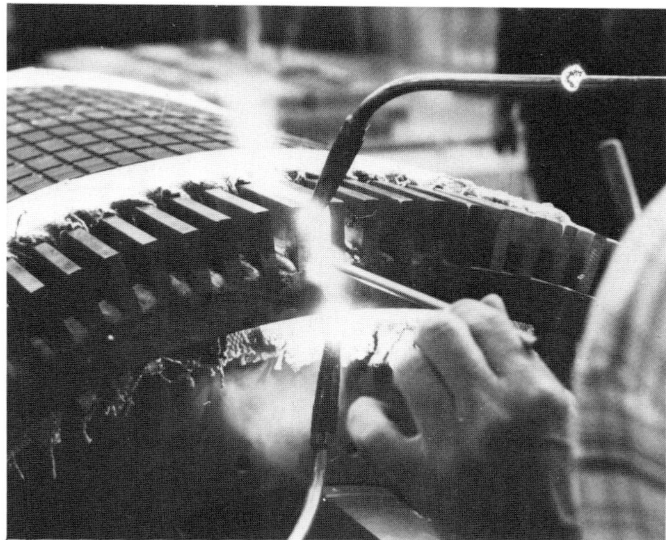

Figure 5.25 Brazing the short-circuit ring under the rotor bars. (*Courtesy:* Brown Boveri Company)

Small motors will employ a deep-groove ball-bearing at both drive and non-drive shaft ends. As the motor size increases, a roller bearing may be used at the drive end to cater for radial thrust of a drive belt, the ball-bearing being used at the non-drive end to locate the shaft axially. More ideally and on larger motors, two roller bearings are used and an additional ball-bearing provided at the drive end, which is free to move in a radial direction. This bearing is solely to locate the rotor axially and can withstand a moderate amount of axial thrust such as that produced by a fan in an axially cooled machine.

The use of rolling bearings is standardized and well-documented in reference books provided by bearing manufacturers. Standard life to be expected, according to speed and loading, is readily calculated and the maximum allowable rotational speed of the bearing for the lubrication chosen is listed.[15]

It is normal to choose bearings which will give a life of 50 000–100 000 h running, or even 200 000 h on the larger motors. Figure 5.26 shows a typical layout of rolling bearings at the drive end and at the non-drive end. Note the radial clearance of the ball-bearing at the drive end.

5.5.2 Sleeve bearings

Examination of typical bearing manufacturers' books[5] will show that there is a definite limit to the rolling bearing rotational speed as bearing diameters increase. For this reason and for extended life, the larger motors of high speed employ sleeve bearings.

Sleeve bearings will be metalled with either a tin or lead-based white metal, the former being the standard of British, and the latter of German, manufacturers.

During the Second World War, Germany was unable to obtain supplies of tin, which was then the conventional material to use, but developed as a substitute a lead-based metal which for normal motor speeds has proved equally reliable and less costly.

As sleeve bearings are used in motors using axial ventilation as well as those with balanced radial ventilation, it is normal to provide a locating bearing at the drive end to take axial fan thrust. Figure 5.27 shows such an arrangement.

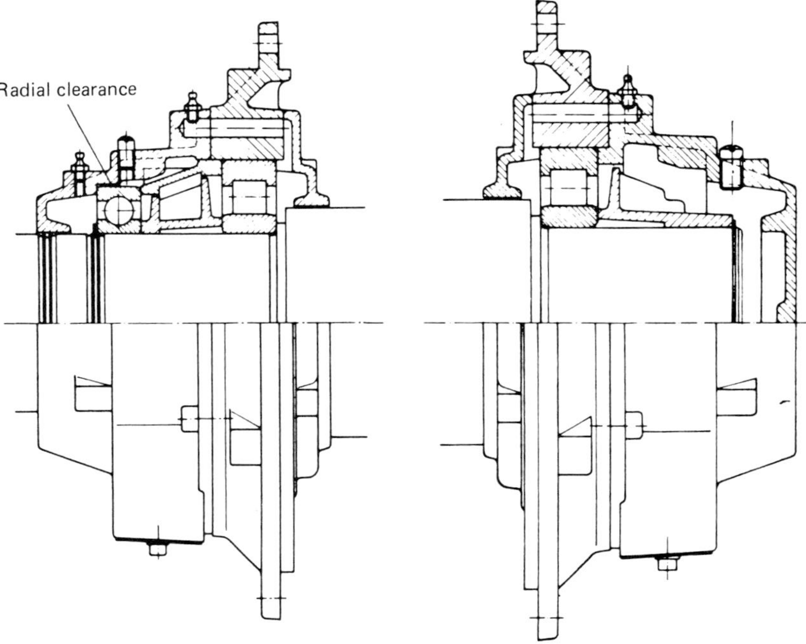

Figure 5.26 Typical rolling bearing. (a) Drive end. (b) Non-drive end

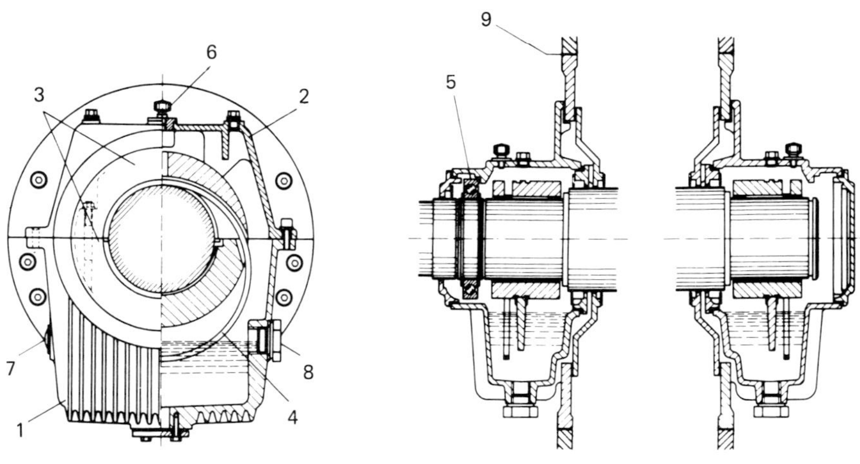

(1)   Bottom housing with cooling fins
(2)   Top housing (cover)
(3)   Bearing shells
(4)   Lubricating ring
(5)   Axial guide bearing
(6)   Vent
(7)   Oil sight window
(8)   Plug
(9)   Bearing insulation

Figure 5.27 Sleeve bearings with locating ball-bearing at drive end

Sleeve bearings may have plain or spherical seats between their shells and the bearing housing. A spherical seat has the advantage of correcting any out-of-true in machining and will ensure the shells are seating against the journals on assembly.

Once it has been assembled and the bearing top housing tightened to the lower housing, the spherical seat becomes rigid and the spherical seat has no influence.

As bearing seals have limited clearance and are not spherically seated, the spherical seating of the shells can only cater for a small degree of misalignment.

5.5.3 Lubrication

Rolling bearings are grease-lubricated, except for a limited use of oil lubrication where high speeds are involved. A grease-thrower will be fitted to the shaft which will pump used grease to a reservoir. It will be possible to empty the reservoir either with the machine running or when stationary. On larger motors it is usual to include the type of reservoir which can be cleaned with the machine running, the system being described as 'grease relief'. New grease can be applied whilst running and the old grease scooped away either by hand or with a specially designed scoop. It should be noted that during the application of new grease to a rolling bearing, the shearing of this grease by the rollers can result in an increase of bearing temperature. The temperature will return to normal after a short period of running.

Sleeve bearings will be oil-lubricated and, depending on the peripheral speed of the journal and to a small extent on the journal pressure on the shell, cooling may be accomplished in the following ways:

(1) Natural cooling with heat radiated from the bearing housing surface. This is possible up to about 13 m/s peripheral speed.
(2) Between 13 and 15 m/s, water cooling by means of a tube cooler in the bearing sump is satisfactory.
(3) Above 15 ms, flood oil will be provided.

It should be mentioned that the main cause of sleeve bearing damage is lack of lubricating oil, and for this reason some customers prefer to fit a flood oil system if the operating conditions make it extremely difficult to see if there is sufficient oil in a bearing housing. An example of this is in a power station where the coalmill motors become coated in coal dust. Flood oil may be fitted to these bearings to ensure oil is present, even though oil-cooling is not essential.

Oil is transferred from the sump to the bearing shells either by disc or oilring. An oilring rotates at a slower speed than the shaft and has lower losses than the disc. The oil enters the bearing shells from machined wedges in the shells on the bearing centre line, and the hydrodynamic operation of the bearing pumps oil from this region to the underside of the shaft from where the hot oil discharges from the bearing ends. This hot oil will be made to pass over the bearing cooler or circulate over as much bearing housing as possible before reaching the area from which the ring or disc is supplied. In this way the oil is cooled by radiation from the bearing housing surface, which may be provided with cooling fins.

Generally, the use of specially shaped sleeve bearing bores is not necessary for the standard ranges of normal speed motors. Providing normal bearing pressures are used, i.e. from 0.5 to 2 N/mm^2*, circular shell bores are satisfactory for speeds up to two-pole rotational speeds. Standard lubricating oil will have a viscosity of 3.7 E or 25 mPa at a temperature of 50° C. A turbine-type oil is often used in view of its low oxidation properties. Use of oil with different viscosity is often permissible but typically an increased temperature rise of up to 5° C may be experienced.

*Pressure is calculated from the projected bearing area of length × diameter.

5.5.4 Bearing insulation

When the design of the flux-carrying component of the motor has some interruptions which are not symmetrical with the flux pattern, an e.m.f. will appear across the shaft ends. This e.m.f. will cause current to flow through the motor frame via the bearings, unless the circuit is broken permanently at some point. To ensure that this is so, it will be necessary to test the insulation periodically, and for this reason it is recommended to insulate both bearings. Usually the drive-end bearing insulation is subsequently shorted by an earthing strap of copper to prevent build-up of static electricity on the rotor. An earthing brush may be used on the drive-end bearing to confine any generated e.m.f. to the motor, and to ensure any static electricity is grounded immediately.

Insulation of the bearings is provided commonly when a stator core is made from segments, and on other designs where the slot combination or rotor core density has been shown by experience to create a shaft e.m.f. Laminations having non-circular shape or non-symmetrical ventilation holes have also been known to cause shaft e.m.f. and to require insulated bearings. Strictly a six- or twelve-pole motor with six lamination segments per circle does not require bearing insulation as the breaks between the lamination segments are in a similar pattern to the main flux. However, insulation is usually provided for the reason that it is standard for other pole numbers.

5.5.4.1 Application of insulation

Insulation used to be almost always applied to the component carrying the bearing shells. In the case of pedestal bearings, the insulation was placed under the pedestal and the holding down bolts were insulated with oversize washers and tubes of insulating material. In the case of end-shield bearings, the insulation was either applied to the seating of the bearing housing in the end-shield or the end-shield was split into two concentric parts and a thin insulation material placed between them (see item (9), Figure 5.27). In all cases it was necessary to take especial care that the insulation was not shorted-out by, for example, temperature detectors or oil piping. In one case I recall, plastic covered water coolant pipes concealed a steel reinforcement which caused a lot of damage before it was located.

More recently, a considerable simplification has been achieved with the insulation being applied to the bearing shells, either where the shells seat in the housing or between a steel addition to the shell and the shell itself. This system is much preferred as components placed on the bearing housing after erection do not now need to be insulated. Temperature probes will have insulation provided by the motor manufacturer and will be tested in the manufacturer's works.

With insulated bearing shells, it is preferable to use insulated bearing seals if these are the floating type which may contact the shaft.

5.5.5 Shaft currents

The damage done to shafts and bearings where electrical current has been flowing is described very adequately in the lubricant supplier's handbooks.[16]

The ball and roller bearings sustain damage which has been likened to a washboard or to ripples on wet sand. Sleeve bearing damage is a series of small craters usually extending in a line 5–25 mm long. Figure 5.28 shows very bad bearing damage due to electrical current. A relatively small amount of current can cause such damage that it will eventually result in lubrication failure and cause the bearing to wipe. In the case of rolling bearings, a steady disintegration of the rolling elements will take place and the bearing will be noisy. It is usually accepted that a shaft voltage in excess of 250 mV requires that the bearings be insulated.

Figure 5.28 Journal damage due to bearing current. (*Courtesy:* Brown Boveri Company)

5.6 Fans, cooling and coolers

Some motors are designed with more heat capacity than is required for full-load running, e.g. when the starting duty is limiting, but the majority are designed from an optimum costing consideration with near full temperature rise under continuous full-load conditions.

The electrical design calculations are commonly programmed on a computer and the components of the equivalent circuit are calculated with satisfactory accuracy. The design engineer, operating between customer and final specification for manufacture, uses the programme to select an optimum machine size largely based on his or her experience and judgement.

As well as the accurate solving of the equivalent circuit for electrical performance, it is quite feasible to incorporate in the programme calculations of airflow, resistance to airflow and temperature rise of the windings. By incorporating this in one programme, the total effect of any changes made to the design can be determined.

5.6.1 Programme outline

Firstly, it is necessary to simulate the heatflow in a small element of the motor, this element being repeated to form the complete motor. The heatflow is solved by analogy to an electric circuit and the temperatures of the iron and copper determined with a given airflow.

The airflow will depend on the resistance of the air circuit and the head developed by the fans and rotating ducts of the rotor. Both the resistance and rotating duct head are dependent on the basic electrical design parameters such as the stator and rotor slot widths, and the rotor diameter.

As an example, a series of results for alternative designs are tabulated in Table 5.1. All can be considered practical designs though, for the sake of comparison, the copper width has been increased beyond what is normally considered ideal. To retain coil flexibility it would be split into two strands in the width.

Table 5.1 Summary of calculation results for a 1500 kW, 11 000 V, twelve-pole air-to-air induction motor

	Slot width (mm)	Copper width (mm)	Air quantity (m³/s)	Temperature rise (K)	Air rise off cooler (K)	Copper mass (kg)	Full-load current (A)
(1)	14	8	2.05	54.2	9.0	589	96.5
(2)	15	9	2.04	53.6	8.7	676	96.9
(3)	16	10	2.02	53.9	8.5	759	98.0
(4)	17	11	1.99	55.4	8.4	846	100.5
(5)	18	12	1.94	59.5	8.4	939	106.3

The table illustrates the effect that widening the stator slot has on the total internal airflow. Also, it can be seen that as the internal airflow is decreased, and therefore has a higher temperature rise, there is a small compensation in that the air entering the inside of the motor has a slightly lower rise above the air entering the cooler tubes (indicated as air rise off cooler). In this case the first item was actually built with the least amount of stator copper. Its temperature rise on test was 54.4 K.

Two types of coolers are commonly used for totally enclosed motors of IP55 or IP44 enclosures. These are the water-to-air cooler and the air-to-air coolers.

5.6.2 Water-to-air coolers

Water cooler design is usually left to the specialist cooler manufacturer, as the skills required for building water coolers are not commonly found within motor manufacturers. The parameters available for cooler design will be specified by the motor manufacturer, such as available water quality and temperature, total expected motor losses and the allowable resistance head or pressure drop the water cooler may present to the given quantity of motor-cooling air. Additionally, the motor manufacturer may wish to specify the material from which the cooler parts in contact with the water are to be made and he will give limiting sizes of the cooler. Probably the whole issue will be discussed and a compromise reached to arrive at the best solution and an agreed price.

Figures 5.29 and 5.30 show motors with top-mounted water-to-air coolers. Both motors employ the double radial air circuit, but the first is a constant-speed motor with the internal air circulated by shaft-mounted cooling fans. The second is a variable-speed synchronous motor drive where the internal air circulation is supplemented by separate fan units. At any time two opposite corner units are in use, and the failure of either unit is detected by an air pressure switch which switches the alternate pair of fan motors into action.

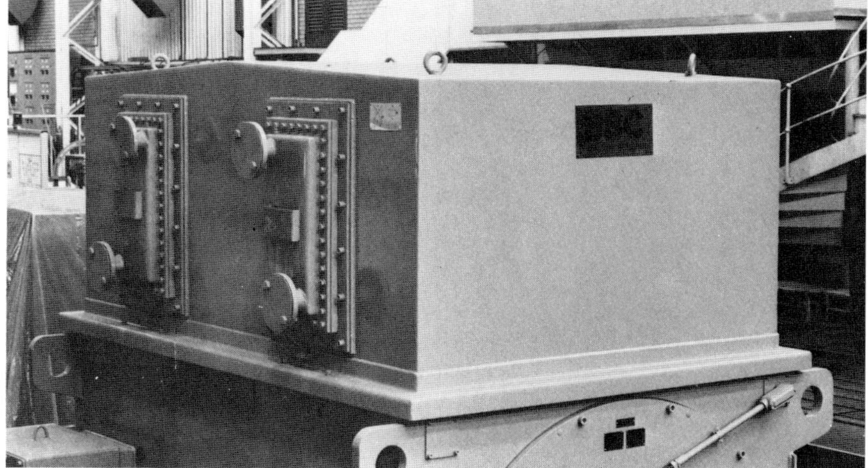

Figure 5.29 Top-mounted water-to-air coolers. (*Courtesy:* Brown Boveri Company)

Figure 5.30 Top-mounted water-to-air coolers with additional fan units. (*Courtesy:* Brown Boveri Company)

5.6.3 Air-to-air coolers

As these coolers are simpler in construction than the water coolers they are often built in the motor manufacturer's factory. Design will be undertaken by the electrical or mechanical motor design engineer. Using a typical design method,[17] a standard range of coolers may be designed for a range of motors. Sufficient cooling tubes will be provided to dissipate the motor internal losses, and as these rows of tubes create internal resistance to the motor-cooling air, various ways are found to minimize this resistance. Figure 5.31 shows an interesting way of presenting a large

Figure 5.31 Air-to-air cooler of low-speed motor. (*Courtesy:* Westinghouse, Texas)

surface of tubes to the internal air whilst restricting the cooler width to the motor width. Figures 5.32 and 5.33 show a single wide horizontal cluster of cooling tubes situated above the angled part of the cooler. This angled part is used to return cool air from the fan end of the cooler to the opposite end of the motor.

In all cases the external air is blown through the cooling tubes either by high-efficiency fans mounted on the cooler (Figure 5.32), or by a shaft-mounted centrifugal-type fan with its air ducted to the cooler tubes (Figure 5.33). It is always

Figure 5.32 Low-speed air-to-air cooler with separate fan units. (*Courtesy:* Brown Boveri Company)

Figure 5.33 Low-speed air-to-air cooler with shaft-mounted fan. (*Courtesy:* Brown Boveri Company)

preferable to have the motor internal air entering the motor after it has left the cool end of the cooling tubes, i.e. the fan end, and this can be arranged readily with a single-ended air circuit of the motor. It can also be arranged with the double radial air circuit by bleeding cool air down the angled cooler sides as already described.

By designing a cooler in this way very efficient heat transfer can be obtained, and air-to-air cooled motors may incur no more than a 10–15° rise penalty over their operation as an open machine.

Figures 5.34 and 5.35 show two typical air circuits. The first is a single-ended axial/radial system and the second a double radial system.

The axial/radial system is becoming popular for all except high-speed motors, as it has the better head generation possibility. The axial thrust of the cooling fan is

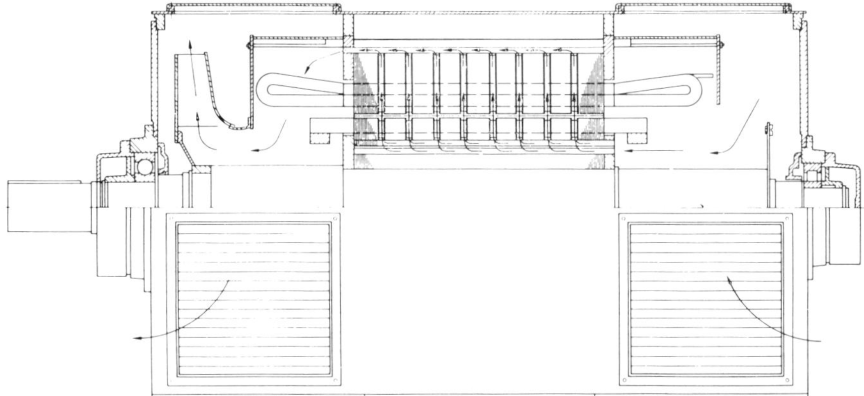

Figure 5.34 Machines with radial cooling, ventilated from one end. The cooling air inlet is at the non-drive end of the motor, the flow being directed axially through the rotor. Warm air is exhausted at the drive end. (*Courtesy:* Brown Boveri Company)

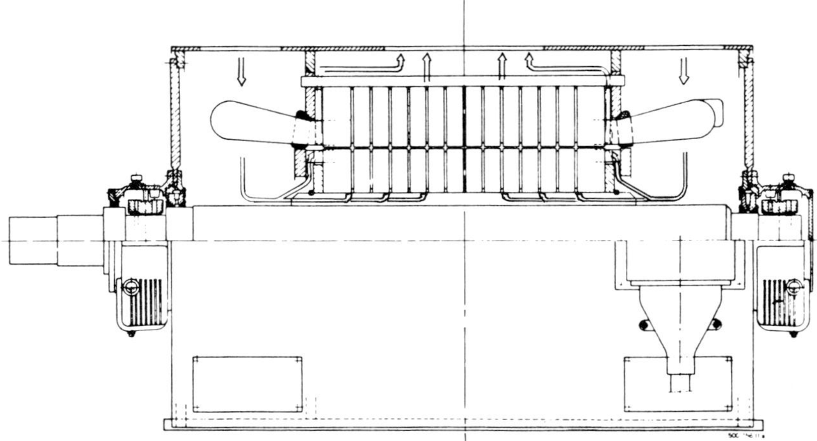

Figure 5.35 Machines with radial cooling, ventilated symmetrically from both ends. Cooling air enters the machine from both ends, the flow being directed through the stator winding overhangs and axially through the rotor. Warm air discharges upwards from the centre of the machine. (*Courtesy:* Brown Boveri Company)

countered by the magnetic centering force of the rotor developed by flux fringing between stator and rotor ducts and core ends, or countered by the locating ball-bearing.

5.6.4 Air circuit calculation

The resistance of the air circuit is proportional to the square of the air quantity and, hence, the resistance curve can be plotted as millimetres of water or pressure drop against air quantity in cubic metres per second, once the resistance factor is known. J.L. Taylor[18] gives a calculation method for the resistance factor K. Using these values for K and the area in square metres, the pressure drop in millimetres of water is equal to $K/\text{area}^2 \times (\text{m}^3/\text{s})^2 \times 1/16$. K/area^2 will be calculated for the major parts of the air circuit as shown below, and summed to a total K/area^2:

(1) Entry to the rotor shaft area.
(2) Entry to the rotor ducts.
(3) Contraction of air at the base of the rotor bars.
(4) Expansion of air at the top of the rotor bars.
(5) Contraction of air on entry to the stator slot area.
(6) Expansion of air at the back of the stator slot.
(7) Expansion as the air leaves the core area.
(8) Cooler resistance.

5.6.5 Fans

The cooling fans will be a combination of actual fans, either radial or angled blade, and the rotating cooling ducts of the rotor. In both cases the head–flow characteristic can be considered in non-dimensional terms where the two are related by the following quadratic equations for radial and 30° backward angle fans.

$$\text{Radian fan: } Y = -3.8X^2 - 0.7X + 0.6 \tag{5.5}$$

$$\text{Angle fan: } Y = -9.2X^2 - 1.2X + 0.55 \tag{5.6}$$

where Y is the head factor and X the flow factor.

In terms of fan diameter, width and rotational speed the head and flow can be calculated as follows:

$$\text{Fan peripheral speed } (F_{\text{ps}}) = \frac{\pi \times \text{fan dia.} \times \text{rev./min}}{60 \times 1000}\text{ , in metres per second}$$

$$\text{Fan area } (F_{\text{area}}) = \frac{\pi \times \text{fan dia.} \times \text{fan width}}{1000 \times 1000}\text{ , in square metres}$$

where the fan diameter is the outside diameter in millimetres and the fan width is the width at the outside diameter in millimetres.

The fan head is given by $F_{\text{ps}}^2 \times 1/16 \times Y$, in millimetres of water.

The fan volume is $F_{\text{area}} \times F_{\text{ps}} \times X$, in cubic metres per second.

If the fan head is plotted against volume, and the resistance head also plotted against volume, the point where the two curves cross will be the actual air quantity. Figure 5.36 illustrates the principle. The calculation is simple to program on a computer or hand calculator. Starting from zero airflow, the fan head is subtracted from the resistance head and, if the result is positive, the calculation is repeated for small increases of airflow. When the result is zero or negative, the calculation ceases and recalls the last value of airflow.

5.6.6 Propeller fans

The design of aerofoil-section fans, or the rolled-steel blade approximation to an aerofoil, is one of the more difficult design arts. Keller[19] gives a good guide to the

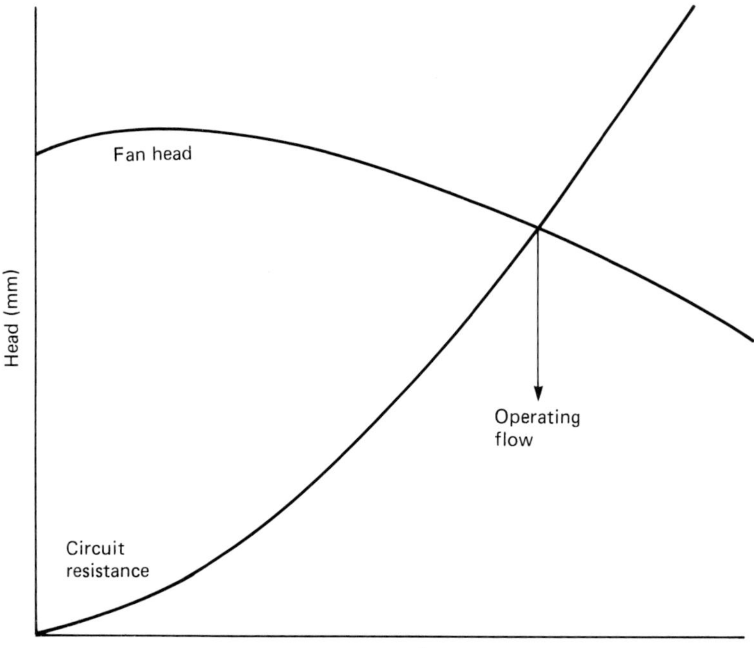

Figure 5.36 Fan and resistance heads against airflow

design of these fans which are only used on two-pole and large four-pole motors. Basically, they require a high peripheral speed to generate the required head.

On one occasion we were having some difficulty meeting both efficiency and temperature rise on a two-pole motor in South Africa. A fan design expert was called in and, after detailed calculations, new propeller fans were made. They did not work as well as they should have done and I think we all felt rather better when a few days later, during practice for the International Formula 1 races at Kyalami, the aerofoils were removed from some of the racing cars. They could not get it right either at the operating altitude and temperature. Since this experience I have always included a correction in the computer program for airflow and cooling to change the air density as a function of altitude and temperature. The density is approximately proportional to $(760 - 0.075 \times \text{altitude})/760$ where the altitude is in metres above sea-level, and is also proportional to the air temperature, i.e. $273/(273 + t)$ where $t =$ degrees centrigrade.

In the calculation of airflow and temperature rise, particularly with air-to-air coolers, this seems to be one of those rare occasions when the results come out better than expected when errors in calculation are made. If the resistance of the circuit is underestimated by 20%, the airflow will only be some 10% below calculation as the resistance is proportional to the square of the airflow. Also, if the internal air is restricted by the cooler more than expected, then the smaller volume of air will actually leave the cooler at a lower temperature than the larger volume would have done. The smaller volume will have a much higher temperature rise through the motor windings and will result in a higher temperature rise than expected, but a large restriction of internal air may result in only a few degrees additional temperature rise of the motor windings.

5.7 Terminal boxes and terminations

Unless the supplier of the motor is responsible for the total installation, his product will be restricted to the motor with a bare shaft end, and a set of main supply terminals.

The connection of the supply to these terminals has led to varying requirements in different countries. The practice in the English-speaking countries is to have a definite break of responsibility at the terminations, and to call for a fault-tested box which will ensure safe operation even under the worst circumstances, whereas the European continent has largely integrated the motor connections with the motor design. This requirement arose from bad experiences of cast-iron boxes exploding upon switch-on to faulty or wet terminals. The object of the fault-tested box is to safeguard life should there be a failure of the interphase insulation in the box or the motor. Insulation failure in the box may be due to moisture, or an object left in the box during connecting to the supply. Motor insulation failure would most likely be due to winding movement during a direct-on-line (DOL) start, a condition which cannot be proved in the manufacturer's works for large motors. The fault-tested box is subjected to two tests:

(1) A through-current test where the fault level current is passed through cables connected to the terminals, and with the terminals shorted in the region of the motor winding. With the r.m.s. fault current flowing for 0.25 s (the normal time for a circuit breaker to open) no damage should occur which might otherwise give rise to a subsequent short-circuit within the terminal box.

This through-current test may be made in two parts. It is advisable to ensure that the peak current does not cause damage due to the high magnetic forces, and also that the heating does not result in damage. For a 45 000 A r.m.s. test, the peak current test should be 115 kA.

Figure 5.37 Fault-tested terminal box. (*Courtesy:* Brown Boveri Company)

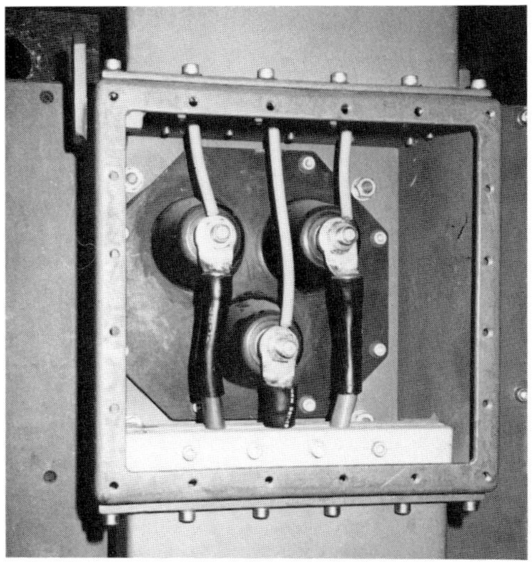

Figure 5.38 Fault-tested box undergoing through-fault test. (*Courtesy:* South African Bureau of Standards)

The test equipment will be calibrated to give this peak current for a few cycles (about 5) and a first through-fault test carried out. A second test will be made for the full 12.5 cycles at an r.m.s. current of 45 kA.

The peak current test is likely to be the most severe.

(2) A second test is made with a short-circuit formed in the terminal box; a piece of stainless steel wire is found to work well and, to maintain the arc for the required number of cycles, arcing horns may be used. In this test the explosion-relief diaphragm should give way in a safe manner, directing the explosion away from personnel who may be near the motor. The rest of the box

may distort but must not release any objects which could be dangerous. Again, the test will be carried out at the fault level current for 12.5 cycles duration.

For a large power station supply the fault level will be 750 MVA on an 11 000 V supply. This gives a fault current of nearly 40 000 A.

Figure 5.37 shows the basic fault-tested box before setting-up for test, and Figure 5.38 the cable distortion after the through-fault test. Figure 5.39 shows the remains of the arcing horns and the diaphragm, after the explosion test had been successfully accomplished and the terminal box cover removed for examination.

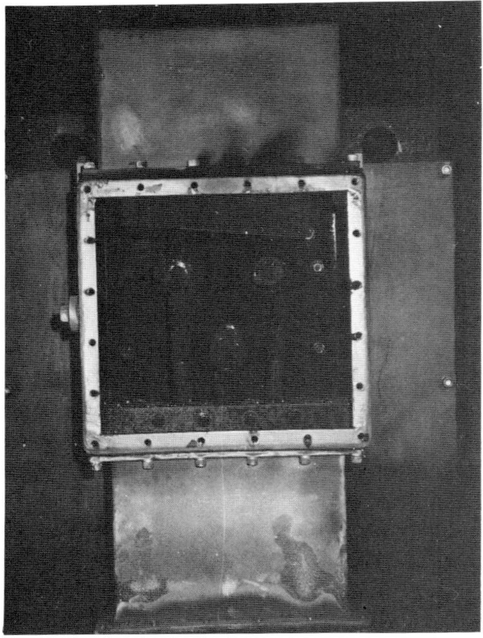

Figure 5.39 Fault-tested box after the explosion test. (*Courtesy:* South African Bureau of Standards)

Any person concerned with the design of terminal boxes should witness such tests to obtain first-hand knowledge of the considerable forces involved. To ensure an internal short-circuit is not formed by tracking due to moisture, a dessicator will be provided. Figure 5.40 shows the fault-tested box with a dessicator screwed into the side of the box. Figure 5.41 shows the same box used as a neutral termination on the other side of the motor.

By terminating cables in this manner, the supply can readily be brought in to either side of the machine. Figure 5.42 shows the same basic design modified to accommodate a voltage surge arrester in an extended box. This box would no longer comply with the fault-tested requirement and would be for use on a lower megavolt-ampere fault level.

When the current is heavy and beyond the normal box design, terminations may be made under the motor and protected by surrounding concrete, or by busbars from the motor side as shown in Figure 5.43 to connect to the sliprings.

5.7.1 Heaters and auxiliaries

Power supply to anti-condensation heaters will be brought into a heater terminal box arranged to give series–parallel connection for different low-voltage supplies.

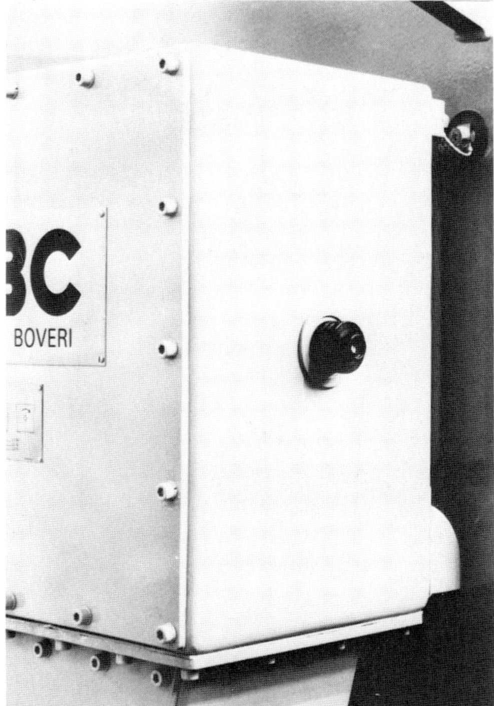

Figure 5.40 Fault-tested box showing the dessicator. (*Courtesy:* Brown Boveri Company)

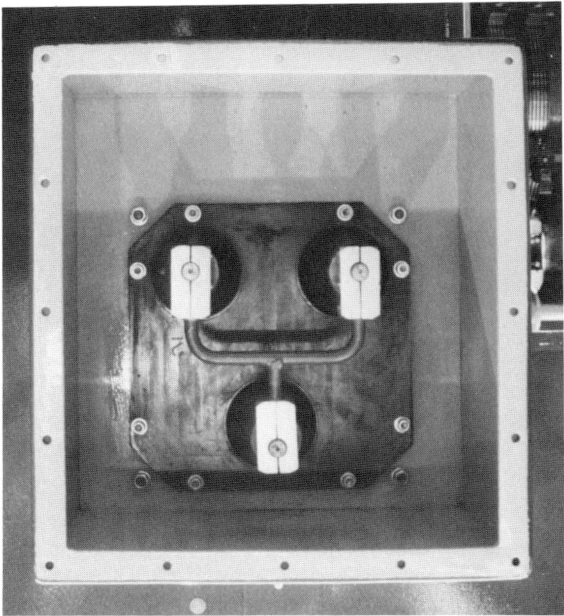

Figure 5.41 Fault-tested box used as a neutral box. (*Courtesy:* Brown Boveri Company)

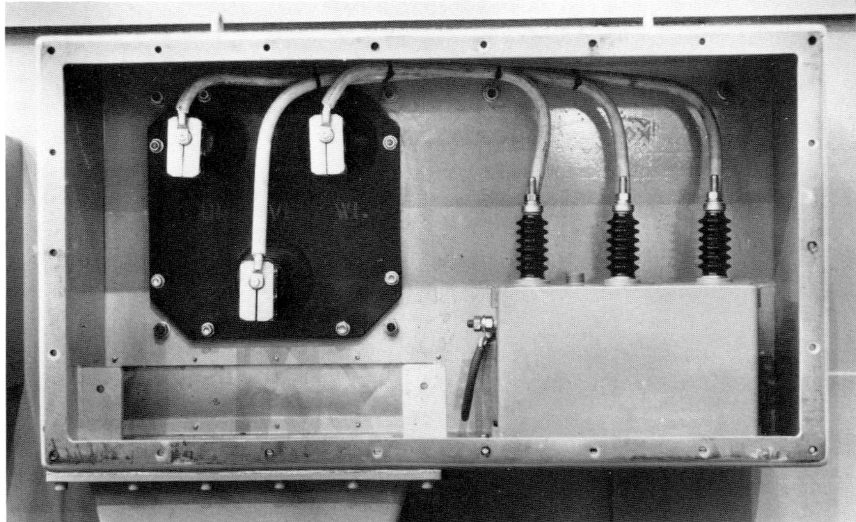

Figure 5.42 Box modified to accept surge arrestors. (*Courtesy:* Brown Boveri Company)

Figure 5.43 Busbar slipring terminals. (*Courtesy:* Brown Boveri Company)

Figure 5.44 shows such a box. Temperature detectors and relays will have their own monitoring terminal box and Figure 5.45 shows the interior of a box containing spark-gap units to ensure earthing of any temperature detector which has come into contact with the winding supply. To the right of the box are provided mounting points for a grid to mount relay units, if required.

The protection of motor windings and bearings is becoming easier as miniature devices are developed. It is now possible to fit reliable temperature detectors to a winding and to operate a relay at a preset temperature in a unit no larger than an old radio valve. New developments will simplify the monitoring aspect in that the switching signal will be given from the motor monitoring unit, which will also store information for subsequent retrieval on the number of starting cycles the motor has undertaken.

Figure 5.44 Heater terminal box. (*Courtesy:* Brown Boveri Company)

Figure 5.45 Monitor terminal box. (*Courtesy:* Brown Boveri Company)

5.8 Heaters and condensation

In the atmosphere, the relative humidity of air is increased and condensation results when air temperature is reduced to the dewpoint, or when sufficient water vapour is added to saturate the air. Saturated air coming into contact with the cold steel surfaces of an electric motor will therefore result in a deposit of moisture.

Where bare terminals are involved, this moisture can cause tracking, ionization of the air and subsequent flashover between terminals of high potential difference.

Moisture was a particular problem with the hygroscopic insulations used in the Class A and Class B systems, as it reduced the insulation level of the insulation material to unacceptably low levels as shown by a Megger test. If low insulation resistance levels are present then it is the practice to dry out the winding by external heating. During this process the insulation level will firstly fall as heat is introduced, but eventually rise as the moisture is driven off.

The non-hygroscopic nature of the Class F epoxide insulation systems has largely removed this moisture problem, though it is still the practice of many users to require heaters in the motor which can be switched on at times when the motor is standing de-energized. The heaters will be designed to maintain the motor at some 5°C above ambient, and thus prevent condensation. Heater ratings will range from 200 W for a 355-frame motor to 1500 W for a large standard frame. Figure 5.46 shows the tubular heater of a 900-frame motor.

Figure 5.46 Tubular heaters on a large motor. (*Courtesy:* Brown Boveri Company)

5.9 Brushgear, sliprings and brushes

The slipring motor once enjoyed much popularity when supplies were weak, as its starting current could be limited. It could also be used for reduced-speed operation by introducing resistance into its rotor circuit. As supplies became stronger, the DOL-started squirrel cage motor displaced the slipring motor, as it was less costly to build and required less maintenance.

The high maintenance factor of the slipring motor was associated with the insulated winding on the rotor becoming contaminated with brush dust, rapid wear of the carbon brushes and, in some cases, of the sliprings themselves.

As modern electronic devices have been developed, particularly the diode and thyristor, a generation of variable-speed slipring motors has again become popular, the variable speed being obtained by extracting slip energy from the rotor via sliprings; sometimes converted to d.c. and inverted to supply frequency, the slip energy is almost totally recovered (see Chapter 4, Section 4.6.3).

The problems encountered with sliprings and brushes are mostly understood, and arise from incorrect design of rings, brush pressure being too high or too low and wrong grade of brush. Atmospheric pollution, e.g. chlorine, or excessively dry air makes for difficult collection of current from sliprings.

If a problem is encountered, the user will be advised to contact the motor manufacturer or one of the brush manufacturing companies who are well equipped to give advice. However, problems are still sometimes encountered which are difficult to explain.

5.9.1 Brushgear

This is the part of the slipring arrangement which carries the carbon brushes and transfers the energy to the rotor terminals. The brushes are held in firm contact with the sliprings or commutator by rigidly mounted brush boxes and springs applying pressure to the top of the brush. The springs will be designed for the grade and size of brush and are usually of fixed pressure, which is constant for the degree of wear allowed of the brush. Earlier brush springs had adjustable pressure, but these have been superseded by the constant-pressure spring.

The brushbox will be honed to size to ensure the brush does not vibrate yet is free to move radially to allow for brush wear.

Figure 5.47 shows the sliprings and brushgear arrangement on a large twelve-pole, 900-frame motor. At the top of the picture is the fibreglass airduct which draws air from the region of the rings and circulates it over the cooling tubes, through washable filters and returns it to the slipring compartment. Air circulation is provided by an auxiliary motor mounted on top of the motor cooler (see Figure 5.32, page 308). This motor is the small vertical-shaft one located above the large cooler motor fan units.

The sliprings are grooved to assist cooling and cleaning out brush dust.

5.9.2 Sliprings

The rings of a three-phase asynchronous motor designed to carry current continuously will be constructed of copper nickel, and will include a spiral groove on those of high peripheral speed. They will be shrunk on to an insulated sleeve which is mounted on the shaft, or could be shrunk directly on to the insulated shaft. Connections from the rings to the rotor winding will be made by rods screwed into and sweated to each of the rings, yet insulated from the rings of other phases. Connections from these rods to the windings will be made of copper strip, insulated from, and firmly clamped to, the shaft. Where rings external to the bearings are used, the connections will be made with cable passing through a bore in the shaft.

This arrangement of copper nickel rings and metal graphite brushes has proved very satisfactory for rings carrying a.c. Higher metal, i.e. copper, content of the brush allows higher current density in the brush. For rings carrying d.c., e.g. the excitation rings of a synchronous motor, an electrographitic grade of brush can be used with steel rings. Steel rings may also be used on an asynchronous motor where the brushes will be lifted and the rings short-circuited mechanically after start-up.

One type of motor still in use in some areas, which often presents difficulty with respect to its sliprings, is the synchronous induction motor. On this motor the rings are firstly used to carry a.c. at start or during periods of asynchronous running, and then d.c. is applied via the same rings to cause the motor to run synchronously. The

Figure 5.47 Brushgear and sliprings of a 900-frame, twelve-pole motor. (*Courtesy: Brown Boveri Company*)

brushes and rings must therefore cater for two different conditions of current density, and this often results in excessive wear of one set of brushes and rings. The situation can be eased by reversing the excitation polarity at regular intervals, as wear will be more apparent on the rings of one polarity.

5.9.3 Commutators

The type of brush used on a commutator of moderate speed is one composed of carbon-graphite. This composition is not affected by sparking and transient overloads such as are experienced on d.c. machines. As a commutator wears, the mica segments separating the copper commutator segments may interfere with the carbon brushes. If that is the case, the mica must be undercut to clear the commutator surface. It is also possible for the commutator to have departed from a perfect circle, as drying-out of the segment and V-ring insulation takes place. In that case the complete commutator must be seasoned and the clamping device tightened. Finally, it will be remachined and the mica cut back from the surface.

On both sliprings and commutators, a smooth light grey patina or skin will be formed after a period of running. This patina is indicative of good operation of the component and its brushes and any rapid brush wear which may have been present during the formation of the skin will now reduce. It is important that such a skin be allowed to remain on the rings or commutator if the whole formation process is not to be repeated. Full information on carbon brushes can be found elsewhere.[20]

5.10 Commutator construction*

5.10.1 Introduction

Commutator motors are used where it is important to have good machine regulation without loss of energy.

The large family of commutator motors includes:

(1) Direct current motors for driving production machines in industrial plants and steelmills, machine tools, papermaking machines, conveying and transport devices, construction machines and vehicles.
(2) Traction motors for driving locomotives from a single-phase 16.66 Hz or 50 Hz a.c. power supply.
(3) Universal motors for driving electrical household appliances and hand tools from a single-phase 50 Hz a.c. power supply.
(4) Three-phase commutator motors for driving spinning and weaving machines and pumps, as well as booster machines for no-loss regulation of three-phase induction motors.
(5) Small generators, e.g. welding machines and lighting generators for road and rail vehicles.

5.10.2 The principle of operation of the commutator

On all these machines the commutator serves the purpose of feeding the current to the rotor. Together with the brushes sliding on its surface it is, in fact, a mechanical frequency converter, converting the frequency of the supplied current to the frequency of the machine rotor. It is for this reason that commutator motors can be fed by current of any frequency, as is demonstrated by the examples of commutator machines listed above. Functional performance of a commutator depends primarily on the quality of its surface. Even the slightest deviation from the ideal smooth, cylindrical form will be detrimental to the contact between the brushes and the commutator surface. This results in sparking, which in turn causes rapid wear or in extreme cases can lead to ring fire, i.e. to arcing around the circumference of the commutator. The higher the electrical loading of the contact between the commutator and the brushes and the higher the circumferential speed and, hence, also the mechanical–dynamic load of the contact system, the more important it is to have a commutator surface as near to perfect as possible.[21] In this respect, commutators on traction motors work under extremely severe conditions.

The unavoidable deviations from the ideal geometrical cylindrical form arising in practice can be split into two kinds:

(1) Out of round, undulations.
(2) Roughness.

The yardstick for assessing the quality of a commutator is primarily the roughness of its contact surface. This is defined by the differences in height between the individual segments, the so-called 'segment steps'.[22] These segment steps are caused by individual high segments not firmly held in the structure. For high-speed machines, segment steps and surface roughness of the commutator in any operating mode may not exceed some few micrometers.

Random measurements on the commutator profile are necessary to check dimensions and manufacture. Non-contacting measuring systems are particularly suitable for this as the commutator can be measured in all its operating modes. Figure 5.48 shows a surface relief measured by a capacitive probe. The accuracy of this measuring method is within 1 μm.

*This section is a translation by Brian Rowntree of an article by W. Heil, AG Brown Boveri et Cie.

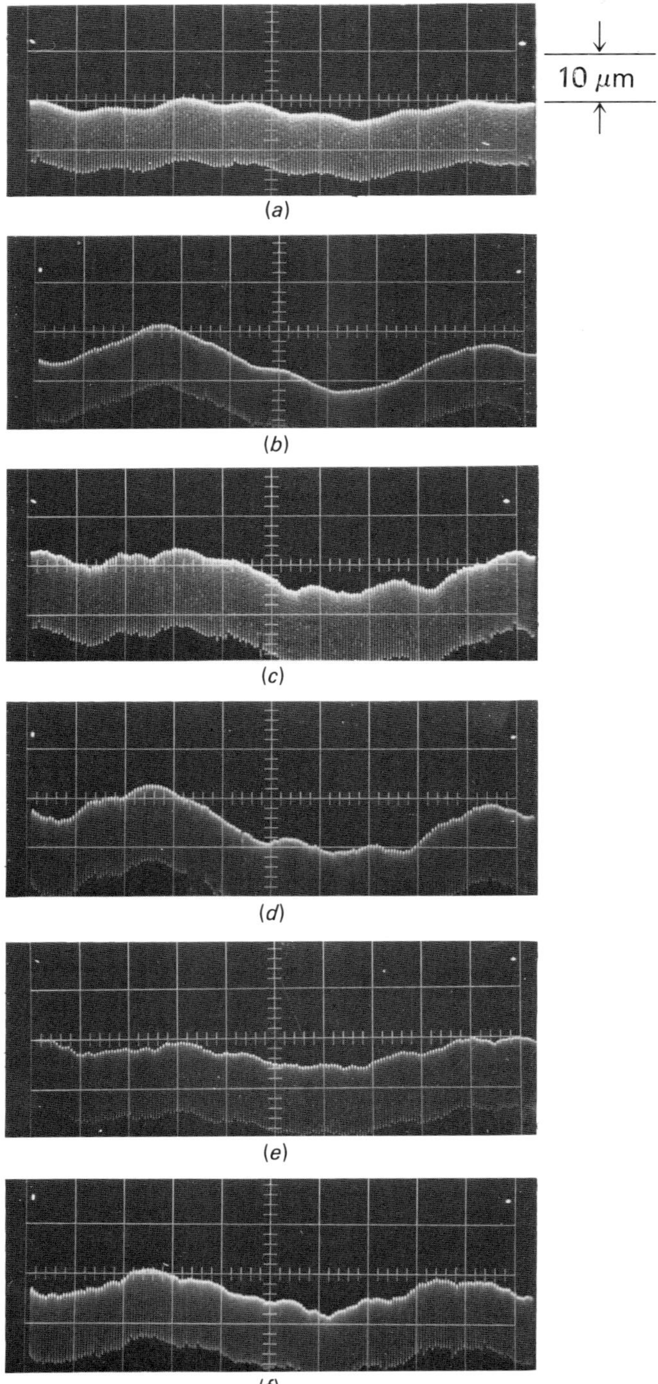

Figure 5.48 Oscillograms of commutator profile. (a) 221 rev./min, 26°C. (b) 3632 rev./min, 26°C. (c) 219 rev./min, 120°C. (d) 3617 rev./min, 120°C. (e) 219 rev./min, 29°C. (f) 3591 rev./min, 29°C. (*Courtesy:* Brown Boveri Company)

5.10.3 Construction of commutators

Though various commutator designs have been used over the years, the basic principles of construction are virtually all the same (Figure 5.49). The wedged-shaped commutator segments (into one end of which the winding ends are soldered) are insulated from each other by compression-resistant micanite or samicanite insulation and held together to form a compact unit by the insulated clamping or shrinkrings.

Figure 5.49 Section through a shrinkring commutator with inner friction rings. (*Courtesy:* Brown Boveri Company)

The commutator segments are preferably made of copper, mainly because copper lends itself to high manufacturing accuracy and has good contact behaviour with the carbon brushes. Another advantage is its high conductivity, though this aspect is of less importance. Resistance to corrosion is also an advantage, and only in sulphurous surroundings is the use of copper not possible because of rapid wear.

Figure 5.50 shows the most important methods of commutator construction, arranged according to their construction and physical features. Up to the present the so-called archbound construction has proven in practice to be superior to all other designs. The segments are pressed radially inwards by the clamping or shrinkrings, whereby they wedge together forming a circular arch.

The tangential surface pressure produced by the wedging effect must be high enough to ensure that the resulting friction prevents individual segments from moving radially outwards due to centrifugal forces or temperature rise. The pressure can be attained and upheld by means of clamping rings fitted into the V-grooves in the segments, or by internally or externally located shrinkrings. To ensure that the arch pressure is attained in the V-ring construction the V-rings must only act on the innermost faces of the grooves in the commutator segments (Figure 5.50 (a(i))).

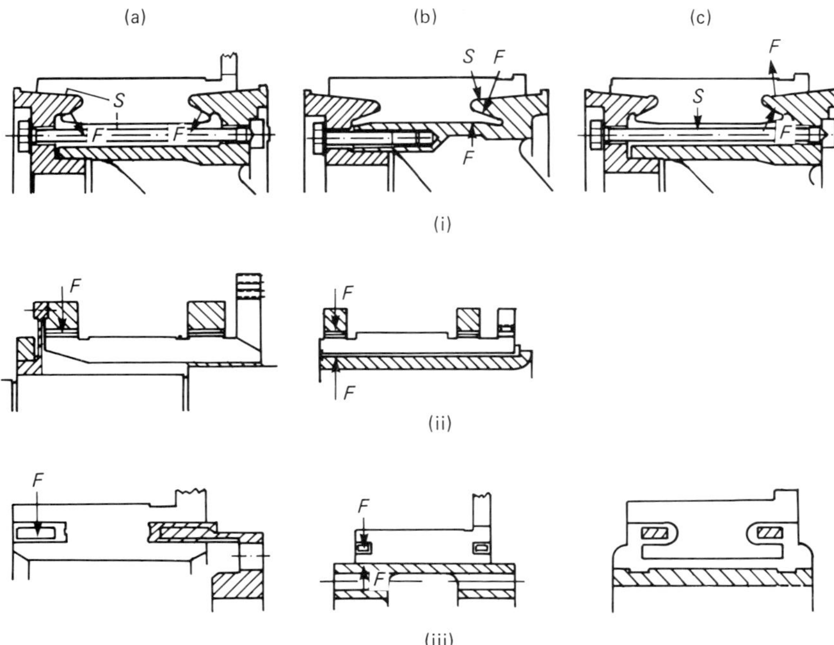

Figure 5.50 Construction types. (a) Pressure on the complete arc of segments. (b) Reduced pressure on the arc of segments. (c) Absence of segment arc pressure. (i) V-ring construction. (ii) Construction with external friction rings. (iii) Construction with internal friction rings. F = force, S = clearance. (*Courtesy:* Brown Boveri Company)

In the case of the shrinkring designs, the commutator should be mounted in such a manner that the arch pressure on the segments will not be reduced (Figure 5.50(a(ii) and (iii))). When the commutator segments are seated on their inner edges the arch pressure will be at least partly reduced (Figure 5.50(b(i), (ii) and (iii))). If the segment assembly is clamped in such a way that the V-rings act on both the inner and outer surfaces of the grooves in the segments, there can be no arch pressure on the segments (Figure 5.50(c(i))). A special design of shrinkring commutator without arch pressure on the segments is the moulded commutator (Figure 5.50(c(iii))). In this design the steel rings are positioned unstressed in the lateral slots in the segments while the commutator is stationary and cold. The assembly of the segments and rings is then bonded to form a whole by the moulding compound.

5.10.4 Manufacturing and dimensioning of commutators

In the mechanical calculation of commutators the circular arch of segments and the clamping devices are considered as being an elastic structure in equilibrium.[23] In this structure the state of mechanical stress does not only result from the forces applied to it (Figure 5.51) but also greatly depends on the manufacturing processes.[24,25] Additionally, the frictional relationships must be considered.[26,27] Figures 5.52–5.54 show various manufacturing stages of a commutator with externally located shrinkrings. First of all the commutator is clamped together by conical segments and rings (Figure 5.52). Then a strong wrapping of binding wire is applied to the free sections between the pressrings to maintain the pressure when the pressgear is removed (Figure 5.53). Next, the commutator surfaces exposed

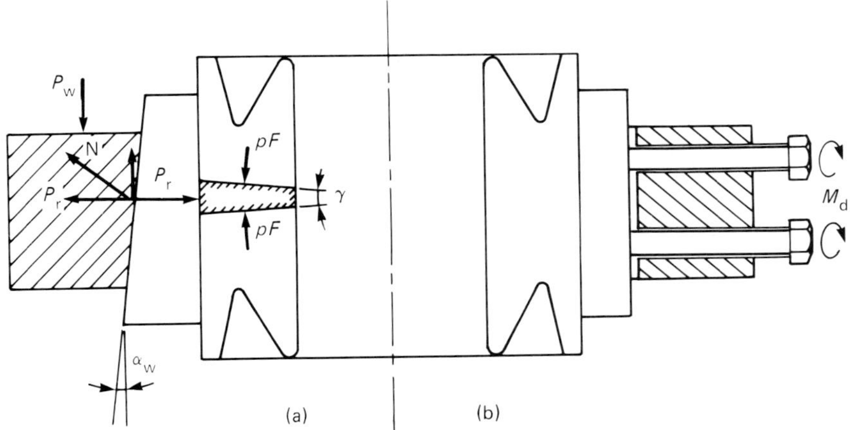

Figure 5.51 Make-up of commutators. (a) With conical tooling under axial pressure. (b) With radial tooling and pressure applied radially by screws. P_w = axial pressure applied, P_r = radial pressure, P_f = pressure on the surface of the segments. (*Courtesy:* Brown Boveri Company)

when the pressrings are removed are machined, insulated, and fitted with the shrinkrings. Figure 5.54 shows the commutator mounted to the rotor shaft.

Each stage of manufacture changes the state of mechanical stress. All the processes, and also the influences of operational stresses can be shown graphically in a stress–strain diagram, in a similar way as for bolted assemblies. The relationships are a little more complicated in the case of commutators with V-rings. Here, the radial forces which produce the arch pressure and the axial forces from the clamping rings acting on the inclined faces of the V-grooves are in equilibrium. The conditions for manufacture and operation can be shown in a force diagram.

It is of great importance that all points of the force diagram lie within the friction cone. If this is not the case, relative movement between the segments of the circular arch and the V-rings can arise during assembly or in operation and would wear away the insulation. This must be avoided by suitable dimensioning. In addition to the forces produced by prestressing, additional forces arise during operation due to the centrifugal effect and the different thermal expansion between copper and steel. The necessary extensive calculations can be carried out quickly and reliably by computer. It is necessary to know as much as possible about the forces arising to be able correctly to design the parts subjected to mechanical stressing. In the main these are the copper segments, the shrinkrings, the bolts, and the clamping-ring or pressring insulation. The highest forces arising in the segments on V-ring commutators is in the sections below the V-grooves and analogously on commutators with internally located shrinkrings in the sections underneath the

Figure 5.52 Commutator restrained by temporary clamping rings. (*Courtesy:* Brown Boveri Company)

shrinkrings. Essentially, the stresses concerned are bending stresses at the transition from smaller to larger section. The classical calculation methods for determining the stresses in commutators,[28] assumes that the lower part of the segment is a beam fixed at one end. However, the effect of the arch pressure reduces the stressing considerably.[29] This influence can also be allowed for in the computer program.

Because of the high mechanical stresses involved, today's commutators are manufactured almost exclusively from silver-alloyed copper. As compared with electrolytic copper this alloy has much superior mechanical properties.

5.10.5 Materials for commutators

The quality and operational behaviour of commutators depends not only on precise calculation and meticulous manufacture but also on the quality of the materials employed and accurate shaping of the individual components. The technical conditions of supply for the commutator segments are prescribed exactly by the

Figure 5.53 Commutator bound with temporary wire banding. (*Courtesy:* Brown Boveri Company)

Figure 5.54 Complete external friction-ring commutator. (*Courtesy:* Brown Boveri Company)

purchaser. In many cases the standard DIN 42963 is used. This standard specifies the permissible tolerances for segment thickness at the top, bottom and in the middle. Deviations from the perfect wedge profile can be present within these tolerances in the following forms: (1) thickness; (2) wedge angle; and (3) surface faults such as local concavity or convexity.

Such deviations can influence the manufacture and operational performance of the commutator in various ways. For instance, in extreme cases, where a large number of very deep segments are involved, thickness deviation can result in a variation in diameter of $\pm 1\%$, whereby this can easily be corrected by grinding down the segment insulation as necessary. Much more serious is wedge angle deviation. If the wedge angle is larger or smaller than the exact nominal value, non-uniform arch pressure will be the result (Figure 5.55). When the angle is larger, the pressure at the outer edge will be greater (Figure 5.55(a)). Should such a commutator wear over the years the mean arch pressure will decrease and the commutator can become unstable. Still more adverse conditions are created by segments with concavity or convexity faults (Figures 5.55(c) and (d)). In such cases, the pressure distribution over the depth of the segment can be so unfavourable that

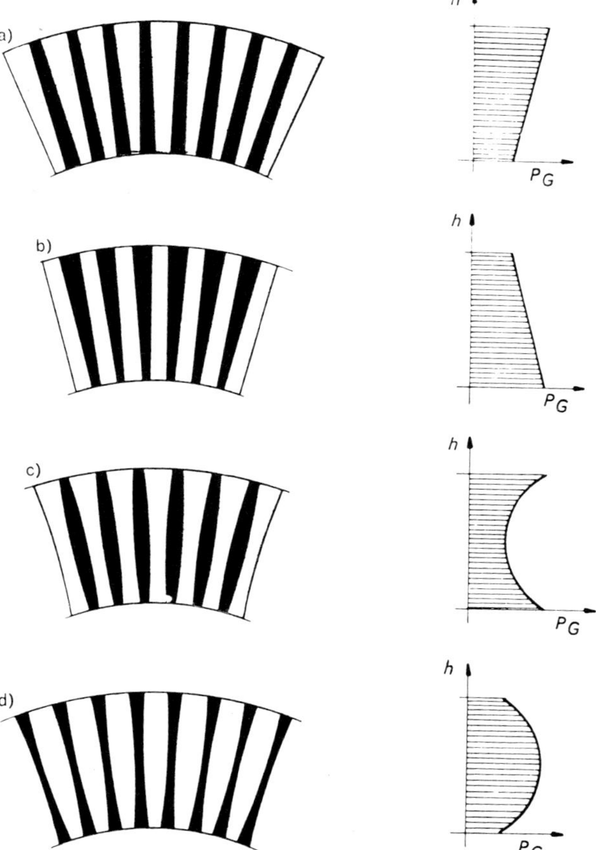

Figure 5.55 Effect of deviation in segment form on the arch pressure. (a) and (b) Wedge angle deviation. (c) Lateral faces concave. (d) Lateral faces convex. P_G = arch pressure, h = segment height. (*Courtesy:* Brown Boveri Company)

the insulation can be locally overstressed and permanently deformed which also causes the commutator to lose arch pressure and become unstable.[30]

Attempts are often made to equalize deviations in segment form by employing a soft insulation material, i.e. one which lends itself to plastic deformation and is then hardened during the commutator-forming process. However, this procedure does not achieve the object. If the insulation can be hardened during the commutator-forming process, the unequal insulation thickness results in a non-uniform pressure distribution as for a non-plastic insulation. If the insulation cannot be fully hardened during the forming process, which is generally the case in practice, then the remaining plastic portion will in time cause a reduction in arch pressure.

From the foregoing it is clear that if a commutator is to be really stable it may only be constructed from segments with the minimum possible wedge angle deviation and surface faults and elastically deformable insulation material. Requirements made with respect to the manufacturing accuracy of the segments can be fulfilled without problem using copper. Moreover, it is possible to manufacture economically copper segments in small numbers. This is an important point because the number of commutator segments differs greatly from one machine to another due to the numerous winding variants. For this reason, it is also not possible to standardize the segment profile. Shellac-bound micanite which, due to preforming and baking only contains a minimum of binding agent, has proven to be a very good segment insulation. In more recent times, epoxy-resin-bound fine-mica, the so-called samicanite, has been employed with equal success. Shellac-bound micanite is also used for the insulation rings in the V-grooves. As opposed to the flat-shaped segment insulation, the rings cannot be preformed so well and therefore contain more binding agent and are consequently plastically deformable within certain limits. Therefore, with this kind of commutator construction it is sometimes necessary to retighten the clamping bolts. In the case of shrinkring commutators the ring insulation may not have any plastic content. On commutators with externally located shrinkrings, an insulation of pure mica is used, preformed with the help of auxiliary shrinkrings. On commutators with internally located shrinkrings polyimide-foil kapton has proven to be excellent for the

Figure 5.56 Rotor of a d.c. motor with a commutator with internally located shrinkrings. (*Courtesy:* Brown Boveri Company)

intended purpose with its high compressive resistance and elastic properties.[31] Figure 5.56 depicts a d.c. motor fitted with such a commutator.

5.10.6 Development trends

For reasons of economics there is the continuous aim to improve the utilization of the volume of all electrical machines. For traction motors, there are also limits imposed on the overall dimensions and weight. These, then, are also the basic requirements for the further development of the commutator. Increasing the specific utilization of the machine means greater heat losses for the same volume. As the permissible temperature limits are fixed, the cooling airflow must be increased and directed more efficiently to the sources of the losses, such as the windings and the core. This can be problematic and is especially so on the rotor because the clamping construction of the V-ring commutator does not leave a lot of space between the commutator and the shaft for the passage of cooling air. From this point of view the shrinkring commutators are much more favourable. Of the two basic design concepts the one with the internally located shrinkrings has the advantage that it requires less space in the axial direction. Moreover, with this design the complicated brushgear necessary for the externally located shrinkring design is not required, which also simplifies erection and dismantling work. With this design concept the cooling air has direct contact with the inner edges of the segments resulting in a better cooling of the commutator. This is important for reducing brush wear and for the operational behaviour of the commutator.

A further reduction in commutator volume and weight is possible by employing a segment material with better mechanical properties than silver-alloyed copper. The employment of steel would be the next obvious step. Considering the low current densities in the segments, the poorer electrical conductivity would be virtually of no consequence. On the other hand, the poorer resistance to corrosion of normal steel would be a disadvantage. Efforts are being made to produce a commutator segment which will be of steel in the inner, highly mechanically stressed zone and of copper in the outer, electrically and climatically stressed zone. This can be achieved by plating or sintering. Another possibility is the employment of stainless steel. However, such steel segments are more expensive to manufacture and it is extremely difficult to maintain the required tolerances. Furthermore, due to the greater tooling costs, drawing steel segments will only be economical when a relatively large number of identical segments are required. Therefore, the development of an economically producible segment material with better mechanical properties than silver-alloyed copper or zircon copper, but with the same good contact and non-corrosive qualities, represents a lucrative challenge for the non-ferrous metal industry.

References

1. CHALMERS, B. J. 'Advances in alternating current machines'. *Electronics and Power*, p. 911 (1977)
2. BONE, J. C. H. 'Influence of rotor diameter and length on the rating of induction motors'. *Electric power applications*. Institution Electrical Engineers, pp. 2–6 (1978)
3. TAYLOR, H. W. 'Eddy currents in stator windings'. *Instn Elec. Engrs J.*, **58** (1920)
4. BRITISH STANDARDS INSTITUTION. *Specification for general requirements for rotating electrical machines*, BS 4999:1976:Part 60 'Tests, Table 60.11.5. (Superseded by Part 143:1987.) BSI, (1976)
5. ALGER, P. L. *The nature of induction machines*. Gordon and Breach (1965)
6. CHALMERS, B. J. 'Alternating current machine windings with reduced harmonic content'. *Proc. Instn Elec. Engrs*, **111**, 11 (1964)

7. RAWCLIFFE, G. H., BURBIDGE, R. F. and FONG, W. 'Induction motor speed changing by pole amplitude modulation'. *Proc. Instn Elec. Engrs,* **105,** Part A, 22 (1958)

8. RAWCLIFFE, G. H. and FONG, W. 'Speed-changing induction motors'. *Proc. Instn Elec. Engrs,* **107,** Part A, 36 (1960)

9. RAWCLIFFE, G. H. and FONG, W. 'Speed-changing induction motors'. *Proc. Instn Elec. Engrs,* **108,** Part A, 41 (1961)

10. RAWCLIFFE G. H. and FONG, W. Close-ratio two-speed single-winding induction motors. *Proc. Instn Elec. Engrs,* **110,** 5 (1963)

11. FONG, W. 'Wide-ratio two-speed single-winding induction motors'. *Proc. Instn Elec. Engrs,* **112,** 7 (1965)

12. RAWCLIFFE, G. H. and FONG, W. 'Two-speed induction motors using fractional slot windings'. *Proc. Instn Elec. Engrs,* **112,** 10 (1965)

13. RAWCLIFFE, G. H. and FONG, W. 'Sum and difference winding modulation with special reference to the design of 4/6 pole PAM windings'. *Proc. Instn Elec. Engrs,* **117,** 9 (1970)

14. RAWCLIFFE, G. H. and FONG, W. 'Clock diagrams and pole amplitude modulation'. *Proc. Instn Elec. Engrs,* **118,** 5 (1971)

15. *SKF Handbook.* Catalogue 3200E Reg. 47. 120 000 (1981)

16. SHELL INTERNATIONAL. *The lubrication of rolling bearings,* p. 154, Figure 259 (1972)

17. HOLMAN, J. P. *Heat transfer* (3rd edn) McGraw-Hill. Chapter 10, p. 321 (1972)

18. TAYLOR, J. L. 'Calculating airflow through electrical machines'. *Elec. Times,* 82 (1960)

19. KELLER, C. *The theory and performance of axial-flow fans.* McGraw-Hill, New York (1937)

20. MORGANITE ELECTRICAL CARBON. *Carbon brushes and electrical machines.* Quadrant Press (1978)

21. REETZ, F. 'About the reasons for commutator damage on traction motors'. *Elektrische Bahnen,* **38,** H3, 54–64 (1967)

22. HEIL, W., LEHNER, L. and SOLHARDT, K. 'Statistical mathematic research into manufacturing methods for commutators'. *Brown Boveri Mitt.,* **54,** H9, 595–602 (1967)

23. WIEDEMANN, E. and KELLENBERGER, W. *Construction of electric machines.* Springer-Verlag, pp. 393–413 (1967)

24. RAUHUT, P. 'Calculation of commutators of the inner friction type'. *Schweiz. Arch. App. Sc. & Technol.,* **12,** 4, 120–132 and 148–156 (1946)

25. RAUHUT P. 'Calculation of the forces in a commutator with arch pressure'. *Elektrische Bahnen,* **22,** H7, 169–175 (1951)

26. HUSZAR, I. 'Concerning the choice of arch pressure commutators'. *Elektrische Bahnen,* **36,** H12, 291–296 (1965)

27. HUSZAR, I. 'Interplay of forces in the swallowtail commutator'. *Sc. Tech. High School Karl-Marx-Stadt,* **8,** H3, 173–177 (1966)

28. LAUB, F. L. 'New fundamentals for the construction of commutators'. *Bull. Schweiz. Elektrotechn.,* **40,** 25, 988–1001 (1949)

29. KELLENBERGER, W. 'Mechanical stresses of commutator segments due to inner shrink rings: a problem of orthotropic forces in cylindrical shells'. *Brown Boveri Mitteilung,* **55,** H10/11, 584–589 (1968)

30. GOELLING, G. 'The layout of the necessary tolerances for commutator segments, and a method of sorting for the setting of large commutators'. *Elektrie,* H3, 109–114 (1966)

31. REETZ (1967) op. cit.

6 Insulation

H. S. McNaughton

Manager, Insulation Engineering and Laboratory,
GEC Large Machines Ltd.

6.1 Introduction

6.1.1 Primary function

The basic function of insulation is to separate electrical circuits at different voltages from each other and from metallic components at earth potential.

6.1.2 Definition of an insulator

From a practical engineering standpoint, insulators can be defined as materials which are poor electrical conductors, i.e. they have high electric resistance. Insulators, unlike metallic materials, have a decrease in resistance with increase in temperature. Between the groups of materials which can be defined as good conductors and good insulators there is a further group which can be defined as poor conductors or poor insulators depending on the application point of view. Insulators also have the ability to hold electrostatic charge. The ratio of the charge-holding capability of a material to that of the equivalent space *in vacuo* is defined as the relative permittivity (Table 6.1).

6.1.3 Function of an insulator

Insulators have three functions; they act as:

(1) Solid dielectric barriers possessing intrinsic electric strength above that of the equivalent air space.
(2) High-resistance creepage surfaces between exposed conductors at different potentials.
(3) Liquid separators between live components. Air and mineral oil are the most commonly used liquid dielectrics.

6.1.4 Electric strength

The ability of an insulator to withstand electric stress without breakdown is one of its most important practical properties. Dielectric strength of insulation is time-dependent. When assessing the electric strength of a material it is essential to know under what conditions it was tested. Electrical engineers need to know the electrical strength of materials when designing insulation systems. Two test values often required are:

(1) Average breakdown value when an increasing rapidly applied voltage is imposed under defined conditions.
(2) Average value which a material can withstand for long periods, e.g. 10, 20 or 30 years.

Table 6.1 Permittivities, i.e. dielectric constants of various materials

Material	Permittivity at 20°C
Air	1.0006
Hydrogen	1.0003
Carbon dioxide	1.0009
Transformer oil	2.2–2.6
Glass	3.5–10.0
Mica (muscovite)	6.0–8.0
Mica (phlogopite)	5.0–6.0
Porcelain	5.5–6.5
Shellac	3.0–4.0
Bitumen	2.0–3.0
Phenolic resin	5.0–6.0
Alkyd resin	5.8–6.2
Epoxy resin	3.3–5.5
Polyester resin	3.0–5.0
Silicone resin	3.0–4.8
Cellulose paper (dry)	2.0–3.0
Aramid paper (dry)	2.2–2.6
Asbestos paper (untreated)	5.0–7.0
Asbestos paper (epoxy-treated)	2.5–3.0
Natural rubber	2.5–5.0
Nitrile rubber	3.5–10.0
Chloroprene rubber	7.0–10.0
Butyl rubber	2.0–4.0
Silicone rubber	2.5–8.0
Chlorosuphonated polyethylene rubber	6.5–7.5
Phenolic-bonded paper	3.0–6.0
Phenolic-bonded cotton cloth	5.0–7.0
Polyester-bonded glass cloth	4.5–7.0
Epoxy-bonded glass cloth	4.0–6.5
Silicone-bonded glass cloth	2.8–4.0
Epoxy-bonded glass-backed mica paper	4.8–5.2
Polystyrene	2.4–2.6
Polyethylene	2.2–3.0
Polytetrafluoroethylene (PTFE)	2.0–2.3
Polyvinylchloride (PVC)	3.5–4.0

Electric strength depends on whether the test is carried out using a.c., d.c. or a defined impulse voltage. Values are also influenced by many other factors including test temperature, electrode shape, humidity, etc. Typical rapidly applied voltage values for some commonly used materials are shown in Table 6.2. Effect of long-term stress on a typical high-voltage a.c. stator winding system is shown in Figure 6.1.

6.1.5 Effect of temperature

Properties of many insulating materials are temperature-dependent. In general, the rate of deterioration, i.e. ageing, increases with temperature. For a material, or system, there is a maximum temperature beyond which it cannot be used if a reasonable life is required. For a particular material, the service life depends on the

Table 6.2 Rapidly applied voltage electric strength of matierals

Material	RAV electric strength (kV/mm)
Resins:	
phenolic	8–14
epoxy	12–24
polyester	16–20
silicone	6–16
Rubbers and elastomers:	
natural rubber	18–24
chloroprene rubber	4–20
butyl rubber	16–32
chlorosulphonated polyethylene rubber	16–24
silicone elastomer	12–28
Papers:	
aramid paper	20–32
asbestos paper (epoxy-impregnated)	12–16
Laminates:	
phenolic-bonded paper	12–20
phenolic-bonded cotton cloth	8–12
polyester-bonded glass cloth	12–16
epoxide-bonded glass cloth	18–24
silicone-bonded glass cloth	6–14
Mica-based:	
epoxide-bonded glass-backed mica paper (rigid)	25–70
micanite sheet (mica flake)–(rigid)	25–50
polyester-bonded glass-backed mica paper (flexible)	12–20
polyester-bonded glass-backed mica flake (flexible)	16–25
Films:	
polyester film	80–120
polyimide film	120–160
Cloths (Flexible–fully cured):	
polyester (isophthalate) varnished glass cloth	30–40
silicone varnished glass cloth	28–36
Plastics:	
polystyrene	20–28
polyethylene	16–20
polyvinyl chloride	18–20
polytetrafluoroethylene	20–28

type of equipment in which it is used. For example, the 'useful lives' of industrial a.c. motors and domestic coffee grinders may be similar, e.g. 15–20 years, although the total running hours in service will differ significantly.

6.1.6 Thermal classification of insulation materials

For many years, thermal classification of insulation materials has been by a system of class letters – Y, A, E, B, F, H and C. Temperatures associated with each class, and basic materials types which fall into each one, are defined in BS 2757:1986 *Classification of insulating materials for electrical machinery and apparatus* (and Table 6.3 of this book). The equivalent IEC document is IEC 85.

Since these standards were issued, development work has shown that the thermal classification of a material can only be used as a guide to maximum permissible

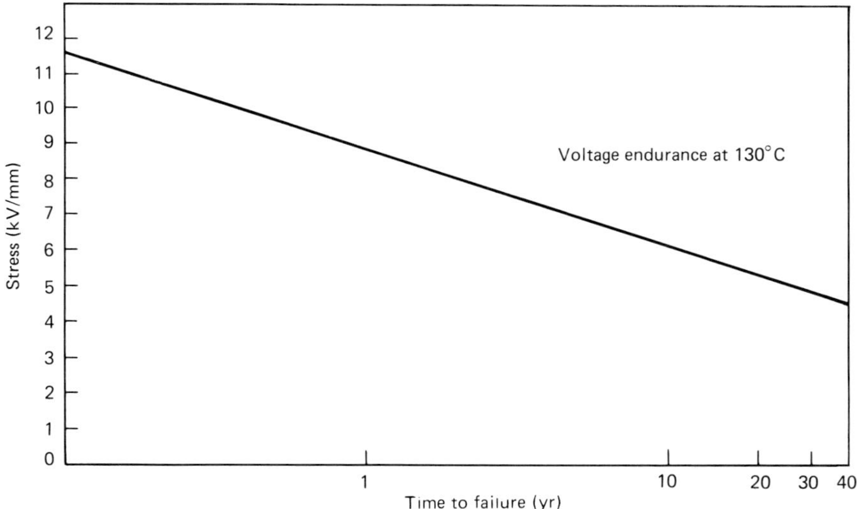

Figure 6.1 Electrical strength vs. time for high-voltage stator bar insulated with resin-rich epoxy bonded mica paper

Table 6.3 Themal classes of electrical insulation listed in BS 2757:1986 and IEC 85:1984

Class	Temperature (°C)	Class	Temperature (°C)	Class	Temperature (°C)
Y	90	B	130	200	200
A	105	F	155	220	220
E	120	H	180	250	250

operating temperature. For most practical applications, where several materials are combined as a system, functional testing of models or complete machines is necessary to establish real service life capability. More recently, new standards have been developed to define the thermal index and thermal profile of individual materials. These indices define precise test methods and end-points for assessing specific material properties.

6.2 Basic types of insulation

6.2.1 Application categories for materials

In practice, when considering materials for use in rotating electrical plant it is often useful to divide materials into the following categories:

(1) Conductor insulation.
(2) Tapes and flexible sheets.
(3) Rigid sheets and laminates.
(4) Sleevings and cables.
(5) Semi-conducting corona shield and stress grading materials.
(6) Bracing tapes, cords, ropes.
(7) Impregnating varnishes and resins.

A brief summary of the major materials associated with each of these categories follows.

6.2.2 Conductor insulation

Covered conductors are usually defined as copper (or alternative conducting metals) covered with a continuous layer of insulation. Both round and rectangular conductors can have either a single or composite covering. The simplest covering is enamel. In the past an oleoresinous material was used but this coating was weak mechanically and had poor abrasion resistance. For applications where the conductors were subjected to mechanical working and possible abrasion, it was common practice to apply an additional covering of cotton, rayon, silk, asbestos or glass fibres. Some 30 years ago a significant improvement was achieved by combining PF with highly adhesive PVA to form an enamel with excellent flexibility and abrasion resistance. Conductors coated with this resin system quickly replaced the earlier enamels; PVA–PF coatings were found to be satisfactory for windings operating up to Class E temperatures. The improved mechanical properties eliminated the need for an additional fibrous covering in many applications. Several alternative thicknesses of enamel are used to meet different electrical withstand levels. Since the introduction of PVA–PF, many other synthetic enamels have been developed. Some of the major ones are shown in Table 6.4. For the majority of small- and medium-sized machine windings where an

Table 6.4 Types of enamel used for covering conductors

Enamel	*Normal maximum operating temperature* (°C)
Oleoresinous	105
Polyvinyl acetal/polyvinyl formal	105
Polyurethane (solderable)	105
Polyamide (nylon)	105
Polyurethane (non-solderable)	130
Polyester (terephthalate ester)	155
Epoxy	155
Acrylic	155
Polyhydantin	155
Polyesterimide	180
Polyesterimide/amide (dual coat)	180
Alkyd-modified silicone	180
Silicone	180+
Polyimide	200
Polytetrafluoroethylene	200

enamel covering is considered adequate, either a single coat of polyesterimide enamel or a dual coat consisting of polyesterimide enamel overcoated with a thin layer of polyamide is used. Both coatings have been shown to be suitable for use on normal industrial machine windings operating up to at least Class F limits. Where the required mechanical toughness cannot be met by an enamel alone, an additional fibrous or film covering is added. Fibrous coatings are usually bonded to the conductor with thermosetting varnish. For large machines where the electric stress between adjacent conductors is low, only a fibrous covering is necessary. Where the electric stress between adjacent conductors is high, e.g. on high-voltage stator coils, a composite covering of either enamel, polyester film, polyimide film or aramid paper with a fibrous covering such as glass lappings or braid is adopted. Conductors for high-voltage machines, i.e. for line voltages greater than 4 kV, are often covered with layers of mica-based tape. These conductors can withstand the high electric stress produced on high-voltage machines in service. Micaceous conductor tapes require a backing to give them adequate mechanical strength.

Typical reinforcements are woven glass cloth, glass fleece, woven polyester cloth, polyester fleece, polyester film, polycarbonate film, polyimide film and aramid paper. Conductor tapes of this type are bonded with resin or varnish which is either fully cured or taken to a B stage. Typical bonding resins are polyesters, acrylics, polyurethanes, epoxides and silicones.

Table 6.5 Typical conductor coverings (excluding enamels)

Covering

Fibrous:
    asbestos lappings
    cotton lappings
    rayon lappings
    silk lappings
    polyamide (nylon) lappings
    polyester lappings
    glass/polyester lappings
    glass lappings
    glass braid

Papers and films:
    aramid (Nomex) paper
    polyester (Melinex, Mylar, Terphane, etc.) film
    polycarbonate film
    polyimide (Kapton) film

Mica-paper based:
    polyester film-backed mica paper
    polycarbonate film-backed mica paper
    polyimide film-backed mica paper
    woven glass-backed mica paper
    woven glass and polyester film-backed mica paper
    unidirectional glass and polyester film-backed mica paper
    double-sided polyester film-backed mica paper

Notes:

[a] All the above coverings can be used on top of an enamel.

[b] Fibrous coverings are normally coated or impregnated with a varnish or resin.

[c] Composite coverings of film or paper plus fibrous lappings are also used.

In some applications, such as for field coils, an enamel- or fibrous-covered conductor is coated with an additional layer of epoxy resin. After application the resin is partly cured to give a dry, non-tacky surface which permits handling of the conductor without difficulty. Conductors of this type are wound into coils which are pressed in a jig or baked in an oven to cure the resin and bond all the turns together (Table 6.5).

6.2.3 Tapes and flexible sheets

6.2.3.1 General

Many components of simple and complex shape associated with rotating machines – such as the slot and end-winding portions of coils and bars, connections, terminal leads, pole bodies, etc. – need to be insulated to a level appropriate for the service voltage. Where the shape is simple, e.g. for a terminal lead or the slot portion of a coil, this can be achieved by wrapping with a wide sheet of material having a dielectric strength several times greater than that of the equivalent air space. Where the shape is more complex, e.g. for a coil evolute, then insulation can be applied more readily using narrow tapes. Many insulating tapes are produced by slitting wide sheets into strips of the required width. Available materials can be divided into two basic types: (1) non-micaceous; and (2) micaceous.

Where the electric stress appearing across the dielectric is relatively low and no discharging is likely to take place in service, non-micaceous materials can be used. Where the electric stress will be high and discharging could occur in service, mica-based materials are essential.

6.2.3.2 Flexible varnish impregnated non-micaceous tapes and sheets

Many standard tapes are based on either woven fabric, fleece or paperlike materials coated with a flexible varnish. Where the base material is a woven cloth, tapes produced by slitting sheet into narrow strips, cut along its length, are defined as 'straight-cut'. Where the sheet is cut at an angle of 45° to the length direction, the tape is described as 'bias-cut'. Many different types of base material, including woven silk, cotton, nylon, polyester and glass cloths, and polyester and glass mats have been used for making flexible materials. Typical impregnating resins such as glyptals, polyurethanes, epoxides, polyesters and silicones have been adopted by various insulation manufacturers. Typical materials are shown in Table 6.6. Thicknesses range from 0.05 to 1 mm.

Table 6.6 Non-micaceous varnished materials

Base fabric[a]		Varnish or resin coating[b]
Woven cotton cloth Woven silk cloth Woven nylon cloth Woven polyester cloth Woven glass cloth Woven aramid cloth Polyester fleece Glass fleece	impregnated with	bitumen shellac oleoresinous alkyd acrylic polyester epoxide polyurethane silicone polyimide

Notes:

[a] Base materials range in thickness from 0.05 to 1 mm.

[b] Resin coatings can be either fully cured or made as B-stage materials which require curing after application. Fully cured materials are sometimes given an additional B-stage coating.

6.2.3.3 Flexible film and synthetic paper tapes and sheets

For many years, the majority of non-micaceous flexible materials were based on woven cloths. This situation changed some 25 years ago with the development of new high-temperature films and fibrous papers. All the materials previously available, such as cotton, glass and polyester cloths and fleeces required to be impregnated with varnish to bring their electric strength up to acceptable levels. The newer materials, however, had inherent electric strength. Sheets based on polyester and polyimide film, and aramid paper, have proved invaluable as slot liners on low- and medium-voltage machines. Aramid paper tape, which maintains its integrity up to + 180° C, has been used widely for conductor and main ground insulation on low-voltage machines (Table 6.7).

Table 6.7 Flexible films and synthetic paper
tapes and sheets

Films:
   polyester
   polycarbonate
   polyimide

Papers:
   cellulose
   polyester
   aramid
   asbestos
   glass
   ceramic

Composites:[a]
   cellulose paper/polyester film
   cellulose paper/polyester film/cellulose film
   polyester paper/polyester film
   polyester paper/polyester film/polyester film[b]
   aramid paper/polyester film
   aramid paper/polyester film/aramid film[b]
   aramid paper/polyimide film
   aramid paper/polyimide film/aramid paper[b]

Notes:

[a] Papers are often impregnated or coated with epoxide, polyester
or silicone resins.

[b] Commonly known as D/M/D, N/M/N and N/K/N respectively.

Table 6.8 Typical types of adhesive tapes

Basic type[a]	*Adhesive*	*Thermal class*
Paper	Rubber resin	A
Crêped paper	Rubber resin	A
Acetate film	Rubber resin	A
Acetate cloth	Rubber resin	A
Polyester film	Rubber or acrylic	B
Woven glass cloth	Acrylic	F
Woven glass cloth	Silicone	H
Polyimide film	Acrylic	F
Polyimide film	Silicone	H
PTFE film	Silicone	H
Vinyl/PVC	Rubber	Y

Note:

(a) Thicknesses range from 0.05 to 0.30 mm.

6.2.3.4 Flexible adhesive-coated tapes and films

By coating various woven cloth tapes and films with resins having good adhesive
and tack properties, a range of products has been produced which is used for many
minor applications in machines. These tapes play small but important parts in many
insulation systems. Self-adhesive tapes based on thermoplastic materials, such as
PVC, butyl and silicone rubbers are also used for specific purposes (Table 6.8).

6.2.3.5 Tapes and sheets bonded with B-stage resins

By treating woven glass cloth with a liquid epoxide resin and then baking the material to cure the resin to a dry flexible stage, a so-called B-stage product is produced. Such materials can be wrapped around various components which are then compressed under heat and pressure to produce rigid insulation. Materials of this type, based on woven cloths, random mats and unidirectionally arranged fibres, are used on many components (Table 6.6).

6.2.3.6 Flexible micaceous tapes and sheets

Tapes and sheets based on mica are of paramount importance in high-voltage windings. To use mica splittings, it has been a normal manufacturing procedure to build up sheets of insulation in the following manner. A layer of backing material, such as woven glass cloth, is coated with liquid resin. On to it is laid a layer of mica splittings. Each splitting, or flake, overlaps those adjacent to it so as to form a continuous sheet of mica. If a thicker material is required then extra varnish and layers of mica splittings are added as necessary. After curing the varnish, the sheet material is used either as a wrap or is cut into narrow tapes. In the past, the majority of standard mica tapes were made by this process. In France in the early 1940s a major advance was made when a process was developed for taking mica flakes of any size and shape and turning them into 'mica paper'. This technique uses vigorous chemical and mechanical action to break down mica splittings into small platelets. A slurry of mica and water is produced which is neutralized and then fed into a modified papermaking machine which reconstitutes the mica into a sheet of paperlike texture. 'As made' mica paper is mechanically weak, but by adding a backing layer of woven glass cloth or polyester fibre it can be handled readily and impregnated with most types of natural and synthetic resins. Using this technique, flexible sheet materials can be produced in the same way as for the older mica-flake tapes (Table 6.9).

Table 6.9 Micaceous materials

Carrier or backing material[a]	Mica	Resin or varnish bond or impregnant

Cellulose paper
Woven polyester cloth
Woven glass cloth
Polyester film
Polycarbonate film
Polyimide film
Polyester fleece
Glass fleece
	+ { mica flakes or mica paper } +	(1) *Fully cured*[b]
		  bitumen polyurethane
		  shellac polyester
		  copal epoxide
		  alkyd silicone
		
		(2) *B-stage – high bond content, i.e. resin-rich*
		  epoxide
		  modified silicone
		
		(3) *Low bond content, i.e. dry tapes for vacuum pressure impregnation*
		  epoxide
		  polyester
		  silicone

Notes:

[a] Micaceous materials are backed either on one or both sides with a carrier to give them mechanical strength. Double-backed materials often use two different carriers, e.g. glass and polyester film.

[b] Materials in which the impregnant is fully cured can be rigid, semi-flexible or flexible depending on the type of resin used. Rigid and semi-flexible materials are normally used in sheet form rather than as tapes.

6.2.3.7 Micaceous tapes and sheets bonded with B-stage resins

In a similar manner to that used to manufacture B-stage resin-treated woven glass cloths (see Section 6.2.3.5) it is practicable to produce mica paper, or mica flake, sheets backed with various materials such as woven glass cloth, polyester mat and polyester film, etc. Materials of this type form the most important range of products used for insulating high-voltage resin-rich windings. Several different types of bonding resins are in common use. The most important group, for industrial machines operating up to full Class F temperature limits, are the epoxides. Tapes and sheets based on such resins, backed with woven glass cloth, have been used for over 30 years on all sizes of machines and for all voltages up to 30 kV. Materials backed with polyester film are also used for windings operating up to 13.8 kV (Table 6.9).

6.2.3.8 Low bond content micaceous tapes and sheets

Of equal importance to the resin-rich tapes described in Section 6.2.3.7 are materials made in a similar manner but bonded together with only a few per cent of resin. These are the so-called 'dry mica tapes' used as the main ground insulation for vacuum pressure impregnated windings. Materials of this type are manufactured with several different backings depending on the application process and the type of resin used for the final impregnation (Table 6.9).

6.2.4 Rigid sheets and laminates

6.2.4.1 General

In the manufacture of most electric machines there is a need for components made of solid insulation. Slot wedges, separators, field coil flanges, end-winding packing blocks, terminal supports, etc. are used on medium or large machines and these are normally made from rigid material.

6.2.4.2 Early materials

In the early days of machine manufacture the only readily available rigid insulating materials were hard wood, porcelain, quartz, glass, mica and slate. Following the development of phenolic resins, a whole new range of materials was introduced which gave electrical machine designers many opportunities to improve the quality of the components built into their machines. By taking a material such as paper, cotton or linen cloth, treating it with varnish and then partially curing to a dry stage, a 'pre-preg.' is produced. If several layers of this pre-preg. are consolidated together under heat and pressure a rigid laminate is formed. Materials of this type are manufactured in thicknesses from 0.1 to 150 mm. From these sheets components can be cut as required.

6.2.4.3 Modern materials

During the last 50 years many new base materials and resin systems have been adopted for producing new rigid sheets. The thermal, mechanical and electrical properties of currently available materials are extensive. Some indication of the range of laminates available today is given in Table 6.10. Although the majority of the rigid materials used in large electrical machine applications are based on thermosetting resins, especially epoxides and polyesters, there are also a number of thermoplastic sheet materials which have proved useful for making components for small- and medium-sized machines.

Table 6.10 Rigid laminates and mats

Base materials	Resin systems
Cellulosic paper Asbestos paper Aramid paper Mica paper Cotton cloth Silk cloth Linen cloth Glass cloth Asbestos cloth Aramid cloth Polyester cloth Glass mat Polyester mat Unidirectional glass	impregnated with { phenolic resin melamine resin polyester resin alkyd resin epoxide resin polyimide resin silicone resin

6.2.5 Sleevings and cables

In many machines there is a need to connect several separate coils together to form a complete winding. On low-voltage windings, after the coil-to-coil joints have been made, insulation of the connections can often be achieved using a length of varnished glass or polyester braided sleeving. Where it is not practicable to make coil-to-coil connections by jointing leads together, it is common practice to use lengths of insulated cable. To connect the ends of windings to machine terminals lengths of flexible cable are normally used. Typical types of sleeving are listed in Table 6.11. Commonly used cables are shown in Table 6.12.

6.2.6 Semi-conducting materials

6.2.6.1 Corona shield and semi-conducting slot packing materials

On high-voltage a.c. stators there is a requirement that the outside surface of coils should make good electrical contact with the laminated core. This is necessary to prevent electrical discharging taking place in any airgaps between the coil surface

Table 6.11 Flexible insulating sleevings

Basic yarn-braided or knitted	Varnish impregnation or coating	Thermal class
Polyester fibre	Polyurethane	B
Glass fibre	Polyurethane	B
Polyester fibre	Isophthalic	F
Glass fibre	Isophthalic	F
Polyester fibre	Acrylic	F
Glass fibre	Acrylic	F
Glass fibre	Silicone elastomer	H
—	Extruded silicone elastomer	H
Glass fibre	Silicone	C

Notes:

(a) Bore sizes range from 0.5 to 25 mm.

(b) Wall thicknesses range from 0.2 to 1.0 mm.

Table 6.12 Cables and coil-end flexibles

Elastomeric and similar coverings:
  Polychloroprene (PCP)
  Nitrile
  Silicone
  Chlorosulphonated polyethylene (CSP)
  Butyl/CSP
  Ethylene propylene rubber (EPR)
  EPR/CSP
  Polyvinyl chloride (PVC)
  Cross-linked polyethylene (XLPE)
  Fluorinated ethylene propylene (FEP)
  Ethylene tetrafluoroethylene (ETFE)
  Polytetrafluoroethylene (PTFE)

Taped coverings:

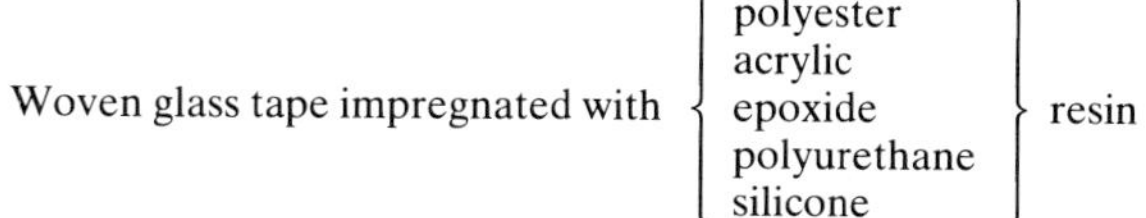

Woven glass tape impregnated with $\left\{\begin{array}{l}\text{polyester}\\\text{acrylic}\\\text{epoxide}\\\text{polyurethane}\\\text{silicone}\end{array}\right\}$ resin

Woven glass or polyester film-backed resin-impregnated mica paper

Aramid paper
Polyester or Polycarbonate film
Silicone elastomer-impregnated woven glass

Note:

(a) Cables are often finally covered with glass or polyester fibre braid to improve their resistance
   to mechanical abrasion.

and core. Although in practice only coils operating at, or close to, the line-end are stressed electrically at a high enough level to produce slot discharging, it is normal practice to coat the external surface of all coils with a suitable semi-conducting medium. In practice, this 'corona shield' can take the form either of a coat of graphite-impregnated varnish or a layer of graphite-impregnated polyester, glass, asbestos or other fibrous material. Today the most commonly used corona shield consists of a layer of graphite-impregnated polyester fleece. This material is used either as a sheet wrapped around the coil-side or as a butted, or half-lapped, layer. An alternative approach is to paint the coil-side with a coat of graphite-loaded varnish. Until a few years ago, woven asbestos tape impregnated with graphite-loaded epoxide resin was used as the corona shield on many large high-voltage windings. Due to possible health hazards associated with asbestos fibres such materials are being phased out (Table 6.13).

Although corona shields are sometimes fitted to 3.3 or 4.16 kV coils, it is more common, at the present time, to fit them only on coils for windings operating at 6 kV and higher voltages. To be effective the corona shield surface resistance needs to be in the range 1–30 kΩ square.

In order that coils or bars make good contact with the core it is necessary for them to be packed tightly in their slots. To minimize airgaps in the slot width, U- or L-shaped liners or packers of graphite-impregnated material are fitted. Although various paperlike materials coated on both sides with graphite varnish were used in the past, it is current practice to fit liners made from graphite-impregnated polyester fleece (Table 6.13).

Table 6.13 Conducting and semi-conducting materials

Corona shield materials:

(1) *Applied by brush:*
Graphite-loaded resin or varnish

(2) *Applied as tape:*

Graphite-loaded resin-impregnated
- woven glass tape
- woven polyester tape
- woven asbestos tape
- woven aramid tape
- polyester fleece

Conducting slot packings:

(1) *Flexible:*

Graphite-loaded resin-impregnated
- woven glass cloth
- woven polyester cloth
- polyester fleece

(2) *Rigid:*

Graphite-loaded resin-impregnated
- woven glass cloth
- woven polyester cloth
- polyester fleece (pressed)

Stress grading (semi-conducting) materials:

(1) *Applied by brush:*
Silicon carbide-loaded resin or varnish

(2) *Applied as tape:*

Silicon carbide-loaded resin-impregnated
- woven glass tape
- woven polyester tape
- woven glass/polyester tape
- woven aramid tape
- polyester fleece

6.2.6.2 Stress-grading materials

On high-voltage a.c. stators, the extensions of the line-end coils beyond the core are subjected to high electric stress. Action must be taken to reduce this stress, otherwise electrical discharging will take place in the air close to the ends of the core. To prevent this problem it is normal practice to apply a stress-grading material to the slot extensions of coils designed for 6 kV or higher voltages. Ideally, the stress-relieving medium needs to have non-linear voltage and current characteristics. In practice, silicon carbide-loaded epoxide resins are used either as a varnish or in a carrier such as woven glass or polyester fibre tape (Table 6.13).

6.2.7 Bracing tapes, cords and ropes

In the construction of most coils and windings, mechanically strong materials are required to hold or brace various insulated components firmly in place. Many of the flexible fibres used in their construction are of the same basic types used as carriers for the varnish-impregnated tapes described in Section 6.2.3. The four most common applications for these materials are:

(1) To act as a cushion over which the main dielectric material, either in the form of sheet or tape, can be applied without any danger of damage being imposed on the component being insulated. For example, it is common practice to apply

a layer of mechanically tough fibrous tape over a brazed or soldered joint prior to applying the main insulation.

(2) To act as 'pull-up' tapes where a component is being insulated with many layers of insulation, e.g. where a component such as a coil for a high-voltage winding is being insulated with many layers of tape it is common practice to intersperse a strong pull-up tape between every four or five layers of the main dielectric tape.

(3) To act as a final binder on the outside of a coil or insulated component. In addition, such binders also act as carriers for the final varnish coating applied to complete windings or components.

(4) To act as a bracing medium for holding components firmly in place so that they cannot move in service. For example, there is a need to restrain high-voltage stator coil end-windings to enable them to withstand the high mechanical forces produced during direct-on-line (DOL) starting. The forces produced can be greater than 100 times those occurring during normal operation. Where there is a requirement that the layers of bracing are bonded together, it is common practice to use materials which have been impregnated with epoxide resin and partially cured to a B-stage condition. Alternatively, the bracing materials can be brushed during application with either hot- or cold-setting synthetic resin. Typical bracing materials are woven, or unidirectional, tapes, ropes, cords or sleevings based on aramid, polyester, polyester/glass, glass or other fibres. The actual material used will depend on the component being braced and the maximum service operating temperature. Typical materials for the four applications are shown in Table 6.14.

Table 6.14 Materials for overtaping and bracing windings and components

Pull up, final binder and bracing tapes

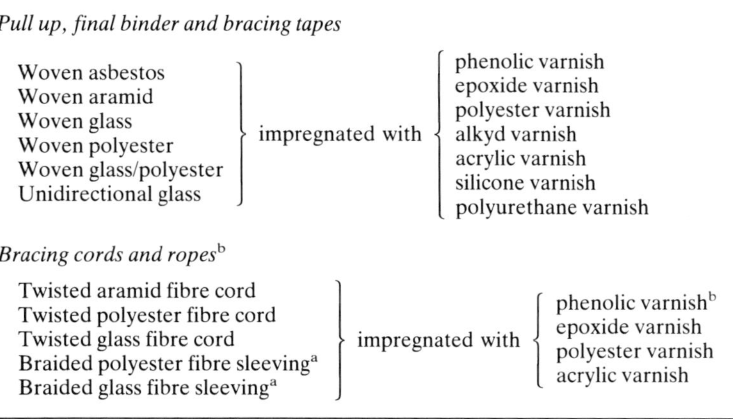

Bracing cords and ropes[b]

Notes:

[a] Large-diameter sleeving can have a central filling of unidirectional glass or polyester fibres to give it additional mechanical strength.

[b] Bracing cords are often impregnated with resin and partially cured to a B-stage. After application the resin is cured using heat so as to develop its ultimate mechanical strength.

6.2.8 Resins and varnishes

6.2.8.1 Definition

Resins are basically high molecular weight natural or synthetic organic solid or semi-solid materials. Varnishes are solutions of resins in appropriate solvents.

6.2.8.2 Development of natural and synthetic resins

In the early days of machine manufacture, natural materials such as asphalt, shellac, various vegetable oils and oleoresinous products were used to produce resins and varnishes. These materials were used to coat and impregnate conductor coverings, insulating tapes, coils and complete windings. Since the development of the first phenolic–aldehyde synthetic resins in the 1870s, there has been a steady and continuous increase in the number of available manmade chemicals with properties which are eminently suitable for producing varnishes and resins. In the 1930s, alkyd resins were developed and followed in the next decade by many new materials including the large classes of epoxides, polyesters and silicones. These latter three resin types have revolutionized the insulation systems used on electrical machine windings.

6.2.8.3 Classification: thermoplastics and thermosets (Table 6.15)

Resins are divided broadly into two basic types: (1) thermoplastics; and (2) thermosets. Thermoplastic resins are those materials which undergo no permanent chemical change on heating. They soften, or even melt, when warmed. They can be moulded into shapes which they retain on cooling back to the solid state. Thermoplastics are often referred to as linear polymers. Thermosetting resins in their initial 'unreacted' condition can be either solid or liquid. They normally consist of two or more components, e.g. base resin, hardener, catalyst, etc. blended together at a low temperature. On initial heating, the resin either melts (if originally solid) or becomes lower in viscosity (if originally liquid). On continued

Table 6.15 Major classes of synthetic resins

Thermoplastics:	*Thermosets:*
cellulose acetate	phenol formaldehyde
cellulose nitrate	aniline formaldehyde
polyamide	cresol formaldehyde
polycarbonate	melamine formaldehyde
polyester (linear)	urea formaldehyde
polyethylene	epoxide
polypropylene	polyester (cross-linked)
polystyrene	polyurethane
polytetrafluoroethylene (PTFE)	alkyd
polyvinyl chloride (PVC)	glyptal
polyvinylidene chloride	silicone
polyvinyl acetate	polyimide
polyvinyl butyral	diallyl phthalate
polyvinyl fluoride	
ethylenetetrafluoroethylene (ETFE)	
Elastomers:	
acrylonitrile rubber	
butadiene rubber	
chloroprene rubber	
polysulphide rubber	
silicone rubber	
butyl rubber	
chlorosulphonated polyethylene (CSP)	
polyurethane rubber	
ethylene propylene rubber (EPR)	
fluorinated propylene copolymer	
chlorinated polyethylene (CPE)	

heating, the viscosity of the liquid is raised by chemical interaction between the basic components until after a time, which can vary from minutes to hours, it sets to form a solid. Subsequent heating of a thermoset resin does not cause the material to liquefy. This setting is accomplished by the chemical process known as polymerization. Thermosetting materials are generally referred to as cross-linked polymers. Further comments on the three groups of polymers of particular relevance are given below.

Polyester resins.   Polyesters as a group are a series of resins in which the key links in the molecular chain are formed by ester groups. Two fundamentally different types are recognized: (1) saturated; and (2) unsaturated polyesters.

Saturated polyesters can be formulated with molecules having either medium or long chains. The latter are known as 'superpolymers' and are of particular importance as they are the basis of Terylene, Dacron, Mylar, Melinex and many other well-known fibres and films. When produced as fibres, or when extruded as a film, saturated polyesters have very high tensile strength.

Polyester fibres, when formed into a mat, are used as slot liners and for other applications where a tough flexible insulation is required. An additional feature, that polyesters can 'wet' glass, has been utilized in producing unidirectional resin-treated glass tapes for banding and bracing of machine windings.

Unsaturated polyester resins are typically liquids, although a few formulations have been produced as solids. The inherent capability of forming cross-links between molecules under heat to form hard, firm solids, i.e. thermosetting, has permitted their ready adoption in electrical applications as impregnating varnishes, resins and enamels. As a group, the majority of available resins are rated Class B. There are, however, a number of polyesters, some of which have been modified chemically, that have been evaluated thoroughly and shown to be capable of continuous operation at Class F temperatures for the largest machines and Class H temperatures for medium and small units.

Epoxy resins.   Epoxy resins were first produced on a commercial basis in the mid 1940s and since that time the growth rate has been phenomenal.

Epoxy resins are characterized by having molecules with terminal groupings of the epoxide type, through which cross-linking takes place on curing. Although there are only a few types of epoxy bases – bisphenols, epoxynovolaks and cylcoaliphatics being the main ones – there are numerous different hardeners or curing agents. By selecting the correct resin–hardener combination, products can be formulated to give particular electrical, mechanical, thermal and chemical properties. Many of those of particular interest for electrical insulation systems use epoxy resins which include latent catalysts. The latter are materials which when added to a resin–hardener mixture do not have any significant effect on the cross-linking or curing action below a specific temperature. However, when the resin–hardener mixture is heated above that temperature then rapid cross-linking takes place. One of the most useful features of many epoxy resin systems is that partial curing of the mixture can be carried out. By this means, many liquid resin–hardener systems can be partly cured such that they change from a liquid to a solid in the so-called B-stage. Resin in the B-stage can remain in this condition for many weeks or months if stored at a low temperature, i.e. $< 5°C$. On subsequent reheating, the epoxy resin–hardener–catalyst mixture will cure to produce the final, fully cured, thermoset resin.

The most important uses for epoxy resins are as adhesives, surface coatings, potting compounds, bonding of mica and fabric tapes, and for treatment of complete windings using dipping, rolling, trickle and vacuum pressure impregnation techniques.

Silicone resins. The substitution of the siloxane group SiO for carbon in the molecular chain of many organic compounds gives rise to chemicals of the generic types called silicones. Such materials generally exhibit enhanced temperature stability compared with organic compounds and have resulted in the development of an entire family of chemicals including oils, varnishes, greases, rubbers and foams. The application of these materials in many industries, not least of which are those associated with electrical machine manufacture, has been significant.

Varnishes formulated with silicones containing methyl and phenyl groups, dissolved in a suitable solvent, are used for impregnating individual coils as well as complete windings. Advantages of the cured products include good flexibility, high thermal stability up to Class H temperature limits and excellent electric strength. Disadvantages are the relatively high cost, low abrasion resistance and susceptibility to chemical attack by many common cleaning solvents. Although silicone resins are used for impregnating many mica and fibrous tapes, great care is needed in their selection and use because of their relatively poor mechanical strength. Silicone elastomers have, in the past, been shown to be weak mechanically and damaged easily. For this reason many of the commercially available silicone rubber insulated cables have an overbraid of either glass or polyester fibre to give them additional mechanical protection. In more recent times, new grades of silicone elastomer have become available which offer greater mechanical strength. Silicone elastomers, and some resins, produce volatile low molecular weight decomposition products during ageing, which prevent formation of a stable surface skin on commutators and sliprings.

6.3 Insulation systems

6.3.1 Operational requirements

In designing electrical machines, many different criteria must be met. Apart from machine bearings and brushgear when fitted, no other major item common to all electrical rotating plant is affected to any significant degree by service ageing apart from the insulation. If a reliable, trouble-free service life is to be achieved, then the insulation system must be designed to meet all foreseeable operating requirements. Consequences of failure of a vital unit of capital plant can be serious both in economic terms as well as in its effect on associated equipment. The most important feature of any insulation system is its reliability.

6.3.2 Effect of stresses in service

In service, windings must withstand the combined effect of stresses set up by the electrical, thermal and mechanical loading imposed during both normal and abnormal operating conditions. Atmospheric contamination to which a machine is subjected can affect adversely the performance and, hence, the service life of the system. Environmental stressing can accelerate appreciably the rate at which ageing takes place. Examination of these stresses will help to highlight the significance of the various properties required from an insulation system (Table 6.16).

6.3.2.1 Electric stress

A winding must be able to withstand the electric stress imposed by the normal operating voltage as well as by any transient stresses set up by impulses reaching the machine terminals from external sources. Normal operating voltage defines the thicknesses of the main ground and interphase insulation. On multi-turn coils, the

Table 6.16 Stresses applied to insulation systems in service

Type of stress	Causes
(1) Electrical stress	Normal operating voltage Impulses Discontinuities in the system
(2) Thermal stress	Normal operating temperature Overloads Stopping and starting
(3) Mechanical stress	Vibrations Short circuits and impulses Direct-on-line starting Centrifugal force
(4) Environmental stress	Abrasive and conducting dusts Hygroscopic deposits, e.g. salt Hazardous atmospheres: oil vapour, steam, solvents, etc

level of impulse expected in service is used to determine the thickness of the interturn insulation. On high-voltage a.c. machines, there is a need to ensure that the slot insulation of line-end coils is substantially void-free otherwise electrical discharging will permit erosion of the organic materials used in the system.

6.3.2.2 Thermal stress

The operating temperature for which a winding is designed defines, to a major extent, the basic materials which can be used (see Table 6.3, page 335). Further thermal stressing is caused by overloads, starting and stopping, as well as by thermomechanical effects produced in the winding during other abnormal conditions.

6.3.2.3 Mechanical stress

Under steady load a winding is subjected to mechanical vibrations set up by the electromagnetic forces acting on the current-carrying components. Stresses are also imposed by thermomechanical expansion and contraction effects on the conductors and associated insulation during starting, stopping, short-circuit and other overload conditions. Direct-on-line starting, short-circuiting and impulses also cause electromechanical stressing of windings and associated bracing systems. On rotating windings, allowance must be made for the effects of centrifugal force.

6.3.2.4 Environmental stress

Up to the early 1960s the majority of large machines were considered as capital equipment. For this reason, whenever a dirty or hazardous atmosphere was likely to exist around a machine, an enclosed air-cooled or water-cooled unit was specified.

In more recent times the change in attitude of many heavy-plant users – particularly in the oil and chemical industries – in considering the overall economic life of specific plants to be not more than 7–10 years, has meant that the cheaper open-type machines have been specified. These allow hazardous materials in the environment to reach the machine windings. In the last few years the consequence of this approach has become evident by the number of installations, particularly on offshore oilrigs and in hot desert conditions, where the number of motor failures has increased substantially. As a direct result of these service problems, many major plant users have reverted to specifying either enclosed machines or fully sealed windings capable of withstanding a water immersion, or spray, test.

Although many users indicate in their specifications the operating conditions expected in service, it is not unusual to find that once a machine has been installed the contamination is very much more severe than was originally specified. Manufacturers need to know both the erection and final operating environments in order that they can ensure that the correct type of insulation system and final treatment are used. The relative importance of each of these stresses varies from one type of machine to another as well as on the planned service duty.

6.4 Insulation technology and winding techniques

6.4.1 Basic winding types

Many different types of winding and insulation systems are used in electrical machines. Each individual equipment manufacturer develops, over a period of many years, systems best suited for the designs and markets into which he sells his particular products.

The windings most commonly in use today can be divided into three basic groups as follows:

(1) *Armature windings.* Windings made up from a number of coils or bars wound into slots in the stationary or rotating core. These include all the major windings associated with both a.c. and d.c. machines (Figure 6.2).
(2) *Field windings.* Windings made up as an assembly of coils wound on to individual poles. These include the major excitation windings used in both a.c. and d.c. machines (Figure 6.3).

Figure 6.2 Typical large d.c. armature. (*Courtesy:* GEC Large Machines Ltd)

Figure 6.3 Typical large d.c. frame. (*Courtesy:* GEC Large Machines Ltd)

(3) *Other windings.* Includes miscellaneous windings such as equalizer and compensating windings on d.c. machines, and cage and pole-face windings on a.c. rotors.

Descriptions of the two major groups are given below. Other windings are described in Section 6.5.

6.4.2 Armature windings

6.4.2.1 Basic design – single- and double-layer windings

Windings wound into slots in core assemblies can be either of the single- or double-layer type. In the former, each of the two coil-sides, making up a coil, are assembled in two slots (approximately one pole pitch apart) such that they fill almost all the available space for the current-carrying conductor. In the latter, each of the two coil-sides making up a coil is assembled in two slots (approximately one pole pitch apart) such that one coil-side fills the bottom half of one slot while the other coil-side fills the top half of the other, associated, slot.

6.4.2.2 Basic types of armature winding

There are three main types of armature winding: (1) mush-wound; (2) form-wound; and (3) concentric-coil windings. Basic details are as follows:

(1) *Mush windings.* Most mush windings consist of a number of simple coils, wound at random with enamel-covered round wire (Figure 6.4). Size and number of turns depend on the voltage, speed and rating of the relevant machine. Mush windings can be either single- or double-layer. Stator slots can be either open, semi-closed or offset. Mush coils are normally wound into slots lined with flexible insulation.

(2) *Basket or diamond-shaped coil windings.* Most a.c. and d.c. machine armatures with medium- and high-power ratings have two-layer windings made up of either single or multi-turn diamond-shaped coils wound with rectangular copper (Figure 6.5). Thickness and type of conductor insulation will depend on the specific design. In low-voltage machines, coils are either wound into open slots lined with flexible ground insulation or have it applied prior to coil insertion. In high-voltage coils, the main insulation is always applied to individual diamond-shaped coils prior to winding as the thickness and dielectric integrity cannot be achieved using a slot-liner system.

For some windings, where the conductor section is large and manufacture of full diamond coils could present problems, each coil is made as two separate coil-sides or bars. After winding, each pair of associated half-coils or bars is joined together to form diamond-shaped coils. This type of construction is especially suited to windings made up of single-turn coils.

A modified type of bar winding, having one end-winding formed to the required shape and the other end left straight, is used for armatures with

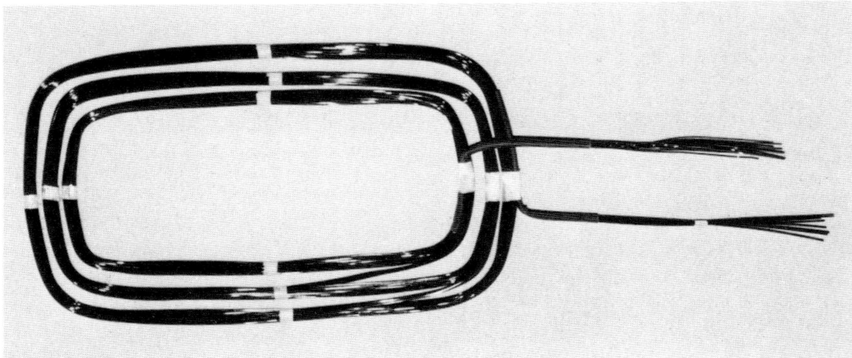

Figure 6.4 Mush coil

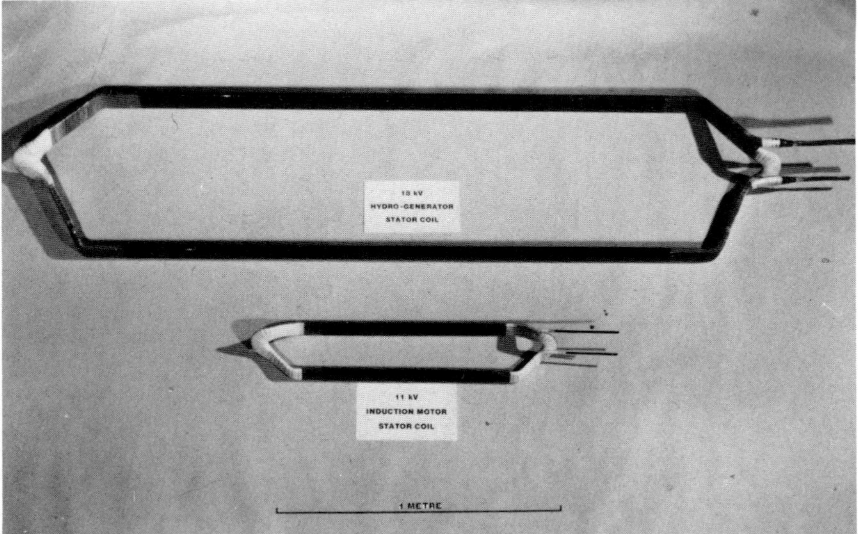

Figure 6.5 Diamond coil

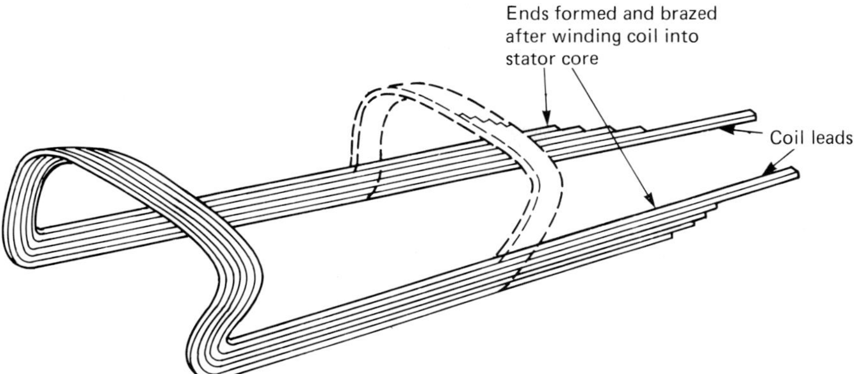

Figure 6.6 Concentric coil

semi-closed slots. After bar insertion, the straight ends are formed to shape. This winding method is expensive because of the amount of forming work required.

(3) *Concentric windings.* Concentric windings are normally of the single-layer type made up with hairpin-shaped coils inserted into semi-closed slots in the armature core. After coil insertion, ends are formed to shape and individual conductors joined together to produce the characteristic concentric shape (Figure 6.6). Individual hairpins are formed from the specified number of insulated rectangular conductors. Main ground insulation is applied as appropriate for the machine line voltage. Concentric windings can also be manufactured with fully formed and insulated coils wound into open armature slots.

6.4.3 Field windings

Low-voltage field windings for both a.c. and d.c. machines are wound either on formers or directly on to individual pole bodies. Where small section conductors are required, either round or rectangular insulated copper is used. Where the cross-sectional area is relatively large, bare copper is used. Bare copper is either wound on the flat or on edge. Many varied designs are used to suit specific design requirements. On low-voltage d.c. machines several different coils are often fitted on the main poles. Main ground insulation can be applied either to the pole body or to each individual coil.

Field coils for rotating components must be designed to withstand the maximum centrifugal force.

6.5 Insulation of machine windings

6.5.1 Alternating current stators: low-voltage

6.5.1.1 General: types of winding

Three types of winding are commonly used, depending on machine rating and number of poles: (1) mush windings are used for small- and medium-sized stators; (2) at higher ratings, multi-turn diamond-shaped coils are adopted; and (3) on the largest stators, single-turn coils and bar windings are used.

Mush windings.   Both single- and double-layer windings are used. Coils are wound with the specified number of turns of round wire. The majority of small- and medium-sized stators have conductors insulated with polyesterimide enamel. Coil leads are insulated normally with a layer of varnished woven glass or polyester fibre sleeving. Common practice is to wind several coils as a series group with sleeved intercoil leads. Coils are wound into slots lined with an appropriate thickness of tough insulation. Liners extend beyond the ends of the core. Cuffed, or reinforced, liners are sometimes adopted to prevent tearing at slot exits. During winding, phase groups are separated by barriers. Slot separators and closures are normally made from either thick slot liner material or strips of laminate (Figure 6.7).

End-windings may be laced together with a running band of woven fibre tape or cord. Joints between coils are either brazed or soldered. Series and pole-to-pole joints are insulated either with varnished sleeving or tape. Terminal cable leads are usually insulated with synthetic rubber of plastic material. Windings are finally treated by dipping in varnish, by trickle impregnation or by vacuum pressure impregnation in synthetic resin (Figures 6.8 and 6.9).

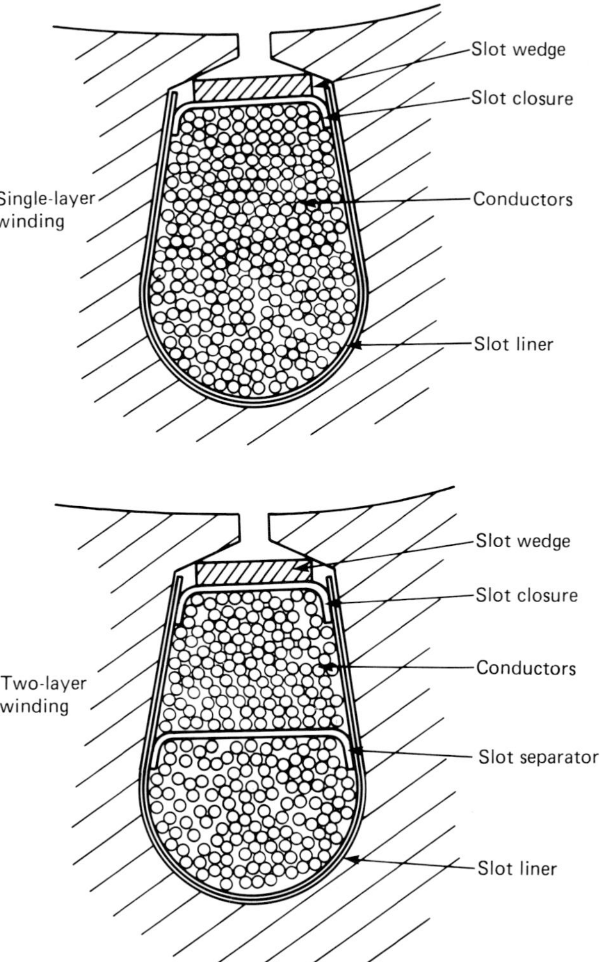

Figure 6.7 Slot build ups for low-voltage windings

Figure 6.8 Low-voltage mush winding

Multi-turn coil windings.   Multi-turn coils for low-voltage stators, i.e. up to 650 V, are usually wound with enamel or enamel and glass-covered rectangular copper. Coils are wound as flat loops and then pulled out to shape. Coil leads are insulated with sleeving or layers of varnished tape. End-windings are taped with one or two layers of insulation for machines designed to operate at voltages greater than 450 V. Lower voltage stators only require a binding tape to hold the coils in shape. Coils are wound into slots lined with an appropriate thickness of flexible insulation. In the past, micaceous sheet was often used on Class B windings. Present practice is to use laminated polyester or aramid paper. Separators and wedges of appropriate thicknesses are usually fitted in each slot. In some designs, top and bottom coil-sides are wound in their own individual liners. Alternatively, a supplementary liner may be fitted in the top half of the slot along with a slot separator (Figure 6.10). After coil insertion, joints are formed and either soldered or brazed. Insulation is then applied to series joints and pole-to-pole connections as appropriate for the stator voltage. Varnished tapings normally are used. Terminal leads and final treatments are similar to those used for mush windings.

Single-turn windings.   When the current loading per slot is high, windings are made up of half coils (or bars) or single-turn coils. Either large-section bare or small-section insulated rectangular copper is used depending on the specific design. On some stators, the coils or bars are formed from one single solid copper conductor strip. Where the required conductor width is large and could present forming problems, two or more strips of thin bare copper are used to achieve the required cross-sectional area. Where the effect of eddy currents flowing in a single deep conductor would give rise to significant losses, conductors are split in the slot depth. Each of the several sub-conductors making up a single-turn coil or bar is insulated with either an enamel or a fibrous covering. To reduce eddy current losses still further, coils or bars may have their sub-conductors transposed either along the slot portion on the end-windings.

Figure 6.9 Low-voltage mush wound stator. (*Courtesy:* GEC Large Machines Ltd)

Ground insulation, based on either varnished fabric or aramid fibre tape is often used for medium-sized stators. For large units manufacturers use mica-based main ground insulation. Both resin-rich and vacuum pressure impregnated insulation systems are used. After application of the main insulation, coils or bars are wound into the core. Slot separators and wedges are made from synthetic resin-bonded laminate (Figure 6.11). After winding and jointing, end-winding taping, fitting of terminal leads and final treatments are carried out in a similar manner to the methods used for multi-turn coils.

6.5.2 Alternating current stators: high-voltage

6.5.2.1 Winding types

Most windings are of the two-layer, multi-turn type with pulled out diamond-shaped coils wound into open slots. Single-layer windings made up of either hairpin or concentric coils are used by some manufacturers. Single-turn windings consisting of half-coils or bars are used on stators with the largest ratings.

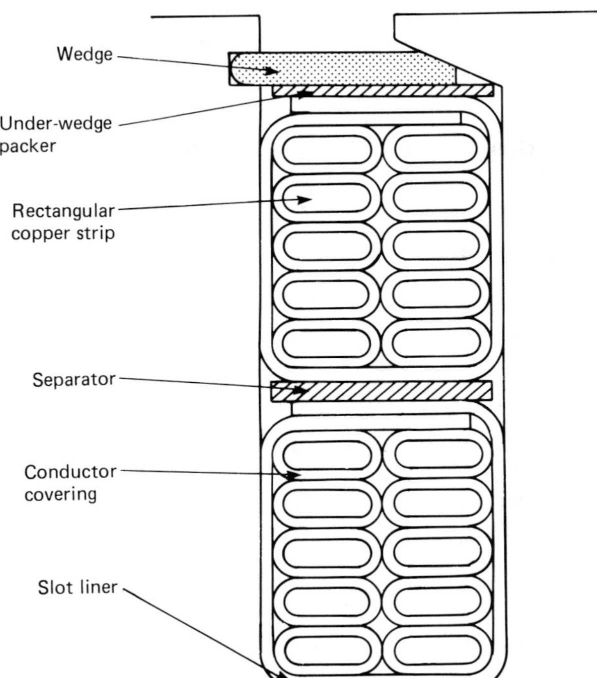

Figure 6.10 Slot build up for low-voltage multi-turn coil

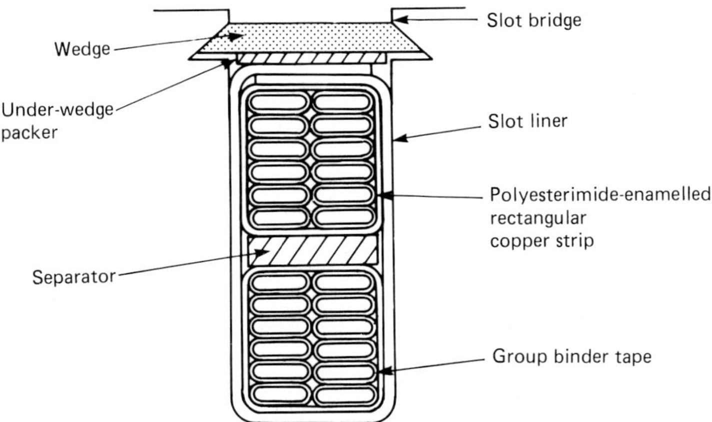

Figure 6.11 Slot build up for low-voltage single-turn coil

6.5.2.2 Basic manufacturing procedure

The first stage in manufacturing multi-turn coils is to wind flat loops of the required number of turns using one or several rectangular conductors. Covered conductors are normally used. Where the copper cross-sectional area per turn is large it is common practice to subdivide it into two or more sub-conductors which are then wound in parallel. By using copper with a small cross-sectional area, stresses applied during loop winding are reduced. Moreover, where the conductor is sub-divided in the slot depth, the overall eddy current loss per turn is also reduced. A typical loop winder is shown in Figure 6.12.

Figure 6.12 High-voltage stator coil – winding the coil loop

Loops are next pulled out into the well-known 'diamond' shape and, depending on the design inter-turn voltage, may have additional insulation applied to each turn. Subsequently, main ground insulation is applied to coil leads, end-windings and straight slot portions (Figure 6.13). After final processing, as required for the specific insulation system, coils are wound into the stator core and connected up. During winding, separating strips, slot liners and/or packers are fitted into each slot. Individual slots are wedged and end-windings braced and blocked to prevent movement in service. Finally, the stator is treated using one of the several alternative processes available for impregnating and/or sealing the winding.

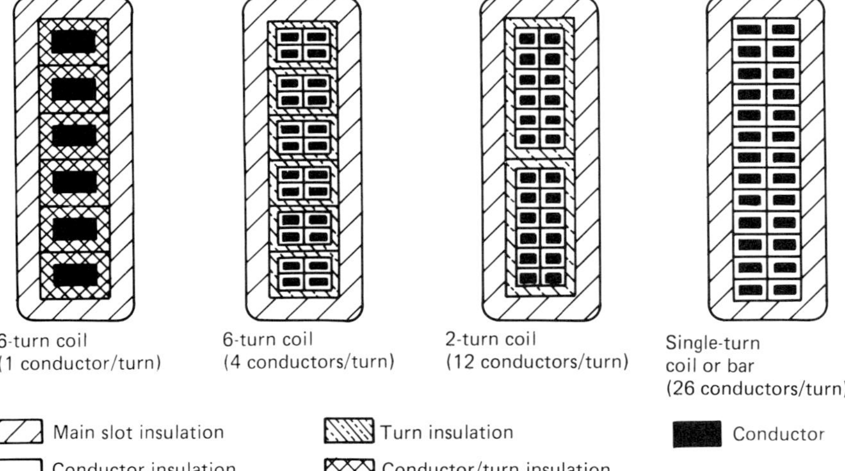

Figure 6.13 High-voltage stator coil – typical slot build-ups

6.5.2.3 Basic insulation systems

Early materials used for insulating high-voltage stator coils and bars consisted chiefly of natural materials. Conductors were covered with cotton, silk or asbestos fibres bonded together with natural resins. Inter-turn and main insulation was made up almost universally from tapes and sheet materials consisting of mica flakes bonded together with natural resins such as shellac, copal or bitumen, carried on a backing of silk, cotton or paper. Natural materials performed satisfactorily in service provided they were used within their definite limitations. Much work was carried out some 30 years ago to improve the technical performance of windings. As a result, all major manufacturers now use insulation systems based on micaceous materials bonded with thermosetting resins. The two basic types which have grown up side-by-side are the resin-rich and the vacuum pressure impregnation processes.

6.5.2.4 Resin-rich systems

Resin-rich systems are, in several aspects, a logical development of the processes used in the past. Coils are wound first as flat loops. Typical conductor coverings in use today are glass, enamel and glass, and mica tapings. After forming, straight slot portions may be consolidated under heat and pressure to specified dimensions (Figure 6.14).

Coil loops are then pulled out into the diamond shape (Figure 6.15). Thereafter, depending on the machine line voltage and number of turns per coil, additional tape may be applied around each turn. The interturn insulation is increased to a level capable of withstanding the maximum steep-fronted surge which the coil/winding will be subjected to in service. Main ground insulation is next applied to the slots and end-windings. Depending on the method of insulation application, the systems can be described as either continuous or discontinuous. In the former the same basic material is applied throughout the coil, i.e. to leads, coil evolutes,

Figure 6.14 High-voltage stator coil – consolidation of conductor stack

Figure 6.15 High-voltage stator coil – pulling out a diamond-shaped coil

end-windings and straight slot portions. In the latter, different materials are applied to the slots and to the end-windings, leads and evolutes.

Both systems perform satisfactorily in service provided the manufacturing procedures have been carried out correctly. On industrial machines, slightly less insulation is applied to the end-windings compared to the slot portions. This is an acceptable design practice because, in service, the voltage appearing between adjacent end-windings does not exceed the machine line voltage except under transient conditions. End-windings are only subjected to approximately 80% of the voltage seen by the ground insulation. For machines with line voltages up to 4–5 kV, ground insulation thickness is normally dictated by the mechanical requirements for producing coils which can be handled and wound without being damaged. For higher voltages, ground insulation thickness is usually decided by electrical requirements. Typically, the electric stress used at the 4–5 kV voltage level is only about 1 kV/mm whereas at higher voltages, e.g. 13.8 kV it is not usually less than 2 kV/mm.

After the main insulation has been applied it is normal practice to consolidate the slot portions under heat and pressure to specified dimensions for coils or bars for machines with line voltages greater than about 5 kV. This operation cures the B-stage resin in the mica paper and produces ground insulation with good electrical, mechanical and thermal properties. Slot portions of coils for voltages greater than about 5 kV have the quality of the dielectric checked by measuring the loss tangent values up to line voltage. (For details of this test technique and recommended limits see Section 6.6.) A typical loss tangent–voltage curve for an epoxy-bonded mica paper coil-side is shown on Figure 6.16. For machines operating up to about 5 kV, because of the lower operating electric stress, there is less need to consolidate the ground insulation. Hence, unconsolidated insulation is often used up to this voltage. Coils so designed are wound into the stator core in the uncured state. Subsequently, curing is achieved by baking the completed stator in an oven.

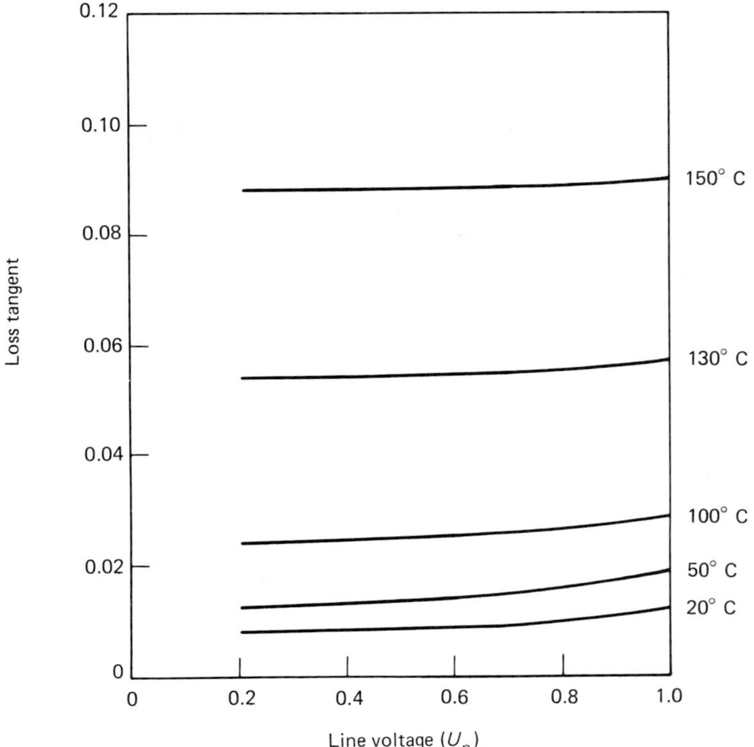

Figure 6.16 Typical loss-tangent–voltage curve for resin-rich epoxy bonded mica paper

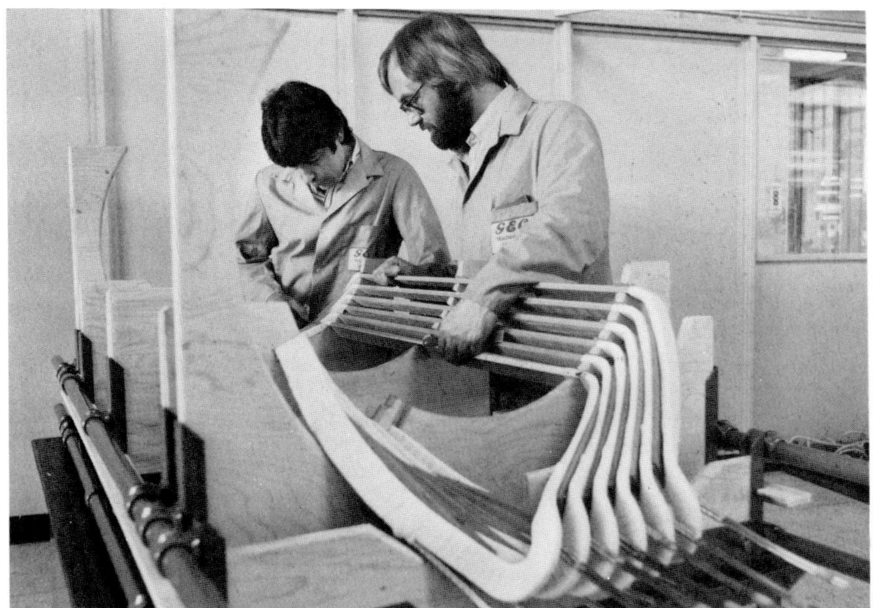

Figure 6.17 High-voltage stator coil – checking the coil shape

Slot portions of coils for voltages over about 5 kV must make good contact with the stator iron. This is achieved either by painting with conducting varnish or by taping with a conducting fibrous material (see also Section 6.2.6). Painted-on corona shields are liable to damage during coil insertion. Fibrous corona shields, such as graphite-impregnated polyester fleece are resistant to abrasion. Coils operating at voltages over 5 kV often need special treatment to prevent breakdown of the air just beyond the end of the core. This is achieved by either painting or taping the slot extensions with a semi-conducting material. The length of this stress grading depends on the current and voltage characteristics of the material used, the machine line voltage and the ground insulation thickness. After coils are completed they are tested prior to assembly. Descriptions of these tests, and how they are carried out, are given in Section 6.6. Dimensional checks are also made to confirm the coils are correct to drawing. A typical checking gauge is shown in Figure 6.17.

Prior to winding, the stator end-winding support or 'surge' rings are usually fitted at each end of the core. In the past, taped metal rings were adopted. Today, resin-bonded filament-wound glass, or segments of laminate material, are used to support the coil overhangs. Computer programs based on finite element techniques are used to calculate forces and, hence, the requisite number and position of rings.

When mounting coils in slots, it is important to ensure that clearance between coil-sides and core is not excessive. Several means of keeping the clearance within acceptable limits are used. The most common method is to wind coils into slots lined with a layer of flexible material. Clearances can be controlled by adjusting the liner thickness. For stators up to 5 kV, liners made of fibrous material such as polyester or aramid fibre paper are used, either alone or in combination with polyester film. Above 5 kV, slot packers or liners are normally made from conducting material such as graphite-impregnated polyester fleece. Separating strips of laminate material are placed between top and bottom coil-sides. Coils are normally retained in their slots by laminate wedges. During wedging, additional packing strips may be fitted to minimize clearance in the depth. For highly rated machines where the bar bounce force is high, alternative means of retaining coils may be necessary, e.g. resilient packers or ripple type springs.

End-windings are braced to prevent their movement in service. The type of bracing system and materials used for a specific machine depends on a number of factors. The proposed service duty is important. For example, the level of bracing required for a DOL-started motor which starts and stops 20 times a day for 30 years needs to be more substantial than that required for a generator which runs continuously for periods of 6 months at a time. Typically, bottom end-windings are braced to support rings either individually or in pairs separated by a packing block of laminate material. Top end-windings are usually braced together in pairs with laminate packing blocks between them. Several rows of blocks and bracing may be used, depending on the overhang length. For machines up to about 4 kV, top end-windings are often archbound. For applications where end-winding forces are low, bracing is often carried out using woven glass, glass/polyester, or polyester tapes. For severe duties, e.g. DOL starting, stronger bracing materials such as unidirectional glass tape or cord impregnated with B-state epoxy resin are used. A typical arrangement is shown in Figure 6.18.

Series joints, pole-to-pole and terminal connections are made and insulated using materials compatible with the relevant insulation system. Stator joints are either soft-soldered or brazed. For Class F windings, brazed joints are essential to cater for the higher operating temperature. Where flexible cables, or coil-end flexibles, are fitted these must be insulated with material suitable for the service temperature. Today, most stators are fitted with cables insulated with EPR/CSP or silicone rubber. Series joints, evolutes, pole-to-pole connections and terminal leads are held rigid with bracing tapes, cords and ropes as necessary.

After winding, most stators are given a final varnish treatment. Normally,

Figure 6.18 High-voltage stator winding – end-winding bracing. (*Courtesy:* GEC Large Machines Ltd)

solvented thermosetting varnishes, such as polyesters which produce tough but semi-flexible coatings, are used. All B-stage materials in the coils, connections and bracing system are cured during the baking operation.

Completed stator windings are high-potential tested at the voltage specified in the appropriate specification. Polarization index, loss tangent and dielectric loss analyser tests are often carried out on large units. Measurements can then be repeated, after a period of service operation, to check the integrity of the winding (see Section 6.6).

6.5.2.5 Vacuum pressure impregnation systems

Vacuum pressure impregnation has been used for many years as a basic process for thorough filling of all interstices in insulated components, especially high-voltage stator coils and bars. Prior to development of thermosetting resins, a widely used insulation system for 6.6 kV and higher voltages was a vacuum pressure impregnation system based on bitumen-bonded mica-flake tape main ground insulation. After applying the insulation, coils or bars were placed in an autoclave, vacuum-dried and then impregnated with a high-melting-point bitumen (asphalt) compound. To allow thorough impregnation, a low viscosity was essential. This was achieved by heating the bitumen to about 180°C at which temperature it was sufficiently liquid to pass through the layers of tape and fill all interstices around the conductor stack. To assist penetration, the pressure in the autoclave was raised to 5 or 6 atmospheres. After appropriate curing and calibration, the coils or bars were wound and connected up in the normal manner. This system performed satisfactorily in service provided it was used within its thermal limitations. In the late 1930s and early 1940s, however, many large units, principally turbine generators, failed due to the inherently weak thermoplastic nature of bitumen compound. Failures were due to two types of problem: (1) tape separation; and (2) excessive relaxation of the main ground insulation (Figures 6.19 and 6.20).

Much development work was carried out to try to produce new insulation systems which did not exhibit these weaknesses. The first major new system to overcome these difficulties was basically a fundamental improvement to the classic vacuum pressure impregnation process. Coils and bars were insulated with dry mica-flake tapes, lightly bonded with synthetic resin and backed by a thin layer of fibrous material. After taping, the bars or coils were vacuum-dried and pressure-impregnated in polyester resin. Subsequently, the resin was converted by chemical action from a liquid to a solid compound by curing at an appropriate temperature, e.g. $\simeq 150°$C. This so-called thermosetting process enabled coils and bars to be made which did not relax subsequently when operating at full service temperature. By building-in some permanently flexible tapings at the evolutes of diamond-shaped coils, it was practicable to wind them without difficulty. Thereafter, normal slot packing, wedging, connecting up and bracing procedures were carried out. Various versions of this basic vacuum pressure impregnation procedure have been used for almost 30 years by many manufacturers for producing their large coils and bars. The main differences between systems have been in the types of micaceous tapes used for the main ground insulation and the composition of the impregnating resins. Although the first system available was a styrenated polyester, many developments have taken place during the last two decades. Today, there are several different types of epoxy, epoxy–polyester and polyester resins in common use. Choice of resin system and associated micaceous tape is a complex problem for the machine manufacturer.

Although the classic vacuum pressure impregnation technique has improved to a significant extent, it is a modification to the basic process which has brought about the greatest change in the design and manufacture of medium-sized a.c. industrial machines. This is the global impregnation process. Using this system, significant increases in reliability, reduction in manufacturing costs and improved output can be achieved. Manufacture of coils follows the normal process except that the ground insulation consists of low-bond micaceous tape. High-voltage coils have corona shields and stress grading applied in the same way as for resin-rich coils, except that the materials must be compatible with the vacuum pressure impregnation process. Individual coils are interturn and high-potential-tested at voltages below those normally used for resin-rich coils because, at the unimpregnated stage, the intrinsic electric strength is less than that which will be attained after processing. Coils are wound into slots lined with firm but flexible sheet material. Care has to be taken to ensure that the main ground insulation, which is relatively fragile, is not damaged. After interturn testing of individual coils, the series joints are made and coils connected up into phase groups. All insulation used is low-bond material which will soak up resin during the impregnation process. End-winding bracing is carried out with dry, or lightly treated, glass- and/or polyester-based tapes, cords and ropes. On completion, the wound stator is placed in the vacuum pressure impregnation tank, vacuum-dried and pressure-impregnated with solventless synthetic resin. Finally, the completed unit is stoved to thermoset all the resin in the coils and the associated bracing system. A typical modern impregnation plant is shown in Figure 6.21.

After curing, stator windings are high-potential-tested to the same standard as already mentioned in Section 6.5.2.4 for resin-rich systems. Loss-tangent

Figure 6.19 Tape separation on high-voltage stator insulation

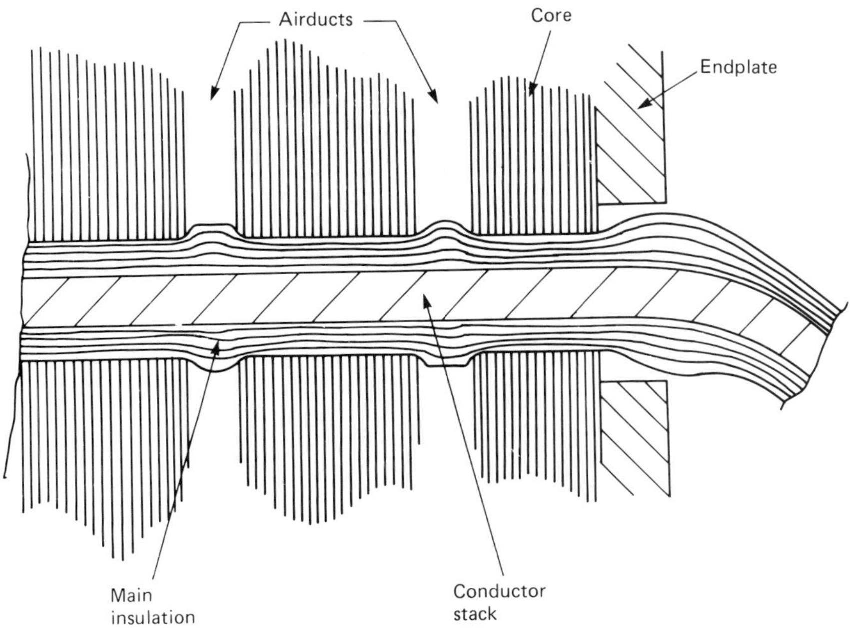

Figure 6.20 Relaxation of thermoplastic high-voltage stator insulation

Figure 6.21 Typical modern vacuum pressure impregnation plant

measurements at voltage intervals up to line voltage are normally made on all stators for over 1 kV. A major difference between resin-rich and vacuum pressure impregnation lies in the importance of this final loss-tangent test; it is an essential quality-control check to confirm how well the impregnation has been carried out. To interpret the results, the manufacturer needs to have a precise understanding of the effect of the stress-grading system applied to the coils. Stress grading causes an increase in the loss-tangent values. To calculate the real values of the ground insulation loss-tangent, it is necessary to subtract from the readings the effect of the stress grading. For grading materials based on materials such as silicon carbide loaded tape or varnish, this additional loss depends, to a large extent, upon the stator core length and machine voltage. A typical set of curves showing the effect

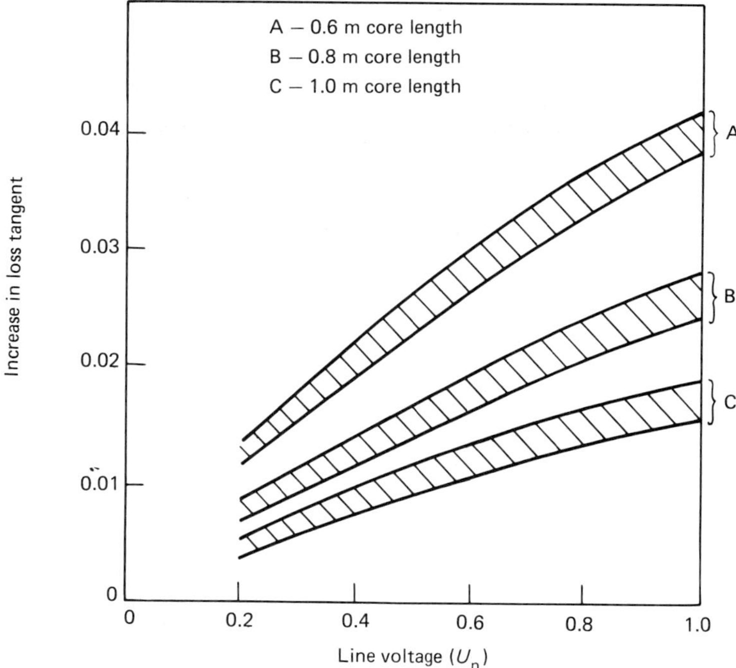

Figure 6.22 Effect of stress grading on loss tangent values

on the loss-tangent values for a specific stress grading system is shown in Figure 6.22.

6.5.3 Alternating current rotor windings

6.5.3.1 General: types of winding

On a.c. rotors, in addition to cage windings, several other basic designs are used. Low-voltage mush coils are adopted for small- and medium-sized wound rotors and for rotor-fed a.c. commutator motors. Fully formed diamond coils and bar windings are used for medium- and large-wound rotors and for pole-face windings on salient-pole synchronous induction motors. On salient-pole machines, wire-wound and strip-on-edge coils are normally used.

6.5.3.2 Mush windings

Mush windings can be either of single- or double-layer design. Up to about 15 years ago the normal conductor covering was PVA enamel. Today, polyesterimide or similar enamels are used because of their increased thermal capability. After winding, coils are normally inserted into slots with either semi-closed or offset openings. Slots are lined with tough insulating material such as laminated polyester or aramid paper. End-windings are supported on flanges. Interlayer insulation is inserted between top and bottom overhangs. Separators of rigid insulation are sometimes placed between top and bottom coil-sides on two-layer windings. Liners are folded over and slots wedged with strips of laminate. To prevent radial movement, end-windings are held down to support flanges by steel or glass binding bands. Coil-to-slipring connections are normally of insulated copper strip clamped to the rotor shaft at appropriate positions.

6.5.3.3 Fully-formed diamond coils and bar windings

Most wound rotors for medium- and large-induction and synchronous induction motors have bar windings fed into semi-closed slots. Each bar is formed in a dogleg shape (Figure 6.23) and fed into semi-closed slots in the rotor core. Bars are normally insulated in the slot portion and on the formed end prior to insertion. After winding, the straight ends are bent to the required shape and insulated. On a few designs, fully formed single, or occasionally two- or four-turn, diamond-shaped coils are fitted into rotors with open slots. Although the operating voltage is relatively low, windings must be designed to withstand the high voltage induced during starting, i.e. up to 4 kV. Main ground insulation for low-voltage rotors often

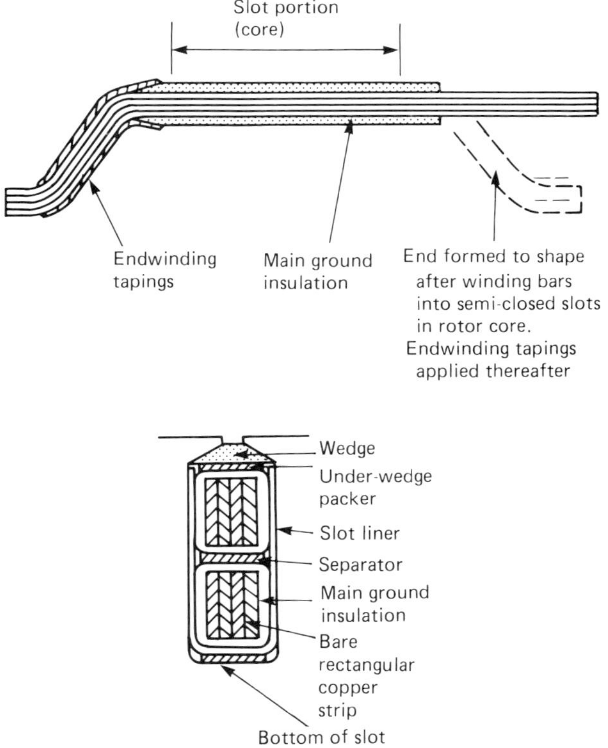

Figure 6.23 Typical a.c. rotor bar

consists of one or more layers of varnished glass or other fibrous tapes. For higher voltages, slot portions are sometimes insulated with B-stage micaceous tapes and consolidated to shape using heat and pressure. On large rotors, bars must be restrained by pulling down the end-windings on to insulated support flanges. Layers of flexible tough insulation are inserted between top and bottom end-windings. Binding bands are applied at high tension to prevent movement during service operation. Completed rotors are sometimes treated by dipping or rolling in a semi-flexible varnish and baked to produce a coating resistant to contamination, dirt and moisture. An alternative approach is to tape bars with dry tapes and then vacuum pressure impregnate the complete rotor.

6.5.3.4 Wire-wound salient-pole field coils

Salient poles are used on synchronous motors and generators with four or more poles. Machines with two poles are usually, but not exclusively, designed as cylindrical rotors.

Wire-wound salient-pole designs use either round or rectangular insulated copper. Coils are often wound directly on to poles wrapped with tough insulation. Top flanges are always fitted between coil and pole tip. In many designs, bottom flanges are also fitted.

Synthetic resin is often brushed between layers during the winding operation. After applying the specified number of turns, coils are clamped along their sides and baked to cure the brushed-in resin. An alternative approach is to use enamelled copper overcoated with B-stage epoxy resin. This resin is cured during the baking schedule. Another approach is to wind and then use a vacuum pressure impregnation process to bond turns together. On high-speed rotors, V-blocks are sometimes inserted between salient poles after assembly on the rotor, to prevent coils bowing in service.

6.5.3.5 Strip-on-edge salient-pole windings

Large synchronous machines have strip-on-edge coils fitted on salient poles. Coils are normally made from bare copper strip wound on-edge or from strips of copper cut into lengths and brazed at the corners.

After producing the basic copper spiral and cleaning thoroughly to remove all loose particles, strips of interturn insulation are inserted. Normally two layers are used to give an effective thickness, after subsequent consolidation, of between 0.20 and 0.30 mm. Interturn insulation can be either B-stage synthetic resin-treated asbestos, polyester, glass or aramid paper. Coils are normally consolidated under heat and pressure. Thereafter, they are either mounted on poles insulated with an appropriate thickness of tough insulation or mounted on insulated spool bodies (Figure 6.24). An alternative approach is to insert dry synthetic paper between turns, clamp the coil to shape and then carry out a full vacuum pressure impregnation treatment.

Completed coils are mounted on the rotor with additional packings as necessary; V-blocks are often inserted between adjacent coils in the interpolar space.

6.5.3.6 Testing

During coil manufacture and after mounting on poles, appropriate interturn and high-voltage tests are carried out to check the integrity of the insulation system (see Section 6.6).

6.5.4 Direct current armatures

6.5.4.1 General

On d.c. armatures, in addition to the main winding, there are several other components whose performance is dependent on the satisfactory operation of their insulation. In addition to the commutator assembly, many medium and large d.c. armatures are fitted with equalizer windings.

In the design of insulation systems for d.c. armatures there is a need to consider the effect of contamination of all possible creepage paths and surfaces by brush dust and other airborne conducting particles. Problems associated with low insulation resistance of armatures in service have impressed upon machine manufacturers the need to produce sealed windings with good final varnish treatments which prevent ingress of contaminants.

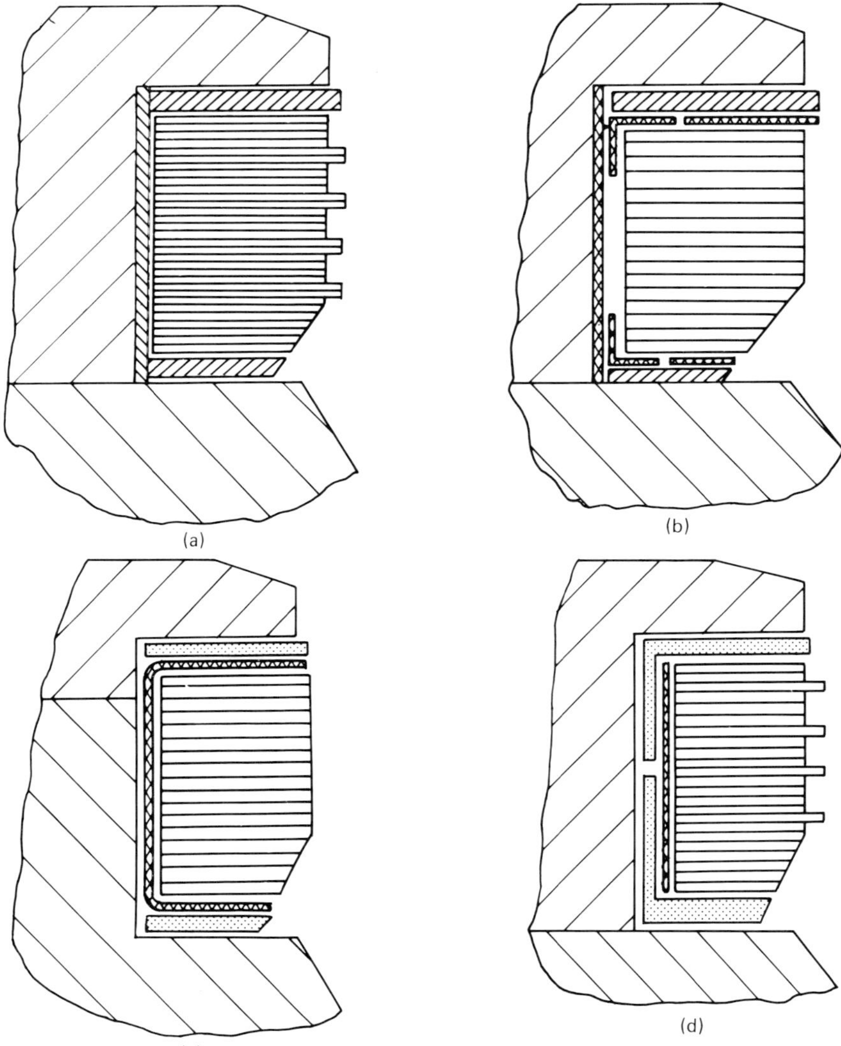

Figure 6.24 Insulation arrangements for strip-on-edge field coils for a.c. salient pole rotors

6.5.4.2 Winding types

Most small industrial d.c. machines have mush-wound armatures. As the rating increases, a point is reached where the design parameters dictate that multi-turn form-wound coils should be used. At higher ratings still, the voltage and current values make the use of single-turn coils or bars the best technical and economic solution.

6.5.4.3 Mush windings

Mush-wound coils for small d.c. armatures are wound using the same type of enamel-covered round wire as used for a.c. mush coils. A common difference is the practice of overtaping coils with a layer of insulation prior to winding into open slots in the armature core. Each mush coil often consists of several conductors wound in parallel. The leads of each of these effectively independent coils are connected to separate pairs of segments. Coils are usually wound into slots lined with tough, flexible material. Separating-strips and wedges of rigid laminate are used to hold them firmly in their slots. End-windings are packed with sheets of semi-rigid material.

End-windings are bound-down to insulated support flanges, where fitted, with several turns of steel wire or unidirectional glass banding tape. Completed armatures are finally treated either by dipping in synthetic varnish, by trickle impregnation, or by vacuum pressure impregnation.

6.5.4.4 Multi-turn coils

In multi-turn windings, individual coils are wound in the normal way as a flat loop which is subsequently pulled out into a diamond shape. Enamel or enamel-and-varnished glass lappings are used as the covering for conductors for Class B and F windings. For Class H machines, polyimide film or aramid paper-taped conductor is often adopted. Where the number of conductors per turn or turns per coil is such that physical handling could present problems, it is common practice to consolidate the slot portions under heat and pressure prior to coil insertion. In some designs, end-windings are taped with one or more layers of insulation before winding into slots lined with an appropriate tough insulation. On other designs, end-winding and main slot insulation is applied prior to winding. Main ground insulation can be either of B-stage resin-rich or low bond content mica tape. Coils are wound into the armature with additional side packings or slot liners as necessary to ensure a tight fit in the slot width. Slot separators and wedges are made from appropriate rigid insulation. After connecting the coil leads to the commutator risers by soldering, brazing or t.i.g. welding, end-windings are packed between top and bottom layers in a similar way to that used for mush windings. Binding bands and final treatment are carried out as for mush windings.

6.5.4.5 Single-turn winding

Single-turn coils are wound as for multi-turn ones except that, because of the larger cross-section, conductors are of bare copper strip. Each conductor is taped with woven glass or similar material. Service interturn voltage is low, e.g. 40/50 V r.m.s. To improve the space factor on medium-sized industrial machines, only alternate conductors are taped. On large armatures, where the copper cross-section can be

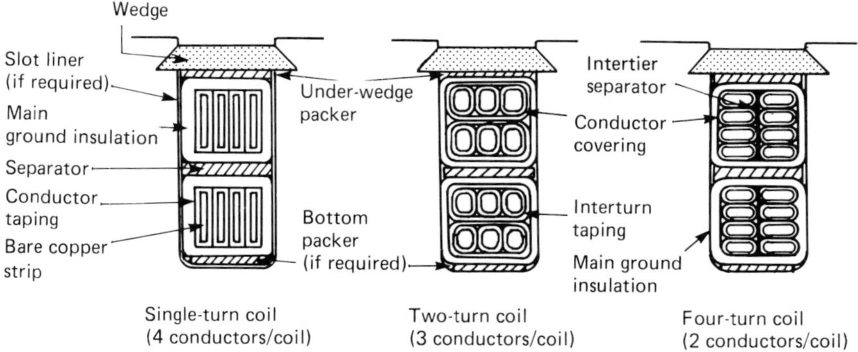

Figure 6.25 Slot build ups for d.c. armature coils

20–25 mm deep by 1–2 mm wide, bar windings are used. Each group of single-turn coils required per slot is normally assembled together and ground insulation applied. Where consolidated insulation is considered necessary for the higher voltage armatures, then tapes based on mica are used. Material can be either of the B-stage resin-rich or low bond content type. Where the former is used, coils or bars have the slot portions consolidated under heat and pressure to specified dimensions prior to winding. Slot liners are not used normally with pre-insulated coils except where there is excess space in the slot width. Where a group of single-turn coils are assembled together with only a single binding layer of tape applied to hold them together, a slot liner system is used. For Class B and F systems, laminated polyester or aramid paper liners are normally used. During winding, slot separators, wedges and end-winding packings are inserted as necessary. Binding bands and final treatments are similar to those used for mush windings. Typical slot build-ups are shown in Figure 6.25.

6.5.4.6 Equalizer windings: general

Three basic types of equalizer winding are in common use: (1) on medium-sized armatures, cable-type equalizers are fitted behind the commutator; (2) on large machines, coil-type equalizers are mounted on a flange close to the commutator and connected to either segments, or risers, whichever is more convenient – in many designs, equalizer coils are assembled on the inner commutator flange, i.e. between the commutator and armature core; (3) an alternative arrangement is to use ring-type equalizers – this type is normally assembled under the armature back-end flange. Copper strip leads on the rings are connected to either the evolutes on coil windings or bar-to-bar connections on bar windings.

6.5.4.7 Cable-type equalizers

This is the simplest type of equalizer winding and consists normally of cable links between commutator segments two pole pitches apart. Because such equalizers tend to be bulky, it is normal practice only to use them where equalization is acceptable every two, three or four slots. Equalizer links are arranged around the inside of one of the commutator flanges and connected to tabs soldered or brazed to commutator segments at appropriate intervals. Groups of cable links are taped together with woven glass or polyester tape to form a harness which is clamped to the flange. The cable covering must be strong enough mechanically to withstand the centrifugal forces produced in service. Cables insulated with butyl/CSP, CSP/ETFE rubbers are used. Cables insulated with silicone rubber are unsatisfactory because the covering tends to creep under load.

6.5.4.8 Coil-type equalizers

Coil-type equalizers are formed as U-shaped hairpins which are pulled out to the specified shape (Figure 6.26). Copper is covered with either enamel, enamel and glass fibre, or some other mechanically tough covering. After forming to shape, groups of two or more equalizers are often taped together as sub-assemblies. Equalizer groups are then mounted on the insulated commutator flange and connected to tabs fitted to the commutator segments or brazed to the risers. In some designs, the equalizer coils are fitted into the bottom of the slots milled in the commutator segments for the main armature coil leads. After assembly on the flange, equalizer windings are packed between top and bottom layers and between adjacent evolute portions with appropriate tough insulation. Thereafter, they are normally banded to prevent movement in service.

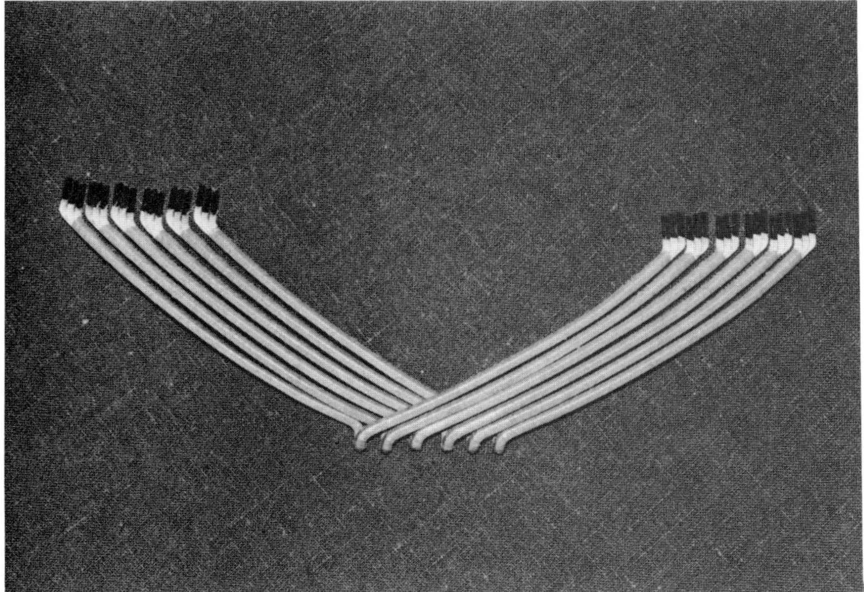

Figure 6.26 Coil-type equalizers for d.c. armature

6.5.4.9 Ring-type equalizers

Ring-type equalizers are often formed from bare copper strip. Copper tab leads are brazed to each ring at the appropriate positions. Each ring is taped with one or more layers of tough insulating tape. Groups of equalizer rings are taped together to form sub-assemblies which are then fitted underneath the back-end flange. Appropriate packings and clamps are next fitted to hold the assembled winding in place and prevent movement in service. The tabs on each ring are either soldered or brazed to the main armature winding at the back-end evolute position. Equalizer-to-coil connections are finally insulated as appropriate. A typical ring-type equalizer assembly is shown in Figure 6.27.

6.5.5 Direct current field windings

6.5.5.1 General

Field windings on d.c. frames are of two basic types: (1) wire-wound; or (2) strap-wound. Which type is adopted for a specific design depends on the voltage and current parameters of the supply. The only significant difference in the design and manufacture of coils for medium and large d.c. machines is one of scale. On small machines, thickness of ground insulation is defined by the electrical stress to which the coil will be subjected: (1) during manufacture; and (2) in service. On large machines the thickness of the main insulation depends to a great extent on the need to make the finished coil sufficiently robust to permit handling during manufacture, installation and service.

6.5.5.2 Wire-wound field coils

Field coils are normally wound on a faceplate winder. Coils are wound on to either a temporary former or spool body fabricated from sheet steel or moulded insulation. Today, the majority of coils are wound with either polyester imide- or

Figure 6.27 Ring-type equalizer for d.c. armature

polyester imide/amide-coated round or rectangular wire. These types of enamel are suitable for continuous operation up to Class F temperature limits for medium-sized machines.

Some grades are capable of operation at 200°C for 20 000 h and, hence, can be used for Class H applications. For large machines where mechanical toughness is important, either a varnished glass or polyester imide enamel plus varnished glass-covered conductor is used. The latter covering is often used for fields supplied from rectified a.c. where voltage spikes may occur in service. Each coil is wound with the specified number of turns of conductor. Coil leads for small machines often consist of insulated cable, whereas for high-current coils a strip of copper is brazed to the start of the conductor. During winding, synthetic resin may be brushed between layers to improve the overall mechanical strength of the finished coil. After winding the required number of turns and fitting the outer cable lead or terminal strip, coils are either clamped and baked on their formers or removed from them and the ground insulation applied. During manufacture of the latter type, lengths of woven glass or glass polyester tape are inserted along the coil sides to hold the turns in place. Where coils will be used in onerous operating conditions, such as on steel mill motors, they are often overtaped with one or more layers of varnished glass or resin-bonded micaceous tape.

Finally, the completed units are either dipped or vacuum pressure impregnated in synthetic varnish or resin, and baked. Alternatively, for less onerous duties, no overtaping is used and only the final treatment in varnish is applied. Wound coils are then assembled on pole bodies insulated with ground insulation consisting of either one or more thicknesses of laminated polyester or aramid paper. Flanges of laminate material of appropriate thickness are often placed on either side of the coil between the magnet frame and pole tip. Suitable means of location is necessary to prevent mechanical movement. On some designs, where coils are not treated prior to assembly, the complete pole and coil unit is treated either by immersion in varnish or using a vacuum pressure impregnation technique.

6.5.5.3 Strip-on-edge and strap-on-flat coils

In general, compole and series coils for medium and large d.c. machines are designed for low-voltage, high-current operation. Such coils are wound traditionally either as strip-on-edge or as strap-on-flat units. Typical interturn insulating materials are polyester or aramid paper, varnished glass or flexible mica sheet. Interturn voltage is low and, hence, thickness is decided on mechanical rather than electrical requirements. Depending on the design parameters, the bare copper width may be made equal to the full depth available for winding. On other designs, the strip may be wound as two, or four, coils side by side with appropriate copper strips brazed on to connect them together in series. Where multiple coils are used, insulation washers are inserted between them. Typical arrangements are shown in Figure 6.28. From a manufacturing point of view, strap-on-flat coils are relatively

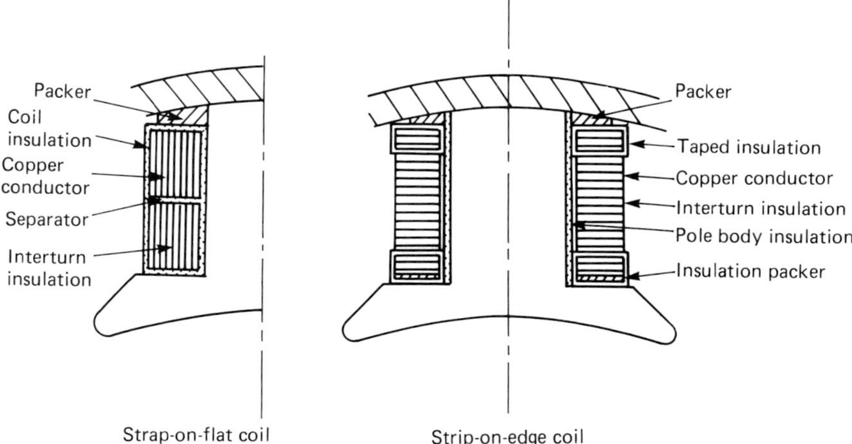

Figure 6.28 Insulation arrangements for d.c. strap-on-flat and strap-on-edge field coils

simple. From a design standpoint they have the disadvantage that the heat dissipation is not as efficient as with strip-on-edge coils. For this reason, in large machines where the output tends to be more critical, strip-on-edge coils are used. The latter can either be wound on edge to produce a coil with rounded corners at the pole ends or made up from strips of copper cut to appropriate lengths and brazed together at the four corners to form a rectangular coil. B-stage treated glass, aramid, polyester, asbestos or mica paper is inserted between adjacent turns. After winding, coils are normally consolidated under heat and pressure.

In some designs of strap-on-flat coils, 'mummified' main insulation is used. In others, and for the majority of strip-on-edge coils, ground insulation is applied to the pole. Coils are then assembled with top and bottom flanges of appropriate rigid insulation. Depending on the final duty, complete coils or pole–coil assemblies may be treated by dipping in varnish or submitted to full vacuum pressure impregnation.

6.5.5.4 Compound coils

In many small and medium machine designs, two or more field coils are required. These can be mounted either side by side or one on top of another. Insulating packers are inserted between them. Typical examples are shown in Figure 6.29.

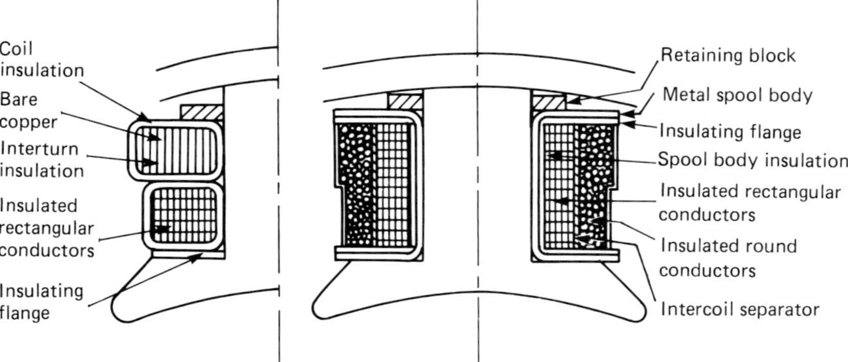

Figure 6.29 Insulation arrangements for d.c. compound fields

6.6 Testing of coils and windings

6.6.1 Functional testing of insulation systems

Before a new insulation system is adopted, it is important to confirm by adequate laboratory testing that it will perform in service for the design life. For all but the smallest machines, life testing of complete windings is too expensive. Assessment of a novel insulation system is carried out on models which simulate all the critical parameters that affect the service life of a specific machine winding. For example, for small a.c. machines the type of model, or functional testpiece, which is used to evaluate the conductor, interphase and ground insulation is known as a 'motorette'. A typical example is shown in Figure 6.30. Functional models are subjected to cycles of thermal, electrical, mechanical and environmental stress as appropriate for the winding. Electrical tests are used to check the integrity of the various components after each ageing cycle. Application of higher than normal stress accelerates the rate of ageing, such that models fail in a relatively short time in the

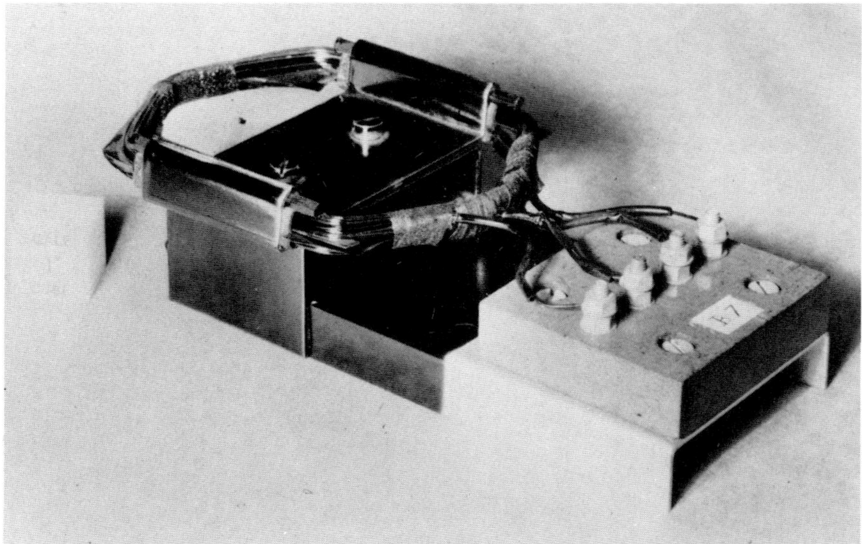

Figure 6.30 Motorette for testing low-voltage insulation systems

laboratory. From knowledge of the effect of the acceleration factors, the equivalent service life can be estimated. Functional testing of low-voltage integral kilowatt windings is well established and the results well proven. For high-voltage windings, i.e. $>1\,kV$, many machine manufacturers have developed their own 'in house' functional test procedures because of the difficulty of designing a universally acceptable model coil assembly which checks realistically all major factors affecting 'service life'.

6.6.2 Routine testing of coils

During production of a set of coils or other insulated components, non-destructive electrical tests are used to confirm insulation integrity. After coils are wound into the machine, similar tests are used to confirm that no damage has occurred during the assembly operation. When coils are tested first as individual units, then after insertion in the core, and finally, after connecting up to form the complete winding, it is sound practice to use test levels in a decreasing order of severity. For example, the final high-voltage acceptance tests for complete windings are defined in national standards. In the UK, these levels are laid down in BS 4999:Part 60. For an 11 kV motor the level is 23 kV for 1 min. Individual coils for such a winding are normally tested at 30% above this value prior to coil insertion. Before connecting up, an enhancement of 20% is commonly used. In addition to the standard high-voltage a.c. acceptance test, manufacturers normally carry out several other electrical checks to confirm the quality of insulated components. Some of the important ones are shown in Table 6.17.

Brief descriptions of these tests are given below.

(1) *Interstrand test.* On multi-turn coils where each turn is made up of two or more insulated conductors wound in parallel, it is necessary to check for short circuits between adjacent conductors. This test is normally carried out at 250 V. The

Table 6.17 Tests carried out on coils/windings

Test	a.c. stators		a.c. rotors		d.c. machines	
Machine type	*Low voltage*	*High voltage*	*Wound rotor*	*Salient pole rotor*	*Armature*	*Field*
Coils and bars:						
interstrand	√	√	—	—	√	—
interturn	—	√	√	√	√	√
loss-tangent	—	√	—	—	—	—
corona shield resistance	—	√	—	—	—	—
high voltage	—	√	—	√	—	√
insulation resistance	—	—	—	√	—	√
Wound stator/rotor:						
interstrand	—	√	—	—	—	—
interturn	—	√	—	—	√	—
loss-tangent	—	√	—	—	—	—
dielectric loss analyser	—	√	—	—	—	—
high voltage	√	√	√	√	√	√
insulation resistance	√	√	√	√	√	√
polarization index	—	√	—	—	—	—
electromagnetic probe	—	√	—	—	—	—
contact resistance (coil-to-core)	—	√	—	—	—	—

test circuit normally includes either a lamp (or buzzer) to give a visual (or audible) indication of shorts. Sub-conductors of single-turn coils and Roebel bars are also checked using the same procedure.

(2) *Interturn testing of high-voltage coils (line voltage $>1\,kV$).* The integrity of multi-turn a.c. stator coils is checked by applying a fast-fronted high-voltage surge across the terminal leads. Several types of surge test equipment are marketed. Most items of this type work on the principle of charging a capacitor to the specified test level and then discharging it through the coil under test. A damped oscillatory voltage with a fundamental frequency defined by the values of the test capacitor and coil inductance stresses the interturn insulation.

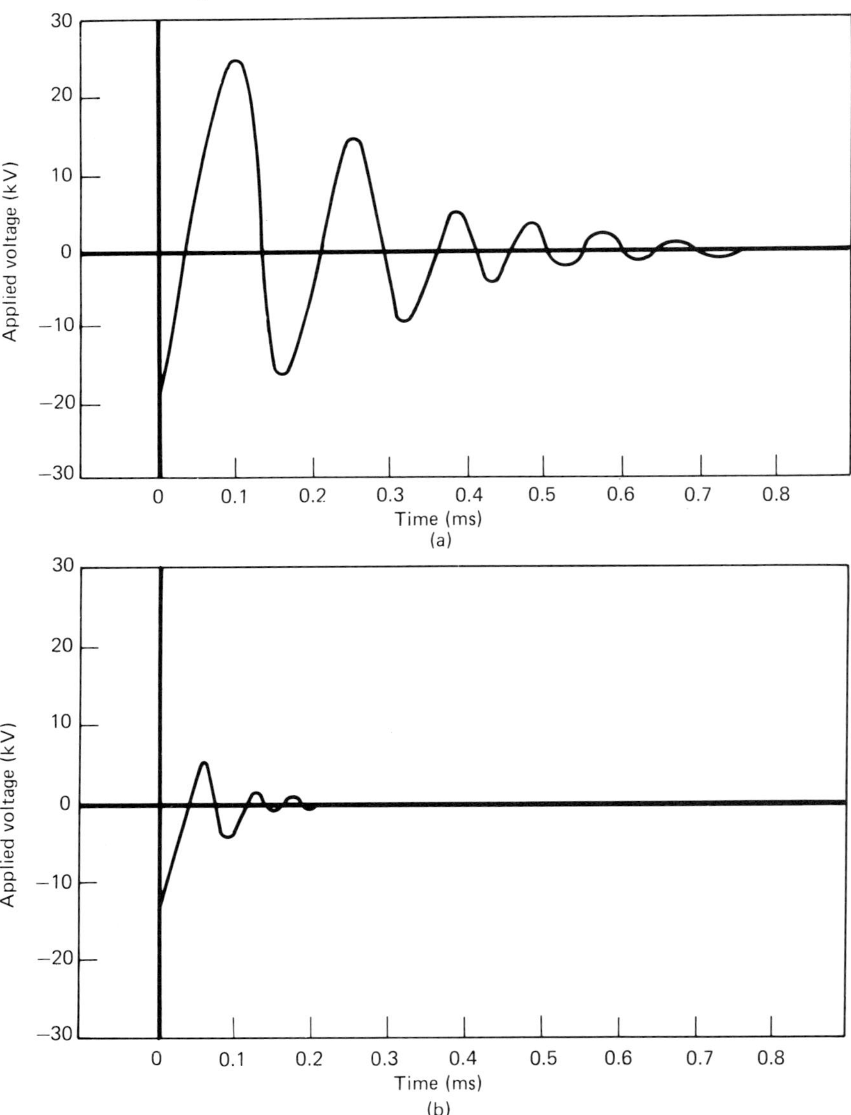

Figure 6.31 Typical voltage traces recorded when surge-testing a.c. stator coils. (a) Trace for a satisfactory coil, i.e. no interturn fault. (b) Trace for an unsatisfactory coil, i.e. with an interturn fault

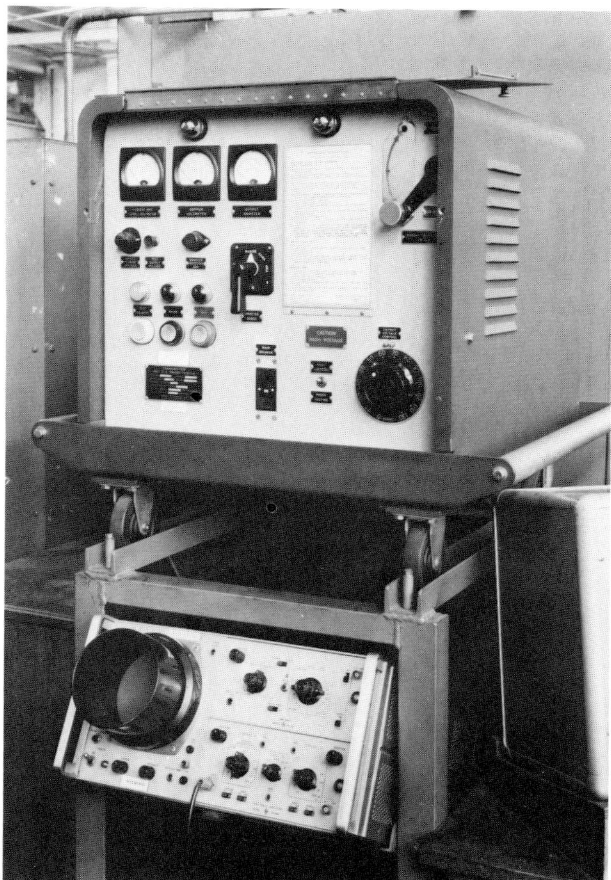

Figure 6.32 Surge tester (30 kV) for checking interturn insulation

Typical voltage traces for coils with and without faults are shown in Figure 6.31. A 30 kV surge tester is shown in Figure 6.32.

In the UK, the normal surge withstand levels for high-voltage a.c. motors are laid down in Electricity Supply Industry Standard ESI 44–5. Motors for the oil industry are specified in OCMA standard ELEC 1. Two levels are defined in the latter document: (1) a 'normal' level which is the same as in ESI 44–5; and (2) a 'special' level which is applicable to motors operating in plants where higher-than-normal surges are expected. The two surge levels for various line voltages are shown in Table 6.18.

Coils designed for vacuum pressure impregnated windings are normally tested at lower voltages than for resin-rich windings because the electric strength of unimpregnated insulation does not develop its full capability until after impregnation. To check that coils designed for global vacuum pressure impregnated windings will meet the levels required for complete machines, additional coils can be mounted in simulated slots, impregnated and then tested at the surge voltages designated in the appropriate standard.

Open-ended hairpin coils for concentric windings are interturn-tested prior to insertion in the stator. A 50 Hz a.c. voltage is applied between adjacent

Table 6.18 Surge and impulse withstand levels for high-voltage a.c. stator windings

Type of surge or impulse *Line voltage* U_n (kV)	*Winding withstand level when subjected to:*		*1.2/50 µs impulses*
	surges with wavefronts between 0.2 and 0.4 µs		
	Normal design level (kV)	*Special design level* (kV)	
3.3	8.1	13.5	18.2
4.16	10.2	17.0	21.6
6.6	16.2	26.9	31.4
11.0	26.9	44.9	49.0
13.8	33.8	56.3	60.2
Design criteria	$3\,\hat{U}_{ph}$ (kV)	$5\,\hat{U}_{ph}$ (kV)	$(4\,U_n + 5)$ kV

Notes:

(a) U_n = r.m.s. line voltage in kilovolts.

(b) $\hat{U}_{ph}$ = peak value of phase voltage in kilovolts, i.e. $\hat{U}_{ph} = \sqrt{2}/\sqrt{3}\ U_n$.

turns. After forming the hairpins into the concentric shape, making the turn-to-turn joints and applying the conductor, turn and main insulation surge testing is often repeated. The test procedure is basically the same as for multiturn coils.

(3) *Interturn testing of field coils.* Integrity of wire and strip-on-edge wound field coils for both a.c. and d.c. machines are checked by applying a medium-frequency a.c. high voltage, e.g. in the range 300–1500 Hz, across the coil leads. By monitoring the current taken by the test supply, coils with and without short circuits can be detected.

(4) *Loss-tangent test on stator coils.* On high-voltage a.c. stator coils, the slot insulation must be well consolidated and substantially void-free to prevent internal discharging taking place at the line-end of the winding in service. In practice, all insulation systems can be considered as imperfect capacitors. Even when completely void-free, a small leakage current flows through them. By measuring the capacitive and leakage currents, the phase angle between them can be calculated. The tangent of this angle, loss-tangent or tan δ, gives an indication of insulation 'soundness'. Most insulation systems are composites of several materials. In practice they, almost always, contain small voids and/or discontinuities. In the simplest example of a coil-side with a single void, the voltage distribution across the dielectric will be non-uniform due to the different permittivities of air and insulation. When a low voltage is applied, a proportion will appear across the void and the remainder across the insulation. As voltage is increased, a value is reached at which the air breaks down and internal discharging takes place. The void is then effectively short-circuited and the full voltage appears across the dielectric. A loss-tangent–voltage curve will show a step change at the voltage at which ionization starts. In practice, real coils contain many small voids. They are often located at different depths through the main insulation. Where many small voids are present, the loss-tangent–voltage curve shows a gradual change in slope with increasing stress. The value of loss-tangent at low voltage and the rate of change of loss-tangent with voltage give indications of the quality of the dielectric. Several specifications exist which lay down acceptance limits. In the UK the two commonly quoted standards are BS 4999:Part 144:1987 and ESI 44–5. In

addition to confirming whether or not voids are present, the loss-tangent–voltage curve also gives an indication of other factors such as degree of curve, presence of moisture and reproducibility of the manufacturing process.

Measurements of loss tangent are normally made on a Schering bridge. On production coils, to ensure that only the loss tangent of the slot insulation is measured, guard electrodes are normally fitted at each end of the slot portion. By using guards, all losses associated with the stress grading system and/or end-winding insulation are eliminated.

Measurements are normally made on resin-rich coils prior to insertion in the core. Typical acceptance limits are shown in Table 6.19.

Table 6.19 Standard acceptance limits for loss-tangent for high-voltage a.c. stator coils and bars

	Maximum acceptance value
Loss-tangent at $0.2\ U_n$	0.03
Change in loss-tangent between 0.2 and $0.6\ U_n$	0.005
Maximum increase in loss-tangent per $0.2\ U_n$ interval	0.005

Notes:

(a) U_n = r.m.s. stator line voltage in kilovolts.

(b) Loss-tangent readings are normally taken at intervals of $0.2\ U_n$ between $0.2\ U_n$ and $1.0\ U_n$.

The effect of the stress grading system can be established by measuring the loss-tangent–voltage curve of a coil-side with and without guards fitted. This depends on the coil voltage, type of stress grading and length of core portion relative to the length of stress grading. For this reason it is not realistic to lay down loss-tangent limits for unguarded coils.

Coils for global vacuum pressure impregnated stators cannot be tested usefully prior to impregnation. In practice, loss-tangent–voltage curves are made normally either on individual phases or complete windings after completing the vacuum pressure impregnation process. Because of stress grading, it is not possible to lay down universal test limits for complete windings. To check the relationship between guarded and unguarded vacuum pressure impregnated windings, additional coils, mounted in dummy slots, can be impregnated and subsequently tested. Results of loss-tangent tests on the dummy coils can be related to the values recorded on vacuum pressure impregnated windings provided the effect of the stress grading is known.

(5) *Corona shield resistance.* Coils for high-voltage a.c. stators, i.e. $>4.5\,\text{kV}$, normally have corona shields fitted. They are usually applied over the full core-plus-support finger length of each coil-side. The surface resistance of the shield is checked by wrapping a band of tin foil, or several turns of copper wire, around the main insulation at the two extremities of the conducting material. A 250-V Megger connected to the two electrodes is used to measure the shield resistance. From knowledge of the coil-side periphery and corona shield length, the surface resistance can be calculated. For satisfactory operation, the shield resistance should be in the range $1\text{--}30\,\text{k}\Omega/\text{square}$.

6.6.3 Routine testing of windings

During winding manufacture it is good quality-control practice to carry out dielectric checks at appropriate stages to confirm that critical aspects of the insulation system meet defined acceptance levels. For example, during the winding of a.c. stators, even although individual coils have been interturn-tested prior to assembly, it is good practice to carry out a further test after coils have been mounted in their slots, wedged and braced. The test procedure is carried out as described already in Section 6.6.2. Whether or not a test at this stage is a good quality-control procedure or is essential depends on a number of factors. Where coils are insulated with tough, resilient interturn materials, an 'in-stator' test may not be essential because the mechanical properties of the system preclude any likelihood of damage occurring during coil insertion. On the other hand, where coils have interturn insulation which is uncured at the assembly stage, such a test is essential. For global vacuum pressure impregnated stators, interturn testing after assembly, but prior to connecting into phases, is essential because of the consequences if an interturn fault should develop at a later stage of manufacture.

The same logic applies to high-voltage a.c. proof tests, where the tests carried out at various stages must of necessity be carried out at voltages greater than the British Standards final withstand level.

On completion, large high-voltage a.c. stator windings normally have loss-tangent tests carried out on individual phases and the complete winding. Measurements are usually recorded at 20, 40, 60, 80 and 100% of line voltage. Additional tests such as discharge energy readings (using a dielectric loss analyser), and electromagnetic or acoustic probe readings on individual slots, are specified often by discerning users of large electrical plant. The results give a fingerprint of the 'as made' new stator. By repeating the same tests every few years, the user can monitor the ageing rate and plan outages for rewinds in a controlled manner.

Appendix 6.1: Insulating materials – suppliers

Relevant section	Component	Supplier numbers (see Appendix 6.2)
6.2.2	Covered conductors	11, 20, 24, 38, 40, 41, 44, 50, 61, 67, 73
6.2.3	Tapes and flexible sheets	3, 4, 6, 9, 10, 14, 19, 23, 26, 27, 30, 32, 36, 38, 39, 40, 41, 42, 43, 46, 47, 51, 52, 54, 55, 57, 65, 66, 71, 72, 74
6.2.4	Rigid sheets, laminates and moulded insulation	3, 5, 6, 34, 35, 38, 40, 41, 42, 45, 48, 49, 69, 71, 72, 74
6.2.5	Sleevings	23, 42, 64, 65
6.2.5	Cables	1, 8, 12, 16, 22, 31, 38, 40, 43, 62, 70
6.2.6	Semi-conducting materials	2, 40, 41, 65, 71
6.2.7	Bracing tapes, cords and ropes	26, 32, 36, 42, 43, 55, 65, 66
6.2.8	Resins and varnishes, putties and adhesives	7, 13, 15, 18, 21, 25, 28, 29, 33, 37, 40, 46, 56, 58, 59, 60, 68, 71, 75

Appendix 6.2: Addresses of insulation materials suppliers – UK

Supplier number	Company	Address
1	AEI Cables Limited	Birtley, Chester-le-Street, Co. Durham, DH3 2RA
2	Acheson Colloids	Prince Rock, Plymouth, Devon, PL4 0SP
3	Anglo-American Vulcanized fibre	Delanco Works, Cording Street, London, E14 6GR
4	Aston Insulating Company Limited	72 Lord Street, Birmingham, B7 4DT
5	Atochem	Colthrop Lane, Thatham, Newbury, Berkshire, RG13 4NR
6	Attwater and Sons	PO Box 39, Hopwood Street Mill, Preston, Lancashire, PR1 1TA
7	BASF Coating and Inks Limited	Hayes Lane, Slinfold, Horsham, West Sussex, PH13 7SH
8	Batt Electrical Company	Church Street, Erith, Kent DA8 1PQ
9	Bayer UK Limited	Bayer House, Strawberry Hill, Newbury Berkshire, RG13 1JA
10	BDF Tesa Limited	Yeomans Drive, Blakelands, Milton Keynes, MK14 5LS
11	BICC Connollys Limited	Ellis Ashton Street, Huyton Quarry, Huyton, Liverpool, L36, 6BW
12	BICC Leigh Elastomeric Cables	Leigh Works, Leigh, Lancashire, WN7 4HB
13	BIP Chemicals Limited	PO Box 6, Oldbury, Warley, West Midlands, B69 4PD
14	Bondina Industrial	Greetland, Halifax, Yorkshire, HX4 8NJ
15	Bostik Limited	Ulverscroft Road, Leicester, LE4 6BW
16	Brand-Rex Company	Viewfield Industrial Estate PO Box 26, Glenrothes, Fife, Scotland, KY6 2RS
17	Britton-Tempo Limited	Sealex House, Ebberns Road, Hemel Hempstead, Hertfordshore, HP3 9RD
18	CIBA-Geigy (UK) plc	Plastics Division, Duxford, Cambridge, CB2 4QA
19	H. Clark and Company	Atlas Works, Patricroft, Manchester, M30 0RR
20	Concordia Electric Wire and Cable Company Limited	PO Box 1, Long Eaton, Nottingham, NG10 3LP
21	Corneluis Chemical	St James House, 27–43 Eastern Road, Romford, Essex, RM1 3NN
22	Crompton Parkinson Limited	Woodlands House, The Avenue, Cliftonville, Northampton, NN1 5BS
23	Croylek Limited	23 Ulleswater Crescent, Coulsden, Surrey, CR3 2UY
24	Delta Enfield Wires	Copperworks Road, Llanelli, Dyfed, SA15 2NH

Appendix 6.2: (Continued)

Supplier number	Company	Address
25	Dow Corning Limited	Avco House, Castle Street, Reading, Berkshire, RG1 7DT
26	Du Pont (UK)	Wedgewood Way, Stevenage, Hertfordshire, SG1 4QW
27	East London Mica Co.	Ringwood Road, London, E17 8PP
28	Emerson and Cuming	1 South Park Road, Scunthorpe, South Humberside, DN17 2BY
29	Epoxylite Corporation	Euroway Estate, Commondale Way, Bierley, Bradford, BD4 6SF
30	Euro-Tech Sales Limited	16 Haywra Steet, Harrogate, Yorkshire, HG1 5BJ
31	Fothergill Cables Limited	PO Box 3, Littleborough, Lancashire, OL15 8HG
32	Fothergill and Harvey	Summit, Littleborough, Lancashire, OL15 9PQ
33	Freeman Chemicals Limited	PO Box 8, Ellesmere Port, South Wirral, L65 0HB
34	GEC Power Engineering	Stafford Laboratory, Lichfield Road, Stafford, ST17 4LN
35	GEC Reinforced Plastics Limited	Mill Lane, Warton, Preston, Lancashire, PR4 1AR
36	Glass Tapes	Tean, Stoke-on-Trent, Staffordshire, ST10 4EA
37	Grilon (UK) Limited	Drumond Road, Aston Fields Industrial Estate, Stafford, ST16 3EL
38	Hi-Wire Limited	Ripley Drive, Normanton, West Yorkshire, WF6 1QT
39	ICI Petrochemicals and Plastics	PO Box 6, Bessemer Road, Welwyn Garden City, Hertfordshire, AL7 1HQ
40	Insulation Systems and Machines Limited	Wharfedale Road, Euroway Estate, Bierley, Bradford, BD4 6SG
41	Insulec (Electrical) Limited	Quarry House, High Street, Dyserth, Rhyl, Clwyd, LL18 6AA
42	Jones Stroud Insulations	Queen Street, Longridge, Preston, PR8 3BS
43	Kensulat Limited	Power Road, Chiswick, London, W4 5PZ
44	Kent Electric Wire Company	Kew Works, Bracknell, Berkshire, RG12 3BG
45	Lantec Limited	Faraday Road, Crawley, West Sussex, RH10 2PL
46	3M Corporation	3M House, PO Box 1, Bracknell, Berkshire, RG12 1JU

Appendix 6.2: (Continued)

Supplier number	Company	Address
47	Marling Industries	Tean, Stoke-on-Trent, ST10 4EA
48	Micanite and Insulators	Westinghouse Road, Trafford Park, Manchester, M17 1PR
49	Permali Gloucester	125 Bristol Road, Gloucester, GL1 5SU
50	Pirelli General	PO Box 4, Western Esplanade, Southampton, Hampshire, SO9 7AE
51	Polypenco	PO Box 56, 83 Bridge Road East, Welwyn Garden City, Hertfordshire, AL7 1LA
52	Primco Limited	Parr Lane, Unsworth, Bury, Lancashire, BL9 8LU
53	Raychem Limited	Faraday Road, South Dorcan, Swindon, Wiltshire, SH3 5HH
54	Rotunda	Holland Street, Denton, Manchester, M34 6GH
55	Rykneld Mills Limited	Bridge Street, Derby, DE1 3LH
56	Sandalin International Limited	63 Market Street, Stourbridge, DY8 1AQ
57	Scandura Limited	PO Box 18, Cleckheaton, Yorkshire, BD19 3UJ
58	Schenechady-Midland	Four Ashes, Near Wolverhampton, West Midlands, WV10 7BT
59	Scott Bader Company Limited	Wollaston, Wellingborough, Northampton, NN9 7RL
60	Shell Chemicals (UK) Limited	Shell Centre, Downstream Building, London
61	FD Sims Limited	Hazelhurst Works Lane, Ramsbottom, BL0 9PL
62	Sterling Greengate Cables Limited	Bath Road, Aldermaston, Berkshire, RG7 5QD
63	Sterling Varnish Company Limited	Fraser Road, Trafford Park, Manchester, MI7 1DU
64	Suflex Limited	Newport Road, Risca, Gwent, NP1 6YD
65	HD Symons	Horace Road, Kingston-upon-Thames, Surrey, KT1 2SN
66	TBA Industrial Products Limited	PO Box 40, Rochdale, Lancashire, OL12 7EQ
67	Thames Wire and Cable Company Limited	Brewery Road, Hoddesdon, Hertfordshire, EN11 8HF
68	Trimite Limited	Arundel Road, Uxbridge, Middlesex, UB8 2SD
69	Tufnol Limited	PO Box 376, Perry Barr, Birmingham, B42 2TB

Appendix 6.2: (Continued)

Supplier Number	Company	Address
70	Vactite	Hawthorne Road, Bootle, Merseyside, L20 6AE
71	Vulcascot Limited	43 Wales Farm Road, London, W3 6XH
72	James Walker and Company Limited	Lion Works, Woking, Surrey, GU22 8AP
73	Webster-Wilkinson Limited	Park Avenue, Madeley, Telford, Shropshire, TF7 5BP
74	BS and W Whitleys (1981) Limited	Pool Paper Mill, Pool in Wharfedale, Otley, Yorkshire, LS21 1RP
75	Winn and Coales	Denso House, Chapel Road, London, SE27 0TR

Bibliography

Insulating materials

GEORGE, H. and METZGER, L. 'Le papier de mica', Samica, *Revue Général de l'Eléctricité*, **59**, 13, 510–524 (1950)

BURNS, R. L. 'Survey of high-temperature polymers for insulation use.' *Insulation*, 67–70 (October 1968)

TRUNZO, F. F. 'Properties of thirteen high-temperature wire enamels,' *Insulation*, 42–51 (February 1970)

PARRISS, W. H. 'Material requirements for epoxy-novolak bonded high-voltage machine insulation,' *GEC J. Sci. & Technol.*, **38**, 4, 157–166 (1971)

ASHWORTH, E. and MURDOCH, C. O. 'Properties of mica paper as an electrical insulation material,' *GEC. J. Sci. & Technol.*, **41**, 4, 107–116 (1974)

FASBENDER, H. and SCHINDELMEISER, F. 'Duroplastic insulating materials in electrical machine construction,' *Kunststoffe*, **62**, 6, 351–354 (1972)

NARAHARA, T., KARASAWA, Y. and MUKAI, J. 'New heat-resistant solventless varnish,' *Hitachi Rev.*, **25**, 12, 417–421 (1976)

NEAL, J. E. 'Development of a Class H (180° C) modified epoxy resin glass-backed mica paper material for use in a.c. and d.c. rotating machines,' *IEEE Symposium on Electrical Insulation*, Montreal, pp.162–165 (1976)

STEINHAUS, R. 'New impregnating materials with a thermal rating of 155–180° C. *Proceedings, 13th Electrical/Electronic Insulation Conference*, Chicago, pp.241–245 (1977)

FERRO, R. H. 'Unsaturated polyester resins for vacuum pressure impregnation,' *Proceedings, 13th Electrical/Electronic Insulation Conference*, Chicago, pp.289–291 (1977)

ANDRES, W. H. 'Mica paper insulations for conductors and cables,' *Proceedings, 13th Electrical/Electronic Insulation Conference*, Chicago, pp.225–230 (1977)

SMITH, J. D. B. and BENNETT, A. I. 'Compatibility studies with mica bonds and solventless epoxy impregnants for form-wound stators,' *Proceedings, 13th Electrical/Electronics Insulation Conference*, Chicago, pp.202–205 (1977)

VIRSBERG, L-G and BJORKLUND, A. 'New type of turn insulation for medium-sized, high-voltage electrical machines,' *Proceedings, 13th Electrical/Electronic Insulation Conference*, Chicago, pp.199–201 (1977)

WAREHAM, W. W. 'Cut motor coil production costs with bondable magnet wire,' *Insulation/Circuits*, 49–51 (August 1977)

MILLER, H. G. 'Evaluation of tapes and wrappers for vacuum impregnation of high-voltage form-wound coils,' *Proceedings, 16th Electrical/Electronics Insulation Conference*, Chicago, pp.413–416 (1983)

ROBERTS, J. W. 'An advancement in mica paper vacuum pressure impregnation tapes,' *Proceedings, 16th Electrical/Electronics Insulation Conference,* Chicago, pp.417–421 (1983)

BERTSCH, X. and JACQUES, A. 'The evaluation of recently developed vacuum pressure impregnated mica paper tapes for high-voltage rotating machines,' *Proceedings, 16th Electrical/Electronics Insulation Conference,* Chicago, pp.422–424 (1983)

LANDRY, L. G. and LEVINE, H. R. 'Sleeving insulation for use on vacuum pressure impregnated form-wound coils,' *Proceedings, 16th Electrical/Electronics Insulation Conference,* Chicago, pp.425–429 (1983)

SMITH, J. D. B. 'Monitoring the storage stability of large volumes of vacuum pressure impregnating resins,' *Proceedings, 16th Electrical/Electronics Insulation Conference,* Chicago, pp.141–145 (1983)

WOJTECKI, P. E. 'New developments in the insulating of electrical conductors by tape wrapping,' *Proceedings, 16th Electrical/Electronic Insulation Conference,* Chicago, pp.131–135 (1983)

JIDAI, E., HASHIZUME, A., HIRABAYASHI, S., IMAI, H., KYUKA, M. and KUYAMA, K. 'Electro-deposited mica insulation systems.' *Proceedings, 16th Electrical/Electronics Insulation Conference,* Chicago, pp.136–140 (1983)

MITSUI, H., INOVE, Y., SEINO, S. and OHTA, S. 'Improvement of rotating machinery insulation characteristics by using mica paper containing aramid fibre,' *Proceedings, 16th Electrical/Electronics Insulation Conference,* Chicago, pp.651–656 (1983)

WOLF, C. J. and FANTER, D. L. 'Environmental degradation of aromatic polyimide insulated electrical wire,' *IEEE Trans. Elec. Insul.,* **E1-19,** 4, 265–272 (1984)

CAMPBELL, F. J. 'Temperature dependence of hydrolysis of polyimide wire insulation.' *IEEE Trans. Elec. Insul.,* **E1-20,** 1, 111–116 (1985)

KIMURA, K. and HIRABAYASHI, S. 'Improved potential grading methods with silicon carbide paints for high-voltage coils,' *IEEE Trans. on Elec. Insul.,* **E1-20,** 3, 511–517 (1985)

MITSUI, H., INOVE, Y. and YOSHIDA, H. 'Influence of mica tape application on insulation characteristics of high-voltage rotating machinery coils,' *IEEE. Trans. Elec. Insul.,* **E1-20,** 3, 619–624 (1985)

SCHMIDLIN, B. and BRANDENBERGER, K. 'Development of an impregnating resin having improved vacuum pressure impregnation properties,' *Proceedings, 17th Electrical/Electronics Insulation Conference,* Boston, pp.100–104 (1985)

INFANGER, F. 'The introduction of high-flashpoint varnishes,' *Proceedings, 17th Electrical/Electronics Insulation Conference,* Boston, pp.265–270 (1985)

LYNN, A. L., GOTTUNG, W. A., JOHSTON, D. R. and LA FORTE, J. T. 'Corona-resistant turn insulation in a.c. rotating machines,' *Proceedings, 17th Electrical/Electronics Insulation Conference,* Boston, pp.308–310 (1985)

TIMMINS, P. A. and MORGAN, J. 'Epoxy trickle impregnation with solventless resin *versus* dip and bake in solvent varnish,' *Proceedings, 17th Electrical/Electronics Insulation Conference,* Boston, pp.368–373 (1985)

SMITH, J. D. B. 'Low-viscosity cycloaliphatic epoxy impregnants for motor and generator insulation,' *Proceedings, 17th Electrical/Electronics Insulation Conference,* Boston, pp.105–109 (1985)

MARKOVITZ, W., GOTTUNG, W. A., McCLARY, G., RICHUTE, J., SCHULTZ, W. JR and SHEAFFER, J. 'General Electrics third generation vacuum pressure impregnation resin,' *Proceedings, 17th Electrical/Electronics Insulation Conference,* Boston, pp.110–114 (1985)

FRAME, R. I. and TEDFORD, D. J. 'Long-term electrical conduction in films of alkyd resin and graphite mixtures,' *IEEE Trans. Elec. Insul.,* **E1-21,** 1, 23–29 (1986)

Insulation systems

MOSES, G. L. 'Thermoalastic insulation improves high-voltage generator coils,' *Westinghouse Engr,* 163–165 (July 1950)

LAFOON, C. M., HILL, C. F., MOSES, G. L. and BERBERICH, L. J. 'A new high-voltage insulation for turbine-generator stator windings,' *AIEE Trans.,* Part 1, **70,** 721–730 (1951)

WOHLFAHRT, O., MORAVEC, M. and DOLJAK, B. 'Micadur, a new insulation for the stator windings of electrical machines,' *Brown Boveri Rev.,* **47,** 5/6, 352–360 (1960)

BLINNE, K., MADDER, O. and PETER, J. 'Orlitherm – a new high-tension insulation,' *Bull. Oerlikon,* 345, 37–60 (1961)

STAFF REPORT. 'Thermoalastic epoxy insulation for large a.c. motors,' *Westinghouse Engr,* 34–38 (March 1965)

JOHNSON, G. L. 'The trickle process method of impregnating motor windings,' *Proceedings, Electrical Insulation Conference,* Chicago, pp.98–99 (1967)

NYLUND, K. and MOLLER, W. 'Oerlikon insulation systems,' *Bull. Oerlikon,* 377/378, 11–23 (1967)

BRITSCH, H. and SCHULER, R. 'Micadur-compact, a fully impregnated stator insulation for high-voltage machines,' *Brown Boveri Rev.,* **54**, 9, 531–537 (1967)

ROTTER, H.-W. 'Micalastic insulation for high-voltage machines of up to medium rating,' *Siemens Rev.,* **36**, 1, 28–33 (1969)

NEAL, J. E. 'Development of resin-rich and impregnated insulation systems and their application to high-voltage electrical rotating machines,' *Proceedings IEEE 9th Electrical Insulation Conference,* Boston, pp.8–13 (1969)

SCHULER, R. 'Insulation systems for high-voltage rotating machines,' *Brown Boveri Rev.,* **70**, 1, 15–24 (1970)

SMITH, T. D. 'Electrical insulation for exacting conditions,' *Laurence Scott & Electromotors Engrg Bull.,* **11**, 3, 13–17 (1971)

NEAL, J. E. 'Machine taping of resin-rich mica paper to high-voltage stator coils for operation in all types of machines up to 30 kV,' *Proceedings, 10th Electrical Insulation Conference,* Chicago, pp.13–17 (1971)

GREENWOOD, P. and McNAUGHTON, H. S. 'Insulation – keynote to machine reliability,' *Energy Intnl,* **9**, 8, 18–22 (1972)

SCHULER, R. 'Experience gained with the micadur-compact insulation system,' *Brown Boveri Rev.,* **72**, 12, 577–581 (1972)

MATSUNOBV, K., ISOBE, S. and MUKAI, J. 'A new insulation experience in large rotating machinery in Japan,' *IEEE Trans. Elec. Insul.,* **E1-7**, 3, 132–139 (1972)

NYLUND, K. and GALLIKER, F. 'Class H insulation systems for traction motors,' *Brown Boveri Rev.,* **72**, 10/11, 514–525 (1972)

ROLSHOUSE, B. and ARNDT, D. 'Vacuum pressure impregnation – the key to void-free coil insulation.' *Proceedings, 11th Electrical/Electronics Insulation Conference,* Chicago, pp.114–117 (1973)

STABILE, S. J. 'Extension of the post impregnation concept to high-voltage windings,' *Proceedings, 11th Electrical/Electronic Insulation Conference,* Chicago, pp.117–122 (1973)

ROTTER, H.-W. and ECKBAUER, L. 'Micalastic insulation for high-voltage machines of small and medium ratings,' *Siemens Rev.,* **40**, 3, pp.99–105 (1973)

ROTTER, H.-W. 'Durignit 2000 insulation for standard-dimensioned motors,' *Siemens Rev.,* **41**, 3, 132–135 (1974)

McNAUGHTON, H. S., ORMEROD, G. and PARRISS, W. H. 'Sealed high-voltage machine insulation for arduous environments,' *Proceedings, BEAMA International Electrical Insulation Conference,* Brighton, pp.25–44 (1974)

AKI, F., MATSUNOBU, K., TANIGUCHI, M., ISOBE, S., KANEKO, K. and SAKAI, M. 'Super high-resin insulation system for stator coils of large rotating machines,' *Hitachi Rev.,* **23**, 3, 94–98 (1974)

NEAL, J. E. 'Current use in UK rotating machines of resin-rich mica paper,' *Proceedings, IEEE Winter Power Meeting,* New York (1974)

NEAL, J. E. 'A review of the resin-rich B-stage materials employed in high-voltage rotating machines with particular emphasis on their application and performance,' *Proceedings, 12th Electrical/Electronics Insulation Conference,* Boston, pp.135–139 (1975)

REJDA, L. J. 'Vacuum pressure impregnation of random wound motors,' *Proceedings, 12th Electrical/Electronic Insulation Conference,* Boston, pp.148–149 (1975)

SCHULER, R., GRIBET, J. and SCHEEL, J. 'Operational experience with high-voltage machines having totally impregnated windings,' *Brown Boveri Rev.,* **76**, 8, 508–512 (1976)

HAKAMADA, T. and MORINO, N. 'New Class H insulation systems for medium-capacity rotating machines,' *Hitachi Rev.,* **25**, 12, 393–398 (1976)

MEYER, H. and PAUFLER, H. 'Micalastic insulation for large d.c. machines,' *Siemens Rev.,* **44**, 12, 559–563 (1977)

SCHULER, R. 'Experience with post-cured insulation systems for large form-wound machines,' *Proceedings, 13th Electrical/Electronic Insulation Conference,* Chicago, pp.220–224 (1977)

ROTTER, H.-W. 'Reliability of low-to-medium-rated high voltage machines with Micalastic insulation,' *Siemens Rev.,* **45**, 11, 479–482 (1978)

ROTTER, H.-W. 'Thermidur – an insulating system for high temperatures,' *Siemens Rev.,* **45**, 11, 483–485 (1978)

GLEW, C. N. and NURSE, J. A. 'Thermalife the new Class F insulation concept for low-voltage stator windings,' *GEC J. Ind.*, **4**, 2, 57–66 (1980)

AMEY, W. and ROTTER, H.-W. 'The Micalastic insulation system in low- and medium-rated high-voltage motors for operation under extreme conditions,' *Siemens Power Engng*, **2**, 5, 136–139 (1980)

McNAUGHTON, H. S. 'Which system – resin-rich or vacuum pressure impregnation? – a review of insulation systems for high-voltage a.c. industrial machines,' *GEC J. Ind.*, **5**, 1, 37–43 (1981)

JONESON, K. 'Micapact II – coils for high-voltage rotating machines,' *ASEA J.*, **54**, 2, 27–32 (1981)

RATHMAN, B. 'Manufacturing of Micapact II coils,' *ASEA J.*, **54**, 2, 42–48 (1981)

McNAUGHTON, H. S. and NURSE, J. A. 'Vacuum pressure impregnation and resin-rich insulation systems for high-voltage industrial machines – a comparison,' *IEE International Conference on Electrical Machines – Design and Application*, London (1982)

NEAL, J. E. and WHITMAN, A. G. 'Insulation of high-voltage rotating machines,' *IEE International Conference on Electrical Machines – Design and Application*, London (1982)

SKIPETRIOV, V. V., FINKEL, E. A. and TIGANINA, T. A. 'Electric-machine insulation based on polyimide film for surface under high-moisture conditions,' *Electrotekhnika*, **53**, 10, 33–34 (1982)

DALAL, M. V. and GANPATE, V. K. 'A new 275 class non-silicone micaceous insulation system,' *Proceedings, 16th Electrical/Electronic Insulation Conference*, Chicago, pp.32–35 (1983)

LISTER, G., LEFEBVRE, K. and KOHN, L. 'Epoxy mica mat Class F stator ground wall insulation by Canadian General Electric – manufacture and quality assurance,' *Proceedings, 16th Electrical/Electronics Insulation Conference*, Chicago, pp.152–156 (1983)

NURSE, J. A. 'Development of global impregnation systems for high-voltage industrial machines giving extended thermal life and enhanced switching surge protection,' *Proceedings, Indian Electrical Manufacturers' Association Conference*, Bangalore, India, Session 1, Paper 4, pp.23–32 (1983)

HUTTER, W., LIPTAK, G. and SCHULER, R. 'Micadur-compact insulation system for rotating high-voltage machines up to medium ratings – behaviour under extreme operating conditions,' *Brown Boveri Rev.*, 6/7, 294–298 (1984)

MULLER, F. 'Vacuband – high-voltage insulation,' *Elin J.*, 3/4, 99–101 (1985)

NURSE, J. A. 'Resivac – an insulation system with extended life,' *GEC Rev.*, **2**, 2, 111–116 (1986)

SECCO, M., BRESSANI, M. and RAZZA, F. 'Progress and development trends in the large induction motor stator winding insulation,' *International Conference on the Evolution and Modern Aspects of Induction Machines*, Turin (1986)

Testing of coil and winding insulation

BHIMANI, B. V. 'Very low-frequency high potential testing,' *AIEE Trans.*, **54**, 148–155 (1961)

VIRSBERG, L.-G. 'Puncture risks when testing large synchronous machines with direct voltage,' *Proceedings, Inst. Elec. Engrs*, **110**, 9, 1637–1639 (1963)

KULL, V. and SCHULER, R. 'High direct voltage for testing the winding insulation of high-voltage machines,' *Brown Boveri Rev.*, **53**, 9, 531–537 (1966)

VIRSBERG, L.-G. 'The high-voltage testing of large electrical machines, particularly at 0.1 Hz/s,' *ASEA J.*, **40**, 5, 71–75 (1967)

OLIVER, J. A., WOODSON, H. H. and JOHNSON, J. A. 'A turn insulation test for stator coils,' *IEEE Trans. on Power Apparatus and Systems*, **PAS-87**, 3, 669–678 (1968)

SCHULER, K. 'Direct-voltage testing of the insulation of windings for rotating machines,' *Brown Boveri Rev.*, **55**, 4/5, 208–214 (1968)

SCHWARZ, K. K. 'Performance requirements and test methods for high-voltage a.c. motor insulation,' *Proceedings, Inst. Elec. Engrs*, **116**, 10, 1735–1743 (1969)

KRANKEL, D. and SCHULER, R. H. 'A method for checking the turn insulation of form-wound coil windings for high-voltage rotating machines,' *Brown Boveri Rev.*, **57**, 4, 191–196 (1970)

SCHULER, R. H. 'Production tests for high-voltage machines with fully impregnated windings,' *Brown Boveri Rev.*, **63**, 8, 513–516 (1976)

PHILLIPPS, W. 'The electrical stressing of interturn insulation on high-voltage motors during switching operations,' *AEG Telefunken*, **67**, 4, 195–199 (1977)

McDERMID, W. 'Review of the application of the electromagnetic probe method for the detection of partial discharge activity in stator windings,' *CEA International Symposium on Generator Insulation Tests*, Montreal, pp. 47–50 (1980)

SCHULER, R. H. and LIPTAK, G. 'A new method for the high-voltage testing of field windings (interturn insulation) on large rotating electrical machines,' *International Conference on Large High-Voltage Electric Systems*, Paris (1980)

CORNICK, K. J. and THOMPSON, T. R. 'Steep-fronted switching voltage transients and their distribution in motor windings: Part 1, System measurements of steep-fronted switching voltage transients; Part 2, Distribution of steep-fronted switching voltage transients in motor windings,' *Proceedings, Inst. Elec. Engrs*, **129**, Part B, 2, 45–63 (1982)

MEYER, H. and POLLMEIER, F. 'Transient stresses on the interturn coil insulation of electrical machines with power ratings above 1 MW,' *Siemens Power Engrg*, **4**, 3, 152–157 (1982)

MEYER, H. and WICHMANN, A. 'Experience and practice with standardized acceptance test procedures for windings of rotating machinery,' *Proceedings, 16th Electrical/Electronics Insulation Conference*, Chicago, pp.146–151 (1983)

WRIGHT, M. T., YANG, S. J. and McLEAY, K. 'General theory of fast-fronted interturn voltage distribution in electrical machine windings,' *Proceedings, Inst. Elec. Engrs*, **130**, Part B, 4, 245–256 (1983)

WRIGHT, M. T., YANG, S. J. and McLEAY, K. 'The influence of coil and surge parameters on transient interturn voltage distribution in stator windings,' *Proceedings, Inst. Elec. Engrs*, **130**, Part B, 4, 257–264 (1983)

CORNICK, K. J. 'Steep-fronted impulse voltage testing of machine stator coils,' *Proceedings, 17th Electrical/Electronics Insulation Conference*, Boston, pp.4–8 (1985)

REYNOLDS, P. H. and LESZCZYWSKI, S. A. 'Direct current insulation analysis – a new and better test method,' *IEEE Trans. Power Apparatus and Systems*, **PAS 104**, 7, 1746–1749 (1985)

RHUDY, R. G., OWEN, E. L. and SHARMA, D. K. 'Voltage distribution among the coils and turns of a form-wound a.c. rotating machine exposed to impulse voltages,' *IEEE Trans. Energy Conversion*, **EC-1**, 2, 50–60 (1986)

SCHWARZ, K. K. 'Transient switching overvoltages and compatibility of rotating machine insulation.' *IEE Power Engrg J.*, **1**, 1, 43–48 (1987)

Ageing of insulation and system testing

MOSES, G. L. Synthetic insulation and the 10° rule,' *Westinghouse Engr*, **5**, 106–107 (1945)

DAKIN, T. W. 'Electrical insulation deterioration treated as a chemical rate phenomenon,' *AIEE Trans.*, **67**, 113–118 (1948)

MAUGHAN, C. V., GIBBS, E. V. and GIAQUINTO, E. V. 'Mechanical testing of high-voltage stator insulation systems,' *IEEE Trans.*, **PAS-89**, 8, 1946–1954 (1970)

WICHMANN, A. 'Functional evaluation of high-voltage insulation on large electrical machines, *ETZ-A*, **92**, 7, 429–433 (1971)

WICHMANN, A. and GRUNEWALD, P. 'Durability investigations on high-voltage insulation – reducing the test period by increasing the test frequency,' *ETZ-A*, **95**, 6, 318–322 (1974)

INSTITUTION OF ELECTRICAL CONTRACTORS. *Guide for evaluation and identification of insulation systems of electrical equipment.* IEC publication No.505 (1975)

WICHMANN, A. and GRUNWALD, P. 'The influence of thermal and mechanical stress on voltage endurance of epoxy-mica insulation,' *ETZ-A*, **97**, 8, 470–474 (1976)

SCHULER, R. H. and LIPTAK, G. 'Long-time functional tests on insulation systems for high-voltage rotating machines,' Paper 15–05, *CIGRE International Conference on Large High Voltage Electric Systems*, Paris (1976)

CARLIER, J., GALAND, J., GRIBET, J., GUYETAND, A., HEROUARD, M., PANNI, S. and PAS DE LOUP, M. 'Ageing under voltage of the insulation of rotating machines – influence of frequency and temperature.' Paper 15–06, *CIGRE International Conference on Large High-Voltage Electric Systems*, Paris (1976)

MAILFERT, R., THORIS, J., RIVIERE, D. and PARGAMIN, L. 'Effect of the superposition of electrical, mechanical and environmental stresses on the fatigue behaviour of composite insulating materials,' Paper 15–10, *CIGRE International Conference on Large High Voltage Electric Systems*, Paris (1978)

KELEN, A. 'Ageing of insulating materials and equipment insulation in service and in tests,' *IEEE Trans. Elec. Insul.*, **E1-12**, 1, 55–60 (1977)

PARRISS, W. H. 'Evaluation of a.c. motor insulation systems by IEEE 117:water immersion or 100% r.h. as the conditioning cycle?' *Proceedings, 14th Electrical/Electronic Insulation Conference,* Boston, pp.90–95 (1979)

KELEN, A. 'Functional testing of Micapact II insulation,' *ASEA J.,* **54,** 2, 36–41 (1981)

MITSUI, H. N., YOSHIDA, K., INOUE, Y. and KALVAHAN, K. 'Mechanical degradation of high-voltage rotating machine insulation,' *IEEE Trans. Elec. Insul.,* **E1-16,** 4, 351–359 (1981)

FUTAKAWA, A., YANDSAKI, S. and KAWAKAMI, T. 'Deformation and fracture behaviour of stator windings subjected to bending loads,' *IEEE Trans. Elec. Insul.,* **E1-16,** 4, 360–370 (1981)

KAKO, Y., KADOTONI, K. and TSUKUI, T. 'An analysis of combined stress degradation of rotating machine insulation,' *IEEE Trans. Elec. Insul.,* **E1-18,** 6, 642–650 (1983)

RENGARAJAN, S., AGRAWAL, M. D. and NEMA, R. S. 'Accelerated ageing of high-voltage machine insulation under combined thermal and electrical stresses,' *IEEE Conference on Electrical Insulation and Dielectric Phenomena,* Bush Hill Falls, pp.129–134 (1983)

FUTAKAWA, A., FUKUNAGA, T., KAWAMURA, H. and MURAKAMI, A. 'Impact wear phenomena of ground insulation in slot section,' *Proceedings, 16th Electrical/Electronics Insulation Conference,* Chicago, pp. 40–45 (1983)

RAMU, T. S. 'On the estimation of life of power apparatus insulation under combined electrical and thermal stress,' *IEEE Trans. Elec. Insul.,* **E1-20,** 1, 70–78 (1985)

RENGARAJAN, S., AGRAWAL, M. D. and NEMA, R. S. 'Behaviour of high-voltage machine insulation system in the presence of thermal and electrical stresses,' *IEEE Trans. Elec. Insul.,* **E1-20,** 1, 104–110 (1985)

PARRISS, W. H. 'Thermal classification of wire-wound low-voltage machines,' *Proceedings, 17th Electrical/Electronics Insulation Conference,* Boston, pp.302–307 (1985)

Miscellaneous

JONES, E. 'Insulation of rotating electrical machinery,' *Proceedings, Inst. Elec. Engrs Pow. Engrg,* **100,** Part IIA, 3, 208–221 (1953)

VERVERKA, A. and CHLADEK, J. 'Corona discharge at the exit of winding slots of rotating machines,' *Elektrotechnicky Obzor,* **47,** 3, 122–129 (1958)

VERVERKA, A. 'Corona protection in slots of electrical machines,' *Acta Technica,* **4,** 6, 459–473 (1959)

ROBERTS, S. C. 'Review of motor insulations with special reference to breakdowns by discharges encountered in the mining industry,' *Trans. South African Inst./Elec. Engrs,* 166–189 (July 1961)

THIENPONT, J. and SIE, T. H. 'Suppression of surface discharges in the stator winding of high-voltage machines,' *CIGRE Report* No. 122 (1964)

SEIDLER, L. E. 'Rotor turn insulation, analysis and development,' *Proceedings, 10th Electrical Insulation Conference,* Chicago, pp.1–3 (1971)

JACKSON, R. J. and WILSON, A. 'Slot discharge activity in air-cooled motors and generators,' *Proceedings, Inst. Elec. Engrs,* **129,** Part B, 3, 159–167 (1982)

NAGANO, S., TARI, M., NAKANO, M., ITO, K. and UTO, Y. 'Early detection of excessive coil vibration in high-voltage rotating electric machines,' *IEEE Trans. Power Apparatus and Systems,* **PAS-101,** 6, 1551–1560 (1982)

CAMPBELL, J. J., CLARK, P. E., McSHANE, I. E. and WAKELEY, K. 'Strains on motor endwindings,' *GEC J. Ind.,* **7,** 1, 13–20 (1983)

STONE, G. C., GUPTA, B. K. and KURTZ, M. 'Investigation of turn insulation failure mechanisms in large a.c. motors. *IEEE Trans. Power Apparatus and Systems,* **PAS-103,** 9, 2588–2595 (1984)

CHABRA, O. P., ANAND, G. L., RAO, A. K. L. and KHARE, A. 'Study on the effect of clearance between adjacent coils/bars in the overhang of rotating electrical machines on discharge inception voltage, *Proceedings, Insulec Conference,* Bombay, Section III, Paper No. 7, pp.49–56 (1984)

SCHERER, K. 'Secure location of stator windings for rotating machines,' *Elin J.,* 3/4, 102–106 (1985)

SHEPPARD, H. R. 'A century of progress in electrical insulation 1886–1986,' *IEEE Elec. Insul. Mag.,* **2,** 5, 20–30 (1986)

Insulation
WHITEHEAD, S. *Dielectric breakdown of solids.* Oxford University Press, Oxford (1951)
CLARK, F. M. *Insulating materials for design and engineering practice.* John Wiley and Sons, New York (1962)
LEE, H. and NEVILLE, K. *Handbook of epoxy resins.* McGraw-Hill, Maidenhead (1967)
SILLARS, R. W. *Electrical insulating materials and their application.* Peter Peregrinus, London, IEE Monograph No. 14 (1973)
BRADWELL, A. *Electrical insulation.* Peter Peregrinus, London, Institute of Electrical Engineers, Electrical and Electronics Materials and Devices, Series 2 (1983)
GALLAGHER, T. J. and PEARMAIN, A. J. *High-voltage measurement, testing and design.* John Wiley, Chichester (1983)

7 Ancillary equipment

D. D. Stephen

Consultant. Formerly Senior Principal Engineer, Marine and Offshore
Division, GEC Electrical Projects Ltd.

Every motor needs some ancillary equipment to enable it to carry out its functions
correctly and safely. In its simplest form this consists of a switch to enable it to be
connected to and disconnected from its electrical supply but, in addition, protection
equipment is normally provided. Some motors require special starting equipment,
transformers, reactors or resistors and synchronous motors require an excitation
system. Other items may be provided for special forms of control, monitoring or
protection.

7.1 Main switch

7.1.1 General

All motors require some device to enable them to be connected to, or disconnected
from, the source of electrical power. The form this device takes obviously depends
on the type of motor, e.g. d.c. or a.c., and its supply voltage and current rating, but
many other factors must also be considered in selecting a suitable device.

7.1.2 Voltage

The insulation level between poles and also to earth must be suitable for the
maximum voltage which can exist in service, including not only machine-generated
overvoltages but also system-generated voltage surges due to normal switching
condition or abnormal conditions such as lightning strikes or cable faults.

7.1.3 Current

The thermal rating must be adequate for rated motor conditions, either continuous
or time-limited overloads or equivalent rating for duty cycle machines, allowing for
any derating resulting from environmental operating conditions such as high
ambient temperatures.

In addition the switch must withstand the mechanical effects of the maximum
current which it can pass in service.

These conditions must both be met while the switch is carrying its rated fault
current.

7.1.4 Transients

These include opening and closing duties, both normal and abnormal due to fault
conditions. Normal closing involves connecting a winding to a voltage and the
prospective initial current flow is not limited by inductance until current has

actually commenced to flow. This phenomenon can produce a very high magnetizing current of short duration and is particularly relevant when transformers are involved. Normal opening of the switch requires the interruption of current flow and the characteristic of the switch is particularly important for this since the system inductance produces a voltage dependent on the current being reduced. Some devices can produce serious overvoltages when interrupting low values of current and this switch characteristic must always be considered.

Abnormal conditions occur when a fault exists at the time the switch is closed or opened and this capability of fault level rating is one of the most important for any switch, and is normally proved by actual tests on sample switches.

Fault interruption is normally initiated by protective relays and, at the time the switch opens, the prospective peak fault current will have reduced slightly. However, when closing on to a fault it is possible that the peak condition will be encountered. The possible fault levels for the system to which a switch is to be connected must be within the proven capability of the switch under the specified operating conditions. The actual power factor of the system viewed from the switch at the time of interruption affects the amount of current which can be handled safely and this factor should always be checked when selecting a breaker fault rating for a given installation.

With some types of switch it is impractical or uneconomic to design it for full system fault level and it is then necessary to provide a back-up device such as the high-rupturing-capacity fuse which will blow in the event of a major fault, thus reducing the prospective fault level on the switch and so preventing the switch being damaged by attempting to interrupt a current beyond its capability.

7.1.5 Control

A switch, being a control unit, must also be capable of meeting all operating requirements. This may require local manual operation or manual initiation of power-assisted closing and/or tripping, either by electrical or mechanical means. Alternatively, it may have to be suitable for remote initiation of closing/tripping. All devices are not suitable for all options and particular control requirements must be considered when selecting a suitable switch.

Frequency of operation also has a direct effect on choice of unit and the starting duty of a motor must be specified to ensure that a suitable switching device is provided. Duties such as inching, reversing and plugging all have an effect on a suitable choice of switch.

7.1.6 Auxiliary devices

It is usual to provide some form of monitoring for all motor drives, and measuring devices such as shunts, current transformers and voltage transformers are usually provided in association with the switch. It is then convenient to include any transducers for use with these devices, as well as meters or instruments to indicate functions such as voltage, current, kilowatts, kilovars, etc.

Protection relays require similar monitoring devices and may also be conveniently included in this package. When the switch forms part of a control circuit, auxiliary contacts to indicate switch status and relay logic are required and this, too, can be included with the switch package.

7.1.7 Installation

Switches can be provided in a variety of enclosures, depending on the application and the environment. Originally, 'open construction' boards carried switches with free access front and back for easy maintenance, but 'dead front' boards are now in common use for safety reasons. This feature is easily obtained by providing front doors to open-type switches or, for example, by using integral 'moulded case' breakers.

To permit easy maintenance, withdrawable/plug-in units are now available in which the switch consists of two parts, a fixed structure with the permanent cable connections and a removable unit containing the actual switch contact unit. The removal may be by means of slide-out rails or a floor-mounted trolley, the former being convenient for units mounted in multi-tier arrangement. Such units may incorporate a partially removed state, which gives main circuit isolation but permits testing of the control system. A circuit earthing function may also be provided, using the switch to connect the circuit side to earth. These units are usually built as a metal-enclosed cubicle which can be integrated into a switchboard or motor control centre.

Although the motor switch can be mounted separately as a single device, such as for use with test machines, it more often forms part of a panel, cubicle or switchboard. It is convenient to use a dedicated motor panel or cubicle where there is a considerable amount of auxiliary equipment involved or where additional starting or excitation equipment is required for a particular motor. This single package can then be fully self-contained and dedicated to the one motor and can be located conveniently near the drive. When environmental conditions prevent such a location, the unit can be installed in a suitable space, e.g. a non-hazardous area, and be remotely controlled from a simple push button station near the drive.

Alternatively, when little or no auxiliary devices are required, it is usually cheaper and more convenient to include the switch as part of a switchboard from which the motor is to receive its electrical power. With such an arrangement, remote control is usually provided for the switch from a conveniently located simple stop/start unit which can be designed to provide essential controls and to be suitable for the environmental conditions where it is located.

Other requirements may, of course, justify other arrangements or groupings, e.g. switches with very high fault level ratings are invariably supplied integral with the associated switchboard, which is designed as a complete unit to meet all the system fault level requirements.

Where several motors are associated with a common drive or process, it may be convenient to group such switches and controls, etc. in their own suite of cubicles or panels, and in practice the optimum arrangement can only be decided after a consideration of the various requirements of the drive and capabilities of the various types of units available for use.

7.1.8 Types of switch

Because of recent trends towards increased safety of both operators and equipment, virtually all motor switches have a fault-interrupting capacity. In a few instances a simple manually operated knife switch is still in use, e.g. in test plants with very low fault levels where the equipment is operated at all times by skilled personnel, but even there it is now customary to have a back-up unit with breaking capacity. Such knife switches can therefore be regarded as isolators or part of a starter unit, and the real motor switch is the device with the interrupting capacity.

Industrial switches may be grouped loosely as contactors or circuit-breakers.

7.1.9 Contactors

These are electrically operated switches and are not designed for fault interrupt capacity. They are particularly suitable for use in factories, etc. where automatic processes require remote control, and they have been developed into very compact units which enable large numbers of drives to be located in a small space such as a

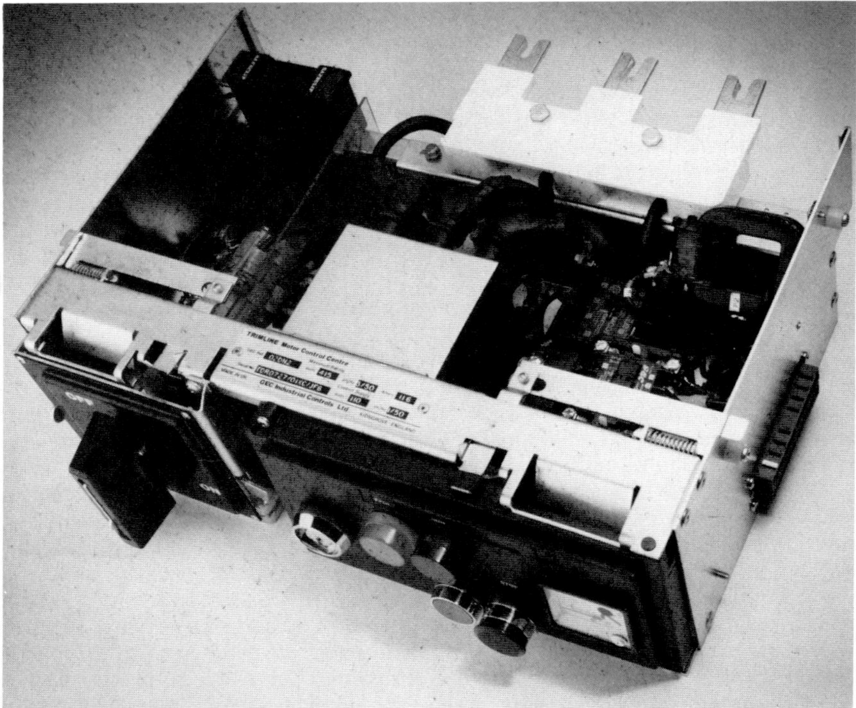

Figure 7.1 Typical motor contactor starter chassis incorporating all control functions. (*Courtesy:* GEC Industrial Controls Ltd)

motor control centre. These are self-contained and obtain their control power from their own electrical supply system (Figure 7.1).

The problem of fault-level protection can be dealt with by providing each motor contactor with suitably rated fuses or alternatively, for higher values of fault level, to provide a back-up switch to protect the complete motor control centre by interrupting the possible fault current. The latter solution is the more economic for small, lower-voltage motors but contactors with integral fault-level limitation fuses are convenient for larger, higher-voltage motors. All contactors are suitable for frequent operations and are basically designed to meet industrial requirements for motor drives (Figure 7.2).

7.1.10 Circuit-breakers

These may be manually, electrically or mechanically operated but must have an interrupting capacity. There is a wide range of type, rating and capability; new designs are currently being developed to cope with modern demands and the optimum choice may depend on several factors. Circuit-breakers are not designed specifically for motor applications, as they are used throughout the power system for generators, transformers, bus section and coupler switches at all levels of voltage. They are not basically intended for frequent operation and, when used with motors, the starting duty can be a significant factor in the choice of a suitable design of switch.

Circuit-breakers for most motors can be closed manually and held closed by a mechanical linkage. Tripping can then be done by releasing the linkage either

Figure 7.2 Typical motor control centre showing installed motor starters. (*Courtesy: GEC Industrial Controls Ltd*)

manually or by an electrical signal operating on a trip coil which acts into the linkage. The electrical signal can be in the form of removal of supply to the coil, i.e. undervoltage release, or by energizing the coil to trip, i.e. electrical tripping. Both systems have advantages and disadvantages and the optimum choice depends on the type of breaker and the application requirement of the motor being started.

When the manual effort required becomes limiting, or on breakers with a very high fault-level rating, either mechanical or electrical closing can be provided with an electrical initiation signal from the operator. Mechanical operation can be performed by charging a powerful spring, which can be compressed by an electrical motor prior to operation, and which is released to provide the closing effort required. Larger switches can be operated by compressed gas but such units are seldom required for motor applications. Electrical closing functions in a similar fashion but a solenoid is energized to provide the closing effort.

The energy required for closing breakers is high and requires a considerable power source. Switchgear batteries are usually provided where several breakers are combined on one board, but transformer/rectifier units can be used for individual breakers if this is more convenient. Separate sources of power are usually provided for closing and for tripping, since the tripping function must be given high integrity for safety reasons (Figure 7.3).

The process of current interruption varies with different types of circuit-breaker. Air-break switches interrupt current in free air and are similar in some ways to contactors except that they are provided with features to enhance their breaking capacities. In many installations this form of switch is acceptable and for low voltages and power it is in common use. With higher ratings and in difficult environments a more compact and efficient switch can be obtained by performing the interruption within a vacuum. This arrangement is available for contactors

Figure 7.3 Fuse-protected vacuum contactor equipment suitable for starting high-voltage motor and including control and protection. (*Courtesy:* GEC Industrial Controls Ltd)

which can have their fault-interrupting capability enhanced by fuses but which, of course, require the fuses to be replaced following a fault clearance operation. However, vacuum interrupters are available which can interrupt high fault levels without the disadvantages of replacing components following a fault clearance operation. Arc interruption in other media has similar advantages and both oil and gas, e.g. SF6, are in common use in circuit-breakers.

In addition to the apparently simple operation of making and breaking current, switches exhibit transient phenomena and effects such as single phasing and current chopping, and voltage restrikes can produce adverse effects on components on some configurations of system with particular types of switch. These factors must also be considered when selecting a suitable type of switch for any particular motor (Figure 7.4).

7.1.11 Conclusion

Manufacturers are continually improving existing designs and increasing limits of operation such as voltage, current, fault levels, etc. as well as producing new products. Thus, it is desirable to keep in touch with all currently available products before selecting a switch for a particular duty.

Figure 7.4 Moulded-case circuit-breaker with magnetic trip. (*Courtesy:* Merlin Gerin (UK) Ltd)

When a motor switch is being added to an existing installation, consideration should obviously be given to utilizing a unit similar to the existing switchboard, motor control centre, etc. but it must not be assumed that this will meet all the requirements of the new drive and all the factors listed above should be checked.

When a completely new project is being considered it is necessary to consider the total installation before deciding on the type and rating of switch to be used and it may be necessary to compromise on some features desirable for the motor application in order to get reasonable performance from switches required for other items of plant. When such a compromise is unacceptable, different forms of switch must be adopted, each selected for its own duty, and the choice of motor switch will involve consideration of all the above items in detail if an economic and safe installation is to be obtained.

7.2 Starting equipment

One of the most important functions of a motor is its starting capability. It must be able to break away and accelerate up to speed its coupled drive, by providing sufficient torque to overcome friction, load torque and inertia effects. It must be capable of doing this under all specified conditions of load and of the electrical supply system.

The inherent starting characteristic of a motor which may be ideally suited to the normal running duty does not always meet the operational starting requirements and in such cases it may be advantageous to incorporate equipment to modify suitably the motor behaviour during starting. The type of equipment used depends on the type of motor and the actual starting duty required but can be grouped conveniently as follows.

(1) *Alternating current motors:*
  (a) *full-voltage starting*, in which the motor windings are connected directly to the electrical supply system. This may be done with the windings connected in their normal running configuration and this is normally described as direct-on-line starting (DOL), which requires only a main switch, or changes may be made to the winding connections to change from starting to running conditions. These latter methods require winding reconnect switches in addition to the main switch.
  (b) *reduced-voltage starting*, in which auxiliary equipment is connected between the supply and the motor, to reduce the voltage at the motor terminals for starting conditions only.
  (c) *rotor-resistor starting*, in which auxiliary equipment is connected in the secondary winding to change the motor impedance during starting.
  (d) *special starting systems*, in which auxiliary devices are connected between the supply and the motor to provide special conditions of voltage and/or frequency to enable the motor to provide optimum starting conditions.
(2) *Direct current motors:* The above list for a.c. motors can be applied to d.c. motors with the exception of rotor resistance starting, but in practice full-voltage starting is not used.

  Normally, single motors used reduced-voltage starting produced by resistors connected in series with the supply; special drives such as large mill motors may employ a special starting system such as a Ward-Leonard or equivalent system.
  (a) *Alternating current motors, other than three-phase.* The vast majority of a.c. motors are for use on three-phase systems and the following descriptions refer specifically to these. The same principles can be used for motors with other numbers of phases, by using appropriate equipment.

7.2.1 Full-voltage starting

7.2.1.1 Direct on line

This is the simplest form of starting an a.c. motor since it only requires a main supply switch but, at the instant of energization, the motor has initially a very low impedance consisting essentially of winding resistance and leakage inductance and draws a heavy current from the supply. This can cause severe voltage drops in the

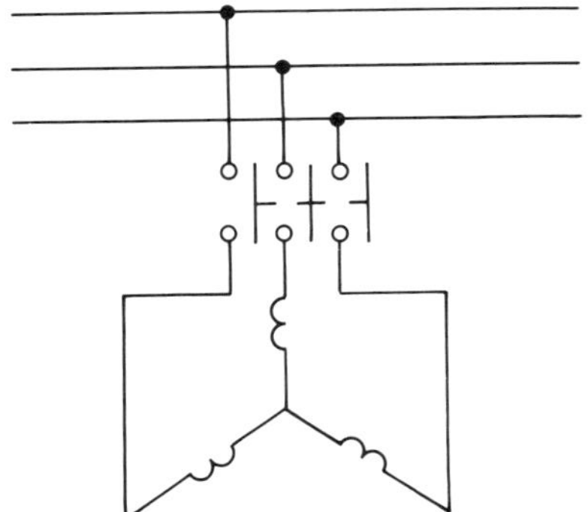

Figure 7.5 Direct-on-line starting

supply system, which may not be acceptable because of adverse effects on other connected plant, or it may impose excessive stresses on the motor windings or on the shaft system and coupled plant.

The voltage drop caused by DOL starting means that, in practice, the motor does not actually start at full voltage and, hence, it takes a lower current and produces less starting torque than it would if the voltage remained at its rated value. At this lower voltage the actual motor impedance may be greater than the actual rated voltage value, owing to reduced saturation of leakage inductances, and this will have a cumulative effect on current and torque. These factors must be included in the design margins of the motor to ensure it meets its starting duty under practical system operating conditions (Figure 7.5).

7.2.1.2 Winding reconfiguration

Star–delta starting.   This is a simple and popular means of reducing the current drawn from the supply at starting but the motor, of course, can then only produce a starting torque appreciably lower than that available with DOL starting.

The motor requires to have six terminals, i.e. both ends of each stator phase available, and must be designed for normal operation when delta-connected. When connected in-star for starting, the motor presents a high impedance to the supply (approximately 1.7 times the delta-connected value) and hence the current drawn at starting is about 57% of the DOL value, giving a torque about 30% of the DOL value. When the motor has accelerated, star-connected, it is disconnected from the supply, reconnected in delta and reclosed to the supply. During this open transition phase the motor windings continue to generate a decaying voltage but, due to deceleration, this does not stay synchronized with the supply and will probably be

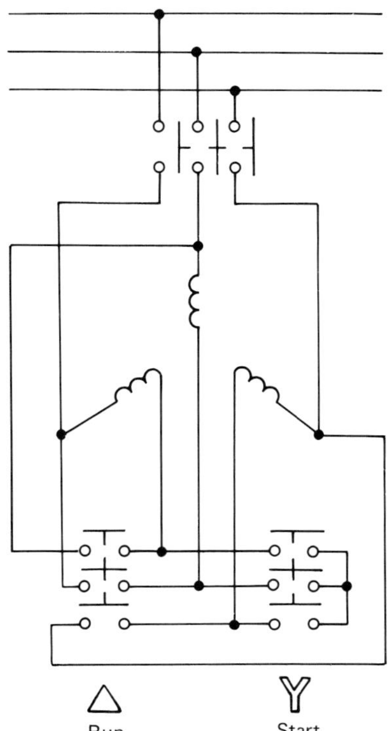

Figure 7.6 Star–delta starting

out of phase when reclosure occurs. This results in a switching surge of short duration but of a magnitude which may exceed the normal DOL switch-on current (Figure 7.6).

Part-winding starting.   When each phase winding has two or more parallel sections, the terminal impedance is increased if some parallel sections are left open-circuited. The ratios of starting torque and current to their DOL values depend on the amount of winding left in circuit; typical values are shown in Figure 7.7.

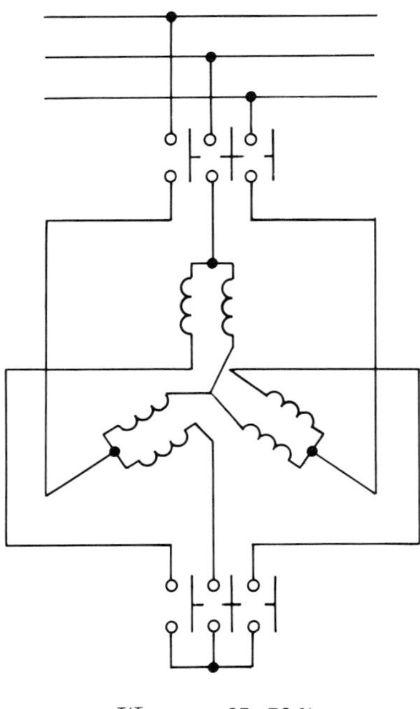

$$I/I_{DOL} = 65\text{--}70\,\%$$
$$T/T_{DOL} = 50\text{--}60\,\%$$

Figure 7.7 Part-winding starting

The starting equipment required is similar to that for star–delta starting in performance, but separate switches are required for each parallel section of winding to be reconnected.

7.2.2 Reduced-voltage starting

It is possible to increase the effective impedance of a motor without changing its winding configuration by either adding external impedance in series with it or by introducing a transformer between it and the supply. The impedance of a motor itself increases as it runs up to speed and is predominantly reactive for all a.c. machines. The use of a series resistor therefore does not give the same effective impedance increase as a reactor and, since it is required to dissipate more heat, it is normally only economic to use it for smaller motors where a cheap, standardized resistor is available which has an adequate rating for the duty, e.g. liquid resistors which are normally used for rotor control with slipring induction motors if they have insulated phase circuits.

Reactors are similar to transformers in construction and enclosure but are, in general, cheaper for the equivalent duty. The significant differences between reactor and transformer starting are that if they are designed to give the same voltage on the motor at standstill, say 50% of line voltage, the starting current drawn by the reactor is 50% of the DOL current but for the transformer it is 25%, while the motor torque in both cases remains the same at 25%. However, as the motor runs up to speed and its impedance increases, the percentage line voltage applied to the motor increases and the percentage of DOL torque also increases, whereas with the transformer the motor voltage remains the same throughout the run-up period (Figure 7.8).

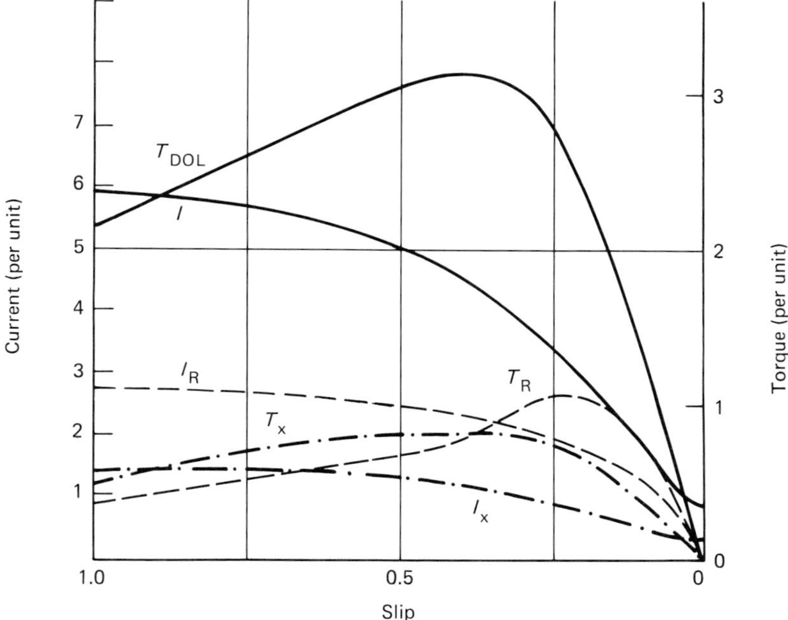

Figure 7.8 Typical comparison of synchronous motor with reactor and auto-transformer starting

Direct current motors utilize resistors in series with their armature circuit to reduce the voltage and can be supplied conveniently in the form of a multi-step faceplate starter or alternatively a drum-type device. Standard units are available for the more usual ratings of motors. The resistance steps should be graded to match the armature resistance to give optimum performance but, unless the starting duty is severe, special grading is seldom required.

7.2.2.1 Reactor starting

It is advantageous to connect the starting reactor at the neutral end of the motor windings as this minimizes the voltage applied to it and also maintains it and the reactor short-circuiting switch at neutral potential during normal motor operation. This arrangement requires a six-terminal motor, which is normal for larger machines, and it provides a simple cabling operation from the motor neutral box to the reactor and from there to the shorting switch (Figure 7.9).

Reactors for starting duties usually require a gapped core and have a non-linear relationship between their impedance and the current flowing through their windings. The actual value is subject to a design tolerance and it is usual to provide

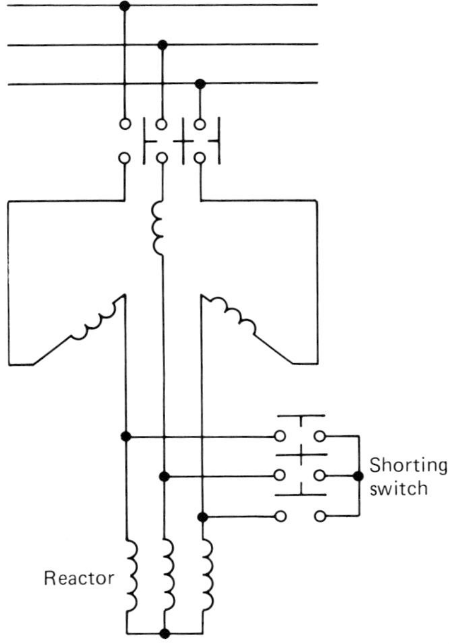

Figure 7.9 Reactor starting

alternative taps to ensure the motor–reactor combination gives the required starting current and torque. When the motor has accelerated to speed, the reactor is short-circuited and this operation produces a transient current dependent on the value of residual voltage across the reactor and the motor subtransient reactance. It is desirable to run up as far as possible using the reactor to minimize this effect and this can be monitored using the line current, which settles at a low value, or the speed which becomes constant. The reactor shorting switch must be rated for the starting current and the shorting current surge and also for any current which might flow if a fault occurred when it is closed in service.

Special multi-step reactors have been used when it was not practical to attain a sufficiently high speed with the full reactor in service, but each step requires a shorting switch and such a solution is not generally economic.

7.2.2.2 Autotransformer starting

When it is desired to use a transformer for motor starting it is not necessary to use a two-winding device and an autotransformer is used as this is more economical. Starting autotransformers are invariably provided with several tapping points to allow for adjustment of starting performance, if necessary. There are two distinct forms of starting: (1) open-transition; and (2) closed-transition.

Open-transition starting is arranged to use the main switch to connect the autotransformer to the supply with the motor already connected to the secondary tapping points. When the motor has accelerated to full speed, the motor is disconnected from the autotransformer and reconnected direct to the supply (Figure 7.10). This operation produces current surges as in star–delta switching, which may not be acceptable. It is also desirable either to disconnect the autotransformer primary from the supply or alternatively to open its neutral point which prevents current flowing unnecessarily. This additional switching is operationally desirable but involves extra cost.

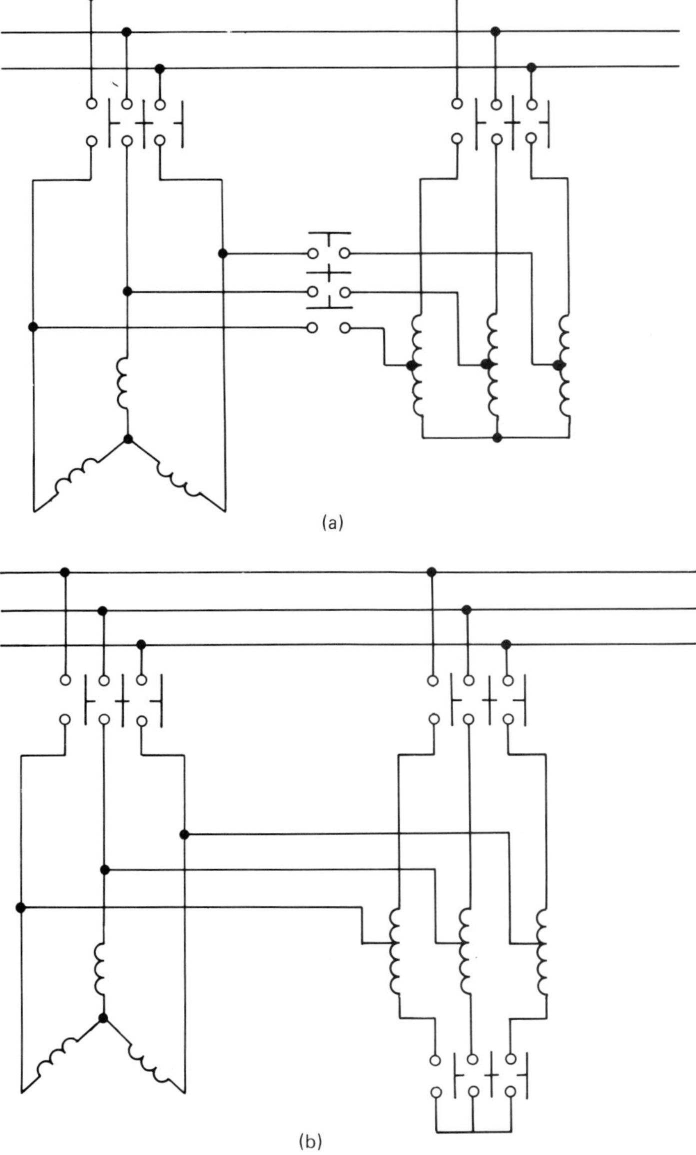

Figure 7.10 (a) Auto-transformer reactor open-transition. (b) Auto-transformer reactor closed-transition

Closed-transition starting (Korndorfer) overcomes some of the above disadvantages. The ideal sequence used is:

(1) Connect the autotransformer tapping to the motor.
(2) Close the autotransformer primary to the supply, so part of its windings act as high-impedance reactors in series with the motor.
(3) Close the autotransformer neutral point and it then functions as an autotransformer and the motor accelerates at a fixed reduced voltage.

(4) When the motor is up to speed, the neutral point of the autotransformer is opened and it reverts to a reactor function.

(5) A short-circuiting switch across the reactor section connects the motor to the line at full voltage.

This ideal sequence requires careful matching of the motor impedance, the autotransformer tap voltage and the reactance of the part winding, to ensure minimum switching surges of acceptable value at each operation. It is possible to buy cheap autotransformers with standard tappings but their impedance characteristics may not match the motor and considerable switching surges may result.

Other special forms of transformer starting may be used for particular conditions. For example, if motors are provided with unit transformers, conventional, tap-changing gear can be used to give reduced voltage starting. Control of the tap change can give reasonably uniform accelerating times and leave the motor running at its required rated value.

Combined reactor and autotransformer units have been used to match special starting duty requirements but these involve increased switching and, hence, costs.

7.2.3 Rotor-resisting starting

Slipring or wound rotor induction motors and cylindrical or salient pole synchronous induction motors have special secondary starting windings connected to sliprings which enable them to have externally mounted resistors connected in series with them. Usually, although not necessarily, these form a three-phase system. Any single value of resistance per phase produces a single motor starting current and torque characteristic and it is advantageous to arrange for the resistance value to be changed to permit improved motor characteristics to be obtained during the whole run-up period. Optimum performance can be obtained if the resistance can be varied continuously over a range from zero (short circuited) to a maximum to give the required switch-on conditions. Multi-step starters can approach the optimum but with cheaper three- or four-step starters quite good matching can usually be obtained.

The most common forms of devices are liquid resistors for continuous variation or contactor–resistor units for stepping. The former exist in several standardized sizes and are very economic, if suitable. However, for automatic systems, the contactor–resistor units are often preferred for operational reasons since they require little maintenance.

7.2.4 Special starting systems

It is possible to provide conversion equipment between the supply and a motor which can be controlled to give satisfactory motor starting capabilities while limiting demand on the supply to an acceptable value. The plant can be rotating machinery or solid state circuitry, the latter now being the more commonly used.

The well-known Ward-Leonard system used a motor-generator set to provide an easily controlled voltage to drive a d.c. motor. The purpose of the system was basically to give a speed control function but the same principle has been used solely for starting large drives or motors connected to very voltage-sensitive systems.

Where reversible hydraulic turbine-generator–motor pump units are used for storage pumping, the same principles can be adopted by using one set as a generator connected to another as a motor unit, which it can accelerate up to speed and synchronize with the supply.

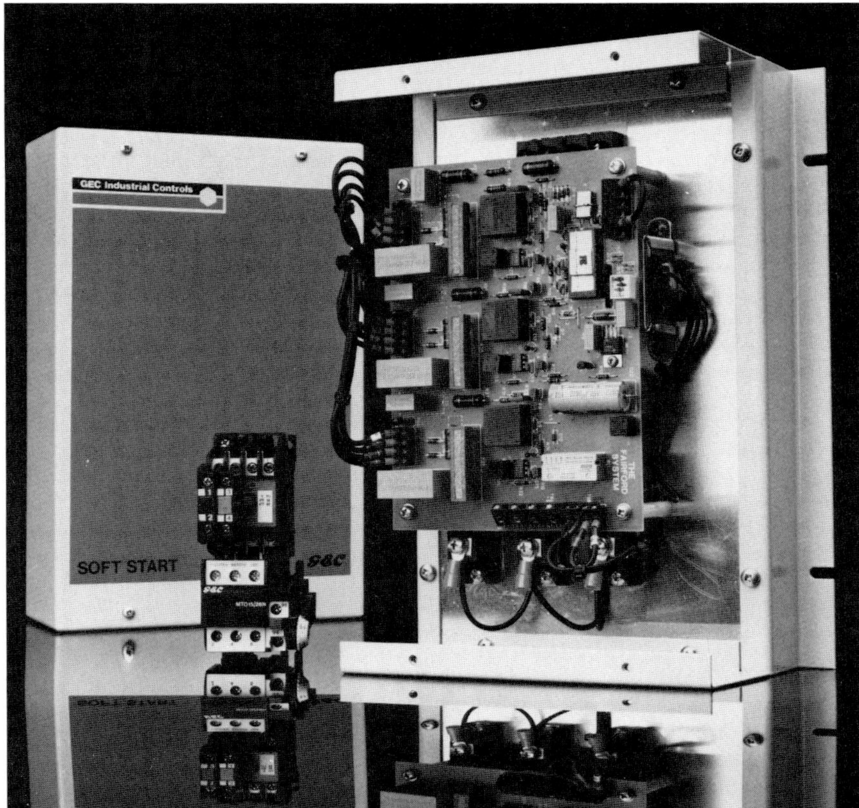

Figure 7.11 'Soft-start' motor controller providing reduced voltage for starting with adjustable ramp rate up to normal. (*Courtesy:* GEC Industrial Controls Ltd)

The increasing availability of larger and more economical static conversion equipment has enabled simple variable-speed drives to be obtained using conventional motors and these, of course, act as special starting units. Even where speed control is not required, similar devices can be used to give so-called 'soft start' systems (Figure 7.11). Several alternative circuits are available, depending on the particular motor characteristics required; the higher the performance required, the greater the complexity of circuits used and, hence, the cost of the unit.

7.2.5 Starting duties

The starting duty of a motor may be more limiting than its steady load rating and it is essential that the machine be designed for the actual starting conditions imposed in practice. Significant factors are:

(1) Torque absorbed by the driven load.
(2) Total inertia of motor plus load (including such factors as entrained water in pump impellers).
(3) System voltage reduction during starting.
(4) Starting time.
(5) Current drawn from the supply.

(6) Actual speed variation with time.
(7) Number of consecutive starts required.
(8) Cooling periods allowed between starts with motor running and with motor stopped.

Some assessment must be made of all these quantities.

In addition to the motor, however, all starting equipment, including switches and protection devices, must be designed to meet these conditions and must be co-ordinated as they comprise, in effect, a single integrated electromagnetic system.

Certain ranges of equipment have standardized capabilities and it is then only necessary to select the appropriate references from the British Standard (or equivalent), but for any special drives individual specification will be required.

7.3 Excitation and power factor control

7.3.1 General

The polyphase a.c. winding on the primary winding of a motor produces a synchronously rotating electromagnetic field, the actual speed of which depends on the number of poles on the winding and the frequency of the a.c. supply. When the secondary (rotor) winding is stationary in this field, it has induced in it a voltage at system frequency; if it rotates, the induced effect is at slip frequency or the difference between the rotating field speed and the rotor speed. If the rotor speed is synchronous, no induced effects appear and to produce torque it is necessary to circulate zero-frequency current (d.c.) in the rotor winding.

The d.c. current provided for synchronous motors is called the excitation current and can be provided in any of a variety of ways, depending on the type of motor and its application. It is usual to provide some means for varying the excitation current to enable the motor characteristics to be adjusted. This may be in the form of a manual control, or a compounded or compensated automatic system, or a full automatic system with either open-loop or closed-loop control.

During starting, slip frequency current will flow in the excitation circuit and winding and the excitation system must cope with this. At the low slip at which the motor will pull into synchronism, it is necessary to pass the d.c. excitation current through the same winding to produce the required synchronizing torque.

There are many alternative types of excitation schemes and equipment but they can be grouped into general categories, as follows.

7.3.2 Slipring motors – d.c. supply

In the past, conventional synchronous motors were provided with sliprings to enable the externally produced d.c. to be connected to the rotor-mounted excitation winding (Figure 7.12). Where a convenient source of d.c. power was available, it was economic to use this for excitation power by providing a two-pole isolating switch, with discharge resistor provided for the winding on shutdown, and a simple series-connected resistor to control the magnitude of the excitation current.

7.3.2.1 Shunt-wound d.c. exciter

It was more usual, however, to provide each motor with its own excitation system and many motors have been provided with their own shaft-mounted d.c. exciter (Figure 7.13). Control is carried out using a small resistor in series with the shunt winding of the exciter. The exciter armature can be connected permanently direct

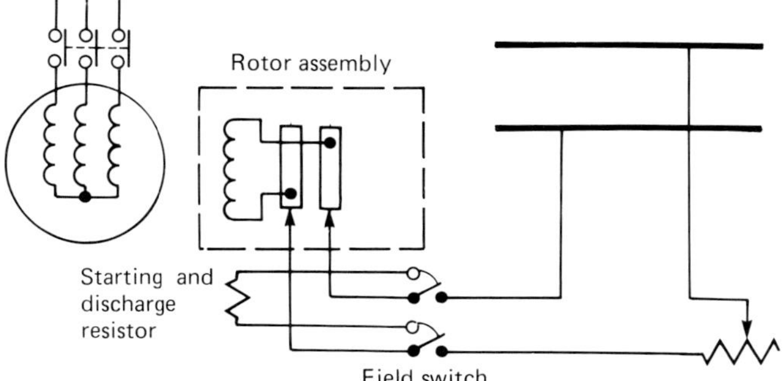

Figure 7.12 Direct current supply excitation

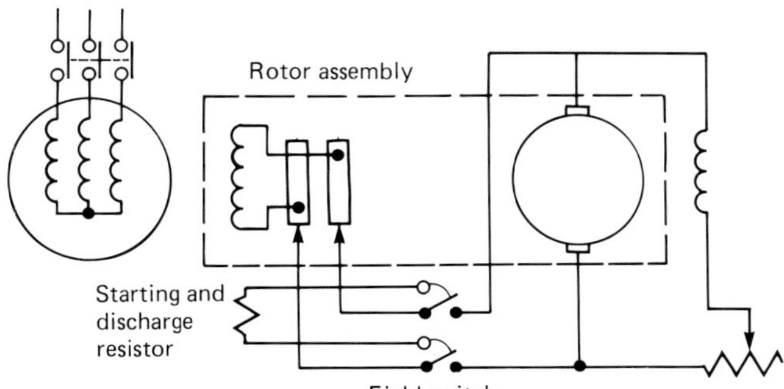

Figure 7.13 Direct current exciter excitation

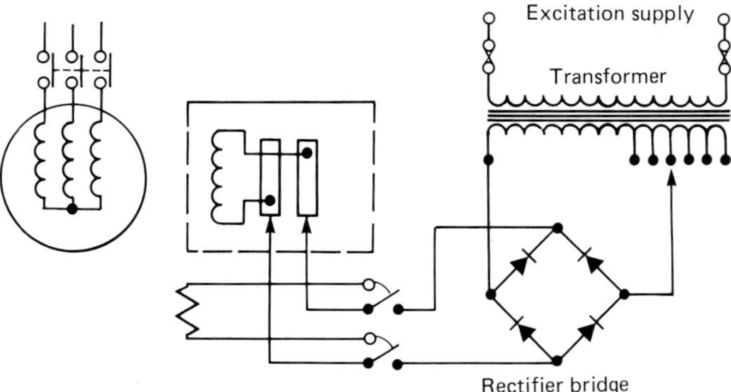

Figure 7.14 Static excitation

to the motor excitation winding via the sliprings but, with this arrangement, slip frequency current appearing during starting had a demagnetizing effect on the exciter and could impair voltage build-up. Most motors with such an exciter are provided with a switch and discharge resistor rated for the particular starting duty. This connects the resistor in series with the excitation winding during run-up, which enables the motor to produce additional starting torque, and only low slip frequency current flows in the exciter during the short time required for synchronizing.

7.3.2.2 Static excitation

Where it is inconvenient to provide a shaft-mounted exciter, a unique source of d.c. can be provided from a convenient a.c. power source by providing a transformer and rectifier unit (Figure 7.14). This still requires an excitation switch and discharge resistor on the d.c. side and needs switching and protection on the a.c. supply side. Adequate step adjustment of a preset fixed value of excitation can be obtained using tappings on the transformer but, if continuous variation is required, a suitable resistor is required in the d.c. circuit.

7.3.2.3 Compensated/compounded excitation schemes

Operation with constant excitation, with manual adjustment available if required, is a convenient and simple arrangement but it is possible to utilize smaller and, hence, cheaper motors for some applications by using automatic variable excitation. This may also have beneficial effects on the supply system by regulating kilovar circulation, or by compensating for voltage variations on the system.

Simple systems used have employed two windings on the exciter, one a conventional shunt and the other fed from a line-current transformer through a rectifier (Figure 7.15). Thus, increase in line current produces increase in excitation current but makes no allowance for a.c. current power factor. A more convenient form of this can be obtained by injecting the line current signal across an adjustable impedance in series with a signal from the motor line voltage. Correct selection of impedance value and variable torque and constant torque phase relationships can give a good compensation over the rated working range. On small motors, this simple compounded excitation system can be used to supply the rotor direct without requiring an exciter.

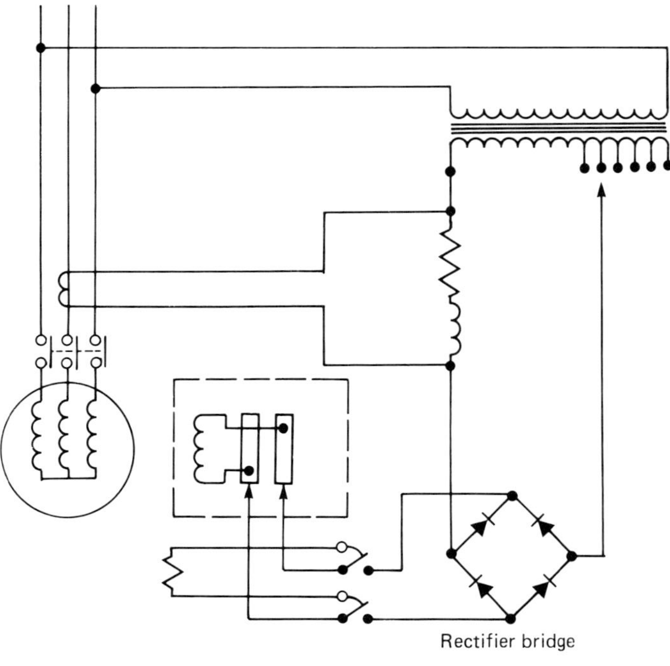

(a)

Figure 7.15 Compounded/compensated excitation

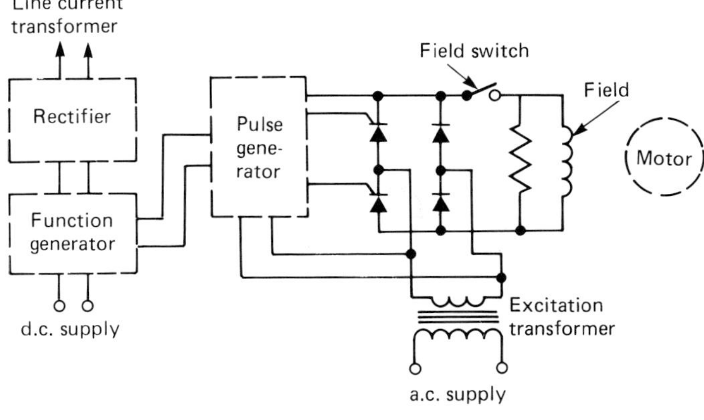

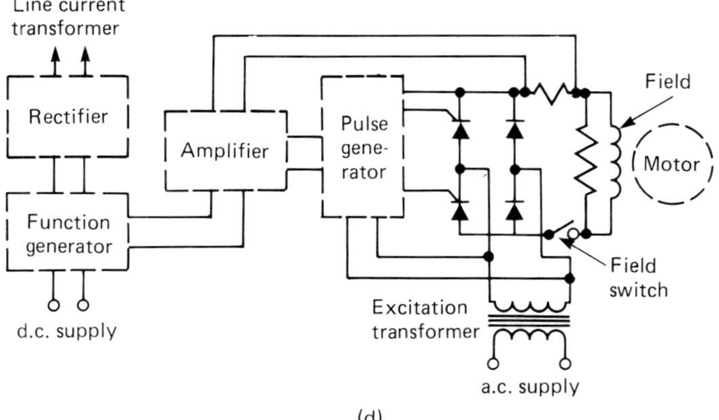

Figure 7.15 (continued)

Rotor assembly

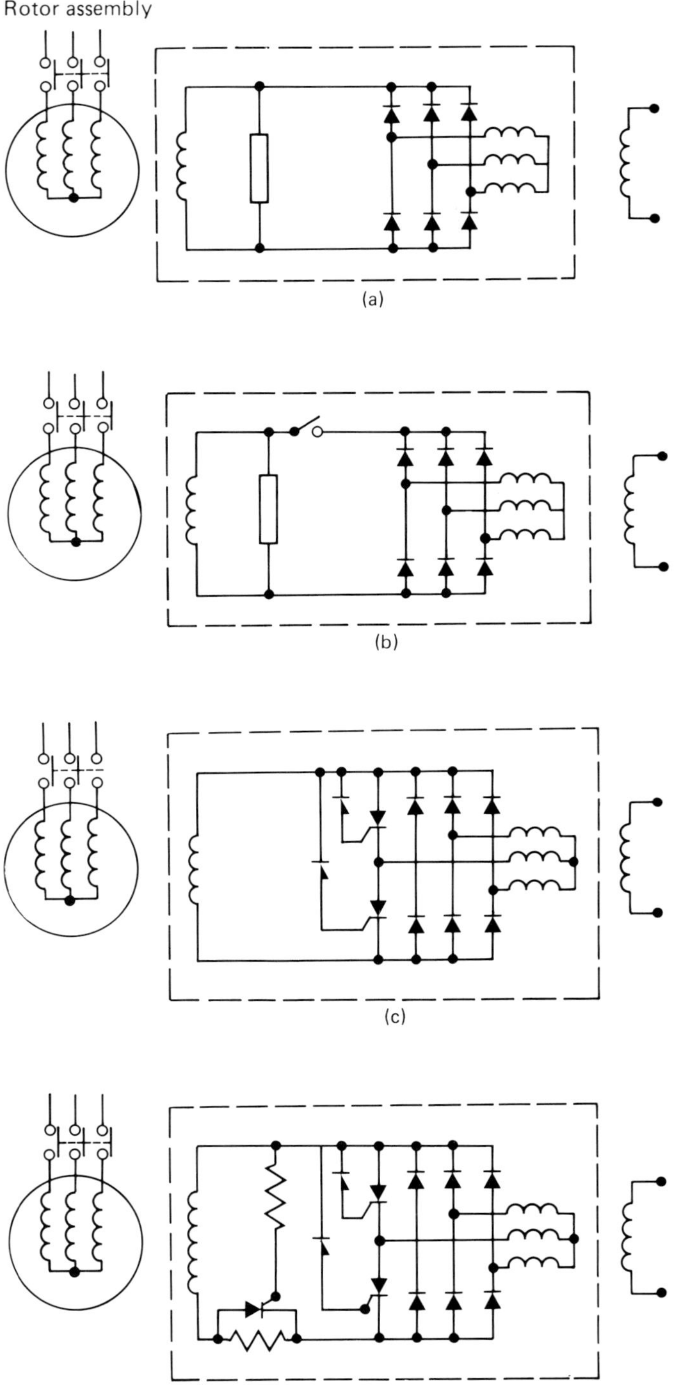

Figure 7.16 (a) Brushless excitation. (b) Brushless excitation switched. (c) Brushless excitation thyristor. (d) Brushless excitation thyristor

The availability of thyristors enabled more sophisticated schemes to be competitive and give much superior results for synchronous operation and also to include accurate automatic synchronizing control to ensure excitation was applied at the correct slip and, if required, at the optimum phase angle (Figure 7.15(b), (c) and (d)).

7.3.3 Brushless motors – a.c. exciters

Some large machines have been provided with a.c. exciters supplying a separately mounted rectifier unit (as described in Section 7.3.2.2) but, with the evolution of suitably rated semiconductors, the concept of mounting the rectifier unit on the motor eliminated the need for sliprings and brushes completely (Figure 7.16). The problems with brushless motors were overcome in the same way as for those previously using d.c. exciters. To produce reasonable starting torque, a starting resistor was connected permanently across the excitation winding on the rotor, to provide torque and reduce duty requirement on the diodes. Superior performance was obtained using a rotor-mounted switch to isolate the diodes up to the synchronizing slip. By using thyristors to act as switches, simple starting was obtained initially by, in effect, short-circuiting the excitation winding during starting and, later, by adding a starting resistor with its own switching devices. As development proceeded, automatic synchronizing at preselected slip and phase angle can be included, all mounted on the brushless motor rotor.

In all these arrangements, steady excitation control is provided by adjusting the a.c. exciter field, which is at a relatively low power level. Any of the forms of control mentioned can easily be applied to this circuit.

7.3.4 Excitation control

Excitation control can be applied to the motor excitation current directly or to the exciter field current, depending on the type of motor and excitation provided but, since the power involved in the excitation field current is much less, it is preferred where practical. It does involve an additional delay during transient conditions, because of the exciter winding time constant, but this seldom presents any serious difficulties.

For it to be possible for a synchronous motor to meet its specified pull-out torque, its excitation must bear a definite relationship to its no-load excitation and

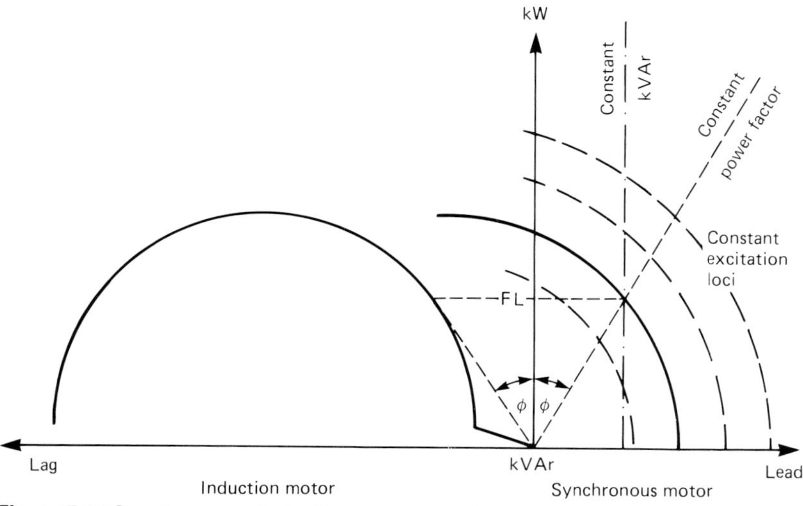

Figure 7.17 Synchronous–induction characteristics

its short-circuit ratio. Fixed-excitation machines are designed such that the rated value of excitation will meet the required condition. In some instances it is advantageous to apply only no-load excitation, or a value less than rated, during synchronizing, and to do this some form of excitation control is required. A conventional rheostatic type of resistor is normally provided where manual control is required, while a simple contactor–resistor unit giving the required preset values can be used for an automatic starting scheme (Figure 7.17).

Simple compensated or compounded excitation systems, as described in Section 7.3.2.3, while providing economic advantages, do not give precise control of excitation and, while they give a more beneficial operating characteristic with varying load than a fixed excitation motor, if accurate control is required, some form of closed-loop control is necessary.

7.3.4.1 Closed-loop control

Such systems can be provided to give control to meet a variety of requirements, either simple or complicated.

Constant kilovar control provides a predetermined amount of kilovars leading up to the limit of the excitation available over the complete range of loads from zero upwards, and gives the same effect as capacitors connected to the system.

Constant power factor requires greater excitation power for the same load but has advantages on systems where voltage drop is severe.

Combinations of the above can be provided, if required, and in some cases control over a limited range, with reversion to constant excitation for higher loads, is more economical.

System voltage control can be provided up to the available leading kilovarage of the motor, in which the motor compensates for lagging kilovarage of other loads which are producing voltage drops.

Nearly every form of sensing circuit and amplifying device has been used in the past, each having its advantages and disadvantages; to obtain the most reliable system requires a careful study of the motor involved, the proposed excitation system, and the motor load and supply system operating conditions.

7.3.5 Power factor control

This term is sometimes applied to synchronous motors, as described in Section 7.3.4, but it is more often used for induction motors. These require a magnetizing current supplied from the electrical system, i.e. they require system kilovars, the actual amount depending on the system voltage and the motor reactance. This has an adverse effect on the system by causing voltage drops and additional losses; improvement of operating power factor, i.e. reduction of lagging kilovar demand, can produce economic savings.

Since it is not practical to change the actual motor power factor, it is necessary to compensate for its lagging demand by connecting capacitors, which take leading kilovars, in parallel with the motor and, hence, reduce the net lagging kilovar demand.

Power factor correction capacitors can be applied in two distinct ways: (1) individual motors can· be compensated by connecting capacitors at the motor terminals and switching both simultaneously; and (2) a separate power factor correction unit can be connected to the system side of the motor switch, i.e. on the busbars.

7.3.5.1 Motor power factor correction

When an induction motor drives a load which is capable of regeneration, a dangerous condition can arise when it is disconnected from the electrical supply, if the capacitor connected to its terminal provides enough leading kilovars to magnetize the motor. Under this condition the machine functions as an induction generator and can generate an open-circuit voltage determined by its own saturation curve. Thus, if the capacitors used provide more than the no-load magnetization kilovars, the voltage will exceed rated value and can damage the capacitors and, in an extreme case, other connected units. With this arrangement, therefore, it is usual to limit individual motor power factor correction by the addition of capacitors giving kilovarage, not greater than the no-load magnetizing kilovarage of the motor. While this gives some correction to the system, the amount may not be adequate to produce real economic savings.

7.3.5.2 Block power factor correction

Connecting capacitors to the system side of the motor switch eliminates the self-excitation phenomenon in all except very special systems. It does require, however, a separate switch for the capacitor with adequate protection devices. Such a unit is only economic where several induction motors are connected to the system and the choice of kilovarage for a single capacitor is difficult when the number of motors in operation may vary. In such an application, multi-step units can be used in which contactors switch capacitors to vary the total kilovarage value. Special units which do this automatically are available, and these can be operated by a signal based on the power factor of the busbar supply to keep this within preset limits, up to the range of capacitors provided.

It is important to check whether any significant magnitude of harmonics can appear on the electrical system, since these can circulate through capacitors producing additional heating and may ultimately lead to their failure.

7.4 Protection

7.4.1 General

The term 'protection' is often used to refer only to the relays included in the control or switchgear associated with a motor but, in fact, it includes all monitoring devices intended to give warning of dangerous operating conditions that arise and also those which activate procedures to reduce damage after a dangerous condition has arisen.

It is possible to monitor all motor parameters but it is not always easy to define exactly when these reach a value requiring corrective action, since many of them are interactive and a hazard may only exist in practice when a particular combination of parameters reaches critical limits. In practice, only a few parameters can be monitored conveniently and economically and it is customary to relate these to likely modes of failure or damage and to use this data as a basis for protection settings.

Since a motor always forms only a part of an installation or drive, it is often convenient to extend the protection philosophy to include some aspects of the drive and also of the electrical system supplying the motor.

7.4.2 Failure modes

Motor failures can be either mechanical or electrical or a combination of these but other factors, such as thermal and chemical, often have considerable significance. The motor design is such as to have considerable safety factors on both mechanical and electrical stress loadings under normal conditions but it is not always economically possible to allow for all abnormal conditions of loadings or environment, such as overtemperature, vibration, shock, etc. or effects due to corrosive gases, liquids or solvents. A machine enclosure is usually selected to minimize such effects as far as possible, and mechanical stresses can be allowed for at the design stage if their magnitude can be predicted. As a result, most failures are of winding insulation and the predominant cause of this is sustained operation at temperatures above the correct operating value. Insulation 'life' is reduced at elevated temperatures and as it deteriorates it becomes more susceptible to damage due to other causes, mainly chemical or mechanical. Thus, the actual failure frequently occurs at some transient condition, such as switching on, when the high mechanical stresses exceed the capability of the thermally weakened insulation. The insulation failure results in the flow of fault current which itself can cause burning and mechanical damage and often obscures the evidence which could have been used to deduce the actual cause of failure.

7.4.3 Overtemperature

Overtemperature of insulation can result from excessive losses in the conductors due to abnormally high currents, or by inadequate removal of heat from the surface because of impaired ventilation, or from a combination of both. Measurement of the external insulation temperature can indicate abnormal conditions without indicating its cause but it is only an empirical value and it does indicate the real temperature of the insulation adjacent to the conductors. It is quite useful, therefore, for low-voltage machines with thin insulation layers, such as small industrial motors. However, it is less satisfactory for higher voltage machines and, on large machines, the presence of a detecting device on the machine windings can produce electrical stress abnormality; when corona charges are likely to occur, they can present a danger. It is common practice to install temperature-measuring devices in winding slots, where the above disadvantages do not apply to the same extent, and usually duplicate devices are inserted at the manufacturing stage so that, should one be faulty or fail in service, the other would remain operative and avoid a major rewind operation to replace the faulty one. On three-phase windings, therefore, it is normal to provide six devices, two in each phase.

There are several alternative temperature-sensitive devices which have been used. Simple thermostats which provide an on–off switching action can be used for direct action, but they are not accurate devices and do not have a high degree of repeatability. They are acceptable for small low-voltage machines of a non-critical drive, where simplicity and low cost are paramount considerations. Thermocouples are simple, robust and accurate temperature-sensing devices and are convenient for mounting on machines or in winding slots. They each only generate a low e.m.f., however, and if their signal has to be transmitted considerable distances it may be necessary to use special compensated leads. If used only for instrument operation, no further complication is required, but if they are to be incorporated into a protection control system it may be necessary to provide amplification units. Standard characteristic thermocouples are now available and proprietary makes of control units are available for use with such devices. Electronic units are available to carry out this function and these are advantageous where several input signals have to be dealt with and computations are required to evaluate such factors as maximum deviation of one signal from the average of several. Such units on large systems can develop into microprocessor units and are then capable of other features associated with the overall protection system.

The other common temperature-sensing device is the resistance temperature detector, which relies on the variation of a standard resistance with temperature. These devices can be very accurate and consistent in performance but require a stable power source to enable accurate temperatures to be deduced. When used for machine-winding protection it is often convenient to use them in some form of bridge circuit, which can be designed to compensate for ambient temperature variation, and can be used on a simple comparative basis to determine when abnormal temperatures are attained without having to evaluate the absolute temperature.

These temperature-dependent devices may be used for tripping machines directly but this is normally only done on smaller machines. On larger machines they are used for alarm or warning only and tripping is carried out by alternative devices. This is also true for overtemperatures other than on windings, e.g. cooling air leaving a heat exchanger can indicate faulty operation of the exchanger, or abnormally high cooling water, or oil temperatures indicating abnormalities which require operator action, but do not in themselves justify shutting down the machine.

7.4.4 Overcurrent thermal devices

In most instances winding overtemperature is related to overcurrent and since it is easy to measure current accurately this forms a convenient means of checking if the heat dissipated in the windings is normal. Under normal ventilation conditions this can be taken as implying that the insulation temperature is normal, and abnormal ventilation conditions would be indicated by one of the warning devices referred to above. Thus, the current which remains within rated value would not automatically shut down the machine but leave the operator to take corrective action. If, however, the current did exceed the normal value then automatic action would be justified. By passing a proportionate current through a thermal model of the machine it would be possible to derive a function representing the machine winding temperature and this could be used to give a warning or trip signal at the required value. An accurate thermal model is not practicable for all machines, because the variation in characteristics is very wide and cannot always be identified accurately for every design. Simple devices with typical characteristics are available at reasonable cost for small machines but it is always necessary to check that the actual machine will be protected adequately by examining the device characteristics. A wide range of thermal relays is available and it is usually possible to select one which matches the requirements of most machines but, in general, the more adaptable the relay the more expensive it is. Such relays can be provided with ambient temperature compensation and have characteristics which differ from 'cold' to 'hot' to match machine heating characteristics more closely. Adjustable operating settings can be provided for machines with different duties and this, in conjunction with tappings on the current input, enables relays to be closely matched to machines whatever their rating may be.

Simple thermal relays are usually designed to protect the primary (stator) winding by matching its characteristics, assuming that other windings will be covered by the same setting. This is true for steady running under normal conditions but is not necessarily true during starting, speed changing or system current or voltage unbalance conditions. It is possible to provide separate relays to match secondary (rotor) winding characteristics, and these are justified on important machines with onerous operating duties. Stalling relays are available for this purpose. These evaluate the integrated I^2t which represents the heat dissipated in the rotor winding and this can be related to an expected temperature rise for standstill conditions, i.e. failure of the motor to break away from rest. A safe design value of I^2t can be obtained from the manufacturer and an appropriate relay setting used to protect against damage. This setting, however, does not discriminate between the motor being at standstill and accelerating and the value of I^2t which will produce the limited temperature rise at standstill would produce a lower value if the motor were accelerating, owing to improved ventilation. A speed switch can be used to discriminate between a stalled motor and an accelerating one, and so permit accurate stall protection with unnecessary restriction on I^2t on a normally accelerating machine.

The effect of system unbalance is to produce additional heating in a machine but it is convenient to analyse this condition into two balanced conditions, a positive phase sequence current system, which can be dealt with by the devices mentioned above, and a negative phase sequence current system which produces significant rotor winding heating. By utilizing a filter network to separate these two components, it is possible to treat each signal separately in relation to the characteristics of the appropriate windings. Special relays are now available which incorporate all these functions in a single device, and with the development of electronic techniques, very compact units are available with a wide range of adjustments to give almost universal adaptability.

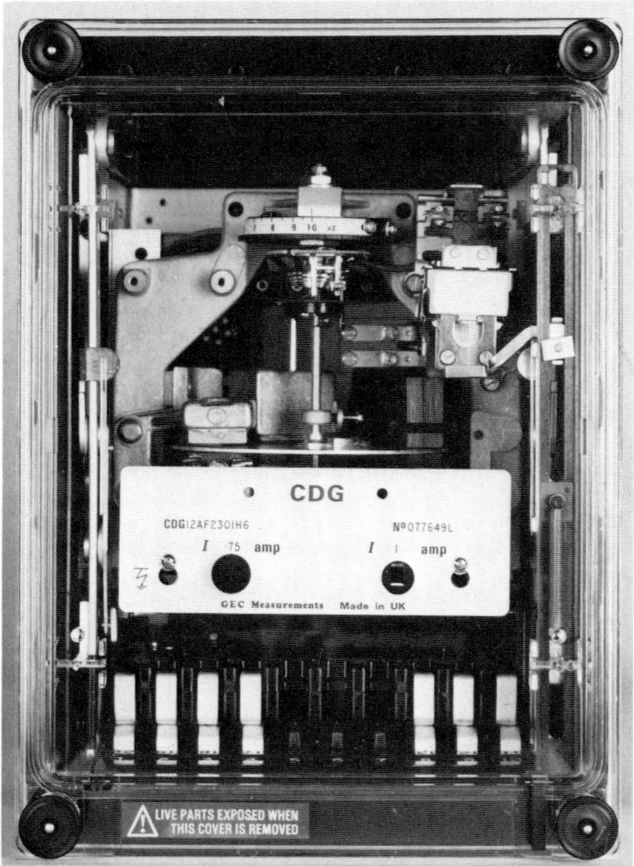

Figure 7.18 Typical induction disc relay providing long inverse definite minimum time–current characteristics. (*Courtesy:* GEC Measurements Ltd)

7.4.5 Overcurrent relays

With some machines, such as synchronous motors, it is possible to deduce the winding temperatures related to primary current and use simple overcurrent characteristics to protect adequately the motor windings against overheating, but it is recommended that thermal relays are used for motors where the effects of motor starting either due to torque, inertia or frequency are significant.

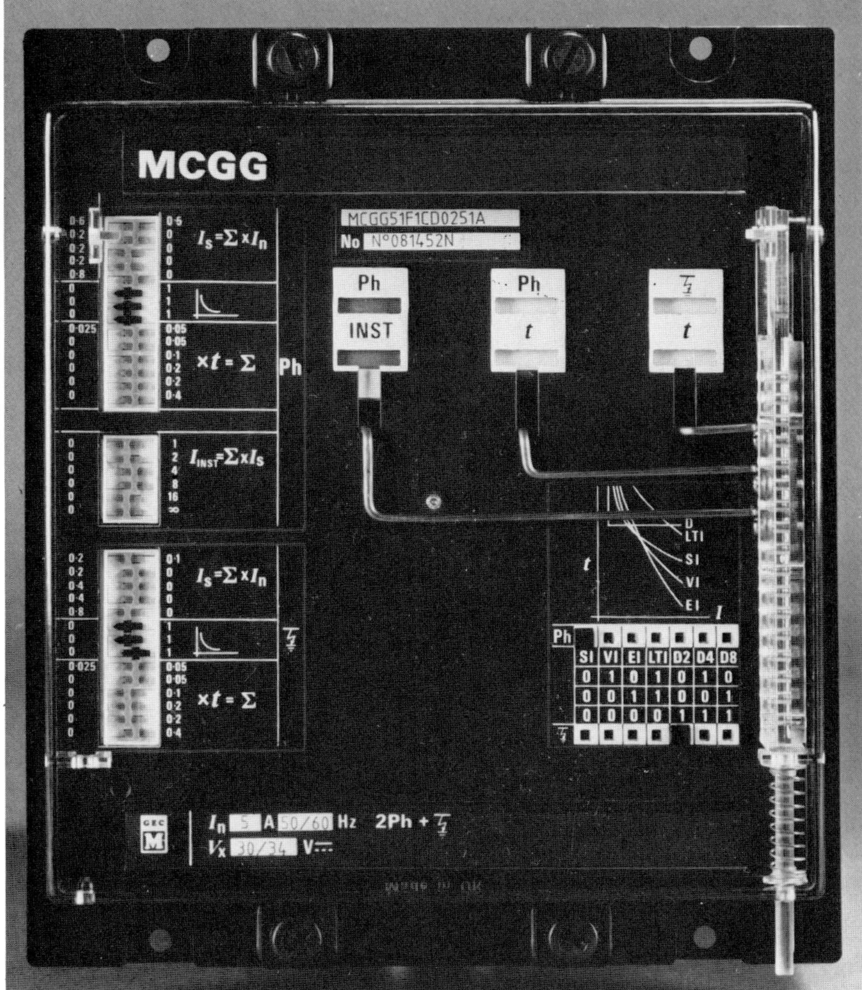

Figure 7.19 Typical solid state microprocessor-based relay providing a wide range of inverse time, definite time, and instantaneous characteristics. (*Courtesy:* GEC Measurements Ltd)

7.4.6 Negative sequence relays

Where severe system unbalance is expected or where the rotor loading may be a critical design limit, it is possible to utilize a relay to measure the percentage negative sequence current loading and relate this to the capability of the machine when running normally. Such a value can be used to give a warning. Alternatively, the relay can be used to measure I^2t, i.e. the equivalent heat input to the secondary winding and either warn or trip when a dangerous condition exists. Such a device should be co-ordinated with a stalling relay, if used, as the functions are related as being relevant to the same winding, but are applicable during different operating conditions.

Figure 7.20 Group of protection relays in modular form. (*Courtesy:* GEC Measurements Ltd)

7.4.7 Compound relays

With the advent of digital relays, it is relatively simple to incorporate several protective functions within one device and several such compound relays are now available. These may be 'general', in that the functions included can be adjusted and co-ordinated for the special requirements of different items of equipment, or may be selected to suit a particular application, e.g. motor protection to include full thermal protection, overload protection, and stall protection, including discrimination for stall or acceleration conditions.

Five types of relays can be seen in Figures 7.18–7.22.

7.4.8 Mechanical factors

Mechanical failure of motors is very unusual since large safety factors are used to ensure generous margins on all stresses under specified operating conditions. However, when abnormal conditions arise, components may reach dangerous stress levels and ultimately fail. Overspeed is an example of such an occurrence, although it is rare for motors. There are some loads which can act as prime movers under abnormal conditions and can cause motors to run away.

Figure 7.21 Solid state motor protection relay providing overload inverse time, stall and single-phase protection. (*Courtesy:* GEC Industrial Controls Ltd)

Whenever possible, the design is such as to prevent the motor failing under all predictable conditions, but sometimes these are not always fully appreciated at the order stage.

Another condition which can lead to failure is vibration, and this can arise due to vibrations transmitted from the supporting structure to the motor, because of abnormalities in the electrical system producing unbalanced forces in the motor, or as a result of unbalances in the driven unit or in the motor itself. Relatively small amplitudes of vibration can cause failure of components if these respond to the particular frequency and resonance occurs. Items such as fan blades or similar rotor projections may be prone to this phenomenon. It is difficult to design all parts of a machine to prevent resonance for every possible frequency but it is usual at the design stage to consider all forced frequencies likely to be significant in service.

The most common source of mechanical failure on rotating machines is the bearings. For economic reasons these must be highly loaded and correct and adequate lubrication must be provided. Any deficiency in lubrication will result in deterioration of the bearing and ultimately can lead to failure. This would be associated with increased losses in the bearing, with consequent increase in operating temperature of the bearing and lubricant. Some types of bearing, such as moderately loaded journal bearings, deteriorate relatively slowly, and operating temperature serves as a good monitor of bearing health. At the other extreme, however, highly loaded anti-friction bearings can deteriorate very rapidly and fail

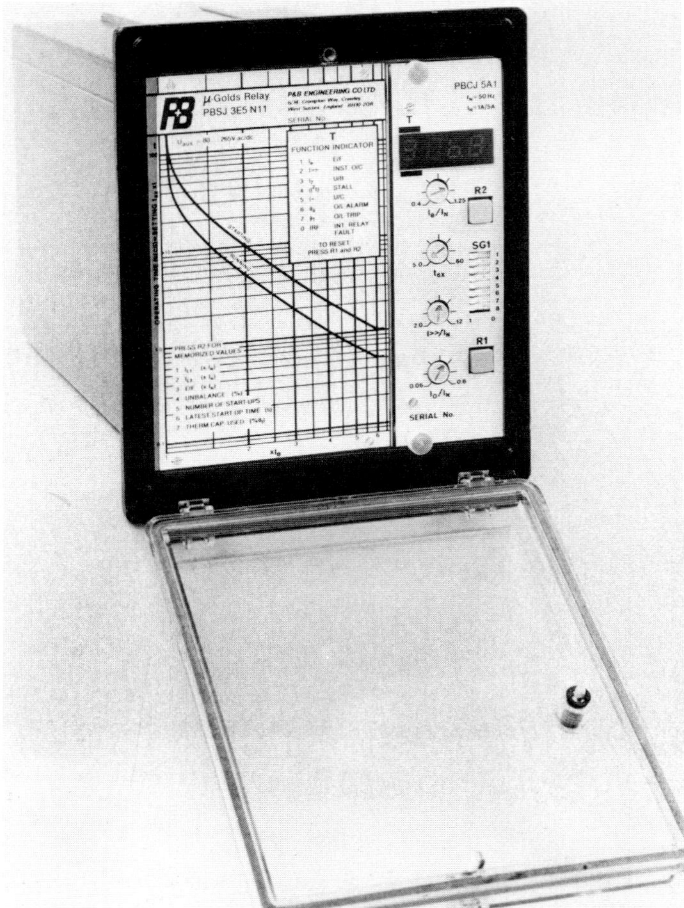

Figure 7.22 Digital motor protection relay incorporating adjustable current–time thermal overload, locked rotor–stall, high set overcurrent and undercurrent, zero phase sequence and negative phase sequence–phase unbalance protection. (*Courtesy:* P and B Engineering Co. Ltd)

before significant temperature rises can be observed and remedial action taken. With such bearings, increased vibration often precedes failure and this effect can be used to monitor impending failure.

7.4.8.1 Monitoring

Protection against mechanical failure, apart from adequate design margins based on all actual operating conditions, consists of measuring bearing temperatures and motor vibration amplitudes. Both methods tend to be used on a comparative basis, i.e. absolute values are not significant in themselves but any change in the values indicates some change in state and justifies investigation to determine what deterioration can be detected. A large variety of devices and methods of application have been used for monitoring and the choice depends largely on the type of motor, its construction, and its operating conditions.

7.4.8.2 Temperature

Bearing thermometers are commonly fitted on pedestal bearings but the term is not truly explicit since the device may be measuring oil sump temperature or bearing liner temperature (usually the shell) and neither gives the maximum temperature at the bearing surface. However, such devices function adequately as comparative devices on moderately loaded bearings. Dial-type thermometers give visual indication and can easily be fitted with adjustable contacts to give alarm or trip signals if required. Thermocouples or resistance-type devices can be fitted similarly, if remote indication or alarm is required. These require ancillary devices to convert the signal into a suitable form for indication or alarm tripping. It is convenient to utilize a selection switch where several signals are involved, to reduce the amount of conversion equipment.

Where bearing lubrication is provided from a separate source, it is usual to provide pressure- or flow-measuring devices to ensure an adequate supply is maintained. Neither device is infallible since pressure can exist without flow if the pipe is blocked and flow will be indicated even if a pipe is ruptured and no oil reaches the motor. Consequently, it is usual to provide some form of bearing temperature measurement as a back-up when a separate lubrication system is being used.

7.4.8.3 Vibration

Vibration detectors are only provided on special motors such as large high-speed units, drives requiring very high integrity, or on motors driving units which themselves require vibration measurements. They require the provision of special mounts, rigidly attached in the correct plane to the bearing, and supply their signal to an interpretive instrument. Devices can be used to measure deflection, speed or acceleration but the simplest device giving adequate information should always be used to minimize cost, complication, and problems of setting up. Should the device be used on a comparative basis, it must be appreciated that abnormal transient conditions due to load changes or system disturbances can produce short-term increases in vibration levels which do not indicate a motor fault condition.

It is possible to use sophisticated equipment for diagnostic purposes to identify particular forms of motor deterioration from the vibration frequency characteristic rather than from simple amplitude change, but this has, as yet, only limited practical application for normal motors.

7.4.9 Special conditions

Some motors may be subjected to unusual operating conditions associated with the supply system which could prove dangerous and might justify the provision of special protection equipment for the motor. Loss of supply, circulating system harmonics and lightning are conditions which fall into this category.

7.4.9.1 Loss of supply

When the power supply to a motor is completely interrupted, the load torque will decelerate it and usually bring it to standstill. Some drives such as pumps may be capable of reverse-power generation and special motor-generator sets can present similar problems which must be dealt with on an individual basis. The problem of regeneration, which appears as reverse acceleration of the motor, can be detected easily by mechanical means and protective devices for this purpose are usually provided by the maker of the driven load which is usually more prone to damage from this condition than the motor.

A more serious condition arises, however, if the motor supply switch does not open automatically following interruption of the supply. This can occur when the switch is mechanically latched, or held closed and requires a separate tripping signal to open it. If this supply fails or the mechanical release mechanism is faulty the motor can be left connected electrically to the system and, when the supply is restored, it will attempt to restart. Mechanical failure can be avoided only by using a tried and proven design of equipment and ensuring that it is maintained and inspected adequately. Continuous monitoring of any tripping supply should also be provided with visual indication of healthy state and audible alarm of any failure.

It is sometimes assumed that switches which are held closed by the supply voltage, or are provided with an undervoltage relay to give a tripping signal, are safe under supply failure conditions, but this is not necessarily so. Any motor will continue to generate a voltage after interruption of the supply and this can maintain the switch closed. This voltage will decay as the motor speed falls and the flux decays but for a finite time a hazard can exist if the supply is restored. The reclosure condition is more dangerous than restoring the supply with the motor at rest because the motor-generated voltage can be out of phase with the restored supply voltage, and this increases effectively the current drawn from the supply above that corresponding to DOL starting at rated voltage.

When there is a possibility of a motor being subjected to sudden restoration of supply it is usual to ensure that the switch is opened on interruption and a voltage relay on the motor side prevents reclosure of the switch if a dangerous generated voltage still exists.

Other simpler forms of protection, such as motor frequency relays, have been used for this condition, but these are not universally satisfactory for all applications and would require detailed analysis of actual conditions for each installation to check if adequate protection were provided.

7.4.9.2 System harmonics

No power system provides a harmonic-free supply to motors and standard specifications state limits which a motor must be capable of dealing with. On systems where these limits are likely to be exceeded, it may be necessary to carry out an accurate analysis of the effect of the actual supply harmonic voltages on the motor harmonic impedance spectrum to determine what harmonic currents could circulate in the motor. The effects of these on losses, torque and other performance characteristics can then be evaluated and, if the motor performance is still within specification, no protection need be provided. Should it limit motor performance, however, then action should be taken to reduce the significant harmonic amplitude. The most effective way of doing this is not always the most convenient or the cheapest but, in general, a tuned rejector circuit can be adequate if only one harmonic is significant.

7.4.9.3 Lightning

When the supply to a motor is carried on overhead lines and the isokeraunic level is significant, it is usual to try to provide some measure of protection against the effect of lightning strikes either direct or induced on the lines producing overvoltages on the motor windings. If the motor terminals are accessible to a lightning strike, it is quite impracticable to protect the winding adequately against overvoltage but, in practice, a length of cable usually exists between the motor and the exposed line, and use can be made of the inductance of this. Fitting line-to-earth capacitors at the machine terminals, in conjunction with the cable inductance, will reduce the wavefront gradient as seen by the motor; hence, the overvoltage appearing between adjacent turns in the winding will be reduced, depending on the wave transit time through the windings and the circuit parameters. It is usual to fit diverters across the capacitors which become conducting at a preset voltage and these will limit the magnitude of the overvoltage. To get close values of overvoltage on such devices, it is necessary to limit the current-carrying capability. To avoid these 'machine-type' diverters, it is necessary to fit higher-current-capacity, but also higher-overvoltage set, 'line diverters' at the overhead line end of the cable.

In the UK, most installations have considerable reactance between them and lightning-prone sections of overhead line, so lightning protection is seldom considered. Should any risk occur, however, a fairly simple calculation can establish whether any protective devices are justified, and what ratings would be suitable.

7.4.10 Fault protection

The other important function of protection equipment is to detect when a fault occurs and take action to minimize ensuing damage. The most common types of fault are as follows.

7.4.10.1 Winding earth faults

The consequence of an insulation failure of the main a.c. winding depends on how the system is earthed, which may be either: (1) insulated neutral; (2) impedance earthed; or (3) solid earthed. Any protection must be selected accordingly.

Insulated neutral system.   The advantage of the insulated neutral system, which is often called the unearthed system, is that without a fault to earth existing no earth fault current flows and it is possible to maintain operation, although the voltage between individual phases and earth can be greater than normal up to a steady value equal to the line-to-line voltage. In practice, this is only approximately true and, while it is generally acceptable for low-voltage systems, the effect of capacitance between the phase conductors and earth becomes significant on higher voltage systems, and such a system can no longer be assumed to be unearthed. A single earth fault on such a system will result in a flow of fault current which may, in time, cause serious damage at the point of earth fault. Depending on the nature of the fault and the characteristics of the system, the phenomenon of 'arcing grounds' may occur which can produce significant overvoltages on the system. Both the nature of the fault and the characteristics of the system can be evaluated to ensure conditions will be acceptable but they can be regarded as disadvantages compared with alternative systems of earthing, depending on the actual operating conditions required.

The truly insulated system cannot be provided with a continuously operating protection system but a monitoring system can be used to check for an earth fault by applying a high impedance connection to earth and measuring any current flow, on an intermittent basis. This does not locate the fault, which can only be done by isolating feeders in sequence until the fault is removed. However, when capacitive current flows in the fault, if it is large enough, it is possible to use a low-sensitivity directional earth fault relay to detect in which feeder the fault exists.

Impedance earthed systems.   With many systems and operating conditions, the disadvantages of both the above systems may be unacceptable and a compromise is used. The advantages of the solid earthed system can be retained partly on higher voltage systems, which have low fault impedances, by earthing through a low value of resistance which will limit earth fault current to about rated line value and will be sufficient to permit simple relays to detect such a fault.

Alternatively, the advantages of the insulated system can be retained partly by

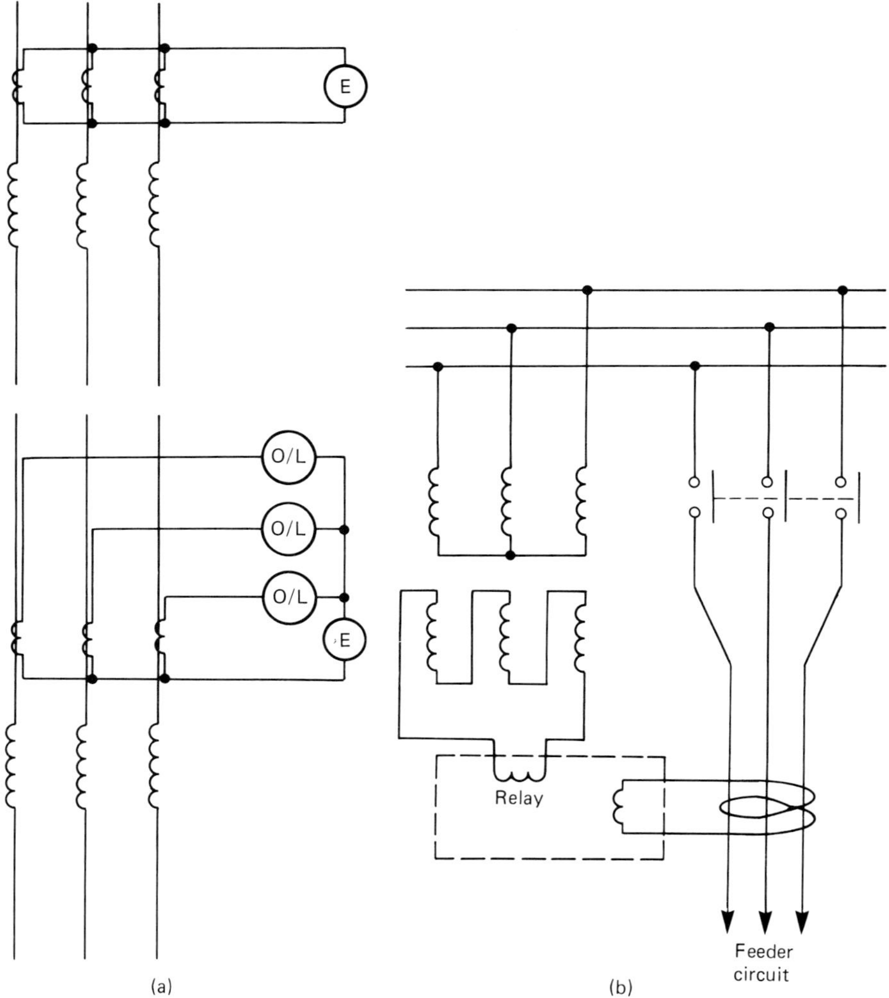

Figure 7.23 Earth fault protection. (a) Residual connections for earth fault. (b) Typical circuit for sensitive directional earth fault detection

earthing through a high value of resistance which will limit earth fault current to the order of a few amps, and which will greatly reduce fault damage and permit operation to be sustained for a reasonable period. A more sophisticated protection system is required to detect such a fault but, since time is not so critical, this is usually a practical solution.

Other forms of impedance earthing may have particular advantages for special systems or conditions but, in general, the above logic applies and the monitoring/protection equipment used is as described above.

Solid earthed system.   Solid earthing employs an alternative philosophy. The advantages are that no significant overvoltage occurs following an earth fault but a large current will flow in the fault and the detection of this can be done easily using a simply relay, which can result in action being taken rapidly. However, the disadvantage exists that some damage will be done at the point of the fault. In practice, on low-voltage systems an earth fault will present a significant impedance and the fault current is not, in effect, a solid earth fault current, but this factor is less significant on high-voltage systems.

Motor neutral.　In most systems any earthing is carried out at the neutral point of a generator or transformer, but very seldom at a motor neutral. This usually precludes the direct measurement of motor earth fault current, and this has to be done by some form of current balance sensing which checks that the net current flow into the motor circuit is zero, i.e. none returns via an earth fault.

Residual connections and a typical circuit for sensitive direction for earth fault detection are shown in Figure 7.23.

7.4.10.2 Winding phase-to-phase faults

An insulation failure can occur which does not involve earth, either by breakdown of insulation between adjacent turns in the same phase (interturn fault) or between conductors in two different phases (phase-to-phase fault). The consequence of either type of fault is similar in that a significant increase occurs in one or more line currents, which then become unbalanced, and local arcing occurs which will produce damage to winding conductors and insulation. The increase in current, and consequent rate at which damage occurs, depends on the location of the fault, but in most cases severe damage is caused in a very short time. For this reason it is unsatisfactory to rely on devices matched to winding thermal requirements, as they would not function fast enough to prevent serious damage to the motor, and would produce possible adverse effects on the supply system.

Fast-acting current relays, described as instantaneous, are available for fault protection and operate satisfactorily on major faults where high line currents occur. However, such devices must be set to a tripping value above the greatest that can occur in healthy operation – usually the motor starting current – and hence do not give protection for a fault current below this value. Very fast relays have an additional disadvantage in that the instantaneous starting current taken by a motor has an asymmetrical component, which can differ for each phase, and the relay sees an appreciably higher current than the nominal r.m.s. a.c. starting current. Smaller motors are usually protected by a relay insensitive to the asymmetry and set just above the starting current. Fault currents of lower value can then continue to flow until the normal overload relay runs out on its time characteristic.

On larger or more important motors, more sensitive fault detection equipment can be provided but this involves additional equipment and cost, and is based on current comparison rather than on absolute value of currents. As the faults being considered occur within the machine windings, there will be a difference between the three line currents at the line terminals and the three appearing at the neutral ends and if these two values are compared for each phase a sensitive form of fault detection is available. Such differential protection is provided for all drives requiring maximum integrity (Figure 7.24).

This system still has limitations on sensitivity, since the characteristics of the current transformers at the line and neutral end of each phase should have identical characteristics for all values of current flowing through them, and this is not attainable commercially. Consequently, the setting of such a relay has to be a finite value of difference current and is usually expressed as a percentage of the winding being protected, e.g. 95% implies that a fault within 5% of the phase winding and the neutral point would not produce a current theoretically capable of operating the relay. This condition is acceptable because this is the region least likely to encounter a fault, the insulation being at low stress levels, and also a fault here will produce minimum rate of damage and would either be detected by other means or

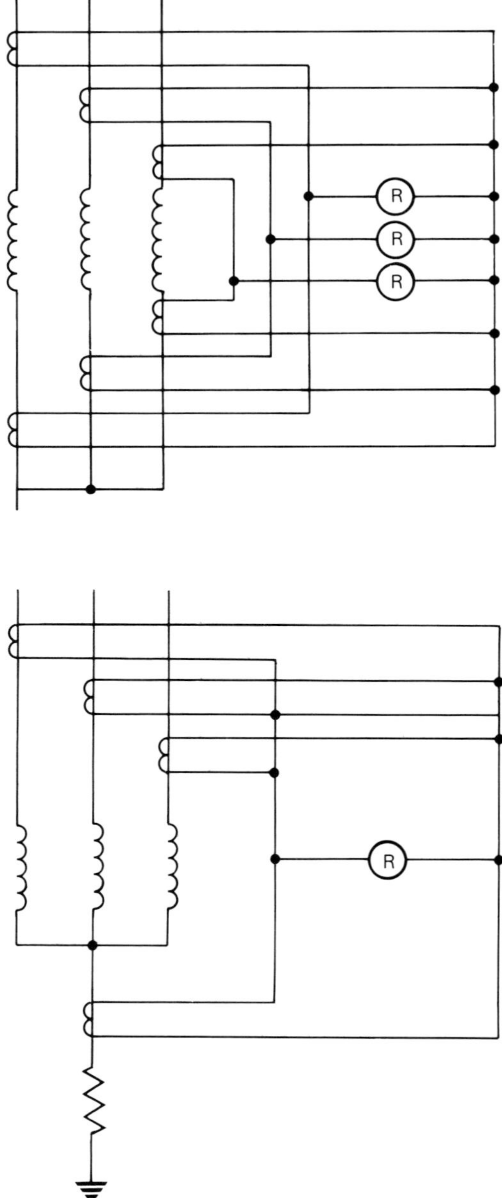

Figure 7.24 Simple forms of primary differential protection. (a) Phase and earth fault.
(b) Restricted earth fault

would degenerate into a more serious fault which would cause the protection to
operate. Various arrangements of differential protection are available depending
on the actual motor, system and operating conditions.

Interturn faults can produce rapid local damage in a winding and will rapidly
deteriorate into a phase-to-earth or a phase-to-phase fault and can be dealt with as

described above. On some very large motors, however, it may be economically justifiable to use some form of interturn fault detection. When there are parallel circuits in each phase winding, an effective system is provided by comparing the currents in each section of winding. This arrangement has the same advantages as the differential system but can detect some faults before they degenerate and cause much more serious damage.

7.4.10.3 Excitation failure

When a synchronous motor is excited it generates a synchronous torque dependent on its reactance and the magnitude of excitation. Thus, loss of excitation or reduction below a critical value can result in the motor failing to meet the required load torque and losing synchronism. As soon as a slip speed occurs, the motor will provide an induction motor torque provided by its starting or winding and, if this is sufficient to meet the load torque, it will continue to run at slip speed. If the induction torque is inadequate, the speed will drop until it meets the corresponding load torque and may reduce to zero, i.e. the motor will stall.

The radial airgap of a synchronous motor is larger than that of a correspondingly rated induction motor and this results in a lower magnetizing reactance and, hence, a lower operating power factor. Thus, when a synchronous machine ceases to run synchronously its power factor will change from unity or a leading value to a lagging value and, in general, its line current will increase for the same kilowatt output. In addition, any residual excitation will produce a synchronous torque superimposed on the induction torque and this will result in a speed pulsation and also a line-current pulsation.

The significant features of excitation failure commonly used for detection purposes are magnitude of excitation, and line power factor. Excitation current under fixed excitation conditions will normally be between the no-load and full-load values and a simple current relay will detect when it is reduced below a safe value. During transient load conditions, a superimposed current will be induced in the excitation winding and this can reduce the instantaneous magnitude to a very low value; any undercurrent relay used must be time-delayed to avoid incorrect operation. With an automatic excitation control system, it is usually impractical to detect failure using an undercurrent relay.

During load transients the motor line current and power factor fluctuate but should the power factor, which is related to the motor load angle, reach a critical value the motor will slip one pole. When this happens, it is theoretically possible that the motor will be restored to synchronism after only one or possibly two pole slips but, in practice, this is unlikely and it is usual to assume that loss of synchronism for any time, even a single pole slip, indicates a fault condition.

It must be realized that this condition also arises when the load torque exceeds the motor healthy torque capability and does not necessarily indicate a failure of excitation, but does indicate a fault condition on the plant.

A suitable device for detecting loss of synchronism is an attracted armature-type relay with coils fed from appropriate line voltage and line current. The relative vectors produced by these can be adjusted to ensure that the locus of normal operation zone does not operate the relay, but any excursion outside this zone, such as by severe power factor swing or current amplitude associated with a pole slip, will operate the relay.

The accurate setting of such a relay requires full information on machine parameters, loading duties and system performance under operating and fault conditions. Non-critical setting can be used on an approximate basis which will give fault detection but may cause inadvertent tripping under marginal but healthy conditions.

7.4.10.4 Single-phasing

When one line supplying a three-phase motor is open-circuited, a synchronously rotating field will still be produced and the motor will continue to run and provide torque. However, a significant negative phase sequence field will also be produced, which will induce extra losses in the secondary winding and produce an additional pulsating torque. Assuming the load remains unchanged, the two remaining line currents will be increased by the order of 50% apart from secondary effects.

The abnormal heating conditions in the motor will cause insulation deterioration and it is usual to trip a motor which has single-phased as quickly as convenient.

Motors provided with winding temperature protection, and those with overcurrent devices, will have a reasonable cover but these do not give close secondary winding discrimination, and some motors are critical in this area. When rapid action is required a negative sequence relay (see Section 7.4.6) can be used.

Digital-type motor protection relays of the compound type (see Section 7.4.7) usually include single-phase protection, and several have facilities for adjustment for setting of this feature independently.

7.5 Relevant performance standards

Motor control and switchgear may have to comply with the latest editions of the following standards, as far as they apply, and are listed in Table 7.1.

Table 7.1

British Standard	Institute of Electrical Contractors equivalent	
88	269	Cartridge fuses for voltages up to 1 kV a.c. and 1.5 kV d.c.
89	51	Direct-acting indicating electrical measuring instruments
142	—	Electrical protective relays
159	—	Busbar and busbar connections
162	—	Electrical power switchgear and associated apparatus
387(1)	—	Miniature air-break circuit-breakers for a.c. circuits
587	—	Motor starters and controllers
775(2)	470	Alternating current contractors for voltages 1–12 kV
923	60	Guide on high-voltage testing machines
2757	85	Classification of insulating materials
2692	282	Fuses for voltages exceeding 1 kV a.c.
3693	—	Recomendations for the design of scales and indices
3938	185	Current transformers
3941	186/A/358	Voltage transformers
4752(1)	157-1/A	Switchgear for voltages up to 1 kV a.c.; circuit-breakers
4941	292	Motor starters for voltages up to 1 kV a.c.

Table 7.1 (Continued)

British Standard	Institute of Electrical Contractors equivalent	
5227	298	Alternating current metal-enclosed switchgear and control gear for above 1 kV
5253	129	Alternating current disconnectors (isolators) and earthing switches for above 1 kV
5311	56	Alternating current circuit-breakers for above 1 kV
5419	408	Air-break switches, disconnectors and fuse combination units up to 1 kV
5420	144	Degrees of protection of enclosures of switchgear and control gear for voltages up to and including 1 kV a.c.
5486	439	Factory-built assemblies of switchgear and control gear up to 1 kV
5490	529	Classification of degrees of protection provided by enclosures
5856(1)	632-1	Direct-on-line a.c. starters for above 1 kV

Other specifications which may apply for marine applications are shown in Table 7.2.

Table 7.2

British Standard	Institute of Electrical Contractors equivalent	
6231	228	Specification for PVC-insulated cable for switchgear and control gear
—	92	Electrical installations in ships
—	92-202	System design – protection
—	92-302	Equipment – switchgear and control gear assemblies
—	92-502	Special features – tankers
—	92-503	Special features – a.c. supply systems 1–11 kV
—	363	Short-circuit evaluation of circuit-breaker installations in ships

8 Works and acceptance testing

E. Spooner

University of Manchester Institute of Science and Technology

8.1 Test categories

Tests on machines conducted at the manufacturer's works fall into two broad groups: (1) formal tests to demonstrate that the machine meets the customer's requirements; and (2) tests to support the development of a range of machines.

The programme of formal tests is a matter for agreement between customer and manufacturer and will normally follow the requirements laid down in specifications such as BS 4999:Part 60. 'Basic' tests are carried out using the first few machines built to a new design to establish that the specified performance is achieved in all respects. 'Routine checks' are carried out on every machine to confirm that it is fault-free on leaving the works. An abbreviated sequence of basic tests conducted to confirm that a motor conforms to the design is called a 'duplicate' set of tests.

During the life of a machine it may undergo further tests for a variety of reasons such as: (1) test following a repair or rewind; (2) to prove the suitability of a different brush grade; and (3) establishing the machine's suitability for a new duty. Some of the more common of these tests are covered in the present chapter.

A formal test programme for a given type of machine might include the tests specified in Table 8.1.

8.2 Test procedures

8.2.1 Preparation

After a machine has been fitted on the test-bed, the engineer in charge must ensure that energizing the machine will not risk:

(1) Damage to the machine.
(2) Damage to the test equipment.
(3) Injury to people.

Factors to consider under (1) and (2) include:

(a) coolant supply;
(b) bearing lubrication;
(c) shaft rotation is unrestricted;
(d) connections have been made correctly;
(e) all brushes have been fitted and are in contact.

432

Table 8.1 Test schedules

No.	Test	Basic	Duplicate	Routine check
(a)	*Synchronous machines (including synchronous induction machines)*			
1	Resistance of windings (cold)	Yes	Yes	—
2	No load losses	Yes	Yes	Yes
3	Locked rotor – current	Yes	Yes	—
	– torque	Yes	—	—
4	Open-circuit secondary induced voltage at standstill (wound rotor)	Yes	Yes	Yes
5	Temperature rise	Yes	—	—
6	Tests to establish efficiency	Yes	—	—
7	Momentary overload	Yes	—	—
8	High voltage	Yes	Yes	Yes
9	Vibration	Yes	—	—
10	Short-circuit saturation	Yes	Yes	—
11	Short-circuit losses	Yes	—	—
(b)	*Induction machines*			
1	Resistance of windings (cold)	Yes	Yes	—
2	No load losses and current	Yes	Yes	Yes
3	Locked rotor – current	Yes	Yes	—
	– torque	Yes	—	—
4	Open-circuit secondary induced voltage at standstill (wound rotor)	Yes	Yes	Yes
5	Temperature rise	Yes	—	—
6	Power factor and any tests to establish efficiency	Yes	—	—
7	Momentary overload	Yes	—	—
8	High voltage	Yes	Yes	Yes
9	Vibration	Yes	—	—
(c)	*d.c. machines*			
1	Resistance of windings (cold)	Yes	Yes	—
2	No load losses and current	Yes	Yes	Yes
3	Temperature rise	Yes	—	—
4	Tests to establish efficiency	Yes	—	—
5	Momentary overload	Yes	—	—
6	Commutation	Yes	Yes	—
7	High voltage	Yes	Yes	Yes
8	Vibration	Yes	—	—

Note: In the case of d.c. machines intended to operate with a voltage ripple greater that 0.04 per unit peak-to-peak amplitude, special arrangements should be made.

Safety-consciousness is particularly important in a test situation since equipment changes daily and it is usually impractical to make everything inherently safe. The following factors are important, others will arise in particular situations:

(a) no danger of touching the terminals;
(b) fencing installed around hazard areas;
(c) appropriate warning signs are displayed;
(d) rotating shafts are guarded adequately;
(e) flammable materials are removed;
(f) sufficient and suitable fire extinguishers are available.

8.2.2 Winding resistance measurement

This is usually the first test in a series and the results are employed in subsequent thermal tests. The test should be carried out with the machine at a stable, uniform and known temperature. The usual method is to feed a stable direct current into the winding and to measure the voltage developed. The following precautions are worth while:

(1) Use the four-wire method whereby the current source and the voltage measurement connections are made separately. This avoids incorporating an indeterminate contact resistance in the measurement.
(2) Employ a stabilized power supply for the current source to avoid the drift which can occur with most other types of source.
(3) Take the voltage reading as close as possible to the winding itself, incorporating as little connecting cable as possible. Even a short length of cable can have a significant resistance. In later thermal tests in which resistance changes are used to derive temperature, cable resistance can lead to serious errors if uncorrected.
(4) For commutator windings, take the voltage reading at each of the commutator bars touching the brush and take the highest reading. Repeat the measurement several times, moving the rotor to a new position for each one, and average the results.

Winding resistances are conventionally declared for an average winding temperature of 75° C for windings with Class A, B or E . . . insulation and 115° C for Classes F and H. If the measured resistance is R_A at ambient temperature T_A then the resistance R_T at temperature $T°$ C, is given by:

$$R_T = R_A (235 + T)/(235 + T_A) \tag{8.1}$$

In exceptional cases where a conductor material other than copper is used, the figure 235 must be replaced with the inverse of the temperature coefficient of resistivity for the conductor employed.

8.2.3 Dielectric tests

The insulation tests normally carried out on large high-voltage machines ($>5\,kV$, $5\,MW$) are covered fully in Chapter 6 and BS 4999:Part 61. For lower voltages it is usually considered sufficient for the winding to pass a flash test. For three-phase a.c. windings, BS 4999 recommends that the winding survive for 1 min with an alternating voltage applied between the winding and the frame with an r.m.s. value of 1 kV plus 2 times rated line voltage. For other windings and parts of the motor, the test voltages are given in Table 8.2. As an alternative to a 1-min test a shorter duration, higher voltage, can be specified.

Table 8.2 Test voltages for flash test

Motor type	Motor power P (kW)	Voltage (U)	Motor part	Test voltage (U_{test})	Test limits
Any	$P \leqslant 1$	$U \leqslant 100$	Insulated parts	$2U+500$	—
Any	$P \leqslant 1$	$U > 100$	Insulated parts	$2U+1000$	$U_{test} \geqslant 1500$
Any	$1 < P < 10$	—	Insulated parts	$2U+1000$	$U_{test} \geqslant 1500$
Any	$P \leqslant 10$	$U \leqslant 2000$	Insulated parts	$2U+1000$	$U_{test} \geqslant 1500$
Any	$P \leqslant 10$	$U > 2000$ $U \leqslant 6000$	Insulated parts	$2.5U$	—
Any	$P \leqslant 10$	$U > 6000$ $U \leqslant 17\,000$	Insulated parts	$2U+3000$	—
Any	$P \geqslant 10$	$U > 17\,000$	Insulated parts	By special agreement	
d.c.	—	—	Separate ex-field coils	$2 \times U_{field} + 1000$	$U_{test} \geqslant 1500$
Synchronous	—	—	Field coil s.c. starting	$10 U_{field}$	$U_{test} \geqslant 1500$ $U_{test} \leqslant 3500$
Synchronous	—	—	Field coil o.c. starting	$2 \times U_{field} + 1000$	$U_{test} \geqslant 1500$
Slipring induction motor de-energized for reversing			Rotor winding	$2U_{rotor} + 1000$	$U_{test} \geqslant 1500$
Slipring induction motor not de-energized for reversing			Rotor winding	$4U_{rotor} + 1000$	$U_{test} \geqslant 1500$

* U_{field}, U_{rotor} refers to the open-circuit voltage of the rotor winding at zero speed with the stator supplied at rated voltage.

8.2.4 Temperature-rise tests

Motors are usually rated thermally for continuous operation. A minority of motors, notably traction motors, are rated for repeated loading in a specified cycle. Details of alternative forms of thermal rating (S1–S8) are given in Chapter 3 and in BS 4999:Part 30 and recommended limits of temperature rise are given in BS 4999:Part 32.

Basic tests are needed to confirm that winding temperatures do not exceed the limits for safe working of the insulation, during operation at the motor rating. The basic tests may be used also to establish an equivalent short-term rating for the purpose of routine checking of subsequent motors. The thermal time constants of motors can be as much as several hours, so the establishment of steady-state conditions would consume an inordinate time for a routine test. Temperature limits for the various classes of insulation as applied to various types of winding are given in Chapter 6.

Having established a temperature rise for continuous operation, a similar test is carried out at a somewhat higher load and terminated after a specified period. This test may last 10 min for a small motor of, say, 10 kW or as much as an hour for a larger machine of, say, 100 kW or more. This test may be used to establish the basis for acceptance of subsequent motors.

Motors intended for use on a complex duty-cycle may be required to undergo a series of temperature-rise tests of varying load and duration. Alternatively, the loading cycle may be simulated on the test bed if adequate loading, control and monitoring facilities have been installed.

For a basic test on a continuously rated motor, it is run at full load for a sufficient length of time to establish thermal equilibrium, and the equilibrium values of winding and coolant temperature are taken to confirm that the specified continuous rating is achieved. During the course of the test the temperature of coolant inlet and outlet are monitored together with those of bearings, ambient air and any parts of the machine in which detectors have been installed. Also, any windings whose temperature can be found by deriving resistance from observations of operating current and voltage should be monitored. Observation of the changes in these temperatures indicates progress toward thermal equilibrium.

Some temperatures can only be measured by stopping the motor and connecting a set of instruments for a short period. This should not be done too often since it disturbs the progress toward equilibrium. It is preferable that such measurements should be left to the end of a test.

The temperature measurements are usually expressed as winding temperature rise above the coolant inlet temperature or above ambient in the case of naturally cooled motors.

8.2.4.1 Temperature measurement methods

Three methods are in common use for measuring the temperature of various parts of the machine:

(1) Thermometer placed in contact with the component concerned.
(2) Embedded detector (thermocouple, thermistor or resistance element).
(3) Winding resistance measured at the motor terminals.

The thermometer method gives spot readings for accessible surfaces only and gives poor accuracy unless care is taken to achieve a good thermal contact.

Embedded detectors provide accurate readings of spot temperatures. If a sufficient number of detectors are installed then a good picture of the thermal condition of the motor is obtained at all stages of a test. It is only in very large machines, however, that the cost can be justified. Also, it is usually impossible to install detectors within the coil insulation and the readings obtained will not reflect accurately the temperature of the insulation next to the conductor. However, in the commonly used two-layer windings of induction or synchronous motor stators, little heat flows between layers and a detector sandwiched by the two coil sides in a slot will give a good reading of the conductor temperature in that region (Figure 8.1).

Winding resistance gives an accurate measure of the conductor temperature averaged over the length of the winding. In the cases of synchronous motor field windings supplied through sliprings, d.c. motor field, interpole and compensating windings, the resistance and, hence, temperature can be monitored throughout a test using the d.c. carried by the winding. Naturally, misleading results will be obtained if the current is not pure d.c. or if parasitic-induced e.m.f.s are present.

In the case of windings carrying a.c. or rotating windings, the resistance can be measured only after the machine has been de-energized and brought to rest. In the time taken for the rotor to stop and for the measuring instruments to be connected, the winding temperature can change substantially and usually it is necessary to take timed readings and construct a cooling curve which is extrapolated back to the instant of shutdown to give the required figure for operating temperature.

8.2.4.2 Analysis of cooling-curve measurements

A simple thermal model of a motor winding and its associated iron core is shown in Figure 8.2. The thermal capacities of the winding and the core are represented by C_w and C_i. The coil insulation offers resistance to heat flow between the two, represented by R. When the machine is stationary and the airflow through the

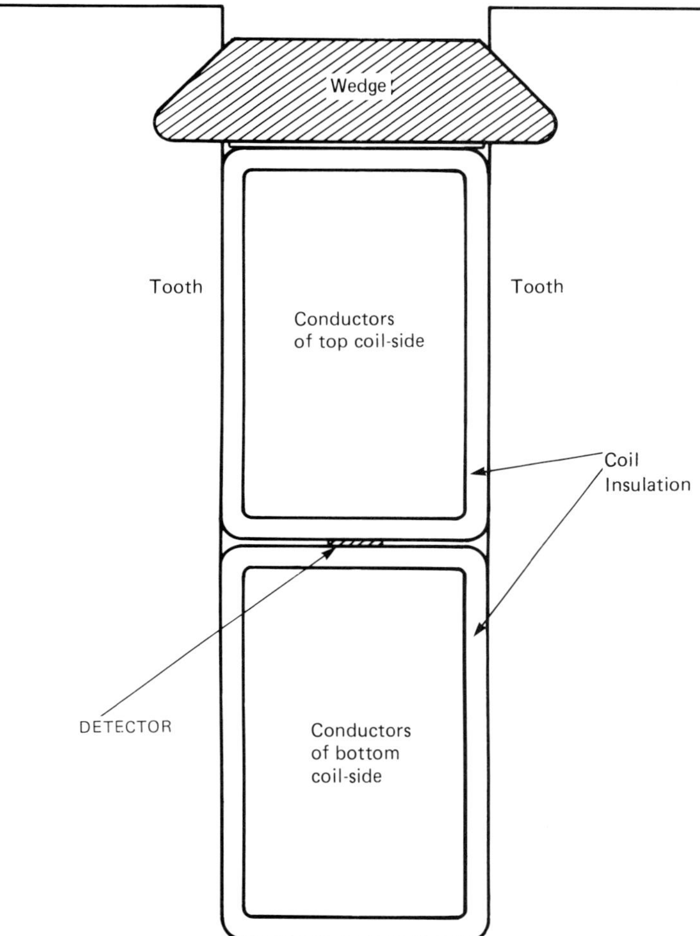

Figure 8.1 Position of an embedded temperature detector to give accurate measurement of conductor temperature

machine has ceased, very little heat will flow other than by conduction between winding and core until they reach thermal equilibrium. Thereafter the machine as a whole cools very slowly (in the still air of the factory) to ambient temperature.

Since the electrical resistance of the winding can only indicate a single figure giving the average temperature, the model of Figure 8.2 representing the winding by a single node is sufficient.

At the instant of shutdown, $t = 0$, the winding and iron temperatures T_w and T_i are T_{w_0} and T_{i_0}. Ignoring the small flow of heat away from the machine, the two temperatures change exponentially toward and equilibrium value T_e according to:

$$(T_w - T_e) = (T_{w_0} - T_e) \exp(-t/\tau) \tag{8.2}$$

and

$$(T_i - T_e) = (T_{i_0} - T_e) \exp(-t/\tau)$$

Taking the natural logarithm of each side of Equation (8.2):

$$\ell n(T_w - T_e) = \ell n(T_{w_0} - T_e) - t/\tau \tag{8.3}$$

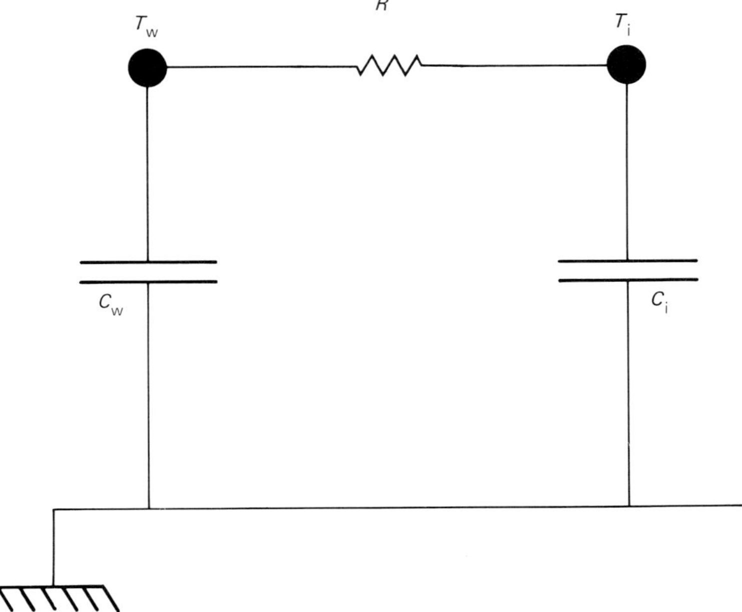

Figure 8.2 Simple thermal model for analysis of cooling curve measurements

A graph of $\ell n(T_w - T_e)$ against t should therefore be a straight line with slope $-1/\tau$ and intercept $\ell n(T_{w_0} - T_e)$. Unfortunately, we need to know T_e to construct the graph from the measured values of T_w. However, knowing that the graph should be a straight line allows us to find numerically the appropriate value of T_e from the data. Figure 8.3 gives a flowchart for a simple process using bisection; a programmable scientific calculator is sufficient for the purpose. Alternatively, a better result for T_e will be obtained if a computer with a program for least-squares fit to a straight line is available.

Having found T_e, the graph can be drawn and the intercept y_0 at $t = 0$ leads to the desired value T_{w_0}, the average temperature of the winding at the instant of shutdown:

$$T_{w_0} = T_e + \exp(y_0) \tag{8.4}$$

Table 8.3 Cooling curve measurements on a 255 kW d.c. traction motor (voltage measured across 19 commutator bars) equivalent resistance at $0\,°C = 8.84\,m\Omega$

Time t (s)	Voltage (mV)	Current (A)	Resistance (mΩ)	Temperature T (°C)
35	71.9	4.92	14.61	153.4
65	70.4	4.92	14.31	145.4
95*	69.2	4.92	14.07	139.0
125*	68.3	4.92	13.88	134.0
155	67.5	4.92	13.72	129.7
185	66.8	4.92	13.58	126.0
215	66.2	4.92	13.46	122.8
275*	65.0	4.92	13.21	116.2
335*	64.1	4.92	13.03	111.4

* These four test points are used in the determination of the equilibrium temperature.

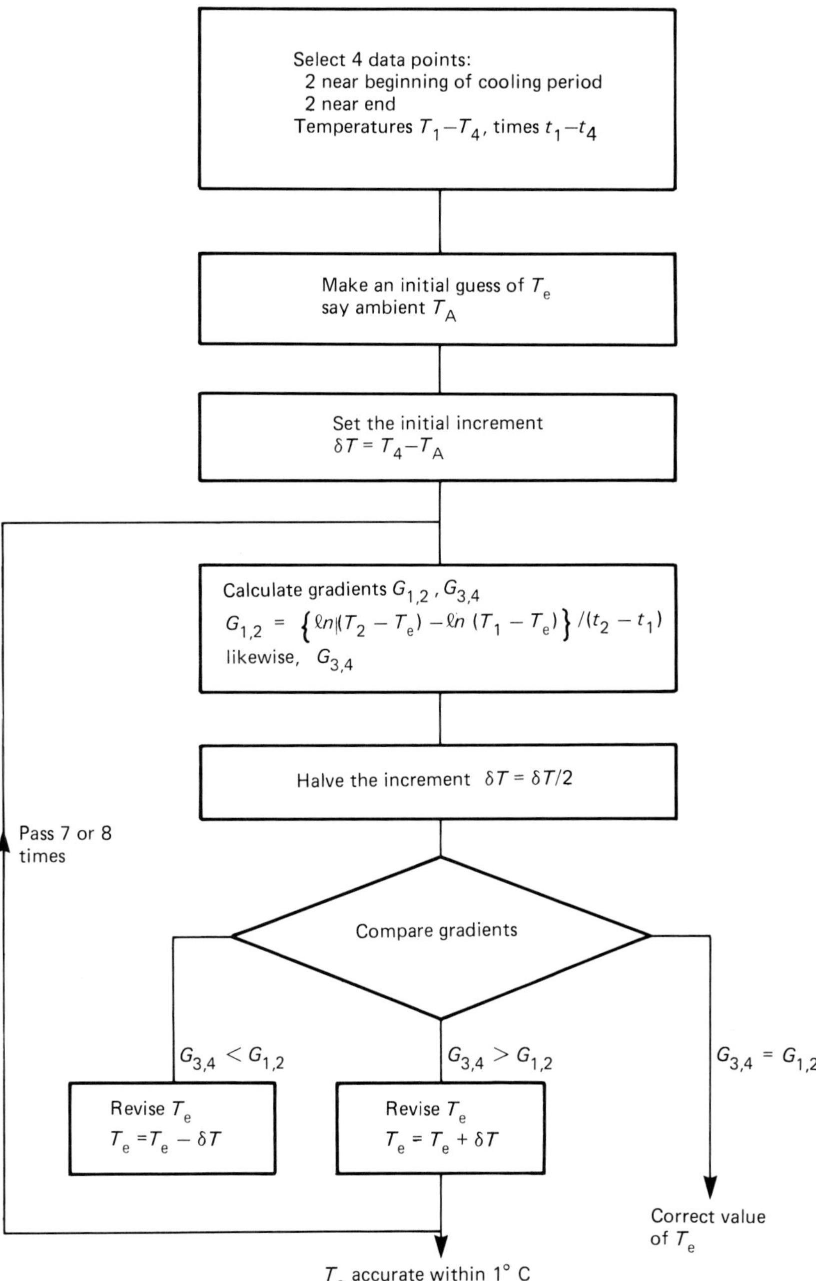

Figure 8.3 Flowchart for computing equilibrium temperature

Example 8.1

To illustrate the above process, a set of readings taken during a test programme on a d.c. traction motor is given in Table 8.3. The resistance values refer to a part of the winding to which connection was made at the commutator by means of a probe with two adjustable contacts, as shown in Figure 8.4. Applying the process of Figure 8.3 to the points indicated in Table 8.2 reveals an equilibrium temperature of 94.2°C. The cooling curve plotted in Figure 8.5 shows the straight-line form expected and gives the shutdown temperature of 161°C from the intercept.

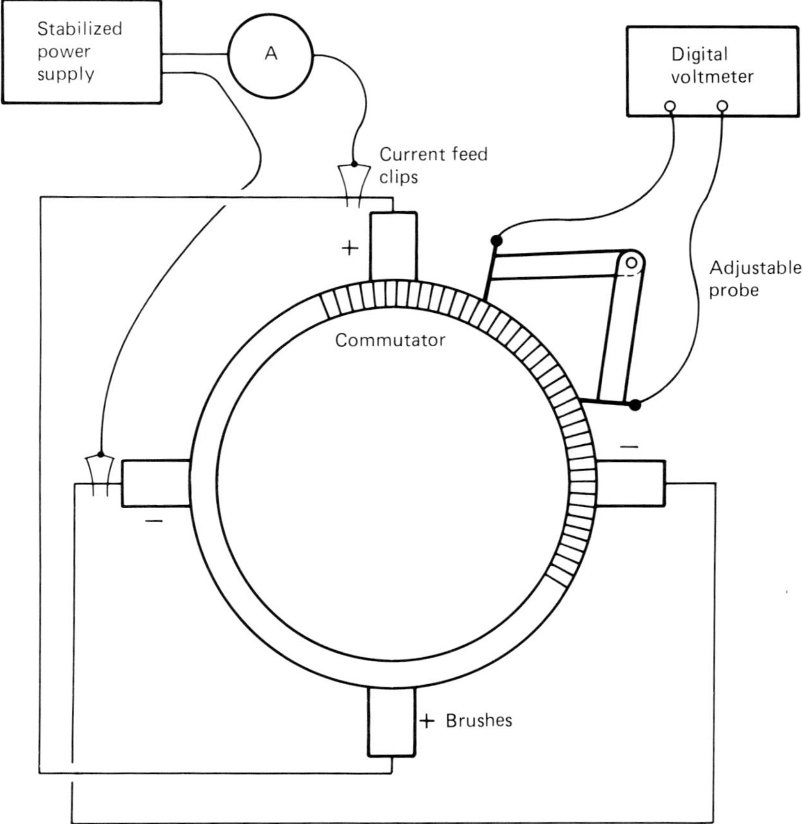

Figure 8.4 Test arrangement for cooling curve measurement on a four-pole d.c. traction motor

8.2.4.3 Additional information

The cooling curve of Figure 8.5 reveals, in addition to the required temperature at shutdown, a value for the thermal time constant for the winding. The slope of the curve is equal to $-1/\tau$. It is readily shown that for the thermal model of Figure 8.2 the time constant is equal to $R \cdot C_w C_i/(C_w + C_i)$. From the known masses of the winding and core we can estimate C_w and C_i with confidence, so that the slope of the graph can be used to find the thermal resistance between the two members.

This information is valuable during the development of a machine and is not readily predicted except by empirical formulae.

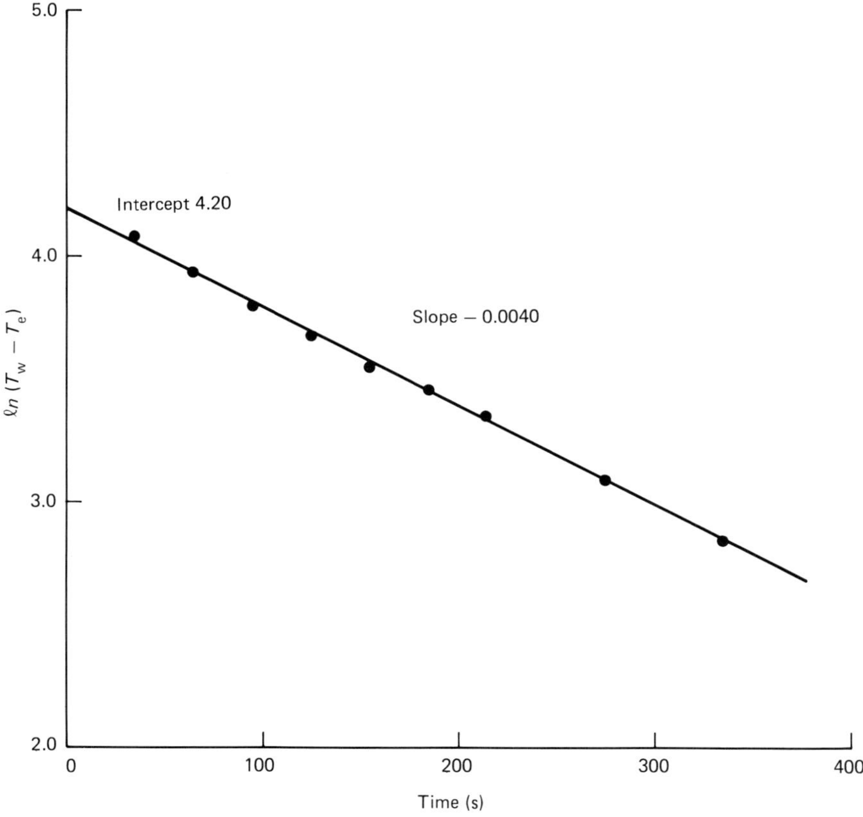

Figure 8.5 Cooling curve plotted on logarithmic scale

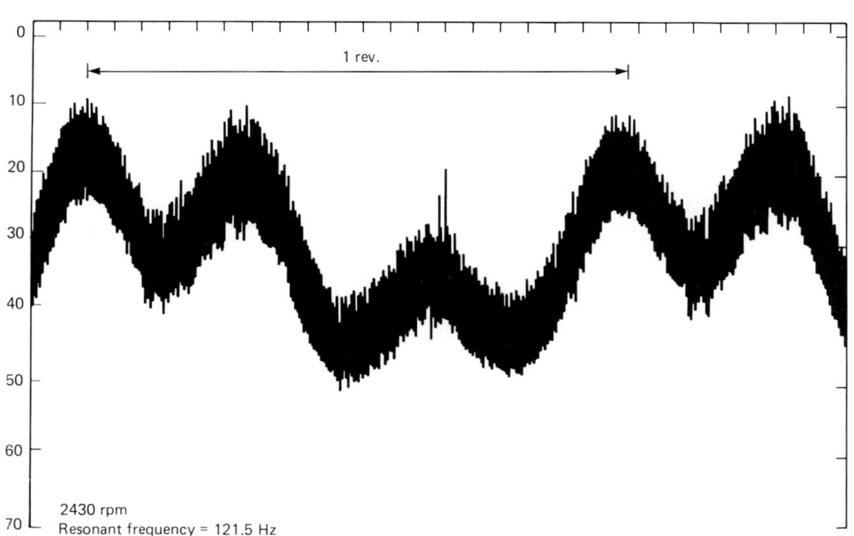

Figure 8.6 Dynamic record of commutator surface profile

For the case of this traction motor example, the masses of winding (including its insulation) and core lead to values of 32 000 and 186 000 J/K for C_w and C_i so that the slope of Figure 8.5 indicates a thermal resistance of 0.0092 K/W. This particular machine was insulated using the vacuum pressure impregnation method which is claimed to achieve better thermal properties than other methods. From the known dimensions, a value of 0.0054 K/W would be expected. The ratio of achieved:expected is about 1.6; similar machines using a resin-rich system have been found to yield a ratio of around 3. In terms of its thermal properties, the vacuum pressure impregnation system does appear to be superior as claimed.

8.2.5 Commutation tests

The majority of d.c. motors of 1 kW or above are fitted with interpoles. The basic form of commutation test is carried out to determine the optimum adjustment of the interpole gaps – both front and rear (see Figure 8.6) and to ensure that the axis of the brushgear coincides with the magnetic neutral axis of the field system.

The tendency for sparking to occur at the brushes depends upon many factors including:

(1) Contact pressure.
(2) Vibration.
(3) Commutator mechanical stability.
(4) Atmospheric pressure and humidity.
(5) Armature current.
(6) Armature speed.
(7) Voltage.
(8) Armature-to-field (m.m.f.) ratio.

On the test bed, the contact pressure is set to the nominal value. The mechanical condition of the commutator is routinely checked using a mechanical recording gauge (the Feinpruf recorder) which indicates the commutator surface profile as the armature rotates slowly past. If suitable equipment is available, a measure of the profile at high speed can be obtained using a capacitive or inductive proximity sensor. A typical record is shown in Figure 8.6.

No amount of adjustment elsewhere can compensate for a bad commutator. If the bars show any indication of even slight radial movement than the test programme should be interrupted for repairs.

The only vibration likely on the test bed is due to the out-of-balance forces of the motor itself or of its loading machine. These should all be small, whereas in service the load might induce severe vibration. Likewise, the test bed is likely to be housed in a warm, dry factory – a luxurious environment for a motor which may be destined to live out in the cold, rain, snow and fog. It will be appreciated that spark-free operation on the test bed is no guarantee of spark-free service. Adding to this the fact that the measurement of sparking is both empirical and subjective, it is clear that commutation test results need to be assessed with great caution.

8.2.5.1 Commutation in d.c. machines

Direct current motors and generators of more than a few kilowatt capacity employ interpoles (sometimes called compoles) for the purpose of inducing the appropriate e.m.f. in those armature coils whose current is reversing. The interpole coil first neutralizes the armature m.m.f. in the airgap under the interpole coil. Secondly, it sets up a reverse m.m.f. which causes a reverse magnetic flux to pass into the armature in that zone. Armature coils entering this field experience a rotational e.m.f. which tends to reduce the current being carried. The strength of the interpole field is adjusted to be just sufficient to cause complete reversal of the armature coil current during the period in which the coil is shorted by the brush–commutator contact. In large d.c. motors, the function of neutralizing the armature m.m.f. is partly filled by a compensating winding set in slots in the face of the main pole.

The e.m.f. needed to reverse the armature coil current during the period of short circuit is proportional to the coil current and the armature speed. The induced e.m.f. is proportional to armature speed and to the flux density in the interpole region.

By the simple expedient of connecting the interpole in series with the armature, the appropriate e.m.f. is induced for all speeds and currents. Unfortunately, magnetic saturation of the interpole and the gathering together of several coils within each slot upsets the matching of e.m.f. induced to e.m.f. required. A certain amount of mismatch can be accommodated by a redistribution of current across the brush contact area. Increasing mismatch results first in tiny localized sparks being drawn between each brush and the commutator bars connected to the most mismatched coil of each slot. This leads eventually to burning of commutator bars in a regular pattern of one per slot, say, one bar in three. Further mismatch leads to more intense sparking which erodes both brush and commutator. Severe mismatch, as might occur during a rapid load change, can lead to sparks extending far enough to initiate a complete flashover from brush to brush.

8.2.5.2 Brush position

In some designs of d.c. motor the brush boxes are carried on a mounting which can be turned relative to the rest of the motor frame and locked in position. It is important to ensure that this mounting is set correctly. The black-band test can be used to check this. For the majority of motors having magnetic symmetry (either designed for bidirectional use or for both motor and generator action) the correct brush position is exactly on the magnetic neutral axis and the simple 'flick' test can be used instead as a routine test for brush position.

The flick test is a null test. It involves applying a voltage momentarily to the field winding with the armature stationary and looking for any voltage induced in the armature circuit (causing the meter to flick). Only if the brushes are aligned with the magnetic axis will the induced voltage be zero. The brush mounting is turned until a null is obtained, and is locked in position.

During the life of a motor, the commutator wears and is periodically skimmed. It is necessary to adjust the radial position of the brush boxes accordingly. Often commutation problems arise after a period in service. It is worth checking both the radial clearance of the brush boxes from the commutator, and their relative spacing around the commutator, as well as conducting a flick test before setting up the motor for a full black-band test. In addition, it is advisable to consult the specialized literature on brush and commutator operation published by the brush manufacturers.[1-3].

8.2.5.3 The black-band test

The black-band test is used to explore the range of tolerance to interpole mismatch. A general form of circuit for the test is shown in Figure 8.7.

For a number of different values of armature current, the interpole current is modified by means of a buck-boost generator until sparking is detected. A range of interpole current is determined for which spark-free (or black) commutation is achieved. The results, expressed as a percentage increase or reduction from the normal current, are plotted in terms of armature current as shown in Figure 8.8.

The buck-boost generator shown in Figure 8.7 needs a very high current capacity at low voltage. For example, a rating of 1000 A, 30 V would be typical for a test

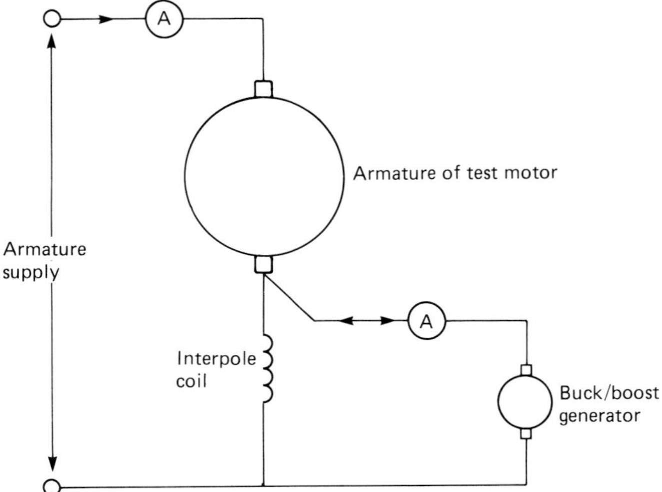

Figure 8.7 Basic circuit for black-band test

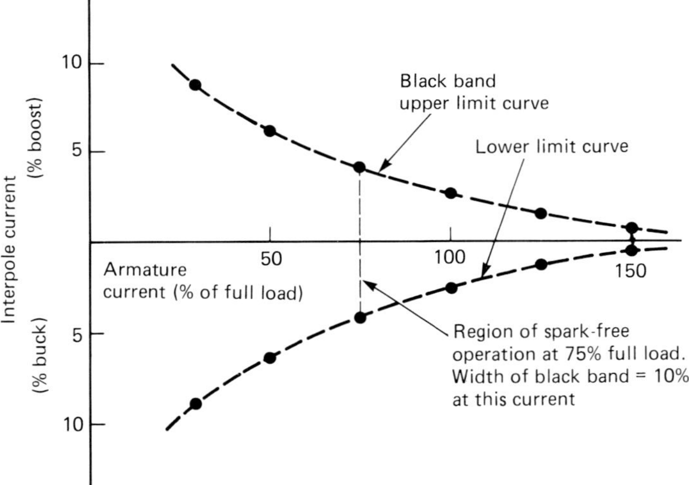

Figure 8.8 Idealized form of black-band test results

installation dealing with motors of up to around 1 MW rating. An alternative arrangement convenient for 'trouble-shooting' applications is to use a lead–acid battery and a variable resistor.

The detection of sparking and the assessment of its severity is a particularly difficult aspect of the black-band test. The usual procedure is to station observers to monitor the brushes visually. The shortcomings are obvious. The method is subjective, it allows only one or two brushes to be monitored, and it is somewhat dubious from the safety point of view. Alternative methods have been tried from time to time employing television cameras or photodetectors coupled by fibre optics. Some success has been achieved and it is to be expected that, in future, such electronic methods will supplant the visual method on installations where black-band tests are frequently carried out.

8.2.5.4 Interpretation of black-band test results

The upper and lower limit curves shown in Figure 8.8 illustrate the form of results
to be expected from a perfectly adjusted shunt-wound motor running at constant
speed. Such a motor would give identical results for either direction of rotation,
provided that it is designed as a bidirectional motor. A series-wound motor
operated from constant voltage would exhibit a much smaller change in the width
of the black band because the lower speed at high current makes for easier

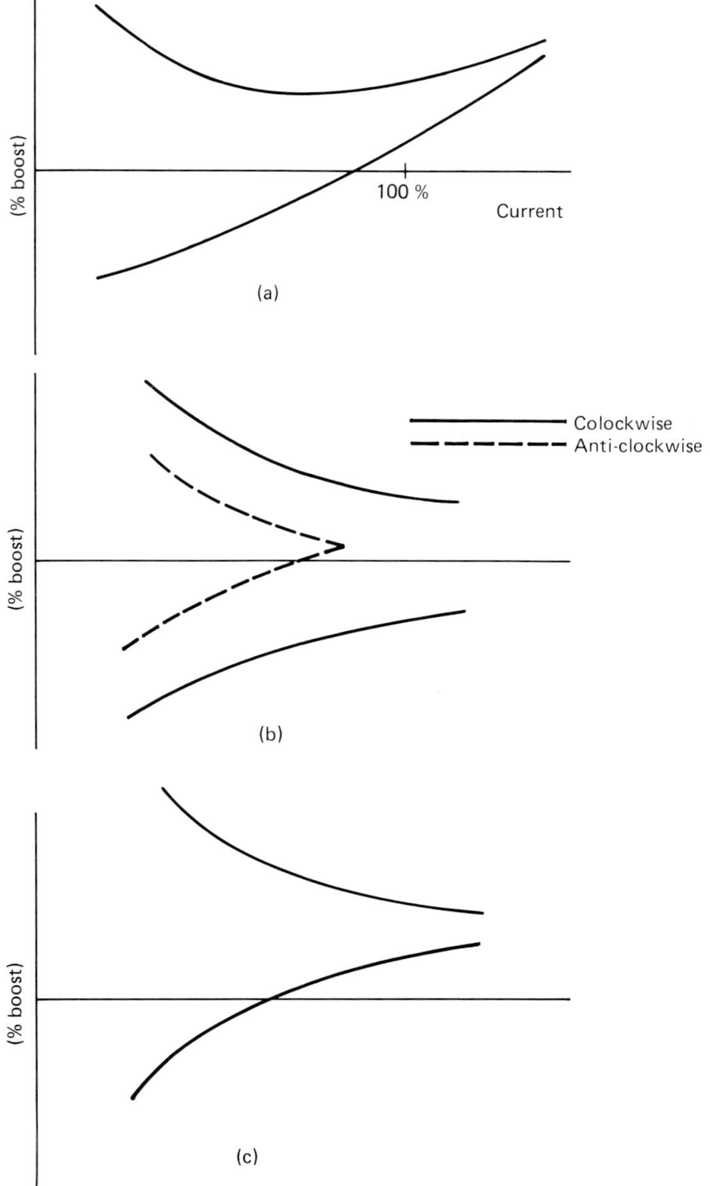

Figure 8.9 Forms of departure of black-band test results from the ideal

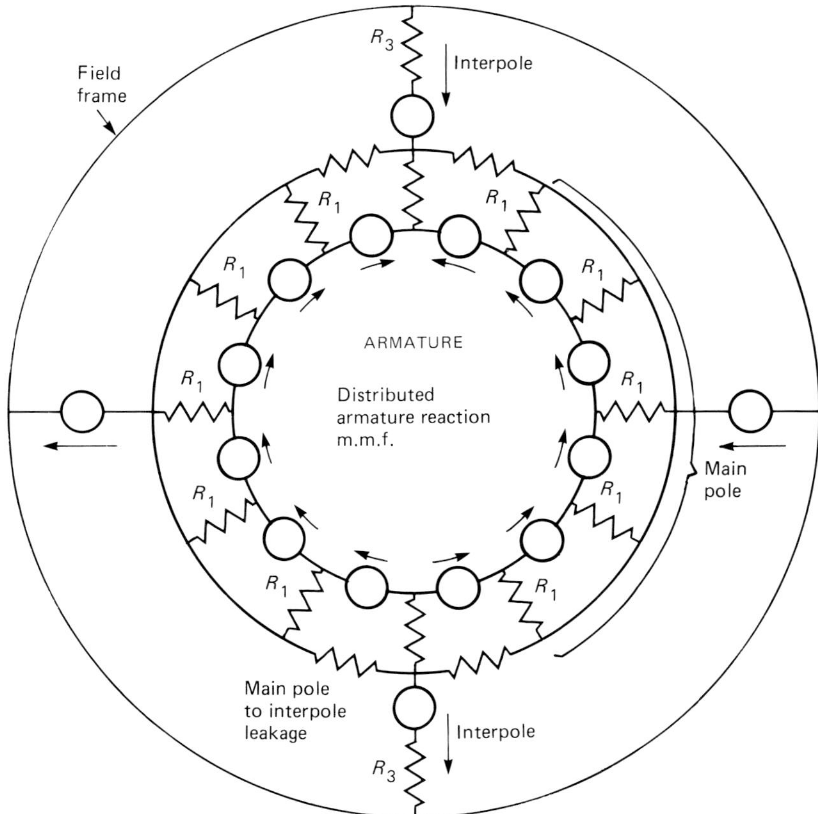

Figure 8.10 Magnetic circuit of a two-pole d.c. machine

commutation. A series-wound motor fed from a constant-power source, e.g. a traction motor for a diesel-electric locomotive, would exhibit a black band widening with increasing current.

In practice, the results depart from the ideal and indicate any modifications or adjustments necessary to give the greatest range of spark-free operation, and the least chance of flashover under transient conditions. Figure 8.9 shows several types of departure from ideal for a shunt motor and the causes and remedies are given below. A similar set of interpretations applies for other motor types.

(1) The curves bend upward for high values of armature current; this results from saturation of the interpole magnetic flux path and is unavoidable. Except under severe overload, this effect should not lead to serious sparking in a properly designed and adjusted motor.
(2) The width of the black band differs for the two directions; this is caused by incorrect brush position. It is readily corrected if the position is adjustable and can easily be checked using a flick test.
(3) The black band is displaced from the ideal centralized position in either the buck or boost direction. This is caused by an incorrect combination of interpole airgap and number of turns. The most convenient remedy is usually to alter the thickness of the non-magnetic shim located at the root of the interpole core.

If the best commutation is achieved with the interpole current boosted by an amount x (e.g. 5% boost corresponds to $x = +0.05$) then the thickness of non-magnetic shim required behind the interpole core is found from an analysis of the magnetic circuit of the motor (Figure 8.10). Thus:

$$R = \frac{-x(z-1)R_1R_2+(x+z)(y-1)R_2R_3+(z-1)R_1R_3}{(1+x)(z-1)R_1+(1+x)zR_2-x(y-1)(R_2+R_3)-x(y-1)^2R_2R_3} \tag{8.5}$$

where R is the required reluctance of the shim (including R_3 below) and is equal to:

$$\frac{\text{Thickness}}{\mu_o \times \text{area}}$$

R_1 is the airgap reluctance of the main pole airgap:

$$\frac{\text{Main pole airgap}}{\mu_o \times \text{pole face area}}$$

R_2 is the airgap reluctance of the interpole airgap:

$$\frac{\text{Interpole airgap}}{\mu_0 \times \text{interpole face area}}$$

R_3 is the reluctance of any existing interpole shim;
y is the main-pole leakage coefficient; and
z is the ratio of interpole to armature amp turns.

In the vast majority of cases, x will be plus or minus a few per cent so that a simpler equation may be used:

$$R \simeq R_3 - xR_2 \frac{(z-1)R_1 - (y-1)R_3}{(z-1)R_1 + z(y-1)R_3} \tag{8.6}$$

8.2.6 No-load tests

All types of machines undergo no-load tests as routine checks. In its simplest form, this test involves running the motor at its rated speed and voltage and checking that the current and power drawn from the supply are within prescribed limits. This simple check will reveal a range of faults such as winding inter-turn short circuits, rotor asymmetry such as might be caused by broken bars in a cage rotor or unsound connections, and mechanical faults, particularly in bearings and seals.

A comprehensive no-load test is a basic test carried out to determine the magnetization characteristic and the constant losses. The necessary arrangements differ between motor types, as follows.

8.2.6.1 Induction motor no-load test

A variable-voltage supply at the rated frequency is needed. If an electronic means is used to vary the voltage, care should be taken to avoid harmonics being introduced to the motor.

The motor is run up to speed and the voltage is set to about 10% above the nominal value. Readings of slip s, input current I and power P are taken over a range of progressively smaller voltage settings V, extending down to about 25% of the rated value.

The magnetization characteristic can be plotted using the voltage and the magnetizing current I_m found from:

$$I^2_m = I^2 - P^2/V^2 \tag{8.7}$$

The input power includes a small copper loss I^2R, mechanical losses, core losses and parasitic losses. The I^2R may be separated by calculation, the friction loss may be determined by extrapolating the curve of $(P - I^2R)$ to the vertical axis as shown in Figure 8.11.

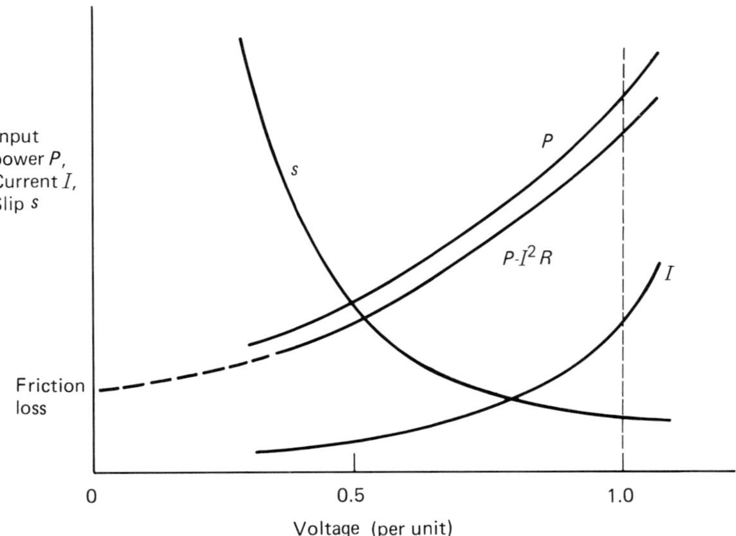

Figure 8.11 General form of no-load test results for an induction motor

The major source of error is in the measurement of power which is carried out at very low power factor.

8.2.6.2 Synchronous motor no-load test

If the motor can be started from the supply, the test can be carried out in a simple manner. The motor is run up to speed and synchronized. The voltage is varied in steps and at each setting the field current is adjusted to bring the armature current to a minimum. In this condition the phasor diagram for the machine is as shown in Figure 8.12(b). It can be appreciated, by comparing the phasor diagrams of Figure 8.12(a) and (b), that setting the field current to give minimum armature current I establishes a close correspondence between the induced e.m.f. E and the terminal voltage V and also makes the armature power factor unity. The magnitude of the in-phase current I_p has been exaggerated in Figure 8.12(a) and the correspondence will be very close in practice.

Measurements of armature current and voltage and field current are used to construct the curves shown in Figure 8.13(a).

Synchronous motors which cannot be started from the normal supply are uncommon. Generators, however, are not usually designed for electrical starting. In either case the no-load test may be carried out with the machine coupled to a small motor which acts as a prime mover. If the generated frequency is the same as the supply frequency, the machine may be synchronized and the test carried out as described above. Alternatively, it may be necessary to carry out a generator

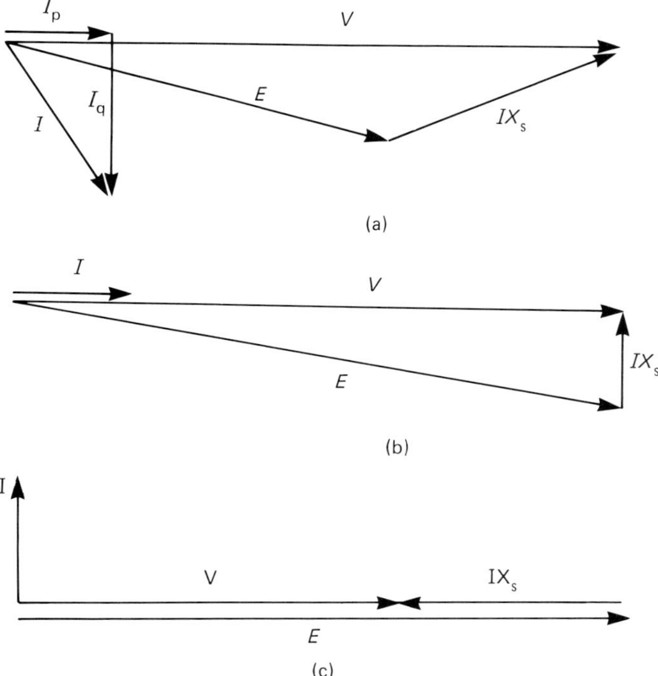

Figure 8.12 Phasor diagrams for synchronous motor operation. (a) Under-excited, drawing lagging current component I_q. (b) Minimum current condition $I_q = 0$. (c) Over-excited, no-load, drawing leading current (zero power factor test)

open-circuit test instead. In this case the field current is applied and the generated e.m.f. is measured directly. Power measurement can be made at the terminals of the drive motor.

8.2.6.3 Direct current motor no-load test

The field and armature should be supplied separately and at variable current and voltage. The supplies should be pure d.c., since pulsating voltages can cause disproportionate pulsating currents and lead to spuriously high loss measurements.

Armature voltage, current and speed measurements are taken over a range of field current settings. The armature voltage is adjusted at each setting to bring the speed as close as possible to the rated value.

The e.m.f. is found by subtracting the IR loss and an allowance for brush contact drop from the measured voltage. The magnetization curve is obtained by plotting the ratio e.m.f./speed against field current.

8.2.7 Locked-rotor test

Induction and synchronous motors undergo a locked-rotor test as a basic or duplicate test. It is analogous to the short-circuit test carried out on transformers. This test gives a measure of the total leakage reactance in the case of an induction motor and of the subtransient reactance of a synchronous machine.

The test is carried out at low voltage to avoid excessive currents. Consequently, flux densities are low and saturation is unimportant. A single set of measurements is adequate because of the absence of saturation.

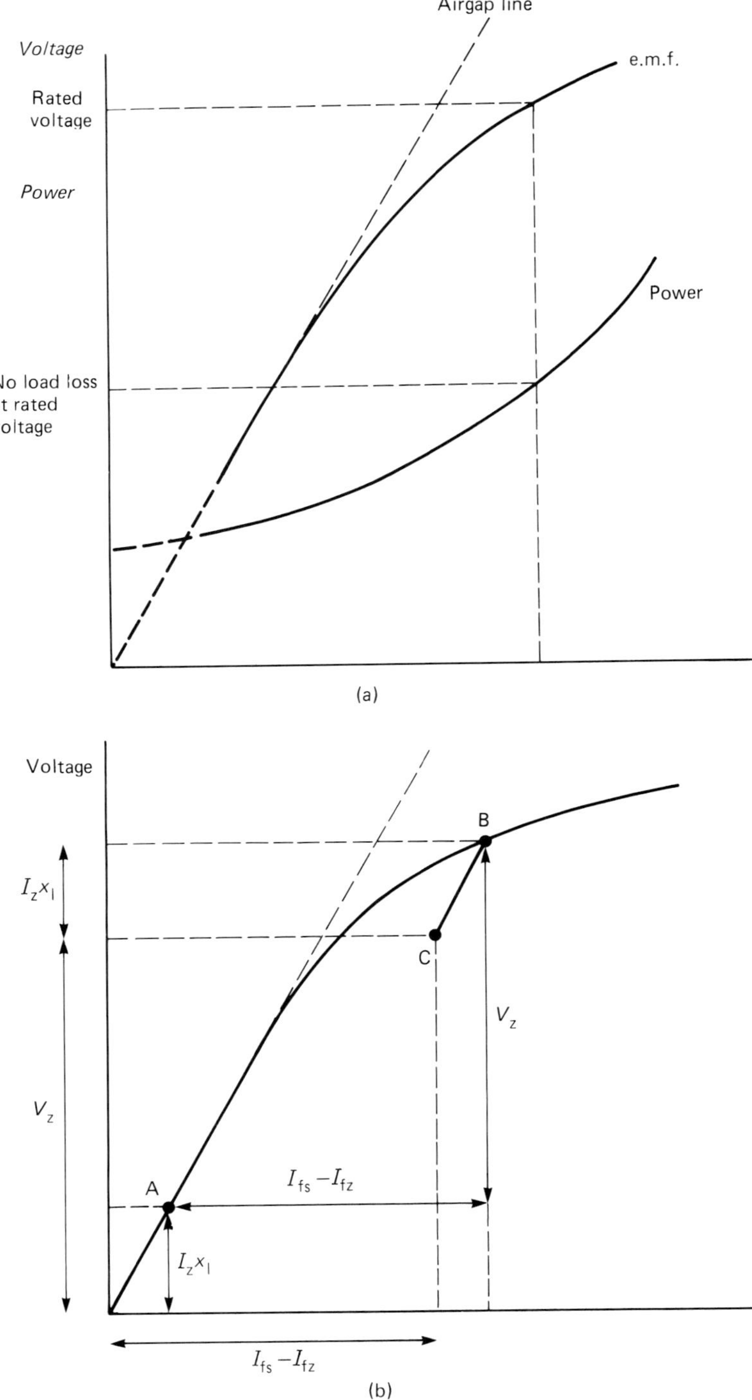

Figure 8.13 No-load test results for a synchronous motor. (a) Field current. (b) Equivalent field current

The rotor should be held firmly in position and, for a basic test, there should be means provided for measuring torque. Usually a torque arm and load cell would be used for restraining the rotor. The secondary terminals of a wound-rotor induction motor must be connected together whilst, for a synchronous motor, the field winding should be left open-circuited. In each case the measured voltage and current yield reactance directly. The torque measurement is required only for a basic test. It relates to the important design feature of starting torque.

8.2.8 Synchronous motor reactances

The leakage reactance of a synchronous motor can be found by using the effect of saturation. Two tests – a short-circuit test and a zero-power-factor test – are needed to provide the required data which, along with the results of the no-load test enable the leakage reactance to be found. A value for the magnetizing reactance is also obtained.

For the short-circuit test the motor needs to be driven by a coupled low-power motor but close control of speed is not necessary. The armature terminals are short-circuited with provision made for current measurement and the field is energized. A few readings of armature current up to full rating are taken along with readings of the corresponding excitation current. Under short-circuit conditions the flux density is very small and saturation is unimportant. The results of this test may be plotted as a straight line through the origin on a graph of armature current against field current.

The net m.m.f. driving flux around the magnetic circuit is the difference between the armature and field m.m.f.s and is just sufficient to drive the short-circuit current through the armature impedance. The armature impedance is mostly the leakage reactance, so that we can write:

$$I_a\,(x_1 + X_{m_0}) = E \tag{8.8}$$

E being the e.m.f. which would have been induced under open-circuit conditions, and X_{m_0} being the magnetizing reactance under the condition of no saturation.

E can be found from the no-load test results by multiplying the slope of the airgap line (Figure 8.13(a)) by the field current measured in the short-circuit test. Hence, the sum of leakage and unsaturated magnetizing reactance is found from Equation (8.8).

If the machine were now to be used as a generator supplying an inductive load, then a larger airgap flux density would be required but the field and armature m.m.f.s would remain in the same phase relationship as for the short-circuit test. Figure 8.12(c) shows the phase relationship, V being the terminal voltage which is zero in the case of the short-circuit test. Of course, it is not necessary to use a physical inductor, simply running the motor on no-load and overexcited ensures that the current drawn from the supply leads the voltage by 90° (very nearly) which is the desired zero-power-factor condition.

If the armature current and voltage in the zero-power factor test are I_z and V_z and for the same airgap flux the no-load test showed a voltage of V_n, then:

$$V_n = V_z + I_z x_1 \tag{8.9}$$

from which we can determine $I_z x_1$.

V_n can be determined by using the results of the short-circuit and no-load tests as follows.

The total m.m.f. driving flux around the main magnetic circuit is the sum of the field and armature winding contributions. The armature contribution may be expressed as an equivalent field winding current I_z so that the voltage V_n will correspond to a point on the no-load e.m.f. curve (Figure 8.13(b)) with $I_{fz}-I_z$ as an effective field current since the field and armature m.m.f. act in opposition when the machine is overexcited.

Similarly, if a short-circuit test were carried out using the same armature current then the airgap e.m.f. will correspond to $I_{fs} - I_z$.

Now the difference between the effective total field currents is sufficient to increase the airgap e.m.f. by V_z from $I_z x_1$ to V_n and is simply $I_{fs} - I_{fz}$. I_{fs} is obtained from short-circuit test results by scaling, and I_{fz} is a direct measurement.

The operating point on the magnetizing curve is raised from the short-circuit case shown in Figure 8.13(b) by point A to the zero power factor case, point B. Alternatively, the point C can be plotted from known data and the point B derived by drawing the line through C parallel to the airgap line.

Point B derived in this way gives V_n from which the leakage reactance can be determined. This method due to Potier gives a result which contains minor effects due to stator resistance, and is based on the assumption that saturation of the main magnetic circuit has no effect on the leakage reactance. The leakage reactance obtained in this way is called the Potier reactance.

The magnetizing reactance can now be obtained by subtracting the Potier reactance from the sum of leakage and magnetizing reactance found earlier.

8.2.9 Voltage ratio test

With the rotor stationary, a wound-rotor induction motor is tested for voltage ratio. Rated three-phase balanced voltage is applied to the primary winding (usually the stator) and the secondary voltage is measured. This test is carried out as a routine check with prescribed limits for the result.

As part of a comprehensive test programme, an enlarged version of this test may be used to determine the magnetizing reactance and to obtain estimates of the leakage reactances and the core losses.

The power input measured at rated primary voltage with a stationary open-circuited secondary is almost entirely core loss, divided about equally between rotor and stator cores. The small I^2R loss in the primary winding can be allowed for by calculation. In service, with the rotor nearly stationary with respect to the rotating field, its core loss will be negligible. The desired core loss figure is therefore about half the measured input power less the I^2R loss.

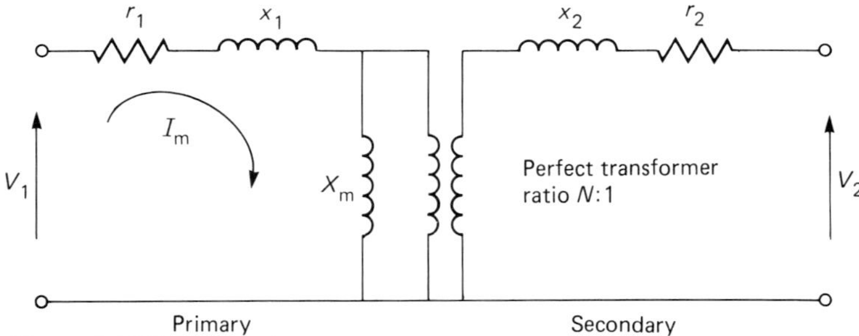

Figure 8.14 Voltage ratio test equivalent circuit

Referring to Figure 8.14, it can be seen that if the ratio N is known from design information, then the magnetizing and primary leakage reactances may be found from V_1, V_2 and I_m:

$$X_m = NV_2/I_m$$
$$x_1 = (V_1 - NV_2)/I_m$$

Feeding from the secondary and making corresponding measurements gives x_2. The resistances are entirely negligible.

If the ratio is not known with sufficient precision then the leakage reactances may be assumed equal in per-unit values and the two sets of measurements will give the reactances and the ratio.

Feeding from the primary (measurement A):

$$NV_{2A} = V_{1A}\, X_m/(X_m + x_1)$$

Feeding from the secondary (measurement B):

$$V_{1B}/N = V_{2B}\, X_m/(X_m + x_2 N^2)$$

The assumption of equal per-unit leakage reactances makes $x_1 = x_2 N^2$, therefore we can find N from the two sets of measurements as:

$$N = \sqrt{(V_{1A}V_{1B}/V_{2A}V_{2B})}$$

The magnetizing and leakage reactances are then found from either set of measurements:

$$\begin{aligned}
X_m &= NV_{2A}/I_m & &\text{referred to the primary}\\
x_1 &= (V_{1A} - NV_{2A})/I_m & &\text{referred to the primary}\\
&= N^2 x_2
\end{aligned}$$

8.2.10 Efficiency and power factor measurement

Measurement of efficiency is a basic test for all machine types, together with the power factor in the case of an induction motor. The method adopted is governed by the means available for applying load.

Motors up to tens, or even hundreds, of kilowatt capacity may be loaded by some form of absorption dynamometer in which the power delivered at the motor shaft is dissipated as heat in the dynamometer. Electrical input power is measured in the usual way at the motor terminals. Mechanical output power is derived from measurements of shaft speed and torque at the dynamometer.

Larger motors require some form of regenerative dynamometer for loading, because of the cost of wasted energy and the difficulty of cooling a high-capacity absorption dynamometer.

Two types of regenerative loading are common. The most straightforward uses a calibrated loading machine, usually a d.c. machine, which accepts the shaft output power of the motor on test and generates into an electrical load. The electrical load may consist of a Ward-Leonard set connected to the mains. The electrical output of the load machine is measured and used to obtain the shaft input power. The load machine must be calibrated beforehand so that its efficiency is known in terms of speed and power. The ratio of loading machine output power to test machine input power is the product of the efficiencies of the two machines.

In the back-to-back method, known as the Hopkinson test in the case of d.c. machines, two identical test motors are coupled, one operates as a motor, one as a generator and power circulates between the two. A power source is needed to make up losses in the two machines and measurement of the input power gives the loss directly and, hence, the efficiency.

8.2.11 Measurement of loss components

Measurement of the individual components of motor loss are normally only carried out for development purposes. The loss components are grouped into constant and load-dependent losses. The constant losses are those which take place when the machine is at speed and energized but unloaded, and the load-dependent losses are those which increase with load power.

Constant losses include:

(1) Bearing friction.
(2) Brush friction and windage.
(3) Hysteresis loss in iron core.
(4) Eddy current losses in iron core and elsewhere.

The power measurements taken in the no-load test give the sum of the constant losses. Also present during a no-load test is the loss associated with excitation. This would take the form of I^2R loss in the field windings of synchronous or d.c. motors. In the case of an induction motor, it is the I^2R loss due to the magnetization current.

The principal load-dependent losses are the I^2R loss in the armature and other windings. In addition, the stray magnetic fields caused by the load current induce eddy-current losses in the windings and in adjacent metal components. These are generally grouped together as stray load loss. A further effect of load current is to modify the field distribution and to increase the 'constant' losses present at zero load.

It is not practical to test separately for each individual loss component but some separation is possible using the tests discussed below.

8.2.11.1 Retardation test

This test can be used for separating the constant losses. It consists of measuring the motor deceleration rate as it coasts to rest following disconnection of its armature supply. The test may be conducted with excitation on or off (Figure 8.15). For induction motors it can only be conducted unexcited.

If the moment of inertia of the rotor is known, then deceleration rates yield torques and powers.

The test results are most usefully taken as speed readings and plotted against time as in Figure 8.15(a). Tangents can be drawn to give deceleration rates and, in turn, torques at various speeds. The torque–speed curves as plotted in Figure 8.15(b) allow the constant loss to be separated into its friction and electromagnetic components. Since the hysteresis torque at a given value of flux density is independent of speed, we can separate them using the torque–speed curve.

To obtain reasonably accurate measurements of loss by this method, it is necessary to ensure that the friction loss is repeatable between tests. The brushes of d.c. motors should be bedded properly and the bearings should be warm enough to allow easy flow of the lubricant. It is advisable to conduct a test with excitation switched off both before and after each test with excitation on, in order to confirm that the friction loss is repeatable.

If the moment of inertia of the rotor is not known from design information, an auxiliary retardation test with a load connected to the terminals can provide its value.

Consider a d.c. motor with a rotor moment of inertia J. Two retardation tests are conducted with the same field current. In the first test, with the armature open-circuit, the deceleration rate at some convenient speed ω is α_1 and the induced e.m.f. is found to be E. In the second test, a load resistor R is connected across the armature terminals and the deceleration rate at speed ω is α_2. R should be chosen so that E^2/R is small compared with the motor-rated power.

In the first test the power consumed in constant losses is $\omega\,\alpha_1\,J$ and in the second test, the power lost in resistance is $E^2/(R+r)$, r being the armature winding resistance.

The total power in the second test is $\omega\,\alpha_2\,J$ which is the sum of electrical loss $E^2/(R+r)$ and constant loss $\omega\,\alpha_1 J$ from which we can find J:

$$J = E^2/(R+r)\,\omega\,(\alpha_2 - \alpha_1) \tag{8.10}$$

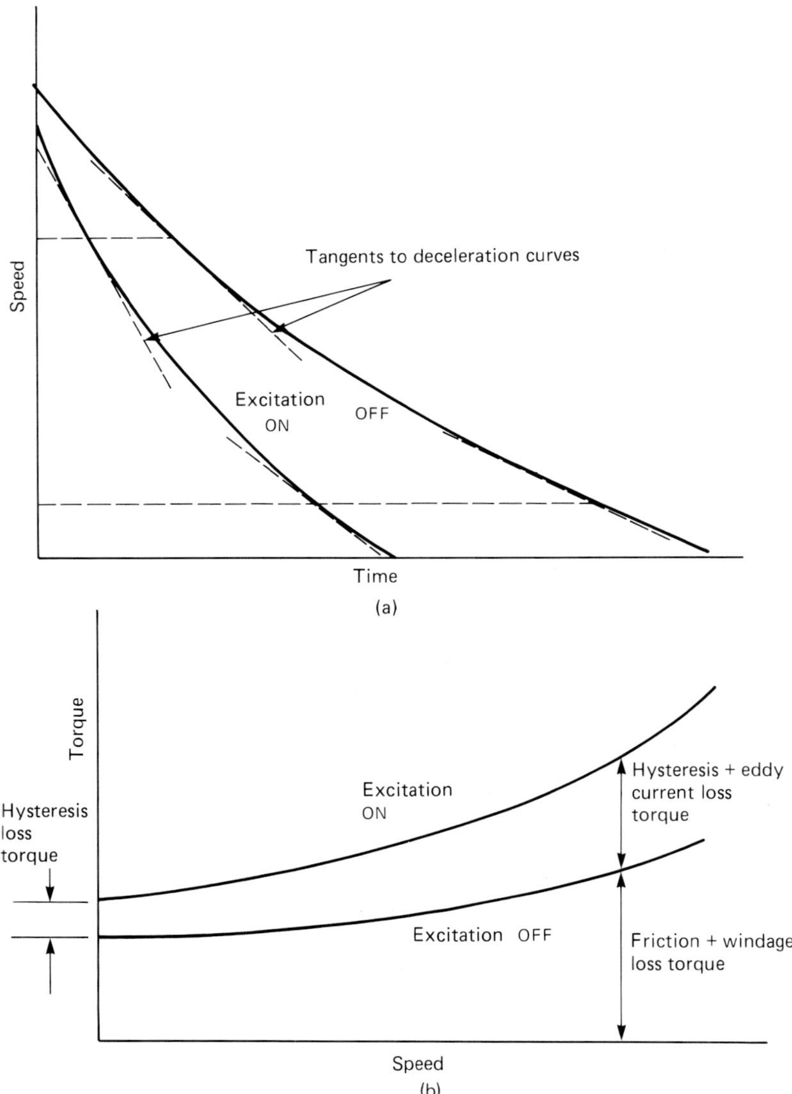

Figure 8.15 Retardation test results. (a) Speed–time curves. (b) Torque–speed curves

8.2.12 Mechanical tests

8.2.12.1 Overspeed test

The overspeed test is a check on the mechanical integrity of the rotating parts of the motor. The test may or may not be incorporated in a test schedule, depending upon the nature of the intended duty. In cases where the motor speed is restricted absolutely, with an induction motor fed at fixed frequency, for example, an overspeed test is of little value. In other applications where occasional abnormal conditions may lead to excess speed, the test is an essential basic test and may even be called for as a routine check of each motor. Traction motors fall into the latter category.

Overspeed test specifications vary according to intended duty. They should always be conducted with the motor hot since this is when it is most vulnerable. Typically, an overspeed test would last for 2 min at a speed of 20–25% above the nominal maximum speed.

A serious failure is obvious! Catastrophic failure on overspeed is rare and results from production defects, since mechanical designs usually carry ample safety margins. More common is distortion resulting from the test. An inadequately seasoned commutator would have an irregular surface following the test; the surface should be checked using a profile recorder as described in Section 8.2. Glass-fibre banding may display signs of stress, where loose threads have lifted from the surface. A particularly vulnerable section of glass banding is the thin section which extends over the commutator risers. Fortunately, a failure of this section can often be repaired quickly and without even withdrawing the rotor.

8.2.12.2 Vibration test

Rotors are balanced as a matter of course, prior to insertion in the stator. The precision of balance specified is a matter for agreement with the customer. Following assembly, the levels of vibration of the machine as a whole may be checked. This is carried out using a set of three orthogonal accelerometers fixed close to each bearing; the vibration amplitude is measured in the two directions normal to the shaft. British Standard 4999:Part 50 gives permitted r.m.s. vibration levels according to motor speed and power for three standards of quality. Table 8.4 of this chapter summarizes these levels.

Table 8.4 Permitted maximum vibration levels for three grades of motor expressed as r.m.s. velocity in mm/s

Grade	Motor speed (rev./min)	Shaft centre height (mm)	Maximum vibration velocity
Normal vibration	600–3600	80–132	1.80
		132–225	2.80
		225–400	4.50
Reduced vibration	600–1800	80–132	0.71
		132–225	1.12
		225–400	1.80
	1800–3600	80–132	1.12
		132–225	1.80
		225–400	2.80
Specially reduced	600–1800	80–132	0.45
		132–225	0.71
		225–400	1.12
	1800–3600	80–132	0.71
		132–225	1.12
		225–400	1.80

For the vibration test, the machine is run on no-load and not coupled to any other machinery. Of course it should be recognized that, in service, the load or the coupling to the load can induce vibration in the motor and it is wasteful to specify a precisely balanced, vibration-free motor without paying careful attention to the driven machinery.

8.2.12.3 Noise

For many applications the noise emitted by a motor is completely swamped by the noise of the driven machinery. For a few critical applications, the noise aspect dominates the specification. Thus, whilst there is no call for any noise measurements to be taken of a coalcutter drive, great pains are taken to reduce and measure precisely and in great detail the noise emitted by any motor destined for use aboard a submarine.

For general-purpose motors, BS 4999:Part 51 recommends maximum emitted noise power according to the enclosure type, the shaft speed and the output power. For a 1.1 kW, totally enclosed, six-pole induction motor, for example, the permitted noise emission is 76 dBA rising to 84 dBA for a two-pole motor of the same power. For a 1.1 MW motor the respective levels are 104 and 112 dBA. Roughly, the permitted noise power is proportional to shaft power.

The units decibel A represent the total power passing through the surface of a reference hemisphere centred on the motor which sits on an infinite plane. A figure of 0 dBA corresponds to 1 pW (10^{-12} W) of noise power. For each 10 dBA increase, the power increases by a factor of 10 so that 60 dBA corresponds to 1 μW of noise power, 90 dBA to 1 mW and 120 dBA to 1 W.

Measurement of noise power consists of measuring the sound intensity levels at a number of points covering a notional surface enclosing the machine and integrating to find the total. Readings should be taken at a minimum of five positions – front, back, left, right and above – at a distance of 1 m from the motor. Sound-level meters give the intensity which refers to the power crossing a surface of 1 m^2. The mean of N readings of intensity L dBA is found using:

$$\overline{L} = 10 \log_{10} \frac{1}{N} \sum_{r=1}^{N} 10^{(L_r/10)} \tag{8.11}$$

In practice, readings are taken not at the surface of a hemisphere but at the surface of an imaginary rectangular box of length $2a$, width $2b$ and height c. An arbitrary conversion is used to obtain the radius of an equivalent hemisphere R:

$$R = \frac{1}{2} \sqrt{(ac + bc)} \tag{8.12}$$

The total sound power is then found using:

$$L_{\text{total}} = \overline{L} + 10 \log_{10} (2\pi R^2) \tag{8.13}$$

The sets of readings are taken over a set of frequency bands each one octave wide and a total emitted noise power determined for each. Since the sensitivity of the human ear is frequency-dependent, weighting factors are applied to the power for each band and the weighted mean is computed using the above formula.

The permitted noise levels are given in terms of the weighted mean for two classes of motor – normal sound power and reduced sound power – the difference being 5 dB across the range.

The operating conditions for taking the readings should be full-speed, no-load and in an environment where the background noise intensity is at least 10 dB below the motor noise level. To avoid the corrupting effect of reverberation, it may be necessary to stand the motor on a set of resilient mounts. It is recognized that more noise will be emitted when the machine is loaded, but the noise generated by loading machinery would confuse test results and the no-load condition is generally used for the purpose of setting acceptance standards.

8.3 Methods of applying load

8.3.1 Absorption dynamometers

The tests for temperature rise, efficiency, characteristics and commutation all need to be conducted on load. For the measurement of efficiency and characteristics, the load must also be measured accurately. A system which applies a controlled and measured mechanical load is called a dynamometer.

Dynamometers may be absorptive or regenerative. Numerous forms of absorptive dynamometers are available, classified according to the mechanism used to dissipate the incoming power. Measurement of torque is accomplished usually by mounting the stationary member in bearings and restraining its movement by a spring balance or load cell. Alternatively, the dissipated power can be measured by appropriate means and used to calculate torque.

Examples of absorptive dynamometers include: (1) rope brake; (2) water brake; (3) magnetic powder brake; (4) d.c. generator with resistive load; (5) eddy-current brake; and (6) hysteresis dynamometers.

8.3.2 Calibrated load machine

The disadvantages of the absorptive-type dynamometers are the cost incurred in supplying energy to be dissipated and the need to remove the unwanted heat from the working area. For powers above a few kilowatts, the extra cost of a regenerative dynamometer can usually be justified.

The most straightforward regenerative dynamometer is a d.c. generator which feeds power back into the mains via a motor-generator set. The net power drawn from the mains is then just sufficient to make up the losses in the test motor, the dynamometer and the motor-generator set. The dynamometer machine is carefully tested before installation and its losses are determined accurately so that electrical measurement of its output power reflects accurately the input power; direct measurement of torque is then not essential.

For higher speeds, the d.c. machine becomes impractical and an alternator–rectifier would be preferable.

For testing d.c. machines using a d.c. supply, the d.c. dynamometer would be arranged to feed directly back into the d.c. supply which need only be of sufficient capacity to supply the losses of the two machines.

8.3.3 Back-to-back testing

When two nominally identical machines are available for test, one may be used as a load machine for the other. The method is only readily applicable to d.c. machines. It was developed for d.c. motors by Hopkinson and improved by Kapp and is often referred to as the Hopkinson test.

Power circulates between the two machines, the loop comprising motor–mechanical coupling–generator–electrical connection. Losses may be supplied either mechanically by coupling a small auxiliary motor to the test set or electrically by connecting one or more generators in circuit with the test machines. The method is in common use in a variety of forms.

8.3.3.1 Synchronous machines

Figure 8.16 illustrates a basic arrangement which may be used for synchronous machines. The coupling must be of a type which allows the relative angular positions of the two shafts to be adjusted. Setting the two field currents determines the working voltage, current and power.

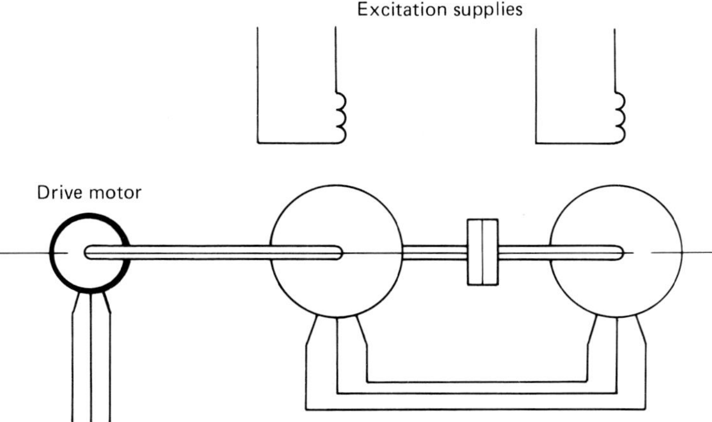

Figure 8.16 Back-to-back arrangement for synchronous motors

8.3.3.2 Induction machines

For an induction machine to deliver power, its speed must differ somewhat from the synchronous speed determined by the supply frequency. Figure 8.17 illustrates one method whereby the slip is accommodated between two back-to-back induction motors by a d.c. link. Adjusting the excitation currents of the two d.c. machines determines the power flow in magnitude and direction.

8.3.3.3 Direct current shunt motors

The basic circuit for two shunt machines is shown in Figure 8.18. The operating currents of the two machines and their common speed are determined by the settings of the two field rheostats. A common variation is shown in Figure 8.19 in

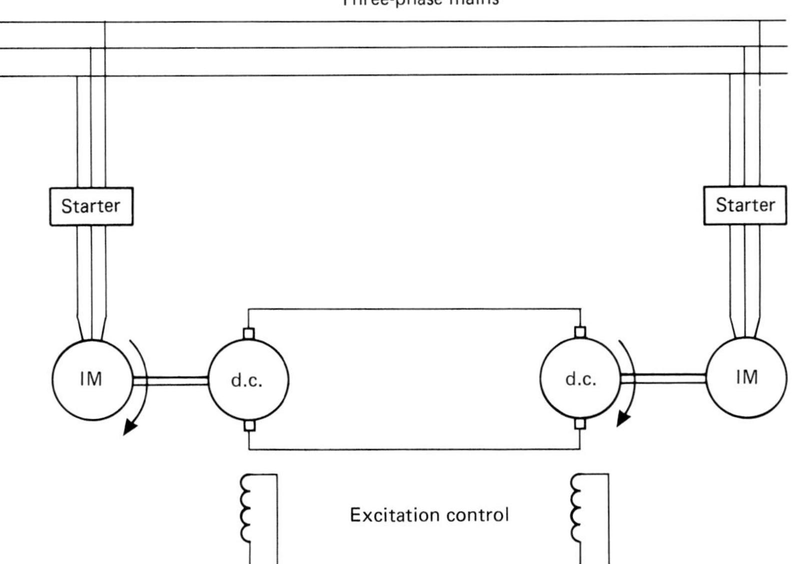

Figure 8.17 Back-to-back arrangement for induction motors

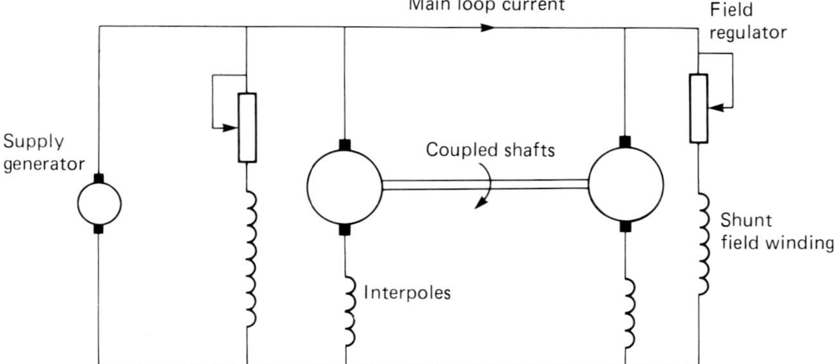

Figure 8.18 Simple back-to-back test circuit for shunt wound d.c. machines

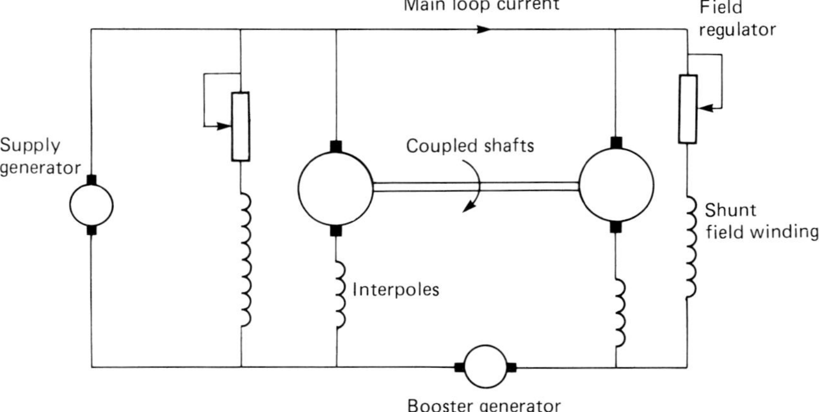

Figure 8.19 Back-to-back test circuit using booster generator. (Field regulators adjusted for equal field currents)

which an auxiliary generator called a booster is used to control the current circulating between the two machines. The field currents of the two motors are set to the same value. The field current and supply voltage determine the speed of the test machines. The main supply provides the no-load losses of the two test motors. In the absence of any e.m.f. generated by the booster, both test machines have the same induced e.m.f. and there is no tendency for current to circulate between the two. As the booster e.m.f. is increased it upsets the equilibrium between the two machines and current circulates around the loop. The magnitude of the circulating current is given by booster e.m.f. divided by the total resistance around the loop; the e.m.f.s of the two test machines cancel.

The main supply provides the no-load losses and the booster provides the load losses, to a close approximation. Accurate loss measurements of the two test machines is therefore accomplished by measuring the total power supplied by the main supply and by the booster.

8.3.3.4 Direct current series motors

Figure 8.20 illustrates the method applied to series-wound motors such as traction motors. The current circulating between the two test machines is determined by the

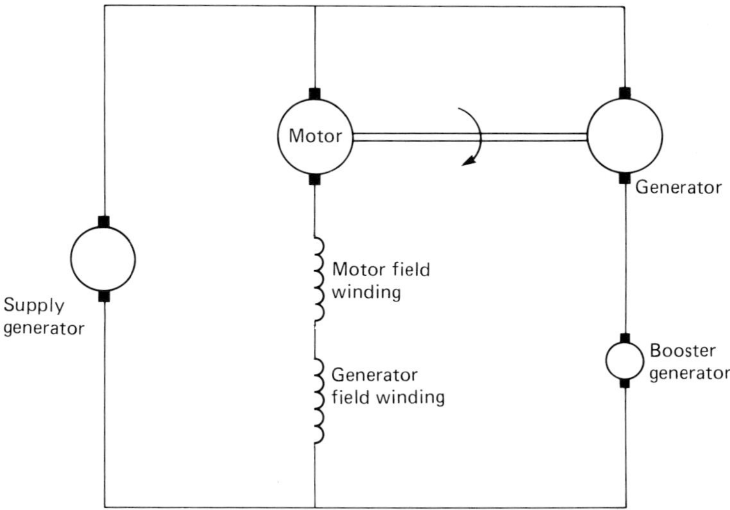

Figure 8.20 Back-to-back test circuit for series motors

booster generator voltage and the total resistance around the loop. The operating voltage is determined by the supply generator. The speed of the machines is fixed by the current and voltage. When bringing the set from rest to a test condition, the booster voltage should be increased first so that the supply voltage does not bring about a runaway condition.

8.3.4 Mixed-frequency test

The arrangement depicted in Figure 8.21 may be used for induction machines in place of the back-to-back configuration. The operation of the system is essentially

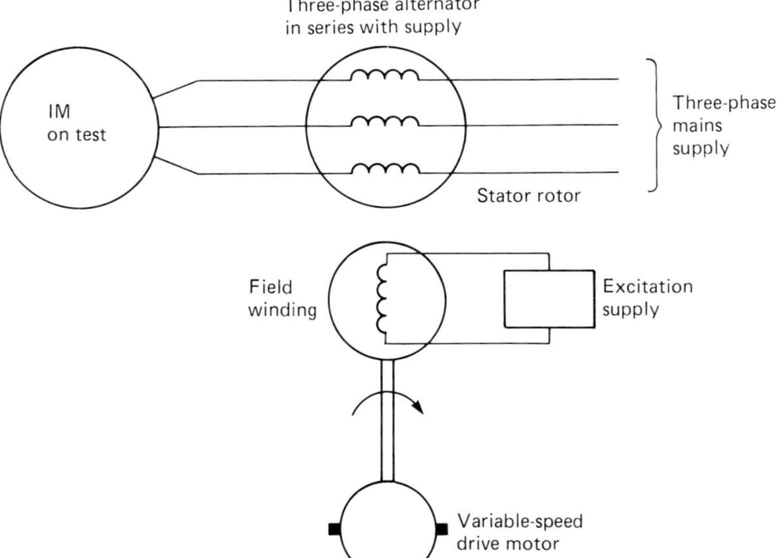

Figure 8.21 Mixed-frequency test for induction motors

to circulate power between a motor and a generator which happen to share a common structure.

The induction motor will run at a speed intermediate between the synchronous speeds for the mains frequency and the alternator frequency. If the alternator frequency can be adjusted to be a few hertz higher than the mains, the induction motor has a negative slip with respect to the mains voltage and generates power into the mains. It has a positive slip with respect to the alternator voltage and absorbs power from the alternator.

Adjustment of the alternator speed and voltage controls the voltage and current at the induction motor terminals, and these may be set equal to, say, their rated r.m.s. values. Since the induction motor is operating on-load electrically, the stray-load losses are present and can be found using measurements of terminal power and subtracing the known constant losses.

References

1. BRITISH STANDARDS INSTITUTION *Specification for general requirements for rotating electrical machines*, BS 4999:1976:
    Part 30: 'Duty and rating'.
    Part 32: 'Limits of temperature rise and methods of temperature measurement'.
    Part 33: 'Methods for determining losses and efficiency from test (excluding machines for traction vehicles)'.
    Part 50: 'Mechanical performance – vibration'.
    Part 51: 'Noise levels'.
    Part 60: 'Tests' (superseded by Part 143:1987).
2. MORGANITE ELECTRICAL CARBON LTD *Brush, commutator and slipring – maintenance handbook*. MEC.
3. LE CARBONE-LORRAINE (GREAT BRITAIN) LTD *A technical guide for brushes for electrical machines*. Le C-B.

9 Installation, site testing and commissioning

P. H. Conceicao

Deputy Chief Electrical Engineer, Generation Department, Kennedy and Donkin, Consutling Engineers

9.1 General – storage facilities

Depending on their size and configuration, motors are dispatched from the motor manufacturer's works in a variety of forms:

(1) Fully assembled and packed in crates lined with moisture-resistant materials.
(2) With the stator and rotor in separate crates or protective timber packing.
(3) Fully assembled and loaded on trailers or low loaders for delivery by road direct from the manufacturer's works.

It is necessary, therefore, to establish at an early stage exactly how the motors should be packaged and delivered. Depending on site weather conditions and transportation difficulties, it may be necessary to create a stores area under roofing long before the motors are required for installation. It is necessary to ensure that adequate facilities are made available on site to handle safely the largest or heaviest package that may be anticipated.

If the motors or components are not installed immediately in their operating positions, they should be stored in a dry, clean storage area at normal ambient room temperature. The rooms should be free from rodents, termites and insects which damage electrical insulation materials. Loose components and auxiliary equipment should be covered with tarred paper or heavy-duty polythene sheets to provide protection from dust, falling oil from overhead cranes, moisture and condensation.

With large motors it is advisable to energize the built-in space heaters which may be provided. The storage area should be provided with the necessary power outlets. It is necessary to ensure that there is no danger of ignition of any of the covers or sheeting protecting the equipment.

In all cases where storage for a considerable length of time is envisaged, the manufacturer should be consulted. On a green-field site it may be necessary to set up a secure power supply from emergency diesel generator sets supplying the site or the housing colony. Heater supplies are usually specified for a 220 V, single-phase or 380 V, three-phase, 50 Hz power source. It is essential in the specification stage to try to select a single common space heater voltage for all motors on a site, preferably a single-phase power source.

Before placing equipment into storage, the machined components should be carefully inspected. Bearings and shaft surfaces are normally covered with a corrosion-preventative barrier. Shafts are often wrapped with a treated fabric cover to prevent mechanical damage and corrosion. If the protective coatings have been dislodged or damaged during transit, the component should be cleaned, made free of rust or corrosion by-products and a fresh protective coating applied. Under no circumstances should rust or corrosion be merely covered over as further damage can occur.

Brushes for sliprings and commutators should not be installed and allowed to rest on the collector surface during long storage because of the risk of surface damage due to corrosion. Brushes should be either lifted or stored separately.

After placing in storage, equipment should be checked at regular intervals. Attention should be given to the humidity and temperature of the storage area. On larger, high-voltage motors, the insulation resistance should be checked and a log kept. Any drastic changes in the polarization index and the 1 min insulation resistance value should be investigated immediately and remedial action carried out.

It must be emphasized that long storage of electrical equipment should be planned carefully and should result in shorter drying-out times and reduce installation and commissioning times.

9.1.1 Receipt of equipment, inspection and handling

Electric motors and their component parts are packed carefully by the manufacturer for shipment, taking into account the site conditions, transportation difficulties and mechanical handling limitations. Before any equipment or package is accepted on site, it should be inspected very carefully for damage incurred during transit.

The crates may have been turned end-over-end despite the various warning marks. The waterproof crate liner or packing may have been opened for customs inspection and improperly resealed, or not resealed at all. The crates may even show signs of crushing due to heavy loads being stored on them.

Stator frames are usually fairly robust. Damage, however, often occurs to terminal boxes and other projections on the casing or frame. Stator packages have frequently been dropped due to slings slipping, and it may be quite difficult to ascertain visually if damage has occurred. Where there is a known instance of droppage having occurred, additional care should be given and tests carried out to determine the integrity of stator cores, clamp plate fastenings and windings.

Rotors transported on cradles are particularly prone to damage if not restrained properly. The use of cables tensioned over the shaft is a prime source of damage.

The cable, if incorrectly padded, can cut into bearing or shaft surfaces forming stress risers. Alternatively, should the rotor slide axially on its cradle, the cables have been known to damage the rotor end-windings causing short circuits and ultimate total destruction of the winding.

If damage is noted or suspected, it may be necessary to unpack the shipment to determine the full extent of the damage. The shipping or transport agent should be notified immediately as also should the manufacturer and relevant insurance agency. Wherever possible, the damage should be recorded, photographed and witnessed. This procedure is extremely helpful to personnel remote from site who have any involvement in assessing the damage and the feasibility of effecting repairs, especially if on site.

The handling, lifting and moving of larger electric motors and components should always be carried out by skilled personnel. Some large machines may be handled as a complete unit. It is necessary, therefore, to consult the assembly drawing giving details of the lengths and lifting capabilities of any slings, eyebolts or shackles. Only certified or tested equipment should be used. Frayed or damaged slings should never be used.

Where rotors have to be threaded into stators, the appropriate sequence drawings and instructions should be obtained from the manufacturer. Where special lifting beams are required, these should preferably be provided by the manufacturer. Slings should never be wrapped around laminated rotor bodies without a protective wrap of a tough flexible material such as leatheroid or equivalent flexible pressboard. Spreader beams should be used to avoid crushing the top edges of packing crates or rotor windings. In all cases where slings are wrapped around machined shaft sections, adequate protective wrapping should be used.

When the lift actually starts, the component should be checked to ensure that the lift is level and within the line of its centre of gravity. Where heat exchangers, for example, are mounted on top of the motor frame, care should be taken to ensure that:

(1) The correct lifting lugs are used for lifting the motor.
(2) Spreader beams are used to avoid crushing the light sheet steel enclosure.

When moving and setting-down large rotors, care should be taken that where V-blocks or trestles are used, pressboard or soft lead sheet should be interposed between any support and the shaft. If it is intended to turn the rotor on its temporary supports, it is recommended that the leatheroid lined support be used with a liberal coating of tallow or similar lubricant.

Figures 9.1–9.4, show some of the stages required for threading a typical large induction motor.

Step 1: Wrap two layers of thick vulcanized fibre round the centre of the rotor body. Wrap a similar protective shroud over the threading end of the rotor to cover the fan blades and the rotor end-rings, as shown in Figure 9.1.

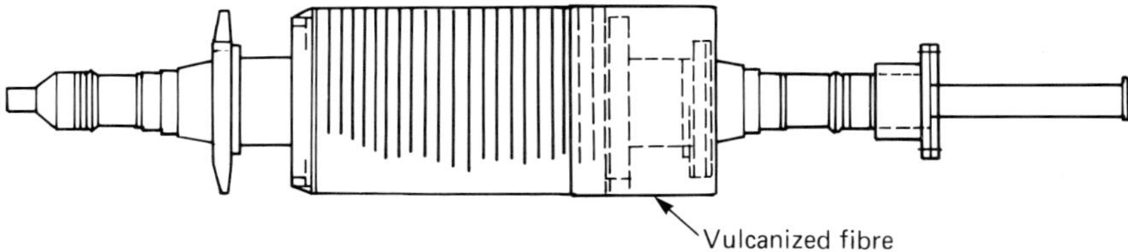

Figure 9.1 A rotor with protective wrap in position. (*Courtesy:* NEI Peebles Ltd/Peebles Electrical Machines)

Step 2: Wrap two turns of each of two identical slings round the rotor on top of the fibre wrap, at approximately the centre of the rotor body as shown in Figure 9.2. Take up initial tension in the slings by means of the crane, then slacken and adjust the position of the slings until the rotor hangs horizontally, checking this by means of a spirit level.

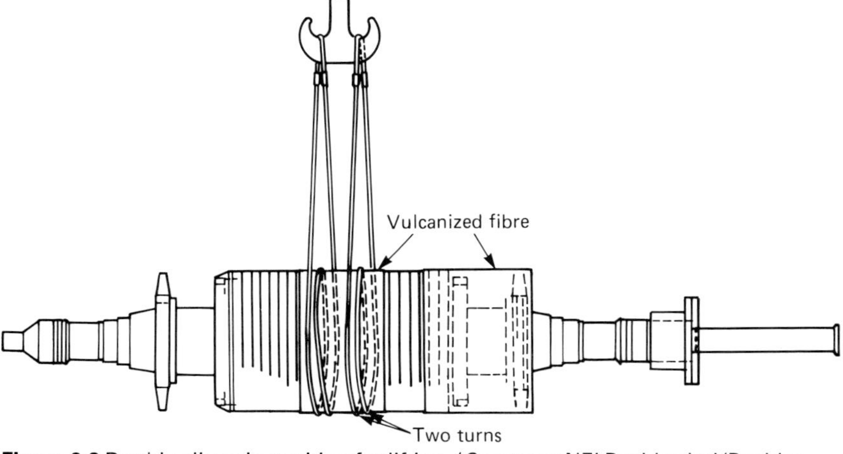

Figure 9.2 Double slings in position for lifting. (*Courtesy:* NEI Peebles Ltd/Peebles Electrical Machines)

Step 3: Offer up the rotor to the stator bore and thread carefully, ensuring that the vulcanized fibre shroud does not snag on the stator wedges or on the end of the core. Ensure that the rotor is central and parallel with the stator bore and proceed with the threading until the slings are adjacent to the winding overhang, as shown in Figure 9.3.

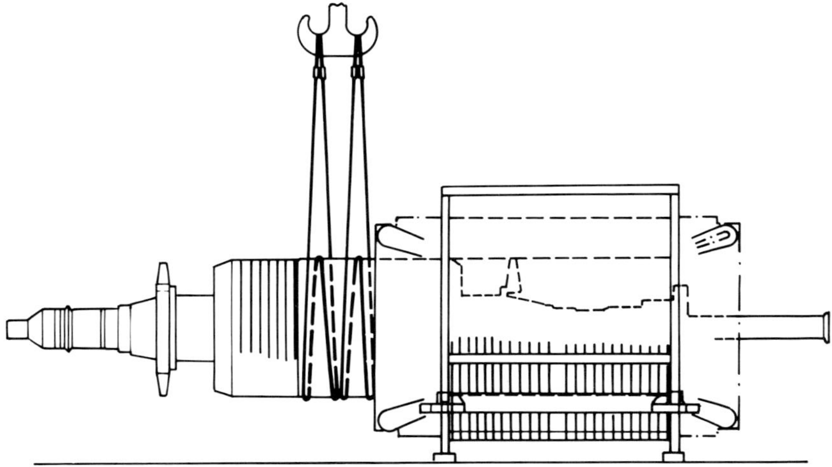

Figure 9.3 Rotor partially threaded into stator. (*Courtesy:* NEI Peebles Ltd/Peebles Electrical Machines)

Step 4: Rest the rotor on blocks with shims used to level up. It is advisable to use a V-shaped top block to prevent the rotor rolling off, as shown in Figure 9.4. Recheck that the rotor is level and parallel with the stator pack bore and remove rotor slings.

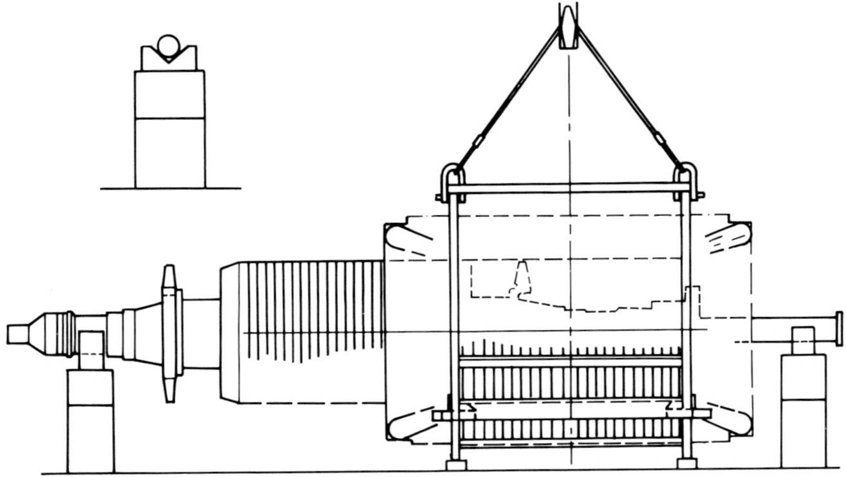

Figure 9.4 Rotor supported on V-blocks with slings removed. (*Courtesy:* NEI Peebles Ltd/Peebles Electrical Machines)

The completion of the threading may be carried out by moving the stator over the rotor. Other methods of completing the sequence depend on the position of the stator on the workfloor or in its final location.

9.1.2 Preparation of foundations

As part of the exercise of establishing the type and disposition of any foundations, it is necessary to examine closely the layout of the combined plant. It will be necessary to work to the national standards or codes of practice relevant to the country of installation. (In the UK, this would be BS 5304:1975: *Code of practice for safeguarding machines*.)

Attention should be given to ensuring that there is adequate access for operation and maintenance of plant. Floors should be of an approved non-slip type and free from projections and other obstructive stops such as minor abrupt changes in floor levels. These should be gently graded.

Adequate free spacing should be provided around the composite motor and its driven member to facilitate maintenance. A minimum of 0.75 m should be provided between any wall and the end of a machine to give all-round access after any necessary guardrails have been inserted.

Gangways should be provided between rows or columns of plant, wide enough for vehicles or forklift trucks which may be expected to pass the machines as finally installed. These gangways should be in addition to the free space round the machines. A minimum width for a main gangway would be approximately 1.75–2 m, whilst an auxiliary branch gangway would be approximately 1.5 m wide.

Lighting levels usually exist or are specified for plant areas. Where necessary, the lighting should be free from stroboscopic effects. Typical lighting levels are listed in Table 9.1. It may be necessary to provide supplementary lighting for major maintenance or dismantling operations.

Table 9.1 Typical lighting levels

Area	Lux	Glare factor
Turbine hall	150–200	25
Switchgear rooms	150–200	25
Workshops	350–450	22
Battery rooms	100	—
Cableways	50	—

Source: Kennedy and Donkin specification.

It is essential to preplan any oil and water pipes for lubrication and cooling services and to locate cable ductways so that they do not clash with the projected foundations.

Motors and driven machines are normally mounted on foundations which must be solid, rigid and level. The thickness of the foundation depends on the size of the plant and the severity of service. The manufacturer should be consulted where any doubts exist.

Where belt transmission is used, the motor should be mounted on slide rails or on a base which permits the belt to be tensioned by moving the motor. Cast-iron slide-rails should not be used for motors mounted on walls or ceilings. Belts which can become electrostatically charged should not be used in hazardous locations. In cases of doubt, the belt manufacturer should be consulted.

Slide-rails normally are secured to the foundation by means of ragbolts let into holes provided in the foundation and finally grouted into position.

The surface of the foundation should be roughened to provide a key for the grout. The slide-rails should be set parallel and the level should be checked. The rails should be packed with steel packing strips not more than 100 mm apart to provide a clearance of approximately 10–12 mm between the rails and the

(a)

(b)

Figure 9.5 (a) Typical sole plates and levelling shims for an 1800 kW motor. (b) An 1800 kW, 992 rev.min, 11 kV, 50 Hz wound-motor resting on its sole plates. (*Courtesy:* NEI Peebles Ltd/Peebles Electrical Machines)

foundations. A polystyrene frame should be placed round the rails to a level of 10–15 mm above the bottom of the rails. The foundation should be well wetted and a grout consisting of 1 part cement to 2 parts sharp sand should be run in. The quantity of water added should give a very stiff mix which should not flow; at the most it should slump very gradually. Excessive water causes shrinkage and a weaker set strength.

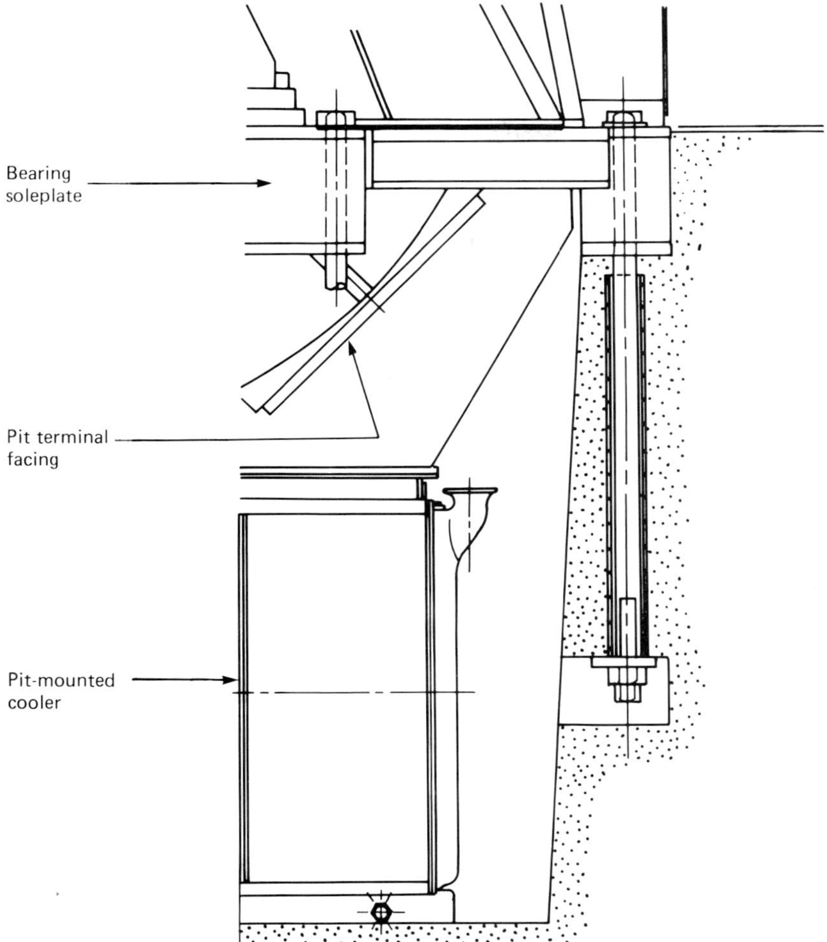

Figure 9.6 Typical foundation arrangement

Direct-coupled machines are usually mounted on bedplates, sub-bedplates and pads or soleplates grouted into the foundations. As unit-constructed large motors become available in sizes up to 25 MW, the machined box base forms a bedplate. Accordingly, it is becoming more common to use pads or soleplates where a combined bedplate for motor and driven machine would have been the normal practice. A typical installation is shown in Figure 9.5(a) and (b).

Care should be taken to ensure that bedplates are level and are not set too high. Low-set bedplates can be shimmed up but a high setting usually requires removal of some of the foundation concrete.

Soleplates provided by manufacturers range from the simple flat slab of steel 50–75 mm thick, with or without ribs on their undersides, to take up horizontal loads to the more rigid I-beam. All taper wedges, packing or thin shims should be of steel; wood or other absorbent materials must never be used. Typical arrangements are shown in Figure 9.6.

When establishing the soleplate dimensions, the loading on the foundation should be no greater than 30–35 kgf/cm^2 to allow for inconsistency in the quality of the grout and its insertion.

Grout may finally be consolidated by tamping-in with fine wire rods. Impact vibrators should never be used for fear of displacing shims and taper wedge packers. Grout should be allowed to set for 5–7 days before the plant is run. In high ambient temperatures it may be necessary to cover the area with damp hessian sacking to control the rate of drying out.

Although the 1:2 concrete-to-sand mixture is commonly used for grouting, there are certain proprietary high-impact-strength, resin-based grouts. These grouts should be used strictly to the manufacturer's recommendations for ambient temperatures, mix and curing times. The permissible surface loadings are vastly greater than the concrete mix, being at least 2–3 times stronger. A change from any specified grout to that of another type should only be carried out with the machine manufacturer's advice.

9.1.3 Drives and couplings

The transmission of power from the motor shaft to the driven member can be achieved in a number of methods which may all be classed broadly in two categories: (1) solid; and (2) flexible.

In general, couplings should be aligned carefully; even where there is a degree of flexibility, the value of good alignment cannot be ignored. Shaft centres should be perfectly in line with their centres coincident. Spigoted couplings are normally supplied when one of the coupled members has a single bearing, usually the motor. The couplings are tightly pressed on, or shrunk on, to the shaft, or are forged integral with the shaft. The single bearing machine has its rotor located on the spigot, the couplings having concentric, accurately mating, male and female spigots. These spigots help in the alignment. When pulling couplings together, the rotor of the single-bearing machine should be supported in the airgap of the machine to get the correct concentricity. It must be possible to insert the same thickness of feeler gauge between the coupling faces all round the periphery before the faces are pulled tightly together. On reciprocating and larger drives, the boltholes are normally reamered in turn, and fitted bolts provided.

Plain couplings are commonly used on plant having a composite bed and where there is no single bearing rotor in the system. The machines are aligned approximately by checking on the coupling rims with a steel straight edge at two positions 90° apart. A spirit-level check on the shaft systems should also be made. Using feeler gauges, a check of the gap between the coupling faces at all points round the periphery should also be carried out. Having set the machines in approximate alignment, the two half-couplings should be fastened together with a single undersize bolt in one hole, using a spacer washer between faces to eliminate axial endplay. Finger-tighten the bolt to permit slight relative movement to allow for misalignment if present. When initially aligned, remove the bolt.

Fix a double-dial indicator clamp on one half-coupling as in Figure 9.7, with the indicators bearing on the periphery and the machined back face (or front face) of the other coupling. The clamp should be on the motor, as this usually can be turned freely. When the gauges are rotated through 360°, neither indicator should show any movement. A motor with dry sleeve bearings would require approximately 55–60 Nm torque applied to the coupling to break away each 1000 kg of rotor weight. Any misalignment can be corrected by moving the complete motor rotor system. Alignment to within 0.025 mm is usually considered acceptable.

Flexible couplings generally should be aligned to the same accuracy as solid couplings. However, flexible rubber-type couplings can be accepted with alignments to within 0.05 mm. In all cases, the manufacturer's setting-up instructions should be followed.

Spline or gear-type couplings generally should be aligned to procedures similar to that of solid couplings. Magnetically mounted dial-type indicators should be used.

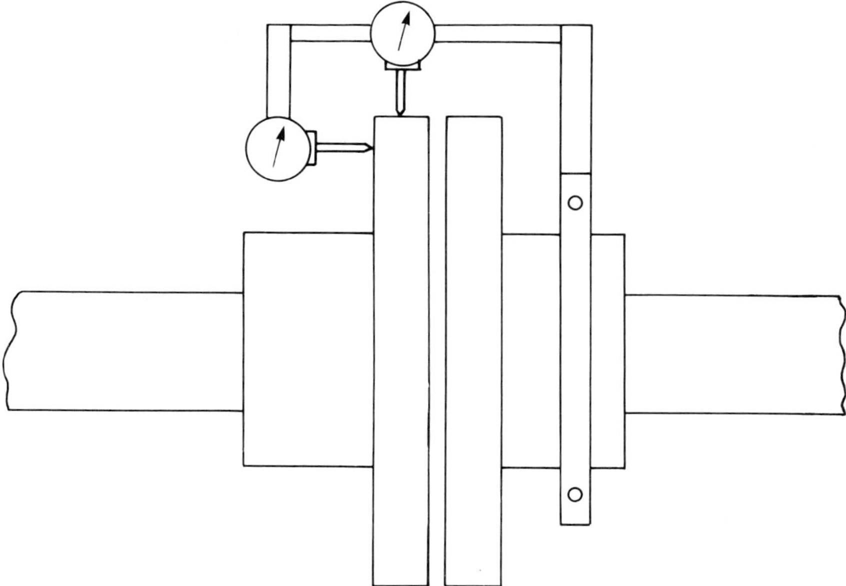

Figure 9.7 Coupling alignment using twin dial gauges

The acceptable offset of shaft centres should not exceed 0.02 mm. Gear couplings should be heated in an oilbath or oven prior to shrinking; gas torches must never be used.

Belt drives, when used, have the motor mounted on slide rails or an adjustable base to permit retensioning of the belts by moving the machine. The drive should be through the lower part of the belt. With fast and loose pulleys, the belt must run close to the motor bearing in the drive position. In general, pulley ratios should not exceed 5:1 with flat belts and the belt speed 30 m/s.

With V-belt drives, the pulley ratio should not exceed 8:1, and the belt speed 23 m/s.

Where cast-iron pulleys are used, these should be checked for soundness after fitting.

9.1.4 Axial alignment

When considering the composite shaft system which may contain a driven member, a clutch or limited end-float coupling and intermediate sleeve and thrust bearing, it is necessary to ensure that combined axial movements are accounted for.

The largest machine should be made the starting-point for lining-up any train of machines. It may be assumed that the largest machine has the largest endplay. This should be checked against the dimension drawings. In very long motors, the axial growth due to thermal expansion must be allowed for by offsetting the stator and rotor in the cold position. This offset, if necessary, will be advised by the manufacturer. The smaller elements should be protected against any end thrusts. The following example, from BEAMA publication No. REM 502-1972, gives an example of a typical complex alignment.

When allowing for thermal expansion of shaft systems, a rough guide is to allow 1 mm for each 1000 mm per 100° K rise. The difficulty is in estimating the average temperature of a long shaft inside a motor enclosure.

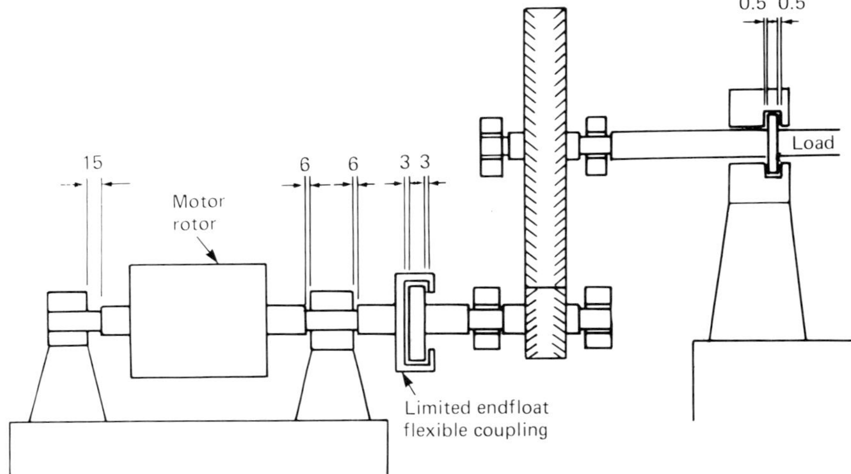

Figure 9.8 Electric motor driving through a flexible coupling and double helical gears. *Source:* British Electrical Appliance Manufacturers' Association Publication No. REM 502 (1972)

A typical drive for an electric motor driving load through a flexible coupling and double helical gear is considered in Figure 9.8.

It is necessary, in all cases where the driven machine (or gearbox) has its shaft axially located, to ensure that there is always residual axial clearance at both ends of the motor bearings under all possible conditions. The minimum initial axial clearance (X) at each end of the motor bearings is given by the following equation:

$$X = A + B + C + D + E + F + G$$

where $A =$ maximum possible thermal expansion between the thrust or locating bearing of the driven machine (or gearbox) and the motor bearing;

$B =$ maximum possible axial displacement of driven machine (or gearbox) shaft relative to the motor bearing due to thermal movement of a stable foundation. If an allowance is specified to cover for settlement or distortion of foundations, then this figure should be added to the figure for thermal expansion and the total used in the equation;

$C =$ maximum possible axial displacement of driven machine shaft due to adjustment or wear at its thrust bearing;

$D =$ maximum possible amplitude of axial oscillation which may appear when double helical gears are used, due to residual errors in the pitch of the teeth;

$E =$ half the total coupling end-float;

$F =$ half the total end-float in the locating bearing in the driven machine (or gearbox); and

$G =$ a residual safe clearance (say 0.5 mm).

Note: Values for B, C and D will vary with the installation and the type of driven machine.

A typical shaft system involving a flexibly coupled horizontal sleeve bearing motor is illustrated in the following example.

Example 9.1

An electric motor about 2250 kW at 750 rev./min is fitted with sleeve bearings, driving a low-speed mill through a double-helical gear. The mill is provided with a thrust bearing having a total axial clearance of 1 mm. The gearbox bearings have greater axial clearances than the mill thrust bearing and do not impose any restriction on axial movement of the shaft. The gearbox manufacturer has stated 0.3 mm maximum possible axial oscillation of one shaft relative to the other due to tolerances on the machining of the gear teeth. The maximum possible thermal expansion of the shaft system between the motor bearing and the load thrust bearing is calculated as 0.6 mm. Separate foundations are involved and 0.5 mm has been specified to take account of any long-term relative movement.

A coupling having limited end-float or elastic axial centering properties is required, and the coupling supplier had offered a coupling having 6 mm total end-float.

It is now necessary to check the minimum axial clearance required at each end of the bearing according to the equation:

$$X = A + B + C + D + E + F + G$$

In this example: $A = 0.6$ mm
$B = 0.5$ mm
$C = 0.0$ mm
$D = 0.3$ mm
$E = 3.0$ mm
$F = 0.5$ mm
$G = 0.5$ mm

Therefore $X = 5.4$ mm

A standard axial clearance sleeve bearing with 6 mm clearance at each end is therefore satisfactory. This bearing would normally have end-faces to limit the rotor float when running uncoupled.

The non-drive end motor bearing could have any value of axial clearance in excess of 6 mm plus an allowance for expansion of the motor shaft, i.e. a large axial clearance sleeve bearing.

A further aspect to be considered when studying the example drawings is the correct evaluation of the effects of vertical growth of the pedestals due to thermal effects. This is particularly important where solid couplings are used with the driven member and the motors have different pedestal heights, the temperatures of the two members being also substantially different.

Problems can also be encountered in the case where the ventilation circuit includes the concrete pit and the motor is axially ventilated. Thermal gradients can exist between the two bearing footings and on either side of the concrete footings. A detailed study of constructional and manufacturing features is contained in CIGRE reports.[1-2]

Vertical motors up to the largest currently manufactured for primary reactor pump cooling circuits are invariably mounted on a spigoted flange or skirt mounted on the pump casing.

Thus, with correct quality control procedures during manufacture, it is unusual to have problems with alignment of vertical motors. Any out-of-vertical alignment usually requires a major machining operation with either the motor or skirt being returned to a machine shop for rectification.

Typically, the eccentricity for solid-coupled large vertical motors tends to be less than 0.025 mm with 0.05 mm being an upper limit. Achievable values for the inclination on the outer diameter of coupling and flange faces tend to be in the order of 0.015 and 0.050 mm respectively.

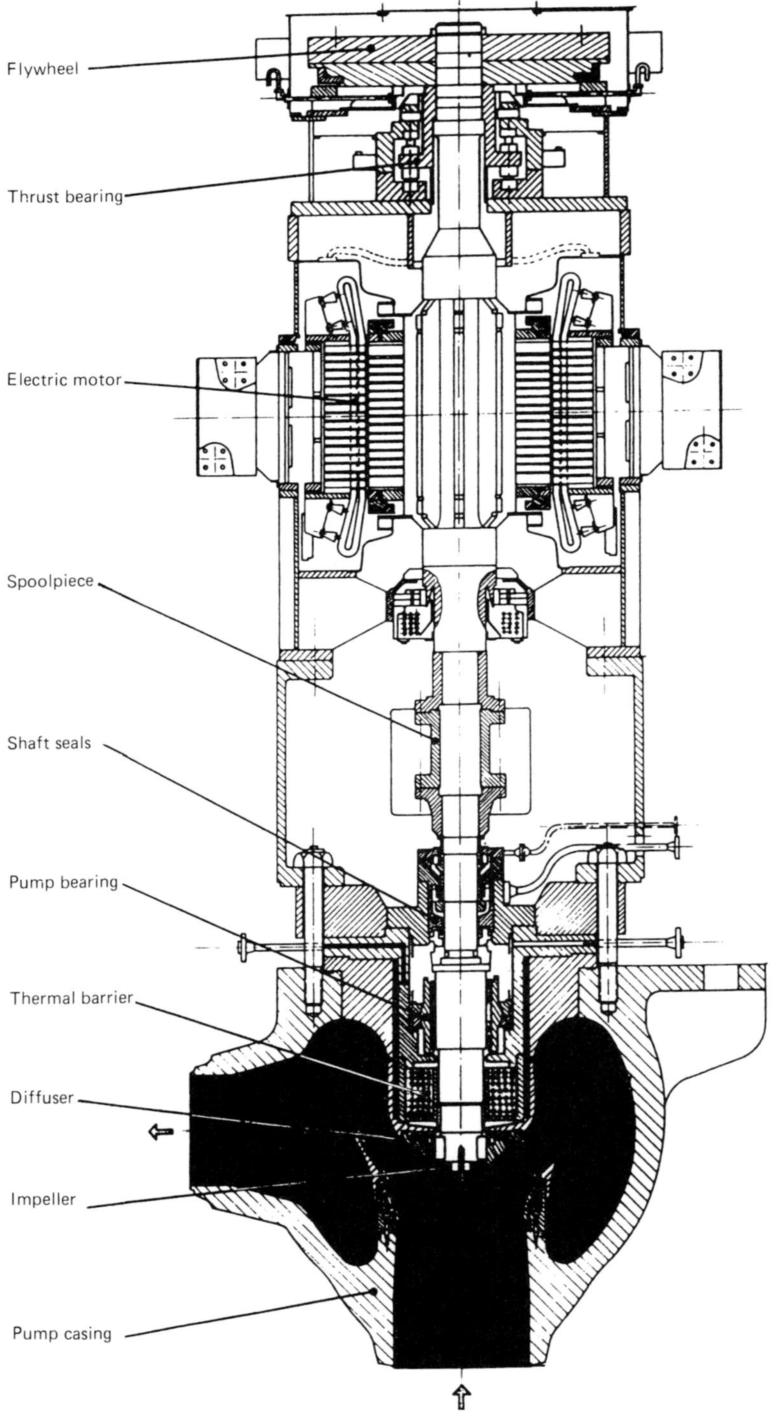

Figure 9.9 A 6500 kW 1485 rev./min reactor coolant motor–pump arrangement. (*Courtesy:* Jeumont Schneider)

On very large vertical motors used for pumped storage purposes, the whole method of construction departs from the conventional vertical motor. It is more usual to find hydro-electric vertical generator practices being used where the stator frame may be located on soleplates and the upper or lower bearings on a bracket which may also rest on soleplates if the stator is located below ground level.

In such installations, there are various combinations of methods, utilizing plumb lines, straight edges, precision spirit levels, pin gauges, special levelling beams and the engineer's level instrument (or theodolite).

The levelling of the soleplate should be carried out by highly experienced staff, preferably employed by the manufacturer of the motor or the pump unit.

Figure 9.9 shows a typical multiple assembly for a nuclear reactor primary motor pump arrangement.

9.1.5 Erection – additional aspects

After the machine foundation plates have been grouted in and alignments achieved, it becomes necessary to carry out a number of complementary stages of checking.

9.1.5.1 Oil systems

On sleeve-bearing machines, the following aspects should be checked:

(1) Bearing insulation.
(2) Bearing white metal surface finish.
(3) White metal bonding to bush.
(4) Bearing seating in its pedestal.
(5) Clearances, radial and axial.
(6) Seals and labyrinths.
(7) Connect and test-bearing temperature indicators.
(8) Check slope and drainage for external pipework.
(9) Clean and scavenge oil pipes/systems, bypassing the bearings.

The pipework should be connected up to the bearings which should then be boxed up and a final scavenge should be carried out. It may be possible to carry out only one scavenge operation, depending on the degree of cleanliness of the components as delivered to site, and if there is no site welding on pipework. This decision must often be made on site. When setting oil drainpipes, a gradient of at least 1:15 is recommended.

Branch pipes from the bearings and main drainpipes should be sized such that with the specified oilflow the pipes are approximately half filled, ensuring that air can be drawn out to the main tank. Large vertical drops, saddles and dips in the pipe runs should be avoided.

9.1.5.2 Cooling systems

Machines provided with air-to-water heat exchangers may have them built either on the machine or provided loose for mounting in a pit or side-mounted in ductwork. It is necessary to ensure that any sheetmetal ductwork provided is supported adequately and sealed at joints. When, finally, the machine is run it sometimes becomes necessary to stiffen with ribs any ductwork which vibrates or is excessively noisy. The plant erector should ensure that cooling-water service pipes are connected to the correct inlet and outlet flanges on header box of the heat exchanger. If, for any reason, the header boxes have been dismantled on site, care should be taken to ensure that any loose baffle plates are correctly reinserted.

All sight gauge glasses on oil and water services should be checked for soundness and freedom from cracks after having being locked into their working positions.

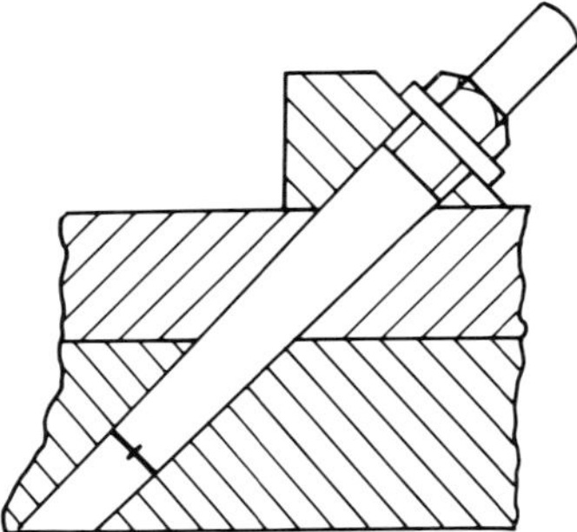

Figure 9.10 Taper dowel pin arrangement

Prior to handing over the machine to the commissioning engineer a final check on the tightness of all holding down and assembly bolts should be carried out. If the alignment is still correct, special dowel pins between the machine and its base should be fitted. There are normally two located adjacent to any two diagonally opposed holding down boltholes. These dowel holes are bored on a slant and are fitted to accept taper dowel pins. A typical pin with threaded shank and nut for withdrawal is shown in Figure 9.10.

9.2 Terminal boxes

Terminations and terminal boxes are invariably finalized at the specification and design stage of any project. It remains only for the site engineer to ensure that the terminals, cables, glands and accessories are correctly made up.

9.2.1 Low voltages up to 1000 V

Low-voltage motors are usually rated up to about 300 kW at voltages below 440 V (1000-V insulation class). Above this output it is more normal to apply a higher voltage.

Low-voltage motors below approximately 100 kW are normally fuse-protected, and above that air-break circuit-breakers are used. Accordingly, there are no serious problems with terminal box explosive forces on fault. It is necessary to ensure compression glands are correctly sized for the cables and that terminal box cover seals are made good; also, that the box is clean and dry before the cover is tightened down. Where star–delta starting is used, it is essential to ensure that the terminals are selected correctly to give the optimum transition from star to delta as recommended by the manufacturer.

9.2.2 High voltages up to 15 000 V

High-voltage motors are generally connected to power systems having correspondingly high-fault power levels. Terminal boxes become more sophisticated, being fitted with pressure-relief devices, dessicators, phase barriers and phase-isolation features.

Pressure-relief devices vary considerably from manufacturer to manufacturer and fall into three basic types:

(1) Rupturing diaphragms.
(2) Specially weakened sections.
(3) Relief gasketing designs.

The first type of terminal box is fitted with a thin foil section which will blow out with the sharp pressure rise generated within the box by an internal fault.

When installing motors with relief diaphragms, ensure that the products of the explosive gases do not blast across gangways or where personnel are likely to be present. It may be necessary to fit deflector hoods, with the manufacturer's agreement.

In the second type (Figure 9.11), it will be seen that a specially weakened cross-line is punched into the box cover, such that the section will lift like leaves without disintegration. One variant is a lead-filled cruciform which blows out on fault. This variant should be fitted with deflectors, if considered to be in an exposed position.

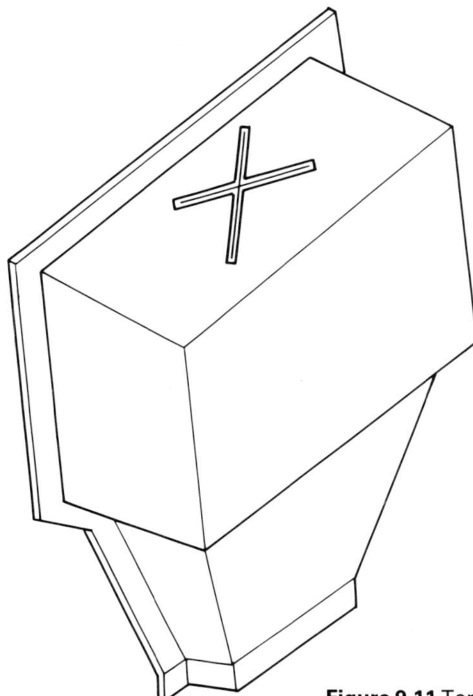

Figure 9.11 Terminal box cover with weakened section

In the third type of relief device, the internal volume of the box is separated from the motor internal volume by a rubber flap gasket. When a sharp pressure rise occurs the flap lifts, voiding the box inwards, again protecting against disintegration.

Where the products of a fault have to be contained, as in a hazardous or flameproof area, the box is of a special heavy construction. This type of box has to be certified and proved.

It is essential to obtain detailed drawings and instructions when connecting-up high-voltage, high-power terminal boxes.

Dessicators are often specified and usually consist of small plugs filled with colour-sensitive silica gel crystals which usually turn pink when moisture has been absorbed. The plugs, which have a tough glass window, are threaded and designed for quick replacement. Used plugs can be dried in an oven and reused.

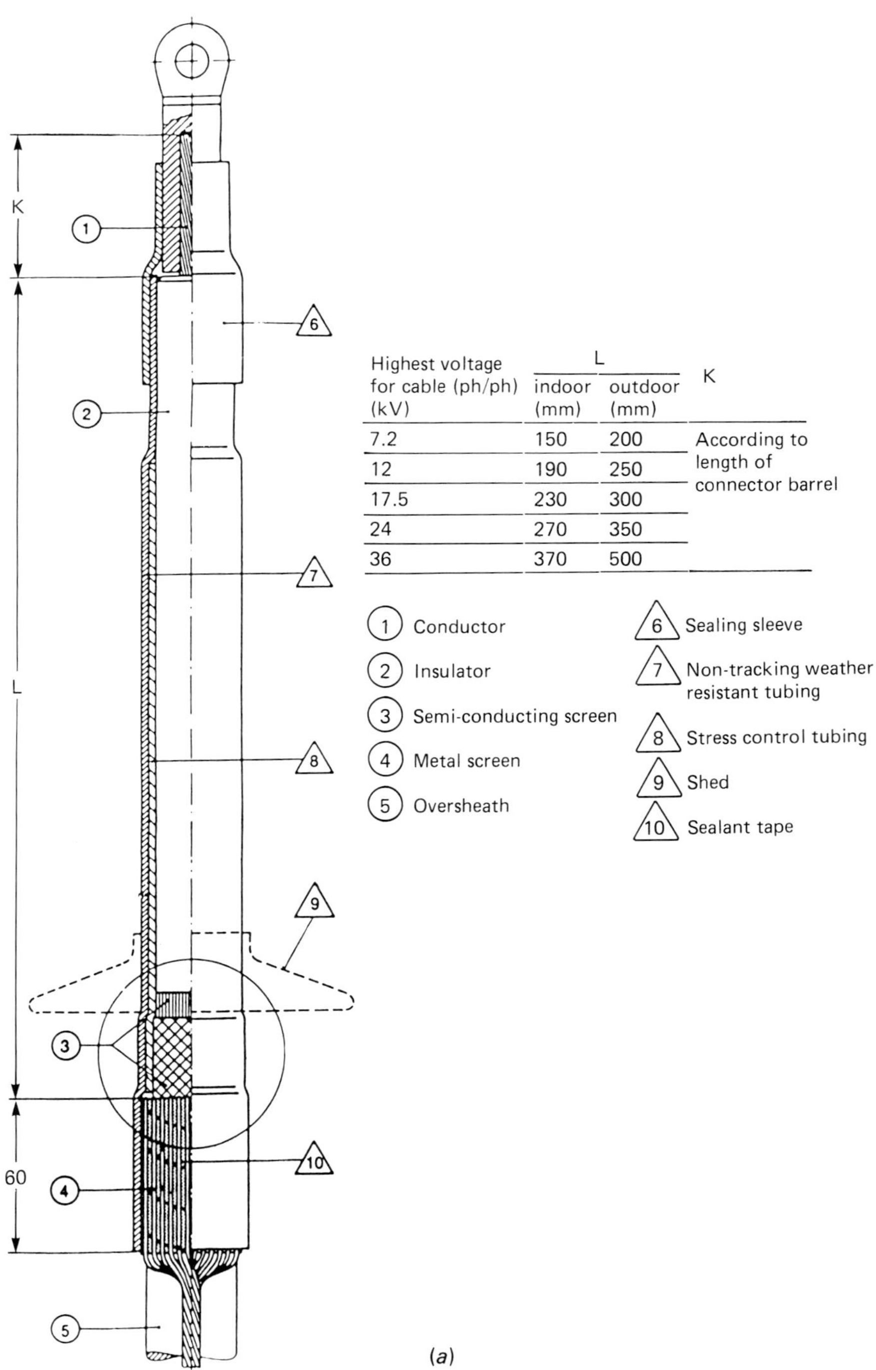

Figure 9.12 (a) Heat-shrink cable terminations. (*Courtesy:* British Insulated Callender Cables Components Division)

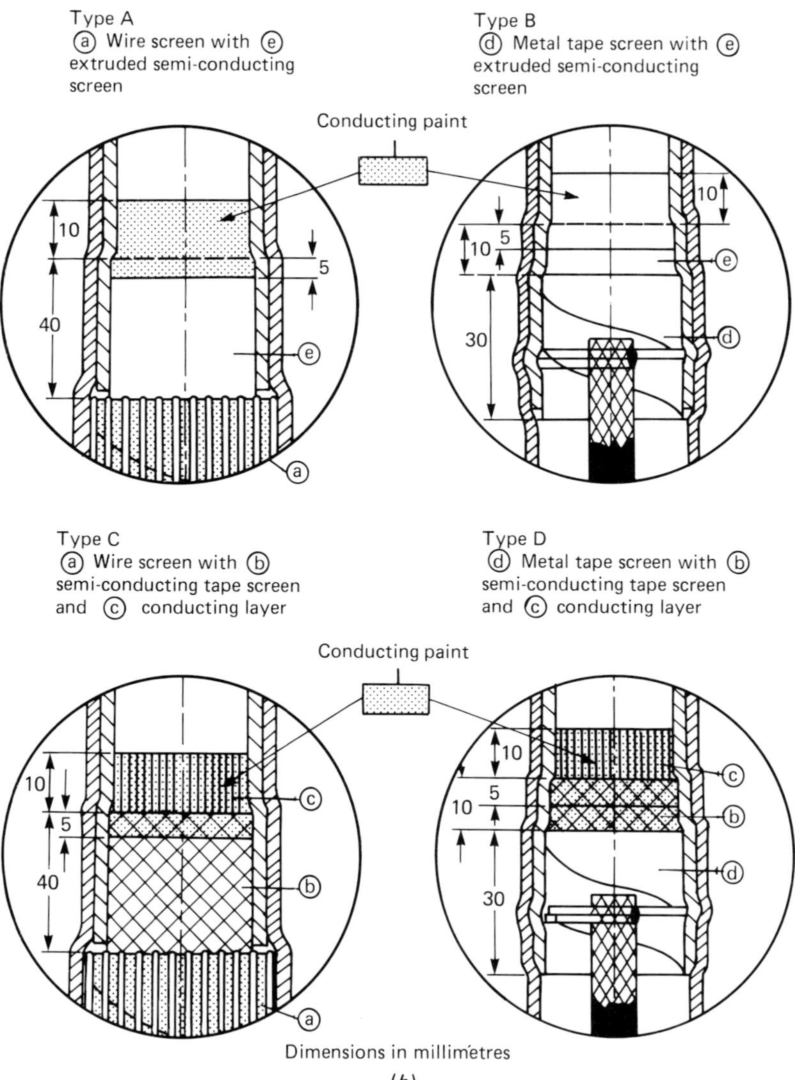

Figure 9.12 (b) heat-shrink cable terminations – details. (*Courtesy:* British Insulated Callender Cables Components Division)

9.2.3 Terminations

There is a certain degree of skill required for terminating high-voltage cables. The older, well-proven, paper–lead-insulated cables used for motor circuits have, in a large number of modern installations, been replaced by the more flexible cables of the PVC EPR and the more modern cross-linked XLPE types. Although considered easier to work with, care has to be taken when stripping back and preparing polymeric cables for use in motor terminal boxes. The recommendations for clearances, bending radii, shrouds, etc. should be followed strictly. Figure 9.12(a) and (b) shows a typical heat-shrink termination suitable for 6.6 and 11 kV motors. The termination is suitable for use up to 36 kV, with the various arrangements of metallic tapes and screens in common use.

9.2.4 Separable connectors

Separable connectors developed in the US to ANSI/IEE Standard 386:1985 for use with flexible polymeric cables were initially used for systems- and distribution-type transformers and their usage spread rapidly to high-voltage electric motors. Figure 9.13 shows the typical features of an Elastimold (Bimold*) connector developed by the AMERACE Corporation. One of the prime advantages is that this fully insulated connector can be tested live without disconnection. A further major advantage is that a formal terminal box is not required. Motor users usually specify a lightly ventilated sheet-metal enclosure round the terminations, more in the nature of a bump guard

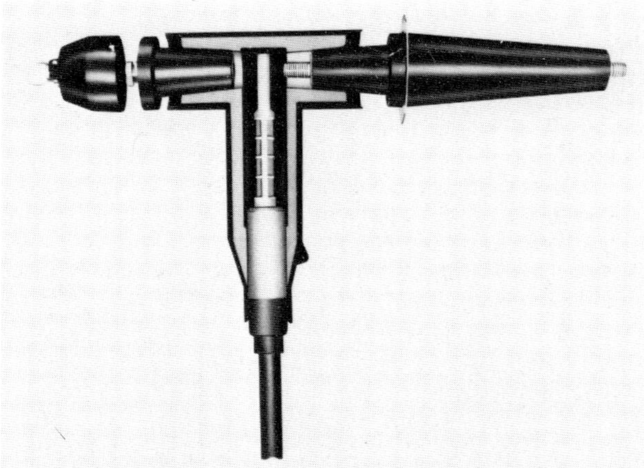

Figure 9.13 Elastimold (Bimold) type 650 LR elbow connector

9.2.5 Pit terminations

Pit terminations are normally used for very high-power motors in excess of 20 MW rating. The winding leads are brought out to insulator bushings mounted in the bottom of the machine. The main advantage is that large terminal boxes are avoided; current transformers, if required, can be mounted in this region and the pit can be designed for security.

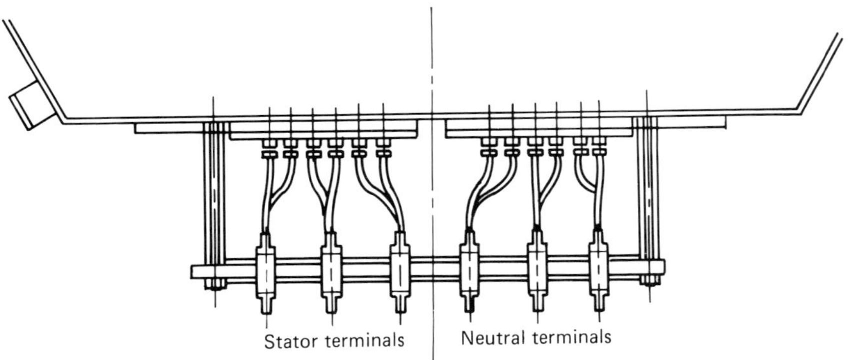

Figure 9.14 Typical pit termination arrangements

*BICC–AMERACE trademark.

Figure 9.14 shows a typical arrangement with line and neutral terminations.

A fully phase-insulated system can be obtained by suitable use of heat shrink sleeves or tapes. This practice is derived from that more commonly used for large aircooled generators with pit terminations, where numerous protective current transformers normally required are mounted at the terminals.

9.2.6 Auxiliary terminal boxes

A large motor may often be provided with a few additional auxiliary (low voltage) terminal boxes. High-voltage motors, particularly, may often be fitted with nine or more temperature detectors embedded in the winding, various air circuit and bearing temperature monitoring devices, vibration monitors and low-voltage space heaters.

It is not considered good practice to introduce live space heater power terminals in a box containing the various sensor terminations. Unfortunately, this occasionally occurs and it is imperative to ensure that such terminals are fully shrouded and fitted with electrical hazard warning labels.

Large wound-rotor induction motors must also be equipped with either a terminal box or terminations connected to the slipring brushgear. These rings can have high voltages across them at start, of the order of 2–3.5 kV. Again, these boxes and cable terminations should be treated with the same care as the primary high-voltage winding terminations. Very large motors may well be equipped with pit-mounted or side-mounted busbar terminations, which should be fully insulated and supported after making up the terminations.

9.2.7 Liquid starters and speed controllers

Wound-rotor induction motors are started by inserting resistances across the sliprings. The resistances can be in the form of metallic resistors or electrodes in an electrolyte.[3–5]

Where grid resistances are used, there will normally be a number of steps shorted-out progressively by contactors with a final running set of contactors which short across the sliprings. It is necessary for the commissioning engineer to ensure that the sequence of operation is correct and that all of the resistance is in-circuit when a start-up sequence commences. It is recommended that leads from the rings to the resistance grids be kept as short as possible and be sized adequately and insulated. It is necessary to keep the last step of resistance plus the leads as small as possible, usually about equal in magnitude to the rotor winding resistance value.

The basic liquid starter or controller comprises a tank containing electrolyte in which three sets of electrodes are immersed, one fixed and one moving per phase. One set of electrodes is shorted together to form the starpoint. The electrodes may be mounted in a single tank or in three separate tanks. The moving electrodes can be driven up or down to vary the resistance smoothly from start through to run up. There are limit switches, timers, geared drives, etc. which require checking for correct operation. A shorting contactor or switch is necessary depending on size.

There is a further type of liquid starter which has no moving parts other than the contactors. The operating principle is based on the difference in resistivity between a liquid electrolyte and its vapour, contained in a fixed electrode chamber. There is a continuous decrease in the resistance value, which is determined by the starting current of the motor. The rotor current initially causes partial vaporization of the electrolyte and the starting resistance is adjusted automatically. A timer finally shorts-out the small resistance which remains in-circuit. These starters normally have their electrolyte temperature controlled thermostatically. As there are no moving parts, erection and commissioning is usually simpler than for the moving electrode type.

There are problems which can occur with liquid starters or controllers:

(1) Local hotspots due to incorrect flow paths or 'scaled-up' electrodes. Localized boiling of electrodes can cause unbalanced rotor currents. Abrupt point descaling can also cause spike voltage to be developed as the scale breaks up. This problem is not evident in the short term and may develop during the guarantee period (normally 1 year after takeover) or even long afterwards.
(2) The electrolyte density may be set incorrectly, giving too-high or too-low values at start. Typically, anhydrous sodium carbonate solution at $30°C$ has a conductivity variance from $0.01\,S/cm^3$ to about 0.045 for solutions of 1–5% strength respectively. The specific gravity could vary from about 1.01 to 1.045 over the same range. The electrolyte should be well mixed and agitated, particularly when the unit has been left standing for very long periods. Circulation pumps are often fitted to large units to circulate the electrolyte before a start can occur, using suitable interlocks or timers which should be checked.

9.3 Commissioning and start-up

Once the motor and driven plant have been aligned successfully, there are a number of checks that have to be carried out before the motor can be energized.

It is necessary to ensure that the switchgear controlling the motor, and any associated metering and protection circuits, have been checked fully. It is imperative to ensure that any trip and emergency shutdown circuits are working correctly before the circuits are energized.

The motor windings should be checked for dryness and also that the insulation resistance and polarization indexes on high-voltage motors have acceptable values as recommended by the manufacturers; details of variation of insulation resistance with ambient temperature and polarization index with insulation type are explained in Chapter 6.

The earth connections to the motor frame and terminal boxes should be checked for tightness and good electrical contact with the site-grounding mat. On motor generator sets, any earthing brushes should have their circuits checked and it should be ensured that brushes are bedded in and placed in their working position.

All auxiliary services such as external oil and water for lubrication and cooling should be checked to ensure that there is adequate flow as specified in the instructions. A check should be carried out to ensure that all hazard warning signs, guards, covers and safety handrails are in position and securely fastened.

The following sections give typical electrical and mechanical checklists. These lists must not be assumed to be final, as for special drives the engineer may have to construct a supplementary checklist.[6]

Where, for example, a motor is fitted with separate motorized fans in place of the more normal shaft-driven fan, it will be necessary to ensure that any interlocks or protective circuits are operative. The same concern is necessary for lubrication services drawn from a separate tank and having motor-driven pumps.

9.3.1 Phase rotation

Phase rotation checks should be carried out on the incoming supply using a phase rotation meter directly on low-voltage supplies and by using secondary terminals on potential transformers for high-voltage supply or phasing sticks. On motors with their terminals marked correctly in the manufacturer's works, the connections can then be made up. If there is any doubt and any risk of damage to the driven machine, the coupling should be split and the motor alone run up.

It is also necessary to ensure that motors fitted with shaft-mounted axial-flow fans have their mechanical rotation compatible with the established phase rotation.

9.3.2 High-voltage dielectric tests and dry outs

Small low-voltage motors insulated with Class F systems normally do not present problems of dampness of the windings. All motors, however, should be checked for dampness before any high-voltage tests are carried out. If a low insulation resistance is obtained, it will be necessary to dry out the machine. All precautions should be taken to obtain a true condition of the insulation on stators or wound rotors. When carrying out any dry out, it is essential that the temperature of the winding be held at a reasonably constant value, or misleading values of insulation resistance will be obtained.[7]

The method chosen for drying out will depend on whether the motor is fitted with anti-condensation space heaters or not, or availability of other supplies and general construction of the motor.

Sufficient heat should be used to obtain an end-winding temperature of 75° C. The rate of temperature rise should initially be slow, to avoid excess formation of vapour and localized heating. Drying out can be effected by any of the following methods:

(1) By using the space heating built into the motor.
(2) By providing temporary space heaters of a black-heat grade placed inside the motor under the end-windings. Care should be taken to avoid charring the windings.
(3) Taking care to make provision for the escape of the moisture by blowing hot clean dry air into the motor.
(4) Placing electric radiators all round the motor. It is necessary to locate the motor in a draught-free room.
(5) By connecting the motor windings (on squirrel cage motors) to a low-voltage a.c. source of approximately 10% of the line voltage. The rotor should be locked to prevent rotation.[8,9]
(6) By passing current through the windings.
(7) Synchronous motor field windings may be dried by circulating current. Care must be taken not to damage sliprings and other components. Clamps which pass currents across the sides of the sliprings should be used. Where direct heating of windings is carried out, the rate of temperature rise is important to avoid excessive temperatures.

The temperature of the windings and insulation resistance should be monitored at regular intervals and the machine should be supervised continually. On initial application of heat, the insulation resistance will drop sharply, then slowly rise until level. On discontinuing the dry out, there should be a further rise in the insulation resistance.

The dry out curve for a typical large motor will have the general shape shown in Figure 9.15. The actual values of time to dry out and insulation resistance will vary considerably, depending on motor size and degree of initial dampness, and on very large motors may take 30 h or more. After successful dry out, the high-voltage tests may be carried out. It is recommended that the test voltage should be 75–80% of the value of the factory tests on a new winding. The exact value is determined by national standards or the manufacturer's recommendations.

9.3.3 Shaft voltage – bearing insulation

Shaft voltages are induced due to variations in the reluctance of the magnetic circuit in an a.c. machine. A pulsating variation in the flux which links the shaft is caused through a circuit consisting of the motor frame, the bearings and the shaft. If this circuit is not broken, a short-circuit path to the induced voltage is formed, causing a current to circulate which may damage the bearing surface. There are many aspects of shaft voltages and currents which require consideration.[10–12]

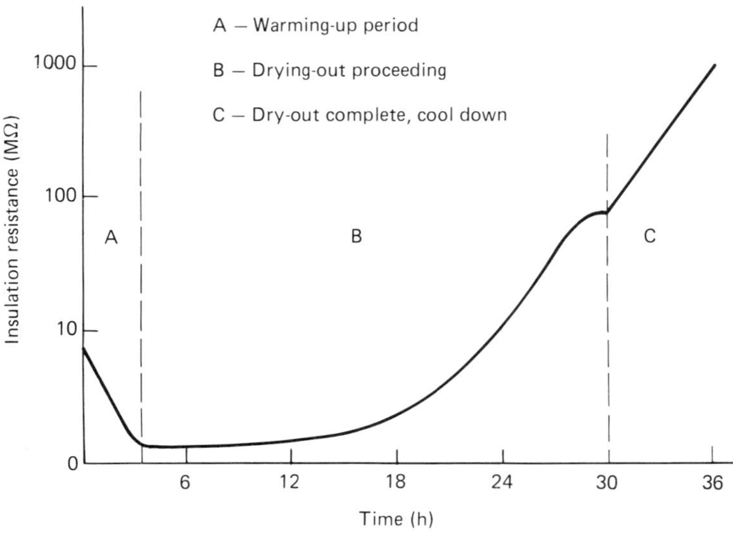

Figure 9.15 Typical dry out curve for a very large high-voltage motor

In normal circumstances it is not economically practical to provide control of the magnitude of shaft voltages; the circuit is interrupted by the insertion of a layer of bearing insulation material immediately under the bearing sleeve or bush or under the feet of the pedestal. It is necessary to ensure that the insulation is not shorted-out by lubrication pipes, spiral metal protection around temperature probe cables and even accidental shorting-out by screwed connections projecting too deeply.

When the motor is first run up, this voltage across the shaft should be measured and recorded as it can be used for diagnostic purposes. Typical values on quite large motors can vary between 1 and 12 V a.c., whilst on smaller motors with anti-friction bearings, values between 50 and 500 mV are encountered. As the majority of motors are fitted with anti-friction bearings with no bearing insulation, many specifications specify a maximum acceptable value before insulation must be fitted. A typical acceptable value is 200 mV for motors up to 250 kW output.

9.3.4 Electrical and mechanical checklist

The combined electrical and mechanical checklists shown in Figure 9.16 have been arranged to give a reasonably logical sequence of events. A user may prefer to separate the electrical and mechanical aspects, but this is not recommended. Special detailed datasheets should be made to record mechanical alignments, airgap measurements, vibration and noise level records, as found to be appropriate.

9.4 Site tests – general

Tests on site usually fall into two main categories: (1) those required for the general purpose of commissioning; and (2) those needed for acceptance guarantee of performance. It is essential to establish at an early stage, preferably the pre-contract stage, exactly what testing is required on site. This is necessary both because unnecessary site testing is expensive and time consuming, and also because

TYPICAL

Electric motors   *Checklist*

Reference ____

____

Plant identification ____

Manufacturer ____

Type ________________________________

Serial no. ________________________________

____

____

Frame size ________________________________

BSS/IEC ________________________________

Kilowatts ________________________________

Volts ________________________________

Phases ________________________________

Hertz ________________________   Rotor volts ________________________________

Rev./min ________________________   Rotor amps ________________________________

F.L.C. ________________________________

Connection diagram ________________________________

Bearings ________________________________

Drive end ________________________   Non-drive end ________________________________

Other nameplate information:

(A) *General checks (applicable to all motors)*
    (1) Motor clean and dry.
    (2) Check motor airgap (not required on motors with ball or roller bearings).
    (3) Check tightness of all nuts and bolts.
    (4) Check dowels fitted if called for.
    (5) Check all safety guards.
    (6) Check lubrication points to specification.
    (7) Check paintwork and finish satisfactory.
    (8) Check air filters/ventilation paths correct.
    (9) Check nameplate correct.
    (10) Check motor identified to plant schedule.
    (11) Check motor earthing satisfactory.
    (12) Check electrical connections are to drawing.
    (13) Check cable box/gland correct.
    (14) Check heater supply available/correct.
    (15) Defect list to be appended.

Figure 9.16 Electrical and mechanical checklists

(B) *Checks on motors fitted with sliprings/commutators*
     (1) Record make and grade of brush as supplied.
     (2) Check brush bedding correct.
     (3) Check brushes are correctly tensioned (record figure).
     (4) Confirm brush stagger correct.
     (5) Confirm brushes set to neutral axis.
     (6) Check brushbox clearance correct.
     (7) Confirm slipring or commutator surface satisfactory.
     (8) Confirm commutator surface satisfactory and correctly undercut.
     (9) Electrical connections correct and secured.
    (10) Electrical clearances adequate.

(C) *High-voltage motors – checks as in (B) plus:*
     (1) Check bearing insulation.
     (2) Check bearing earth straps.
     (3) Check/set magnetic axis.
     (4) Ensure high-voltage cablebox to specification.
     (5) Check diaphragm/pressure relief device.
     (6) Check dehydrating device.
     (7) Check terminal shrouds fitted as appropriate.
     (8) Check cablebox compound level.

(D) *Commissioning checks*
     (1) Take insulation resistance of motor. (*Note:* For high-voltage motors record
         polarization index.)
     (2) (a) defects should be cleared prior to issue of any certification;
         (b) issue appropriate energizing certificate.
     (3) Position 'danger' notices.
     (4) Check phase of the supply is correct and colour true.
     (5) For motors where direction of rotation is important, to avoid damage to the
         machine a '*flick test*' can be done or the motor coupling disconnected.
     (6) Check the motor starts smoothly with satisfactory run-up time.
     (7) Check motor current is within specification.
     (8) Bring motor up to full load and check specification.

(E) *Running checks*
     (1) Vibration/torsionals during run-up acceptable.
     (2) Vibration on no-load satisfactory.
     (3) Vibration on full-load satisfactory.
     (4) Bearings oilflow/temperatures satisfactory.
     (5) Cooling water flow/temperatures satisfactory.
     (6) Differential inlet/outlet air temperature satisfactory.
     (7) Winding temperatures satisfactory.
     (8) All gauges/indicators/recorders satisfactory.
     (9) All auxiliary motors oilpumps, fans satisfactory.
    (10) Defect list to be appended.

(F) *Additional special checks*

Figure 9.16 (Continued)

special instruments, current and voltage transformers may be necessary. Any special equipment is often difficult or impossible to obtain on certain sites. Customs and Excise regulations often cause extreme difficulty to the contractor and can delay the overall programme.

9.4.1 Standard commissioning tests

Where a motor has been fully works-tested, including the heat run with load losses and efficiencies being established, site tests should be minimal.

9.4.1.1 Alternating current motors

Typically, tests on a.c. motors consist of the following:

(1) Insulation resistance.
(2) Polarization index (high-voltage motors).
(3) High-voltage winding tests.
(4) Phase-rotation check.
(5) Light-running tests.
(6) Shaft/pedestal insulation where applicable.

Test (3) causes the largest amount of problems. It is often accepted that with low-voltage motors, of 380–440 V, this test can be waived provided that the insulation resistance was carried out using a 1000 V Megger. This deviation should have been agreed at the contract stage.

When carrying out test (3) on high-voltage motors, it must be remembered that the winding capacitance is dependent on output, speed and voltage. For large motors, a large-capacity transformer will therefore be required.[13,14] Typically, a 2000 kW motor may require between 10 and 30 kVA for the entire windings, depending on motor speed and voltage. At 10 000 kW this requirement could increase from 40 to 100 kVA. A single phase tested to earth, with the other two phases earthed, would require approximately 40% of the values stated above.

Diodes on brushless synchronous motor rotors must be protected from damage when carrying out the high-voltage test. The bridge circuit should be shorted out.

It is fairly common practice on site to carry out the test using a cable test set with d.c. voltage. This should again be agreed at a contract stage. It is recommended that the test should be carried out to comply with ANSI/IEEE Standard 95-1977. A plot of microamperes against kilovolts should be recorded and studied for irregularities.

9.4.1.2 Direct current motors

Motors which arrive on site fully assembled usually provide little difficulty in commissioning.

Typically, tests on d.c. motors consist of the following:

(1) Insulation resistance.
(2) High-voltage winding tests.
(3) Light-running tests.
(4) Commutation check on load.
(5) Load–speed characteristic.

On large d.c. motors which may have to be assembled on site, it will be necessary to carry out partial works tests. The brush neutral position may have to be re-established and polarity checks carried out. The test programme in the case of large mine winders and steelworks drives requires special consideration and should be agreed with the manufacturer.

9.4.2 Acceptance guarantee tests

When manufacturing very large motors, usually greater than 10 MW, it is often impossible to carry out all the prototype acceptance tests in the factory. Thus, it is usually impossible to carry out a full-voltage starting test to confirm that any stringent voltage dip limitations on the system have been fully met. It is then necessary to agree a procedure for recording and evaluating such a test when it is carried out on site.

The full load kilowatt heat run is another test that is often carried out on site. In many instances, full load cannot be achieved due to temporary site limitations. It is necessary again to agree details for an acceptable procedure within the contractual requirements. Should any special measuring equipment be required, this must be agreed. In the planning stage it is often more convenient to specify a higher grade of instrument current and voltage transformers for the motor metering panel to facilitate testing.

It is important therefore to consider the limitations of site testing, guarantees and tolerances at an early stage in a project, and preferably at the planning stage before a specification is issued.

When finally evaluating motor performance, it must not be forgotten that the most important feature of an electric motor is to be able to start, to run up its load and to deliver its rated load efficiently. There can be no compromise on the first two requirements, particularly if the system is weak. Poor electrical performance in respect of efficiency, power factor or slip (where critical) can be evaluated and a penalty involved if necessary.

9.4.3 Compliance with specifications

National standards, such as BS 4999:Part 69, list the various tolerances on motor performance; IEC 34-1 has similar values. When assessing data, consideration should be given to the accuracy obtainable for any parameter determined on site. For very large or important drives it is usual to have a specially tailored motor specification prepared by a specialist. Any tolerances on items considered vital should have been listed.

From experience, it is considered that the following items often have special requirements:

(1) *Starting current:* usually linked with system disturbance which may be such that no deviation from a specification is acceptable. It may be necessary to install capacitors to assist during the starting period.
(2) *Slip:* can be related directly to pump output and evaluated if the output is critical. Apart from a major rotor redesign, this aspect would require agreement to any stringent tolerances.
(3) *Efficiency:* can be directly related to running costs which can be evaluated. Before any decision is made the manufacturer should, for example, be given the opportunity to consider trimming ventilation losses if a motor has a thermal margin due to excessive fan effect.
(4) *Power factor:* usually related to tariff. Power factor correction can be considered, depending on the severity of the shortfall.
(5) *Temperature rise:* as there are no tolerances on temperature rise, this aspect causes the greatest difficulty when evaluating acceptance criteria. Again, the specification should be reconsidered because, for example, a motor insulated with a Class F system, but specified with Class B or any value below a full Class F system can be acceptable subject to discussion. In this respect it must be remembered that the same motor tested two or three times under virtually identical conditions can show variations of temperature rise of 2–3° K. Further, it is often possible to evaluate the insulation life reduction provided the true operational conditions are known.
(6) *Noise:* where noise is critical, especially if personnel are working around the machine, the motor would normally be rejected after applying any specified tolerance. In an extreme case of slot noise, which cannot be silenced acoustically, a new rotor may have to be provided or the stator winding reconnected.

(7) *Vibration and unbalance:* these have been coupled for convenience. It is necessary to get this aspect corrected before acceptance, as the long-term effects of excessive vibration can damage seriously the motor structure and even break-up the foundation. Again, where super silence and smooth running is required, this should be specified in detail.

Generally, the commissioning engineer does not make the final decisions on acceptance apart from shutting down a machine considered to be dangerous. It is necessary to be reasonably appreciative of the problem areas and scope for rectification. Where problems are encountered, it is therefore vital to ensure accurate data are recorded along with any helpful information which may assist the motor manufacturer or user.

9.5 Functional monitoring

9.5.1 Vibration and bearing noise

Vibration, if excessive, can be damaging to both plant and foundations. Where vibration is experienced, it is necessary to try to locate the cause before any rebalancing is attempted. Occasionally, it is found necessary to apply jack struts to selected parts of the installation. The foundations should be checked for any movement, and all foundation bolts checked for tightness. This check should also include all coupling bolts.

Portable, hand-held, vibration detection equipment can quickly establish the severity of the vibration on bearings pedestals and bedplates. British Standard 4999:Part 50 gives a guideline (Table 9.2) for vibration limits, measured on the bearing on a horizontal plane through the shaft centreline of horizontal machines, both in the factory (column 1) and on site (column 2). Experience shows that manufacturers can usually obtain lower values than those listed.

On vertical motors, the top bearing is sometimes found to have values 50% greater than those achieved on horizontal motors.

If the severity of the vibrations persists, the manufacturer should be informed and an in-depth investigation carried out. Rebalancing of rotors can be carried out on site, as provision is usually built into the rotor assembly to enable this. Certain specifications call for the facility to balance high-speed rotors on site by the provision of balance grooves and weights on easily accessible parts on the rotor, such as fan hubs. Under no circumstances should a commissioning engineer attempt to relocate balance weight fixed by the manufacturer without permission; neither should he affix illegal weights on any rotating masses.

On motors equipped with anti-friction bearings, a further diagnostic aid is available to determine the quality of the bearing. Shock-pulse monitoring can be carried out and the bearing noise level taken.

The noise level can be compared with that recorded in the factory (if specified), to determine if bearing damage has occurred.

9.5.2 Cooling medium monitoring

9.5.2.1 Ventilation

When the motor is on full load, it is necessary to check the temperatures of the cooling medium. Where machines are fitted with separate heat exchangers, both the primary and secondary coolants should be checked if possible. It will be necessary to check the differentials of the cooling media and actual site ambients to enable alarms and/or protective trips to be properly set up. Advice from the manufacturer should be obtained if large deviations are noted from the works test data.

Table 9.2 Limits of vibration amplitude (peak to peak in millimetres)

Speed (rev./min)	Column 1	Column 2	Speed (rev./min)	Column 1	Column 2	Speed (rev./min)	Column 1	Column 2	Speed (rev./min)	Column 1	Column 2
3600	0.020	0.025	1800	0.036	0.045	900	0.059	0.074	450	0.085	0.105
3400	0.021	0.027	1700	0.038	0.047	850	0.061	0.076			
3200	0.022	0.028	1600	0.040	0.050	800	0.063	0.078	400	0.090	0.115
3000	0.024	0.030	1500	0.041	0.052	750	0.065	0.081			
2800	0.025	0.032	1400	0.043	0.054	700	0.068	0.085	350	0.094	0.120
2600	0.027	0.034	1300	0.045	0.057	650	0.071	0.089			
2400	0.029	0.036	1200	0.048	0.060	600	0.074	0.092	300	0.100	0.125
2200	0.031	0.039	1100	0.051	0.064	550	0.077	0.096			
2000	0.033	0.042	1000	0.055	0.069	500	0.080	0.100	250	0.100	0.125

Note: Where the actual test speed is not tabulated, interpolation is permitted.

Source: BS 4999:Part 50.

Where there are a number of motors or machines located close together, it is necessary to check that hot air exhausted out of one machine is not recirculated by the machine or sucked in by the air inlet of another machine adversely located. This problem is more likely to occur with long trains of machines such as multiple, open-ventilated, motor generator sets with air blown downwards on leaving the machines.

When setting alarms based on factory test data, sufficient allowance should be made where the site ambient varies by a large amount from summer to winter; IEC 34–1 makes a correction for variation on test and site ambients where the difference of ambients exceeds 30° C.

Where site cooling mediums such as air or water temperatures are found to exceed specified values by a large amount, this fact should be reported immediately to the manufacturer. It may become necessary to provide temporary ventilation to a machine room having insufficient capability for heat removal. This should also be recorded for evaluation.

9.5.2.2 Bearings

Similar checks should be carried out on any bearing cooling water or lubrication oil circuits. The differential temperatures should again be recorded and compared. Where the bearing is monitored, this should also be compared. If a hot bearing is suspected and a very fast rise in temperature of the bearing metal observed, it may be necessary to shut down if the temperatures show no sign of levelling off. Immediately check oil and water flow quantity against specified values. It may be prudent to lift the top half of the offending bearing and push out the bottom bush for inspection, as any abnormal wear or bedding-in pattern may give the solution to the problem.

When checking the bearings visually, it is prudent to check that the bearing was assembled correctly and that the oil is not bypassing the bush giving correct flow but no effective heat transfer. Under this condition, oil exit temperature and oil differential temperature would not be correct. Under no circumstances should bearing alarms or trips be disconnected for commissioning purposes.

9.5.3 Electrical protection

Electrical protection of motors is normally incorporated with the switchgear. However, the commissioning engineer may well have to ensure that the protection equipment has been checked adequately. It is necessary to be familiar with the equipment provided before attempting to start-up plant. Emergency stop buttons, start and stop facilities, dynamic braking (if fitted) and all interlocks should be located and checked. The correct operation of position or limit switches should also be checked.

Finally, it is necessary to ensure that all wiring checks, phase sequencing and protective relaying have been proven and that all trip circuits are healthy before attempting to start up.

9.6 Diagnostic – general

When starting up any motor for the first time, whether it be asynchronous, synchronous or d.c., there is a high probability that spurious trips may occur. It is necessary to inspect any flag or other indicators which may show in which mode the trip originated. A knowledge of the driven plant characteristics and run-up time is essential. On the first run-up the trip may occur instantly the breaker is closed, during the run-up period, or after the motor has apparently settled and is taking up load. It is necessary to establish exactly when the trip occurred during the start-up sequence.

It is advisable after any trip to log the event, and to ensure that a sufficient cooling period is allowed before restarting. Check that all equipment is returned to the 'ready to start' position, e.g. liquid controllers have been wound back. Most common circuits automatically run equipment back if motorized, but on the first commission run it is prudent to take nothing for granted. A similar check should be carried out where an autotransformer, Korndorfer or star–delta starting sequence is involved.

9.6.1 Diagnostic – failure to start: general

9.6.1.1 Events and possible causes

(1) *Nothing happens:* line fuses may not have been inserted or breaker failed to close; check that the incoming circuit has been energized.
(2) *Overload trip:* start-up current greater than setting – check and rectify; if so, restart.
(3) *Starting sequence incomplete – nothing happens:* check interlocks, limit switches, battery voltage for solenoids, low busbar voltage.
(4) *Protective circuits operate:* check the following if fitted; overcurrent, earth fault, single phasing, reverse power.
(5) *After diagnostic (1)–(4) motor still fails to start:* driven load may be locked mechanically or seized up; split couplings and try motor alone.

9.6.2 Diagnostic – failure to run up: a.c. motors

9.6.2.1 Event and possible causes

(1) *Motor crawls and is noisy:* bearing stiction high, insufficient torque, low voltage; rotor rubbing stator – pulled over, unequal airgaps.
(2) *Run-up time too long:* voltage may be low; load excessive – unload drive; check if autotransformer taps too low; electrolyte incorrect.
(3) *After diagnostic (1) and (2) motor run-up time still too long:* the load inertia may be excessive – design error – consult manufacturer.
(4) *Synchronous motor run-up too long after diagnostic (1) and (2):* field excitation may be present throughout run up – especially for separately excited motors.
(5) *Synchronous motor runs up but fails to synchronize or hunts:* field circuit not operating; insufficient excitation for pulling into step; exciter defective – check output voltage; load excessive – reduce load and retry.
(6) *Synchronous motor running satisfactorily, but trips or pulls out of step:* loss of excitation – check sustained line voltage depression – check excessive load fluctuations if driving a compressor load – unload.
(7) *Excitation circuit fails to build up:* short circuit on field windings; open circuit on field winding.
(*Note:* Field circuit faults may often only occur 'at speed', not standstill.)
(8) *Wound rotor motor runs at low speed with starting resistance out of circuit:* generally too-high resistance left in the rotor circuit; cables – check size; brushgear – brushes; sliprings – check for heating.
(9) *Wound rotor motor, excessive brush wear, dusting or sparking:* wrong grade of brush; incorrect brush spring tension; slipring surface rough or spiral grooved incorrectly; brush spacing wrong circumferentially; humidity too low for brush grade chosen; abrasive eccentric rings; high current density, overloads, brush sticking in boxes.
(10) *Synchronous motor d.c. collector rings, excessive grooving:* complete checks as (9) above; negative ring grooving after prolonged running due to electrolytic action – improved by reversal of polarity.

9.6.3 Diagnostic – motor running problems: a.c. motors

9.6.3.1 Events and possible causes

(1) *Stator overheating in parts:* rotor offcentre – misalignment; single-phasing with noisy running; unbalanced voltages on supply; bearing play excessive with rotor offcentre; stator winding coil faults.

(2) *General overheating:* overload; incorrect power factor for synchronous motor; too high or low stator voltage; ventilation circuits fouled; cooling water not flowing adequately; build-up of temperature in machine room; too many starts per hour.

(3) *Mechanical noise vibration – axial bumping – noisy bearings:* short-circuited coils on stator or rotor windings; unbalanced pull – airgap offcentre; defective anti-friction bearings; coupling or foundation bolts loose; core laminations buzzing; rotor fan blades hitting fan guide; stator and rotor axially misaligned and bumping due to insufficient bearing clearance; bent or misaligned shaft.

(4) *Anti-friction bearings running hot:* excessive grease filling; incorrect grade of grease; dirty or insufficient grease; bearing race skidding in housing – also noisy; excessive side loading due to belt pull or misalignment.

(5) *Sleeve bearings running hot:* insufficient oilflow if force-lubricated; insufficient cooling water if water-cooled; oil level low or oil rings not picking up oil; dirty oil or wrong grade; oil rings dragging or incorrectly fitted; side thrust on white metal shoulder; damaged babbit or white metal bedding surface; blistered white metal surface.

9.6.4 Diagnostic – general d.c. motor problems

9.6.4.1 Warning

Successful commutation is affected by various factors such as speed, load cycling, environment, etc. Whilst the commissioning engineer should be able to recognize commutation problems, it is unwise to make unapproved changes. Brushgear is usually set up in the factory and adequate reference, punch marks or dowels are provided to enable correct setting up.

Major changes should be carried out or be supervised by a specialist who is familiar with the motor in operation.

In general, a healthy commutator will have a semi-glossy light tan coloration. If the gloss is excessive, this may be accompanied by signs of brush chattering.

Occasionally, a light mottled surface with a random pattern, both axially and circumferentially, may be noted on a perfectly healthy commutator. If a heavy dull blackish film is observed, this may be due to an oil–dust mixture. The commutator may require cleaning by wiping with a coarse lint-free cloth or canvas fixed to a flat wooden paddle. Under no circumstances should emery cloth or paper be used.

9.6.4.2 Events and possible causes

(1) *Motor fails to start:* check fuses – supply availability; connections healthy; correct to diagram; brushes not lowered to contact commutator; starter arm not contacting studs on faceplate starter; check cable continuity.

(2) *Motor trips on start:* shunt field open-circuit or incorrectly set; main fuses or trips incorrect; load stalled or shaft locked; resistance steps shorted out too quickly; incorrect voltage.

(3) *Motor runs up but overspeeding or runs on low speed:* brushgear rocker incorrectly set; incorrect voltages, armature or field; field windings incorrect; series motor no-load speed high; compound motors unloaded – speed high; overloaded – speed low.

(4) *Motor running, but brush sparking heavy:* brushgear in wrong position; incorrect spacing of brushes; brushes incorrect grade; brushes improperly bedded; brush pressure incorrect; brush stagger incorrect; motor on heavy overload; mica segments proud – undercutting wrong; brush trail or reaction incorrect for direction of rotation – check instruction manual; high commutator bars – commutator shifting; brushes sticking in brushboxes; broken or short-circuited armature winding; partially shorted field coil; loose pole bolts.

9.6.5 Diagnostic – d.c. motor commutator problems: d.c. motors

9.6.5.1 Events and possible causes

(1) *Streaky surface – light metal-to-brush transfer:* light loaded motor – too many brushes; brush pressure incorrect – low; brush too porous or abrasive – atmospheric contamination; fumes and/or abrasive dust.

(2) *Grooving over face of commutator brush width:* abrasive brush grade or abrasive dust; check axial stagger if track grooving occurs.

(3) *Thready surface – fine deep hairlike lines with heavy metal-to-metal transfer:* light-loaded motor – too many brushes; brush pressure incorrect – low; brush too porous; fumes in atmosphere; mixed grades of brushes used.

(4) *Copper drag – with heavy erratic flashing. Obvious on trailing edge:* low brush pressure; heavy vibration on commutator surface; abrasive or incorrect brush grade; harmful gas/fumes in atmosphere.

(5) *Pole pitch bar-marking appearing at pole pitch or twice pole-pitch spacing:* shunt field unbalanced; defective equalizer connections; poor armature/riser connections; low brush pressure; heavy vibration; incorrect brush grade on rectifier-fed motor.

(6) *Slot bar marking – trailing edge marking giving a regular pattern – related to conductors per slot:* incorrect electrical or magnetic settings; unequal pole airgaps; overloading.

(7) *Alternative dark and light bar marking or groups of differently coloured bars forming a regular pattern round the commutator:* due to electrical design causes; incorrect commutation; deficient equalizing; usually requires manufacturer's assistance if persisting after checks on magnetic setting.

References

1. CONFERENCE INTERNATIONALE DES GRANDS RESEAUX ELECTRIQUES. *Report on vertical fixed-speed motors for general industrial use rated 1000 kW and greater.* CIGRE Publication No. WG11.06, ELECTRA (1987)
2. CONFERENCE INTERNATIONALE DES GRANDS RESEAUX ELECTRIQUES. *Report on vertical fixed-speed motors for primary and secondary coolant pumps for use with nuclear reactors.* CIGRE Publication No. WG11.06, ELECTRA (1987)
3. HOTCHAMPS, P. 'Medium and large liquid rheostats'. *ACEC Rev.*, **3**, 4, 45–48 (1966)
4. McFARLAND, G. L. and ALVAREZ, W. 'The liquid rheostat for speed control of wound rotor induction motors'. *AIEE Trans.*, **67**, 603–610 (1948)
5. ELLIOTT, R. L. 'Water rheostats for shipyard use'. *IEEE Trans. Ind. Gen. Applications,* **IA-18**, 5, 568–572 (1982)
6. ERICKSON, C. J. E. and MORRISON, W. G. 'Placing a 20 000 hp motor in service'. *IEEE Trans. Ind. Gen. Applications,* **1A-14**, 5, 397–401 (1978)
7. GEC TURBINE GENERATORS. *Insulation testing, drying-out and cleaning procedures for air-cooled a.c. generators.* GEC Turbine Generators Publication No. 1207–72
8. MOON, H. YUEN. 'Low-voltage heating of motors in refineries and chemical plants. *Inst. Elec. & Electronic Engrs Trans. Ind. Gen. Applications,* **IGA-5**, 3, 300–309 (1969)
9. HOLMQUIST, J. R. 'Motor heater'. *Inst. Elec. & Electronic Engrs Ind. App. Soc. Newsletter,* 3–5 (1984)

10. HUHN, J., KRANEN, H. and WIPPERMANN, M. *High-rating motors for reciprocating compressors* (Schorch Report), pp.69–75 (1977)
11. SOHRE, J. S. 'Shaft currents can destroy turbo-machinery'. *Power,* 96–100 (1980)
12. CAGGIANO, J. L. 'Shaft voltage measurement'. *Turbo-machinery Intnl,* 29–31 (1983)
13. WIESMAN, R. W. 'Capacitance of synchronous machine armature windings determined for high potential test'. *Gen. Elec. Rev.,* 26–30 (1947)
14. *AC motor maintenance manual – HV motors.* ANSALDO Publication No. 1–4016A, pp. 20–27

Standards and publications

BRITISH STANDARDS INSTITUTION *Code of practice for safeguarding machines.* BS 5304. BSI (1975)
BRITISH STANDARDS INSTITUTION. *Specification for general requirements for rotating machines.* BS 4999:1976 Part 69 'Tolerances'. BSI
BRITISH STANDARDS INSTITUTION. *Specification for general requirements for rotating machines.* BS 4999:1976, Part 50 'Mechanical performance – vibration'. BSI
BRITISH ELECTRICAL APPARATUS AND MANUFACTURERS' ASSOCIATION. *Recommendation for the design feature of shaft systems of plant involving electrical machines.* BEAMA publication REM 502 (1972) BEAMA
AMERICAN NATIONAL STANDARDS INSTITUTE/INSTITUTE ELECTRICAL ENGINEERS. *Separable insulated connectors for power distribution systems above 600 V,* Standard No. 386. ANSI/IEE (1985)
AMERICAN NATIONAL STANDARDS INSTITUTE/INSTITUTE ELECTRICAL ENGINEERS. *IEEE recommended practice for insulation testing of large a.c. rotating machinery with high direct voltage,* Standard 95. ANSI/IEE (1977)
INTERNATIONAL ELECTROTECHNICAL COMMITTEE (1983). *Rotating electrical machines:* Part 1: 'Rating and performance.' IEC Publication No. IEC 34-1. IEC

10 Maintenance and failures

S. P. Watson

Formerly Senior Development Engineer, Mather and Platt Ltd.

10.1 Maintenance management and policy

10.1.1 General policy decisions

The overall policy of any profit-orientated organization must be to minimize the cost of operating its plant while still maintaining acceptable levels of safety. Decisions regarding maintenance policy will therefore be influenced by:

(1) The mode of operation of the plant.
(2) The historical likelihood of failure (where such information exists).
(3) The cost of plant outage due to either failure or planned maintenance.
(4) The consequences of outage due to failure.
(5) The cost of planned maintenance.
(6) The cost of standby plant.
(7) Access to the plant.
(8) Replacement plant strategies.
(9) The level of engineering support available from the original motor manufacturer.

If sufficient information is available, an economic model can be produced to test the effects of various alternative policies. Where only limited information is available it is better to use experienced judgement to complete the economic model rather than using experience to decide upon policies.

10.1.2 Safety policy

Like any other item of electrical plant, electric motors should only be maintained by adequately trained personnel under appropriate supervision. Before any work commences on an electric motor which is, or has been, in service, the appropriate safety precautions should be observed. These will almost certainly include mechanical interlocks and/or a system of 'permit to work'. Whatever system is used, it must be certain that at no time can a motor be energized accidentally while it is undergoing maintenance. Certain protective devices such as earth leakage circuit-breakers are suitable protective devices on small low-voltage motors in order to minimize risk to operating and maintenance personnel.

Before the terminals of any electric motor are exposed (for instance to disconnect cables), the supply must be isolated and locked off and confirmed to be at earth potential by the use of a suitable test device supplemented, if necessary, by the use of locally applied earthing straps.

In addition to the dangers associated with working on electrical plant, another safety aspect is the effects of any repair or maintenance on its original design specification. Care must be taken to ensure that any work undertaken does not impair the operating safety of the plant. This is even more essential when applied to motors which are certified to operate in hazardous areas (see Section 10.2.3.2).

The guidelines issued by the Health and Safety Executive should be followed at all times during the course of repair or maintenance of electric motors. It is recommended that during any re-installation following removal from service the relevant sections of the IEE wiring regulations are observed. (These regulations only apply to onshore buildings in the UK.)

10.1.3 Implementation of the policy

Standard plant-operating procedures should include provisions for maintenance or failure, which will ensure that defined policies are implemented as intended. These procedures should define:

(1) The policy in detail, including all safety precautions to be observed and a method of recording actions for future reference.
(2) The head of department, or his nominated deputy, responsible for implementing the defined policy.
(3) The ultimate authority in the event of any problems which may arise.

10.2 Standard cleaning and overhauling practices

10.2.1 General aims

The efficient operation of any motor depends upon correct installation and the subsequent care that it receives in service. The ingress of foreign matter into the motor enclosure or bearings is one of the principal causes of failure. It is therefore essential that cleaning and inspection is undertaken on a regular basis to ensure trouble-free operation. To prevent bearing failure, it is important to check the condition of the lubricant at regular intervals and replace as necessary.

10.2.2 *In situ* cleaning practices

The principal causes of electrical breakdown result from ingress of: (1) moisture; (2) oil and grease; (3) dirt; and (4) carbon/metal dust. The motor must therefore be kept clean and dry both internally and externally (unless designed specifically to operate otherwise, such as motors with hermetically sealed windings). Many simple cleaning operations can be accomplished with the machine in its normal operating position without the need for major dismantling.

The necessary frequency of cleaning will depend upon the operating environment and the type of motor. To remove dirt or dust a vacuum cleaner is the ideal tool, alternatives being blowing out with dry compressed air or handbrushing. Fairly simple access to the windings of ventilated motors may be possible. Where this is the case, all visible dirt and dust must be removed.

If the motor contains brushgear (commutator or sliprings) then all carbon dust must be removed and the brushes checked for excessive wear or overheating.

The internal surfaces of totally enclosed motors generally do not need to be cleaned or inspected unless either the ingress of foreign matter is suspected or if it contains brushgear. If this is the case, it will usually be more convenient to remove the motor from its site to a suitable workshop for dismantling.

Heat dissipation either by convection or radiation occurs from all external surfaces. Dirt or contaminant on these surfaces will lead to a reduced heat transfer coefficient which may ultimately lead to overheating. The external surfaces of all motors must therefore be kept in a clean condition.

Air grills must be kept unobstructed and clean, to maintain the airflow quantities required to cool the motor.

Air-to-air heat exchangers need to be inspected to ensure the free passage of air down all the tubes. If the tubes are becoming restricted or if any are blocked due to the build-up of dirt, these should be cleaned using compresed air or a non-metallic tube brush.

10.2.3 Overhaul procedure

10.2.3.1 General considerations

A motor may be taken out of service for overhaul for a number of reasons such as:

(1) Preventative or planned maintenance.
(2) Failure.
(3) A suspected fault condition developing.
(4) Prior to redeploying the motor on a different driven equipment.

10.2.3.2 Motors certified for use in hazardous areas

If the motor has been designed for use in a 'hazardous area' it will almost certainly be certified by a recognized third party test house such as BASEEFA. In some cases, particularly type-N motors to BS 5000:Part 16, the motor may be certified by the manufacturer (which is implicit in the compliance with the specification). If the motor is to be reused in a hazardous area after overhaul then special care will be required in any work that is carried out to ensure that the certificate of compliance is not invalidated, leading to what could be a potentially dangerous situation following reinstallation.

The BEAMA/AEMT *Code of practice for the repair and overall of electrical apparatus for use in potentially explosive atmospheres* recommends procedures and practices to be followed in connection with the repair and overhaul of electric motors. The following brief extract from this code summarizes the salient features to be considered.

Assuming that repairs and overhauls are carried out using good engineering practices then:
(1) If manufacturers' specified parts or parts as specified in the certification documentation are used in a repair or overhaul then the apparatus should remain in conformity with the certificate.
(2) If repairs or modifications are carried out on apparatus specifically as detailed in the certification documents then the apparatus should still conform with the certificate.
(3) If repair or overhaul is carried out on apparatus in accordance with this code of practice and the relevant Standard(s), although not in compliance with 1 and 2 above, then it is unlikely that the apparatus will be unsafe although it may not conform fully with the certificate.
(4) If other repair or modification techniques are used then it will be necessary to ascertain, from the manufacturers and/or the certification authority, the suitability of the apparatus for continued use in a potentially explosive atmosphere.

If work is undertaken on motors being used in hazardous areas then it will be necessary to refer to the relevant constructional and testing standards which are listed in Table 10.3.

10.2.3.3 Removal of motor from site

Before dismantling, the motor should, whenever possible, be removed to a suitably clean and dry workshop, preferably fitted with appropriate lifting gear. This is particularly important for motors sited out of doors. If it is not possible to move the

Table 10.1 Summary of motor failures, and possible causes and remedies

Failure due to	Possible causes of failure condition	Description of results of failure	Possible remedies
Overspeed	Turbining due to reverse fluid flow through pump or compressor	Damaged rotor core, end-winding and any rotating part due to centrifugal forces, damaged bearing surfaces. Damaged stator due to debris from rotor failure	Rewind rotor. Replace overstressed components. Check shaft for straightness. Check outer diameter of core for concentricity. Check stator/rotor for rubbing. Check bearing condition. Fit overspeed and/or reverse rotation protection
Stall	Seizure of bearings/gearbox/coupling, etc. Jammed mechanism, overload condition, combined with failure or incorrect setting of overcurrent protection devices	Motor will absorb locked rotor current and generally have only minimal cooling at zero speed resulting in one or a combination of: (1) Heat-damaged stator winding. (2) Heat-damaged rotor winding. (3) Damaged leads or terminations.	Check stator and rotor windings and rewind as necessary. Check leads and terminations and replace where necessary. Install or correctly set overcurrent protection devices. Rectify cause of failure condition
Overload	Increased demand on motor due to abnormal load conditions combined with failure/incorrect setting of overcurrent devices and failure of thermal cutout devices when fitted	Depending on degree of overload increased losses, therefore accelerated thermal ageing of stator/rotor insulation systems and slipring/commutator assemblies due to elevated temperature rise leading to premature failure. Accelerated brush shedding and slipring/commutator wear, possibly leading to flash over	Rewind rotor/stator. Clean or replace slipring/commutator assemblies and brushes. Fit thermal cutout devices and/or correctly set overcurrent protection devices

Table 10.1 (Continued)

Failure due to	Possible causes of failure condition	Description of results of failure	Possible remedies
Overvoltage	Wrongly tapped transformer. Wrongly connected three-phase motor (delta-connected instead of star)	Increased losses resulting in same results as for *Overload*	As for *Overload*. Fit overvoltage protection
	Voltage surges due to vacuum contractors, lightning strikes, system faults, etc	Electrical breakdown of insulation resulting in inter-turn and earth faults	Identify and repair damaged coils. Rewind stator/rotor (possibly with a higher system insulation level). Fit surge protection capacitors
Undervoltage	Wrongly tapped transformer. Wrongly connected three-phase motor (star-connected instead of delta)	Increased losses resulting in same results as for *Overload*	As for *Overload*. Fit undervoltage protection
Bearing seizure	Insufficient or excessive lubricant. Contaminated lubricant or bearing surfaces. Old lubricant. Misalignment combined with inadequate monitoring of bearing temperature/vibration levels. Brinnelling from local vibration source, bearing capacity overload, e.g. excessive tension in drive belt. Offloading of ball/roller bearings on vertical motors. Corrosion due to ingress of moisture. Electrical pitting due to shaft currents caused possibly by breakdown of bearing insulation	Motor stops rapidly resulting in possible damage to bearing supports/end-shields, shaft, etc. See also *Stall*	Replace bearings and seals. Metal spray, or plate and regrind shaft. Check bearing supports/end-shields and shaft. Fit bearing temperature/vibration monitors. See also *Stall*. Remove cause of overload. Clean or replace bearing insulation
Single-phasing (three-phase a.c. motors)	Loss of one suppoly line due to system fault. Failure of protection scheme	Loss of torque and asymmetric fault currents in stator and rotor windings, resulting in severe *Overload* condition	See *Overload*. Fit or set protection scheme

Ingress of abrasive dust	Wrong enclosure specified for working environment	Abrasion of exposed insulating surfaces leading to premature failure. Overheating due to blocking of air passages resulting in same condition as failure of cooling system	Rewind as necessary. Change motor enclosure. Resite motor. See *Overload*
Ingress of moisture	Wrong enclosure specified for duty. Failure of enclosure or bearing seals	Electrical breakdown of insulation, resulting in interturn and earth faults	Identify and repair damaged coils. Rewind stator/rotor (possibly using a sealed winding system) Identify and repair damaged seals. Check compatibility of enclosure for duty. Dry out motor throroughly and reseal winding with varnish before reintroducing into service
		Bearing seizure due to corrosion	Replace bearing, bearing seals and possibly lubricant
Failure of ventilating/ cooling system	Burn-out of auxiliary cooling motor. Loss of coolant. Blocked cooler tubes or air passages	Increased temperature rise, resulting in increased thermal ageing of stator/rotor insulation systems and slipring/ commutator assemblies, premature failure	See *Overload*. Rectify cause of fault
Frequent start/ stop duty cycle (Cage induction motors)	Inadequately designed motor for duty cycle. Wrong type of motor being used. Wrong type of starting system being used	Fractured rotor bars/end-rings. Thermally aged stator winding. Damaged stator caused by debris from rotor failure. Stator end-winding movement resulting in stator winding failure. Coreplate burning at rotor tooth tip	Rewind rotor using an engineered design. Rewind stator with a designed end-wiring bracing system. Replace starting scheme (fluid coupling, soft starter) or type of motor

Table 10.1 (Continued)

Failure due to	Possible causes of failure condition	Description of results of failure	Possible remedies
Failure of slipring assembly (wound rotor induction motors)	Failure to replace brushes at specified intervals leading to metal-to-metal contact. Rapid brush wear. Severe overload. Contaminants in surrounding atmosphere. Incorrect brushes. Worn brush holders or brush pressure incorrect. Wrong grade fitted	Metal from sliprings worn or melted away. See *Overload*	Replace sliprings and brush assembly. Identify and rectify original cause of failure
Failure of commutator assembly (d.c. motors)	As for *Failure of slipring assembly*. Severe sparking at brushes due to bad commutator condition, brushes set off-neutral, motor overload or interpoles not properly adjusted, resulting in rapid brush wear and overheating	Metal from commutator worn or melted away. See *Overload*	Replace commutator and brush assembly. Identify and rectify original cause of failure

motor then the area surrounding it must be cleaned as far as is practicable and, again in the case of a motor sited out of doors, a temporary protective enclosure should be erected around it.

When lifting motors ensure that the correct lifting points are used. Do not use the lifting lugs on sub-assemblies such as on the heat exchangers to lift a complete motor unless these have been specifically designed for the purpose.

If motors, particularly larger ones, are being transported, the free shaft-end should be positively locked to prevent the bearings from brinelling.

Before attempting to carry out any work on a motor it should be isolated from all electrical supplies, both main and auxiliary. All terminal leads must be disconnected in the terminal box including heaters, thermistors, thermocouples, etc. and identified to assist future reassembly. In the case of a high-voltage motor fitted with a compound-filled sealing chamber, the sealing chamber will have to be unbolted and left attached to the mains cable.

Any exposed cables or sealing chamber apertures should be protected to prevent the ingress of moisture or dirt. If the sealing chamber is being left in position it should be supported to prevent any damage to the mains cable. The motor can now be uncoupled from its drive system.

If a coupling is fitted, scribe a line across the two halves to ensure correct angular alignment during reassembly. Any holding down bolts and dowel pins should now be removed. If any shims have been used, these should be identified and carefully retained to ensure correct realignment with the drive system.

The motor can now be lifted from its site.

10.2.3.4 Dismantling

Most motor manufacturers produce general or specific maintenance/installation manuals for their particular products. Where such a manual exists this should be consulted before dismantling commences. It will generally be necessary to remove major components (when fitted) in the following sequence:

(1) Silencers, external fan and ducts.
(2) Heat exchanger.
(3) Terminal box(es).
(4) Endcovers and/or bearing.
(5) Slipring or commutator assemblies.
(6) Rotor – this normally requires special consideration especially on larger motors.

   The normal method of removing the rotor is to fit a tube over one end of the shaft (which, in the case of a motor fitted with large-diameter internal fan relative to stator bore, will be at the end opposite the fan). This effectively extends the shaft.

   The tube should be long enough to enable the rotor core to be threaded clear of the stator and also strong enough to support the rotor weight. The rotor is then slung between the extension tube and shaft at the opposite end. It is then lifted carefully so that it is suspended freely without touching the stator corepack. The rotor can be carefully threaded out until the rotor core is clear of the stator. The rotor can then be lifted by a rope slung around the corepack using a different hoist or crane jib and finally lifted clear of the stator.

   During this operation it is normally the stator end-windings that are exposed to the greatest likelihood of damage and a great deal of care is necessary to prevent this. The operation is illustrated in Figure 10.1.

During dismantling the positions of all components should be marked to assist during reassembly. Airgap lengths and the relative position of the stator and rotor cores should also be noted.

Figure 10.1 Rotor being unthreaded from stator

Before cleaning, a quick check for any visible signs of obvious damage (build-up of debris, signs of corona discharge, etc.) will generally prove a worthwhile exercise.

10.2.3.5 Cleaning

After the motor has been dismantled, individual components must be cleaned thoroughly.

When steam cleaning equipment is available, this is ideal for most components, including windings and slipring–commutator assemblies, provided they can be thoroughly dried out (which can be confirmed by measurement of insulation resistance) before reassembly. An exception to this can be motors containing coreplates which have been manufactured with a water-soluble interlaminar surface insulation coating. If the cleaning process removes this insulation then unacceptably high losses are likely to occur in the motor core. The original motor manufacturer should therefore be consulted if any doubt exists. Alternatively, hot water and detergent can be used to good effect. Again, if applied to insulating materials the above comments still apply.

Problems may, however, occur when using water to clean insulating materials. If the insulation system is old or particularly porous then thorough drying out can prove difficult and if any doubt exists a proprietary cleaning fluid based typically on a mixture of trichloroethylene and white spirit should be used. Availability of such cleaning fluids overseas may prove problematic and airline regulations generally forbid their transport. Therefore, for overseas repair or overhaul it may be necessary to ship sufficient quantities of cleaning fluid in advance of the work or to hold quantities on site.

10.2.3.6 Assessing the need for refurbishment or replacement of components

Once the motor has been cleaned completely, individual components require detailed examination to ensure they will operate satisfactorily until the next major

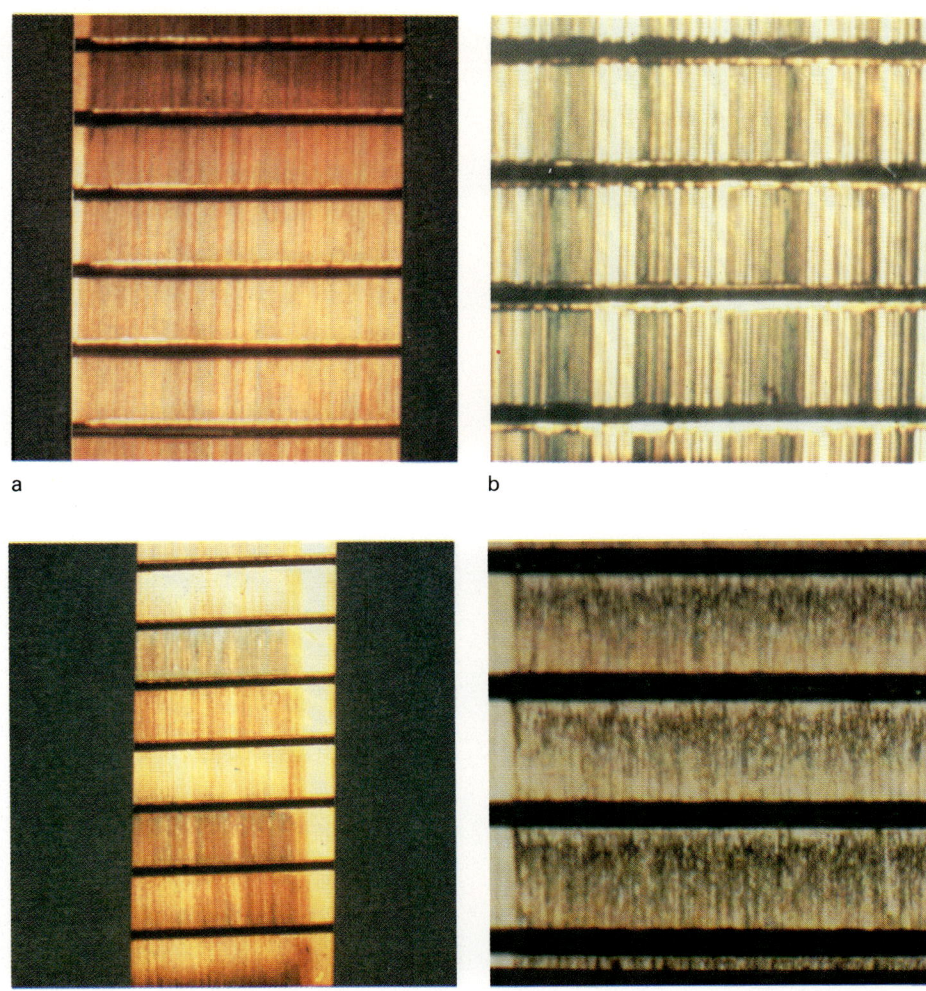

a

b

c

d

Plate 1 *above and overleaf*

(a) Typical uniform colour intensity of a normal skin. The brushgear is performing satisfactorily. **(b)** Raw grooved skin showing bands of raw or very slightly skinned copper. This indicates the metal is being attacked. The most frequent causes are: (1) excess humidity; (2) oil vapour; (3) corrosive gases; (4) under-loaded brushes; or (5) the use of an unsuitable brush grade. **(c)** Dark blotches with sharp or irregular edges followed by lighter areas in alternating fashion with gradual reduction in colour intensity. The most frequent cause is a fault in a bar or group of bars causing radial movement of the brush. **(d)** Bars marked at their centre or edges. The most frequent causes are: (1) defective maintenance; and (2) commutator poorly turned or trued.

e

f

g

h

Plate 1 *continued*

(e) Alternate light and dark bars distributed around the commutator. The most frequent cause is electrical in origin and most probably an armature winding fault.
(f) Metallic erosion (burning) at centre of bars. **(g)** Dark fringes due to high micas.
(h) Commutator with axial profile showing track growing with correct wear. This indicates normal operation after a very long period of operation.

Courtesy of Le Carbon

Plate 2 *above*

Rotor of 4.8 MW, 3600 r.p.m., 13.8 kV sea-water injection pump drives for offshore platform. The 'Maploc' construction accomodates axial expansion and minimizes joint stresses and metallurgical degradation.

Courtesy of Mather and Platt Ltd.

Plate 3 *below*

Stator unit of 7460 kW, 6.6 kV, 60 Hz, 4-pole induction motor for a compressor drive.

Courtesy of Brush Electrical Machines Ltd.

Plate 4 *above*

"Metric" Cage Induction motor to drive a
Stabilizer Interheater Pump in a chemical
works with saliferous atmosphere. Certified
to BS 5000 (Ex)N. CACA Weatherproof
Machine rated at 355 kW, 3.3 kV, 50 Hz,
1475 r.p.m.

Courtesy of Laurence, Scott and
Electromotors Ltd.

Plate 5 *below*

DC motors driving the finishing stands of a
hot strip rolling mill. 5500 kW rating with a
working peak torque of 1.75× nominal.

Courtesy of GEC Large Machines Ltd.

overhaul. Table 10.2 summarizes the type of signs which may be observed. In addition, the following points should be considered.

Stator and rotor.   If a rewind of either the stator or rotor is required then the following factors should be considered:

(1) Is the winding in question vacuum pressure impregnated? If so, stripping the winding may prove difficult and in some cases less economic than replacing a complete corepack (where this is feasible), or even a complete motor, unless purpose-designed release films or agents have been applied at the manufacturing stages to alleviate this difficulty.
(2) What type of insulation coating is applied to the core laminations? Some organic-based insulation coatings totally disappear at temperatures in excess of 200° C. Burn-out ovens which operate at temperatures in excess of 400° C are sometimes used to assist in removing coils. If this is not recognized, the occurrence of these two features together will lead to vastly increased iron losses, resulting in overheating and very rapid failure of the new winding when reintroduced into service.
(3) If a bearing failure has occurred, it is possible that the complete bearing surface may have collapsed. The result of this could be that the rotating rotor has rubbed on the stator resulting in damage to both components. These should be carefully examined to identify any such damage and its extent. If damage is apparent, metallurgical checks may be required.

Bearings.   Prior to opening a bearing for cleaning and inspection, it is important that all outside surfaces are cleaned thoroughly to prevent the entry of dirt into the bearing housing.

Any bearing surfaces or housings which are to be left exposed for even relatively brief periods of time should be covered by a lint-free cloth or a plastic sheet, to prevent dirt from contaminating the bearing.

When using cloths to clean or handle bearings they must always be of the lint-free type.

The decision to replace bearings may well have been taken from the way in which they had been performing whilst the motor had been running in service. Careful inspection may also indicate the need for replacement.

With sleeve bearings, the shaft landing and white-metal bearing surface need to be examined carefully for signs of metal removal, uneven wear or localized markings caused by overheating. This type of bearing is generally used on larger high-speed motors and if there is any defect which is not corrected, it will probably lead to a very rapid bearing seizure with minimal warning.

When bearings do need replacing many are available directly off the shelf. However, purchasing directly off the shelf can sometimes present problems since the type and clearances of the bearings are important factors in their selection and some types of motors use matched-pair assemblies.

With some types of bearing, specific selection is made by the manufacturers such that their bearings are particularly suitable for electric motors. This optimum selection can be important in minimizing vibration levels, so new bearings should always be ordered as 'suitable for use on electric motors'.

If any doubt exists regarding bearing selection then the original motor manufacturer should be consulted.

Sliprings and commutators.   These and their associated brushes and brushgear should be inspected carefully for overheating or undue wear. (Aspects of commutator skins and wear are illustrated in Plate 1.)

Table 10.2 Identifying problems with the motor's major individual component parts

Component	Visible signs, tests or measurements	Cause and remedial action required
Windings	Mechanical damage to insulation	Carelessness in assembly or disassembly – local repairs, e.g. by retaping and sealing with varnish, will generally suffice if the damage is not extensive
Bracing, intercoil packers and winding supports	Looseness	Poor design/manufacture – general ageing. Any movement must be eliminated to prevent insulation failure due to fretting fatigue
Wedges	Looseness	Poor design/manufacture – general wear – possibility accelerated by high vibration levels. Looseness must be eliminated (possibly by rewedging and/or redipping and curing). On vertical motors loose wedges can drop out
Corepacks	Signs of localized overheating on surface	Breakdown of inter-laminar insulation. Increased losses and overheating will occur leading to winding failure. Corepack and winding will generally require replacing
	Excessive tooth movement	Axial movement in corepack teeth may lead to high frequency tooth vibration (caused by magnetic fields) leading to fatigue failure at narrowest part of tooth. Increase pressure on the tooth tips at the extremities of the core – if any of the teeth have broken off, the corepack will generally require replacing
Bearings	Pitting, scratching, scoring, denting, corrosion of bearing surfaces, balls, rollers or cages. Contaminated lubricant	A number of causes can result in the bearing damage. These can range from damage caused: (1) during fitting; (2) brinnelling during transit or due to operation of adjacent machinery on site; (3) inadequate or contaminated lubricant; (4) shaft currents; (5) cummulative errors in tolerances; and (6) metal fatigue due to operating bearing in excess of design life. If any damage is apparent it will be necessary to replace complete bearings in the case of ball and roller bearings and possible bearing shells in the case of white-metal bearings. Shaft seatings should be examined for damage and if any is apparent it will be necessary to remachine

Rotor cage	Broken bars. Bars detached from short circuit ring. Loose bars. Damaged or cracked short-circuit ring. Hairline cracks in short-circuit rings and exposed bars can be identified by using suitable dye pen techniques. Ultrasonic techniques can be used to check integrity of bar/short-circuit ring joints	Rotor design inadequate for motor duty. Motor being misused in service. If any damage to the rotor cage is apparent, fitting new bars and/or short-circuit rings will be necessary. Care must be taken in selecting the appropriate materials. For instance, high-speed motors may be fitted with chrome–copper short-circuit rings, which have been selected for their superior mechanical strength. Simple rebrazing of these will diminish the mechanical strength and could possibly lead to later mechanical failure
Sliprings and commutators	Plate 1 illustrates aspects of commutator skins and wear.	
	Rapid brush wear	Incorrect brush pressure – correct pressure. Incorrect brush grade – fit correct grade. Damaged commutator or rings, e.g. scored or out-of-round. Repair damage or replace. Defective commutation–trace reason and rectify. Brush sparking–trace reason and rectify
	Wear-scoring or pitting of rings or commutator	Chemical attack of commutator rings due to corrosive environment. Review degree of enclosure of commutator and rings – fit rings of different composition immune to attack from particular corrosive elements. Sparking of brushes–trace fault and rectify. Over temperature–trace fault and rectify
	Excess temperature	Blocked ventilating passages. Incorrect brush pressure. Badly-bedded brushes. Motor overloaded. Incorrect brush grade fitted–diagnose and rectify condition

Brush wear is generally expressed in millimetres of reduction in brush length per 1000 h of service. This is only correct when comparing brush wear in sliprings or commutators which are operating at the same peripheral speed. This is because rate of wear is a direct function of the distance travelled by the commutator or ring surface in a given time.

If brush wear appears to be excessive then the cause must be investigated to try and minimize the more frequent maintenance which will be required and to avoid an eventual failure. If brush wear is uneven between brushes, e.g. brush lengths differing by more than 20%, the cause should be investigated.

Mechanical and electrical parameters which may influence brush wear are:

(1) Coefficient of friction c.
(2) Applied pressure p.
(3) Peripheral speed n.
(4) Volt drop across ring (commutator–brush interface v).
(5) Current in brush i.

These parameters are interdependent: c is a function of n and i, and v is a function of p and i. The characteristic curve for rate of wear, as a function of applied pressure, is illustrated in Figure 10.2. From this, it is clear that an optimum band of operating pressure exists and, outside this band, accelerated brush wear can occur. Brush pressure must therefore be checked carefully as must the free movement of the brushes in their holders.

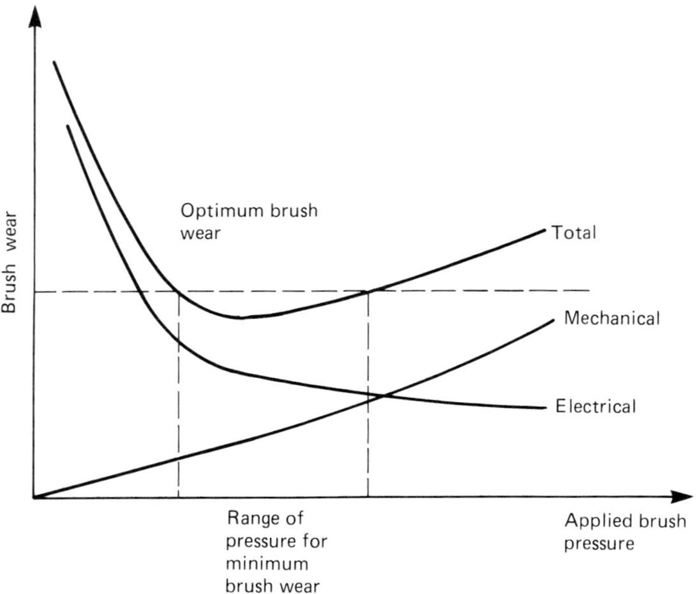

Figure 10.2 Characteristic curve showing rate of brush wear

Other factors which may influence brush wear are:

(1) Operating temperature.
(2) Machine vibration.
(3) Deformation of commutator or rings.
(4) The surrounding environment (relative humidity, presence of corrosive gases, abrasive dusts, etc.).

(5) Circulating currents between commutator bars.
(6) Defective commutation.
(7) Brush rocking in the case of motors operating on a reversing duty cycle. This is illustrated in Figure 10.3, from which it is clear that the contact surface is reduced and therefore the actual applied pressure and probable wear rate increased. This phenomenon is aggravated by increased clearance between brushes and brush holders.

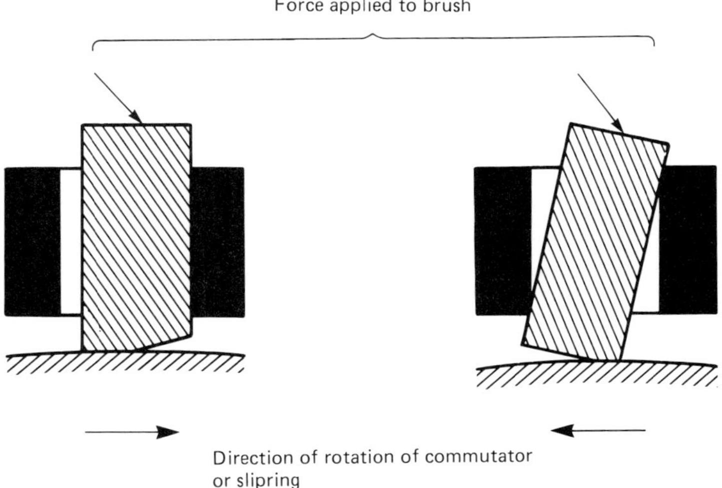

Figure 10.3 Illustration of mechanism of increased brush wear due to brush rocking

Most of these factors will require monitoring in service. Deformation of commutator or rings can be measured whilst the machine is being overhauled and, if necessary, they can be skimmed to true-up (in the case of commutators, the micas will require undercutting) and then be polished.

Fitting new brushes.   When new brushes are fitted they need to be bedded-in such that the brush face is radiused to match the curvature of the commutator or slipring. This is most frequently carried out in the motor and is best done by placing the brush in its holder and sandwiching an abrasive paper or cloth between the brush and the ring or commutator. The paper is then pulled backwards and forwards along the slipring or commutator surface until all contact surfaces are fully bedded. Final bedding can be accomplished, where necessary, by rotating the motor and using a pumice stone on the slipring or commutator surface, the abrasive dust acting as a grinding agent.

After brush-bedding, all brushes should be drawn back in their brush-holders and all dust blown or sucked out. When new brushes have been fitted the motor should ideally be started at reduced load which should be then gradually increased until a skin is formed.

Never mix different grades of brush on the same motor.

10.2.3.7 Reassembly

Reassembly can commence once all component parts have been cleaned, inspected and refurbished or replaced in accordance with the preceding paragraphs.

Assembly follows the reverse sequence to disassembly. Simple checks should be made during the course of assembly to ensure that the motor is building-up correctly. The position of the rotor core relative to the stator should be the same as when the motor was dismantled, usually being axially central and radially concentric.

Good engineering practice should be used when tightening nuts and bolts to prevent distortion of components.

Where gasket material or sealing compound or strips have been used, ensure these can be reusable. If any doubt exists, replace them with new ones. To ensure that the integrity of the motor's enclosure is not impaired, the advice of the original motor manufacturer should normally be sought when replacing sealing strips and compounds.

Table 10.3 Standards and other publications related to the overhaul or repair of motors certified for use in hazardous areas

Number	Title
BS 5345:Parts 1–8	*Selection, installation and maintenance of electrical apparatus for use in potentially explosive atmospheres (other than mining applications or explosive processing and manufacture)*
BS 5501:Parts 1, 3, 5–8	*Electrical apparatus for potentially explosive atmospheres*
BS 229	*Flameproof enclosure of electrical apparatus*
BS 4683:Part 1	'Classification of maximum surface temperature'
BS 4683:Part 2	'The construction and testing of flameproof enclosures or electrical apparatus – type of protection "d"'
BS 4683:Part 3	'Electrical apparatus for explosive atmospheres – type of protection "N"'
BS 4683:Part 4	'Electrical apparatus for explosive atmospheres – type of protection "e"'
BS 5000:Part 16	'Rotating electrical machines (motors – with type of protection "N"')
BS 4761	*Specification for sprayed unfused metal coatings for engineering purposes*
BS 4758	*Specification for electroplated coatings of nickel for engineering purposes*
BS 4641	*Specification for electroplated coatings for chromium for engineering purposes*
BS 4999	*Specification for general requirements for rotating electrical machines*
BS 4999:Part 20	'Classification of types of enclosure'
BS 4999:Part 60	'Tests' (superseded by Part 143 (1987)
BS 5490	*Enclosures (IEC 529) degrees of protection*
SFA 3009:1972	*Type of protection (Ex)s*
SFA 3012:1972	*Type of protection (Ex)i*
BASEEFA Ex-Memo No. 1	*Explanatory memorandum on marking and labelling of certified apparatus and components as used in the UK*

Note: Additional relevant standards are listed in BS 5345:Part 1, Section 7.

Once the motor has been fully assembled the rotor should be turned by hand (or, on large motors, with the aid of a bar) to ensure free rotation. If the motor is in a workshop where suitable test facilities are available the motor should be run unloaded. Prior to this, insulation resistance and continuity of windings should be checked.

When test facilities are not available, the motor should be returned to site and aligned to its load but not coupled to it. Insulation resistance checks should be carried out on windings and, where appropriate, other electrical components. Standards and other publications relating to repair of motors certified for use in hazardous areas are listed in Table 10.3. Recommended minimum insulation resistance values are tabulated in Table 10.4.

Table 10.4 Recommended minimum insulation resistance values

Motor supply (kV)		Recommended instrument terminal voltage (V)	Recommended minimum insulation resistance in megohms at winding temperature of					
Over	Up to and including		10 °C	15 °C	20 °C	25 °C	30 °C	40 °C
	0.66	500	16	11	8	6	4	2
0.66	3.3	1000	40	28	20	14	10	5
3.3	6.6	1000	80	56	40	28	20	10
6.6	11.0	2500	120	84	60	42	30	15
11.0	13.8	2500	160	112	80	56	40	20

The motor can now be wired-up to the mains and run unloaded to confirm everything is operating satisfactorily. The motor can now be coupled to its load and returned to service.

If the insulation resistance readings are lower than recommended in Table 10.4 this indicates that there is probably moisture present in the windings. These should

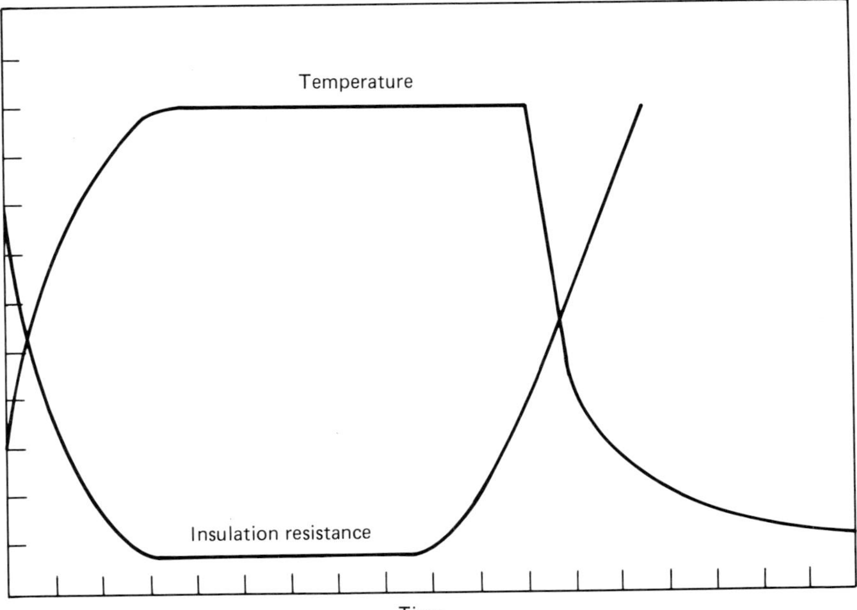

Figure 10.4 Typical drying-out characteristics

be dried thoroughly using an industrial oven, blow-dryer or similar device (but taking care not to exceed the temperature limit of the insulation) before the motor is energized. Figure 10.4 shows typical drying out characteristics.

10.2.4 Standard maintenance procedures

Motor manufacturers should supply standard minimum maintenance procedures for their motors. These will be modified to some extent by the motor's operating cycle and environment and should normally be used as an initial guide. As operating experience increases, the maintenance schedule can be modified by the user to match specific requirements.

A typical minimum maintenance schedule for a cage induction motor is illustrated in Table 10.5.

10.3 Applying effective record-keeping to maintenance procedures

10.3.1 General aims

Collection of reliable data on motor usage and incidence of failure enables:

(1) Optimum maintenance and replacement strategies to be formulated.
(2) A critical review of current operating and maintenance procedures to be undertaken, hence identifying any deficiencies and allowing corrective measures to be implemented.
(3) An assessment of how well the original motor design specification matches the operating requirements.
(4) A critical assessment of various manufacturers' motors to be undertaken.

A cost will be associated with collection of such data, which may consist of a capital investment in terms of computer-based data logging systems or the revenue costs associated with manual systems. These costs should be justified in terms of a reduction in maintenance costs and plant downtime due to failure, combined with an increase in plant life. With computer-based systems, other data may also be recorded such as energy absorbed which may lead to better energy utilization and therefore lower operating costs.

10.3.2 Data recorded

Most motors do not operate continuously at full load, even though many will be nominally rated for this duty. For accurate determination of maintenance intervals, it is important to have detailed information relating to the motor's operation.

Different types of motors will be affected differently by differing types of operating cycle. For instance, on cage induction motors high thermal, centrifugal and electromagnetic stresses are generally associated with direct-on-line (DOL) starting. With this type of motor it is important to know the number and duration of starts, along with the number of hours run, in order to determine maintenance intervals rather than simply using the criterion of elapsed time.

As discussed in Section 10.2.3.6, brush wear is determined by distance travelled. Therefore the maintenance interval for brushes and brushgear will be established by the product of the number of hours run and the peripheral speed of the sliprings or commutator.

Winding insulation life is, in the absence of abnormal phenomena (such as voltage surges) primarily determined by its operating temperature. The relationship between operating temperature and insulation life is exponential and of the form:

$$\text{Life} = A \exp (B/T)$$

where A and B are constants and T is absolute temperature.

Table 10.5 Typical minimum maintenance schedule for a cage induction motor

Motor condition	Maintenance Ref	Task	Period							
			1 Day	1 Week	2 Weeks	1 Month	3 Months	6 Months	1 Year	2 Years
		Read Instruments	X							
Motor running		Verify coolant flow	X							
	1	Check desiccant		X						
	2	Examine cables				X				
	3	Replenish lubricant	See nameplate							
Motor stopped	A	Clean exterior					X			
	B	Change lubricant							X	
	C	Check bearing insulation						X		
	Optional	Check heater continuity								X
		Check winding insulation								X
		Check thermistor circuits								X
		Check security of terminals								X

Therefore, an assessment of when a winding is reaching the end of its useful life due to thermal degradation can be made using the relationship:

$$\text{Fraction of life remaining} = 1 - \sum_{n=0}^{n=x} \frac{t_n}{A \exp\left(B/T_n\right)}$$

where t_n is duration spent at maximum measured temperature T_n and x is the number of different temperature time levels.

This assessment would, of course, require winding temperature detectors and to make this viable, computer logging would be required to perform the integration. The equation given above is, of course, fairly simplistic and ignores many factors. If suitable definitions for the other factors are available, then these should be incorporated.

Any periodic maintenance should be recorded, as should every failure along with the cause (where this can be determined) and any remedial action undertaken.

All available cost information should also be recorded, such as:

(1) Cost of maintenance.
(2) Cost of overhaul.
(3) Cost of repair.
(4) Cost of down-time directly associated with motor outage or maintenance requirements.

10.3.3 Using the data

The data collected can be utilized in a combination of ways to reduce the cost of operating motors, briefly:

(1) If a large enough sample is available a statistical analysis can yield general trends.
(2) Optimal replacement and maintenance strategies can be formulated (see Section 10.5).
(3) Preventative maintenance can be scheduled more effectively.
(4) Most plants will have requirements which vary on a cyclic or seasonal basis. If statistical data on not only the motor but the plant in general is available then the combination of the two can be used effectively to determine the best time for a motor to be taken off-line for any reason.

10.4 Preventative maintenance

10.4.1 General aims

Preventative maintenance is the application of maintenance procedures to operating plant to obviate the possibility of failure and to extend its operating life. This may consist of either simple on-line maintenance or costly off-line overhauls. The use of all available information is essential to ensure optimal preventative maintenance strategies.

10.4.2 Condition monitoring

Various techniques are available to monitor the condition of the various components parts of an electric motor. Examples are:

(1) *Bearings*: shaft proximity probes or housing-mounted accelerometers, temperature detectors, lubricant analysis meters and shock pulse meters.
(2) *Windings*: *Off-line* – insulation resistance, polarization index, tan delta and interturn impulse;
On-line – coolant gas analysis and high-frequency emission (discharge inception).

These indicators can be used to determine the need for off-line maintenance.

10.4.3 Record-keeping

The importance of reliable records is discussed in Section 10.3. Analysis of the information may yield the statistical likelihood of failure of any particular component in the period of time being considered, thus determining the need for off-line maintenance.

10.4.4 Economics

The economic arguments discussed in Section 10.5 should be applied to all preventative maintenance decisions. Because most electric motors are relatively simple devices in terms of construction and operation it will normally be found to be uneconomic to conduct off-line preventative maintenance unless ensuing failure can in some way be predicted with a high degree of certainty. This will be determined by the amount of information available on which to base a decision.

10.5 Economic considerations

10.5.1 General aims

The general aim of an economic policy relating to the maintenance and replacement of electric motors is to minimize their life-cycle costs. If the mortality characteristics of a particular electric motor are known then it is possible to formulate optimal maintenance strategies. If annual motor running costs are known than an economic model can be developed to determine the optimal electric motor replacement interval. For these economic approaches to be valid it is essential that suitable and reliable data are available. The acquisition of such data is discussed in Section 10.3.

10.5.2 Reliability

Reliability can be defined as the 'probability that a device will perform its required task under a given set of environmental conditions for a selected time duration'. The reliability of most devices can be characterized by the typically shaped curve of Figure 10.5.

The sectors of this model are described as follows.

(1) The sector A–B depicts early life mortality due to erroneous design, fabrication and malfunctioning parts. With many electronic components, failure due to early-life mortality is minimized by using burn-in techniques, i.e.

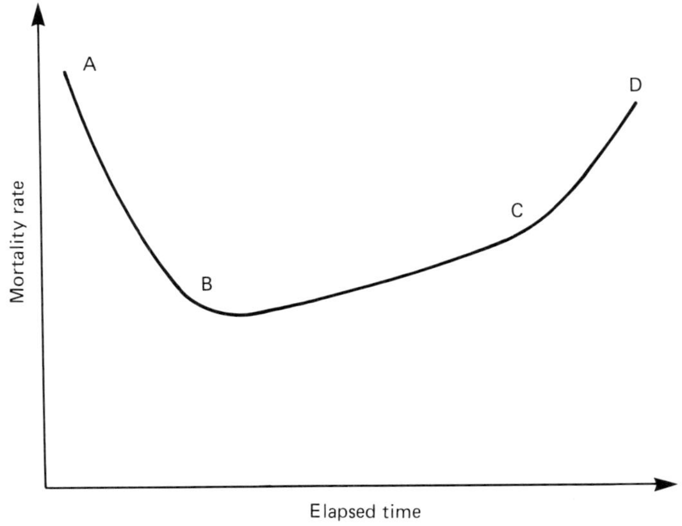

A–B Early life mortality
B–C Random failure
C–D Fatigue and wearout phenomena

Figure 10.5 Characteristic shape of the familiar 'bath-tub' curve

running the devices for a time which takes them beyond point B on the curve during the manufacturing process.

In the case of electric motors (particularly the larger ones), rigorous tests are generally undertaken at various stages of manufacture, hence ensuring the quality of the final product and minimizing the difference in hazard rate between A and B.

(2) The sector B–C depicts chance (random) failures due to the environmental stress exceeding the strain capabilities of the device. Some causes of these types of failures are listed in Table 10.1 (pages 499–502).

(3) The sector C–D depicts fatigue and wear-out phenomena. The point at which this sector is entered will depend upon the complexity of the motor. Typically, a complex d.c. motor with commutator and brushgear assembly will enter this sector much earlier than the rugged and robust cage induction motor.

10.5.3 Optimal maintenance strategies

A vast amount of suitable and reliable data is required to formulate optimal maintenance strategies. The type of data required consists of the results of studies undertaken on several populations of particular makes and types of motors subjected to varying levels of maintenance.

If the net present value of the cost of maintenance is greater than the product of the cumulative probability of failure and the net present value of the cost of failure then the maintenance strategy employed cannot be economically justified. Since, physically, electric motors are relatively simple devices with a small number of operating parts, then only in exceptional cases will any off-line maintenance be

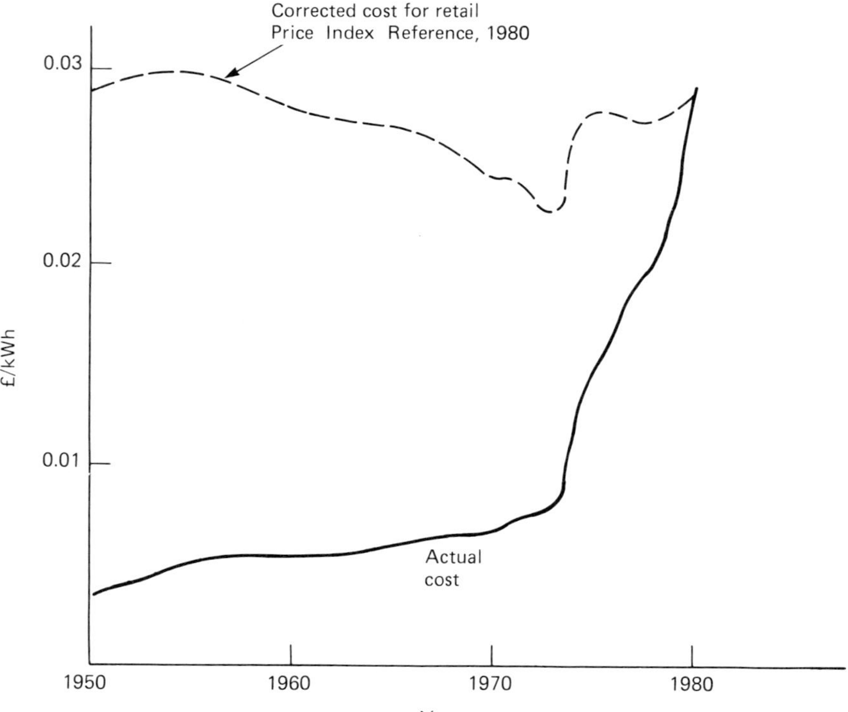

Figure 10.6 Real cost of electrical energy over the 30 years prior to 1980

justified. Routine maintenance as recommended elsewhere in this chapter and by the manufacturers will generally be the most economic solution.

Electric motors, like many other devices, suffer from a diminishing mean time to successive failure due to lack of maintenance ability to return them to the 'as good as new' state each time they are repaired. Such a phenomenon is ideal for deriving optimum replacement strategies.

10.5.4 Replacement strategies

An economic model for optimizing replacement strategies can be developed. The model is simplified by making the following assumptions:

(1) Only those costs which vary with the age of the motor or motor under review are incorporated in the economic analysis. Other costs which do not vary with these considerations are irrelevant to the economic decision. Since in real terms energy costs have not varied significantly in the past (Figure 10.6) for the purposes of the model these can be regarded as constant and should not be included unless comparing motors whose operating efficiencies differ.
(2) The economic rules will be based upon cashflows (not accrued expenses) so that the time value of money can be considered properly in the economic decision.
(3) The routine maintenance policy being used is optimum.
(4) If a motor is to be replaced then it is assumed that its scrap value would equate to its disposal charges. Also during a replacement (or renewal) there is no irretrievable loss of revenue caused by the motor being off-line for a short time. The model can, of course, be modified if these assumptions are considered invalid.
(5) There are no financial or other constraints which would influence the optimization of the economic model.
(6) The real cost of capital to the business considered is assumed to be fixed.

The present value p_n of a motor's costs running for n years is:

$$P_n = K + C_1 + \frac{C_2}{1 + r} + \cdots \frac{C_i}{(1 + r)^{i-1}} + \cdots \frac{C_n}{(1 + r)^{n-1}} \qquad (10.1)$$

where K is the installed cost of the motor; C_i is the annual motor running costs (as defined in (1) above); and assuming that C_i is incurred at the beginning of each year such that the present value of C_i is given by:

$$\frac{C_i}{(1 + r)^{i-1}}$$

where r is the real annual cost of capital to the business being considered, and is assumed to be fixed.

By using real costs (as opposed to inflated 'money' costs) then the cost profiles for each identical successor will be identical and assuming the successor is installed after n years will be:

$$\frac{P(n)}{(1 + r)^n}$$

If we let $Z(n)$ be the present value of the entire series of replacement motors then:

$$Z(u) = P(u) + \frac{P(u)}{(1+r)^n} + \frac{P(u)}{(1+r)^{2n}} + \frac{P(u)}{(1+r)^x}$$

which is a geometric progression which reduces to

$$Z(u) = \frac{P(u)}{1 - \left(\dfrac{1}{1+r}\right)^n} = P(u)k(u)$$

where $k(n) = \dfrac{1}{1 - \left(\dfrac{1}{1+r}\right)^n}$

and is known as the present-value-to-perpetuity factor.

The economic objective is to find the value of n which minimizes the value of $Z(n)$ (see Figure 10.7), i.e.:

$$Z(n) - Z(n-1) < 0$$

$$Z(n+1) - Z(n) > 0$$

Alternatively, it can be shown that if $x = 1/(1+r)$ then the optimal replacement strategy can be written as:

$$C_n < \frac{K + C_1 + C_2 x + C_3 x^2 \cdots + C_i x^{i-1} + \cdots C_{n-1} x^{N-2}}{X + X^2 + \quad \cdots \quad X^{n-2}} \tag{10.2}$$

$$C_n + 1 > \frac{K + C_1 + C_2 x + C_3 x^2 \cdots C_i x^{i-1} + C_n x^{n-1}}{x + x^2 + \quad \cdots \quad x^{n-1}} \tag{10.3}$$

10.6 Failures: causes and remedies

10.6.1 General

If a motor has been correctly installed and commissioned then a major failure will not normally occur during its design life. If a failure does occur then it will probably be as a result of incorrectly set protective devices or inadequate condition monitoring.

Condition monitoring can take many forms and correct interpretation of the results can forewarn of an impending failure. For instance, an increase in bearing temperature rise or vibration level as measured by appropriate monitoring equipment may indicate an impending bearing failure.

When protection equipment is not fitted, then an increase in the noise emitting from a motor, as detected by the operator or maintenance engineer, may well indicate excess vibration or loose components in turn possibly indicating an impending bearing or component failure.

Table 10.1 (pages 499–502) summarizes the most common motor failures, together with possible causes and remedies. This is intended only as a general guide and should not be regarded as a comprehensive fault-finding chart since, in practice, many less obscure reasons for failure can occur.

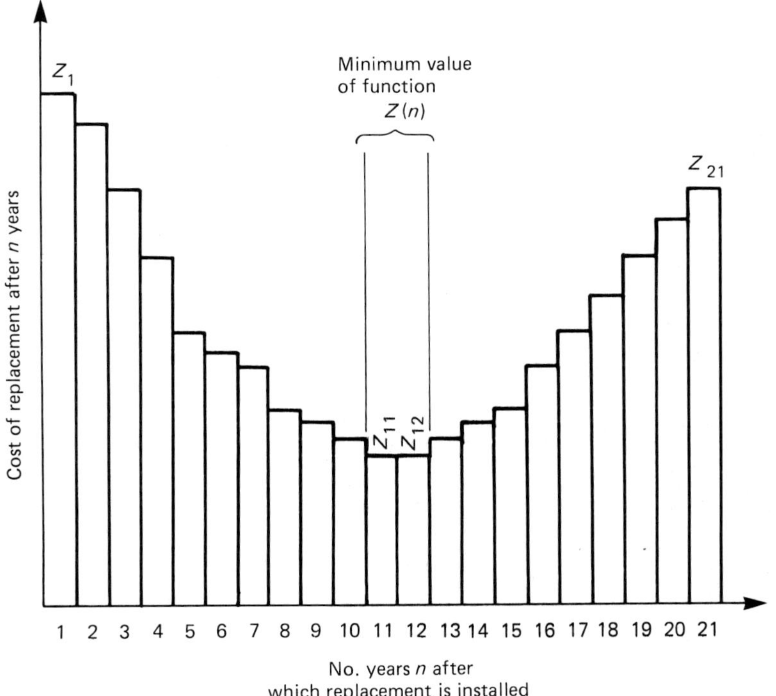

Figure 10.7 The economic objective is to select the value of n which minimizes the function $Z(n)$

10.6.2 Identifying the initial cause of a failure

Identifying the initial cause of a failure can in many instances prove a difficult task, especially if significant damage to other components has resulted, possibly masking the original problem.

Details of all monitored variables immediately prior to a trip-out can prove useful in fault finding. Many protective relays which are now available are capable of storing this information for later retrieval.

It is not normal for a motor which has been in constant service to suddenly fail without prior warning. If this happens, then the mode of operation of the driven equipment and drive transmission should be suspected as the possible cause.

If the motor has been out of service for any time prior to the failure then the conditions under which the motor has been stored will probably be at fault, e.g. ingress of moisture or contaminants to the windings or bearings. It is therefore good engineering practice to undertake basic insulation resistance checks and to inspect the condition of the lubricant before running a motor which has been out of service for an extended period (which will depend on the type of motor and local environment).

If the motor has been overhauled or repaired, or the supply system or protective scheme altered in some way, then this again will give guidance to the possible cause of failure.

Figure 10.8 Stator end-winding with inadequate bracing

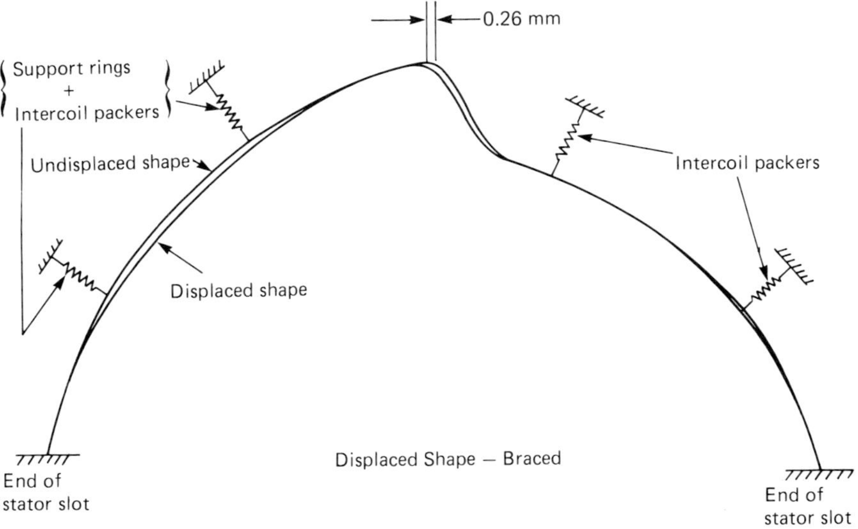

Figure 10.9 Typical presentation of end-winding displacement for 6 MW high-speed cage induction motor

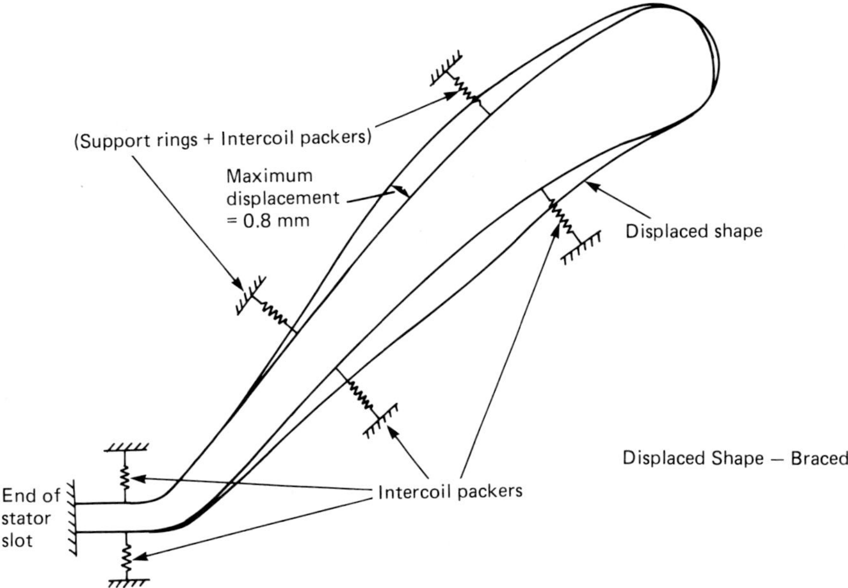

Figure 10.10 Typical presentation of end-winding displacement for 6 MW high-speed cage induction motor

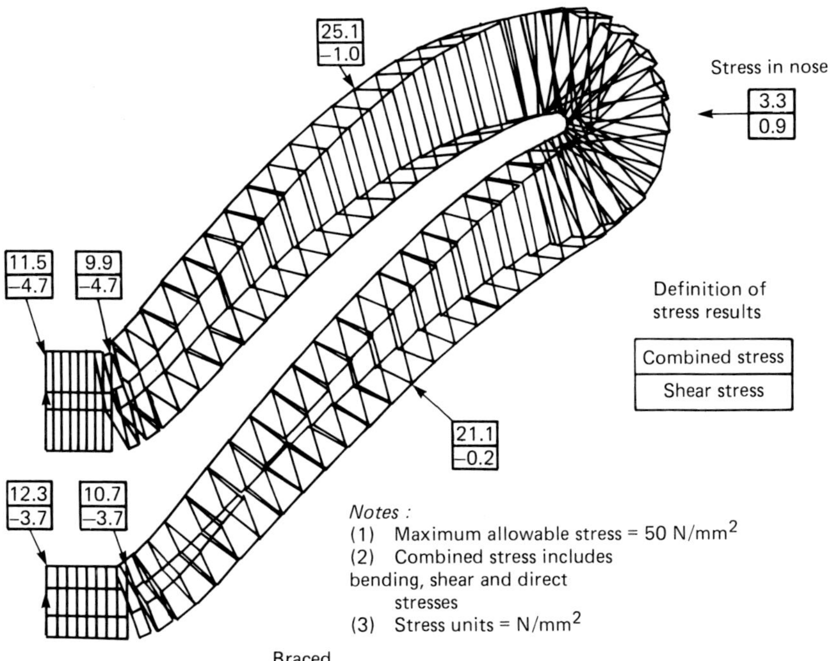

Figure 10.11 Typical end-winding model for 6 MW high-speed cage induction motor

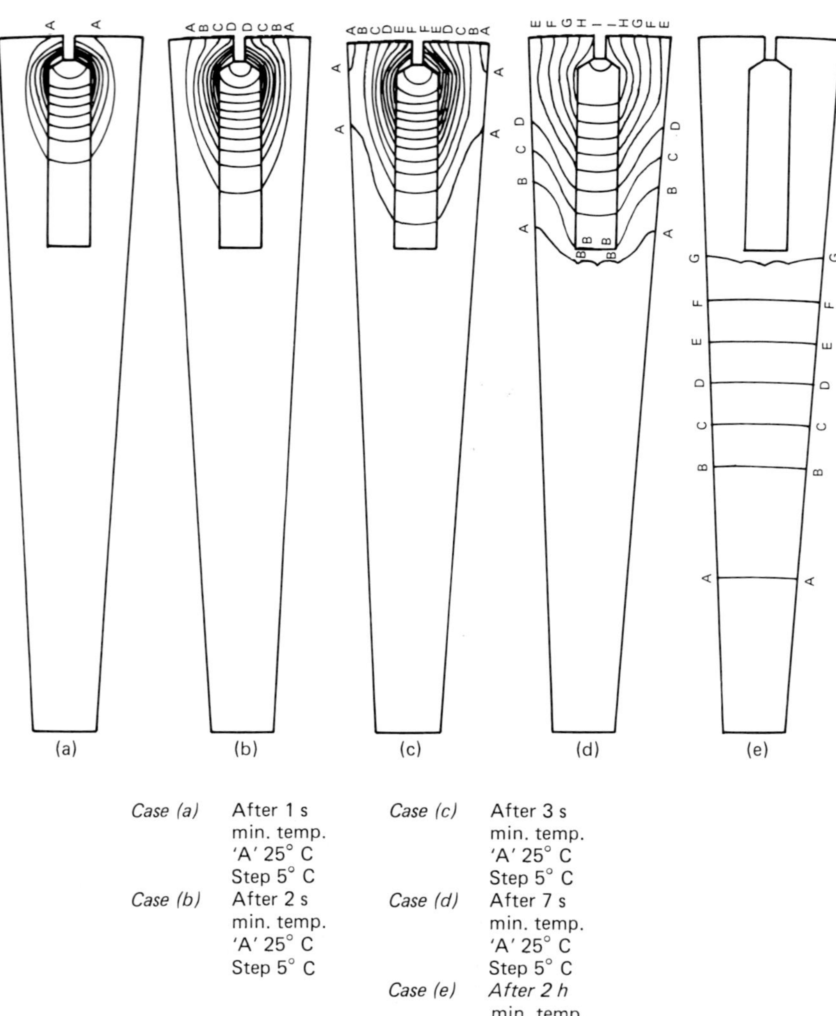

Case (a)   After 1 s
            min. temp.
            'A' 25° C
            Step 5° C
Case (b)   After 2 s
            min. temp.
            'A' 25° C
            Step 5° C

Case (c)   After 3 s
            min. temp.
            'A' 25° C
            Step 5° C
Case (d)   After 7 s
            min. temp.
            'A' 25° C
            Step 5° C
Case (e)   After 2 h
            min. temp.
            'A' 70° C
            Step 2° C

Figure 10.12 Typical single-start isotherms in the region around a rotor bar for a 6 MW high-speed cage induction motor

10.6.3 Implementing remedies

Whenever possible, the initial cause of the fault should be identified to enable a future recurrence to be avoided. Careful inspection of all components which may have been affected by the failure is essential, and any that are dimensionally out of tolerance or show signs of being overstressed in any way should be replaced.

The procedures given in Section 10.2.3 (page 498) for overhaul are equally applicable in the implementation of repairs.

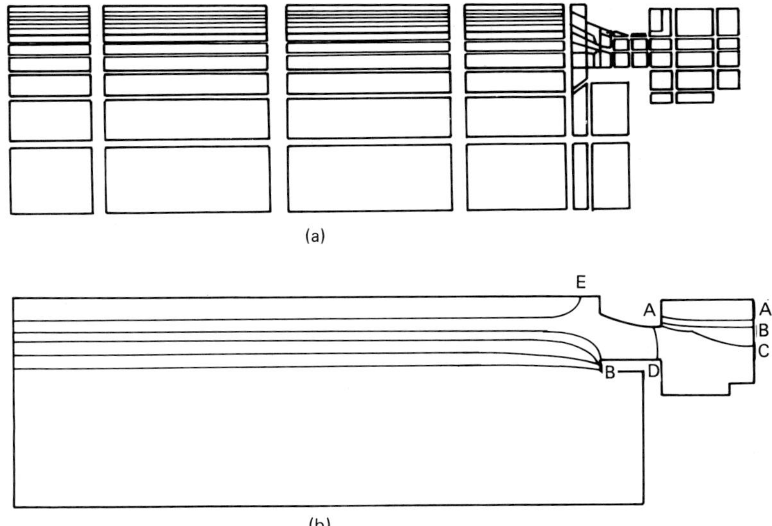

(a)

(b)

Figure 10.13 Typical axi-symmetric analysis for a 6 MW cage induction motor. (a) Axisymmetric. (b) Rotor axisymmetric isotherms after 8 seconds – Minimum Temperature 'A' 30° C, Step 10° C

Figure 10.14 High-speed rotor being balanced prior to assembly

10.6.4 Recurring failures

Where failures occur early in a motor's life, simply restoring the motor to its condition prior to the failure may lead to another failure in an even shorter time. If this does occur it may indicate that the original motor design is in some way inadequate to cope with the operational demands on the motor. Further repairs without taking expert advice will probably lead to a rapid reduction in MTBF. This is illustrated in Figure 10.8 where the stator end-winding bracing and packing has been clearly inadequate for the operational requirements of the motor. If this condition is left unattended, end-winding failure would result.

In the case illustrated, the methods outlined below were used to redesign the system of end-winding packing and bracing and the motor was subsequently rewound.

Methods are available to assess the suitability of designs for various operational conditions using finite element analysis techniques. Cage induction motors are usually subjected to very high thermal, electromagnetic and mechanical stresses during starting. A complete finite element analysis of the stresses involved will determine the motor's suitability for the imposed starting conditions.

A brief extract typical of plots obtained from such an analysis are given in Figures 10.9–10.13. These show the rotor thermal analysis and stator end-winding stress analysis. The particular rotor in question is shown in Figure 10.14. Such an analysis requires detailed modelling of the motor and superimposed load conditions. This is generally expensive but can usually be justified on larger, key motors.

11 Units, dimensions and conversion factors

B. J. Chalmers

University of Manchester Institute of Science and Technology

11.1 SI units

SI is the accepted abbreviation for the 'Système International d'Unités' which is the standard metric system of units adopted throughout the world. It is based on the primary or base units shown in Table 11.1.

Table 11.1 Primary SI units

Physical quantity	Common symbol	Unit name	Unit symbol
Length	ℓ	Metre	m
Mass	M	Kilogram	kg
Time	t	Second	s
Current	I	Ampère	A
Temperature		Kelvin	K
Luminous intensity		Candela	cd
Substance		Mole	mol

Note: The kilogram is the unit of mass, not force.

Other SI units are derived from the primary units as shown in Table 11.2. The dimensions of each unit are given, in terms of the primary units; these dimensions are sometimes useful when it is wished to check the dimensional consistency of terms in an equation.

Since manufacturers of permanent-magnet materials persist in using c.g.s. units, it is necessary to note the relationships between these and the corresponding SI units.

The form of some equations is varied. In particular, for flux density B:

in the c.g.s. system: $B = H + 4\pi I$
and in SI: $\qquad B = \mu_o H + J$
$\qquad\qquad$ or $B = \mu_0 (H + M)$

Table 11.2 Derived SI units

Physical quantity	Unit name	Unit symbol	Unit dimensions
Velocity	Metre/second	m/s	$\mathrm{m\,s^{-1}}$
Acceleration	Metre/second^2	m/s^{-2}	$\mathrm{m\,s^{-2}}$
Angle	Radian	rad	—
Angular velocity	Radian/second	rad/s	$\mathrm{s^{-1}}$
Angular acceleration	Radian/second^2	rad/s^2	$\mathrm{s^{-2}}$
Frequency	Hertz	Hz	$\mathrm{s^{-1}}$
Force	Newton	N	$\mathrm{m\,kg\,s^{-2}}$
Torque	Newton metre	Nm	$\mathrm{m^2\,kg\,s^{-2}}$
Moment of inertia		kg m^2	$\mathrm{m^2\,kg}$
Energy	Joule	J	$\mathrm{m^2\,kg\,s^{-2}}$
Power	Watt	W	$\mathrm{m^2\,kg\,s^{-3}}$
Electric charge	Coulomb	C	$\mathrm{s\,A}$
Electric potential	Volt	V	$\mathrm{m^2\,kg\,s^{-3}\,A^{-1}}$
Capacitance	Farad	F	$\mathrm{m^2\,kg^{-1}\,s^4\,A^2}$
Resistance	Ohm	Ω	$\mathrm{m^2\,kg\,s^{-3}\,A^{-2}}$
Magnetic flux	Weber	Wb	$\mathrm{m^2\,kg\,s^{-2}\,A^{-1}}$
Magnetic flux density	Tesla	T	$\mathrm{kg\,s^{-2}\,A^{-1}}$
Inductance	Henry	H	$\mathrm{m^2\,kg\,s^{-2}\,A^{-2}}$
Magnetomotive force	Ampere	A	A
Magnetizing field	Ampere/metre	A/m	$\mathrm{m^{-1}\,A}$
Current density	Ampere/metre^2	A/m^2	$\mathrm{m^{-2}\,A}$
Permeability	—	H/m	$\mathrm{m\,kg\,s^{-2}\,A^{-2}}$
Permittivity	—	F/m	$\mathrm{m^{-3}\,kg^{-1}\,s^4\,A^2}$

The relationships between the units are given in Table 11.3.

k is the factor by which the value of a quantity measured in a c.g.s. unit must be multiplied in order to find the corresponding value in the SI unit.

Table 11.3 Magnetic units

Quantity	c.g.s. unit	SI unit	k
Flux	Maxwell or line	Weber	10^{-8}
Flux density B	Gauss	Tesla	10^{-4}
Magnetizing field H	Oersted	Ampere/metre	$10^3/4\pi$
Magnetization I	(Gauss/4π)	—	—
$\qquad\qquad J$	—	Tesla	$4\pi/10^4$
$\qquad\qquad M$	—	Ampere/metre	10^3
$(BH)_{\mathrm{max}}$	Megagauss oersted	Joule/m^3	$10^5/4\pi$

11.2 Constants and conversion factors

Values of important constants, in SI units, and some useful conversion factors are given in Table 11.4.

To determine a conversion factor for any particular quantity and system of units, the general relationship is:

$$(\text{Value in unit 2}) = R(\text{value in unit 1})$$

$$\text{where } R = \frac{\text{size of unit 1}}{\text{size of unit 2}}$$

If the size of the unit is increased, then the measure value in that unit is decreased.

R may be evaluated by expressing the dimensions of the unit in terms of the primary units, in the form $(mass)^a$ $(length)^b$ $(time)^c$ . . . as shown in Table 11.2, and then calculating:

$$R = \frac{(\text{mass unit})^a\,(\text{length unit})^b \ldots \text{in unit system 1}}{(\text{mass unit})^a\,(\text{length unit})^b \ldots \text{in unit system 2}}$$

Table 11.4 Constants

Permeability of free space μ_0	=	$4\pi \times 10^{-7}\ H/m$
Permittivity of free space ε_0	=	$8.854 \times 10^{-12} F/m$
Acceleration of gravity g	=	$9.807\ m/s^2$

Conversion factors

Length	1 m	=	3.281 ft
		=	39.37 in
Mass	1 kg	=	0.0685 slug
		=	2.205 lb (mass)
Force	1 N	=	10^5 dyne
		=	0.2248 lbf
		=	7.233 poundals
		=	0.102 kgf (or kilopond, kp)
Torque	1 Nm	=	0.738 lbf/ft
Energy	1 J	=	10^7 erg
		=	0.7376 ft lb
		=	0.2388 cal (1 cal = 4.186 J)
		=	9.48×10^{-8} BTU
Power	1 W	=	$(1/746)$ hp $= 1.341$ '×' 10^{-3} hp
Moment of inertia	1 kg m^2	=	0.738 slug ft^2
		=	23.7 lb ft^2

11.3 Metric multiples and submultiples

Table 11.5

Multiples				*Submultiples*		
T	tera	10^{12}		d	deci	10^{-1}
G	giga	10^{9}		c	centi	10^{-2}
M	mega	10^{6}		m	milli	10^{-3}
k	kilo	10^{3}		μ	micro	10^{-6}
h	hecto	10^{2}		n	nano	10^{-9}
da	deca	10		p	pico	10^{-12}

Index